Advances in Delivery Science and Technology

Series Editor
Michael J. Rathbone

More information about this series at http://www.springer.com/series/8875

Kenneth A. Howard • Thomas Vorup-Jensen
Dan Peer
Editors

Nanomedicine

Springer

Editors
Kenneth A. Howard
Department of Molecular Biology and Genetics
Interdisciplinary Nanoscience Center (iNANO)
Aarhus University
Aarhus C, Denmark

Thomas Vorup-Jensen
Department of Biomedicine
Biophysical Immunology Laboratory
Aarhus University
Aarhus C, Denmark

Dan Peer
Department of Cell Research & Immunology and Department of Material Science and Engineering
Laboratory of Precision NanoMedicine
Tel-Aviv University
Tel-Aviv, Israel

ISSN 2192-6204 ISSN 2192-6212 (electronic)
Advances in Delivery Science and Technology
ISBN 978-1-4939-3632-8 ISBN 978-1-4939-3634-2 (eBook)
DOI 10.1007/978-1-4939-3634-2

Library of Congress Control Number: 2016938513

Printed on acid-free paper

This Springer imprint is published by Springer Nature
The registered company is Springer Science+Business Media LLC New York

Editor Kenneth A. Howard would like to dedicate this book to his sister Jacqueline Teer.

Preface

Nanomedicine is set to meet medical challenges such as early and rapid diagnosis, targeted and effective drug delivery, and novel ways of providing organ and tissue replacement that work towards new personalized medicines. Utilizing properties occurring at the nanoscale for unique medical effects as well as understanding and controlling cellular molecular processes and clinical translation are key elements in nanomedicine. These skills are strongly bound by the threads that connect nanoscience, physics, material science, immunology, molecular biology, pharmacology, pharmaceutics, and medicine. The book *Nanomedicine* sets to capture this broad spectrum of disciplines whilst still maintaining focus on the key elements that establishes it as a *bona fide* unique field. The inclusion of adequate background information for understanding the basic principles makes this text an excellent learning tool for students as well as a useful reference for more senior scientists and other practitioners in the field.

The opening chapter defines the current field of nanomedicine by addressing the constituent key elements and describing major nanomedicine initiatives and programs, as well as product landscape to understand the development process and the impact of nanomedicine. The following three chapters focus on immune recognition with emphasis on the response by the innate immune system, notably complement activation of relevance in removal of nanomedicines as well as in disease pathogenesis. Chapter 2 focuses on complement factor inhibitors developed by structural biology determination. Chapter 3 addresses processes leading to nanoparticle-induced innate immune activation and discusses consequences for nanomedicine-based therapeutics. Chapter 4 focuses on protein polyvalent interactions with molecular recognition patterns at the nanoscale which is relevant in design criteria for the reduction of immune recognition of particulate nanomedicines. The next section focuses on nanomedicine-based diagnostics. Chapter 5 describes microfluidics-based tools for single cancer cell analysis and the use of nanotechnology for improved signal responses. Nanotheranostics and in vivo imaging modalities are addressed in Chapter 6. Chapters 7–11 describe a range of nanoscale nanopharmaceuticals. Chapter 7 gives an overview of a range of nanocarriers used for cancer treatment with a focus on targeting approaches. The ability to influence interaction with the extracellular and intracellular biological environment by particle geometry is the focus of Chapter 8. Nano-drug delivery systems for delivery of protein\peptide drugs to the brain are described in Chapter 9, and polymer-based DNA delivery to dendritic cells for cancer immunotherapy in Chapter 10. The section concludes with Chapter 11 on the use of the biopolymer silk in nanomedicine. The clinical translation of nanomedicines is dependent on safety and regulatory approval, addressed in Chapter 12. The final section describes nanomedicine-based tissue regeneration and cell therapy with the application of nanotechnology for medical implant drug release in Chapter 13, and Chapter 14 describes guided cellular responses by surface cues for nanomedicine applications.

The book *Nanomedicine* is relevant both as a textbook of an emerging field and as an essential read that gives insight and clarity to this interdisciplinary subject.

Acknowledgements

The editors would like to express their sincere gratitude to all the authors for the time and effort they have given in the preparation of the excellent chapters contained in the book. We are grateful to the CRS and especially Series Editor Michael Rathbone for giving us the opportunity to put together a book we feel is needed in the field. Our special thanks also go to our publisher Springer and their staff members Carolyn Honour, Renata Hutter, Sarah McCabe, Jacob Rosati, and Michelle Feng He, all who worked with both timely attention and patience during the preparation of this book.

Contents

Contributors

Jenolyn F. Alexander Department of Nanomedicine, Houston Methodist Research Institute, Houston, TX, USA

Qusai Al Nazer Laboratory of Nanomedicine and Biomaterials, Brigham and Women's Hospital, Harvard Medical School, Boston, MA, USA

Gregers Rom Andersen Department of Molecular Biology and Genetics, Aarhus University, Aarhus C, Denmark

Morten Østergaard Andersen The Maersk Mc-Kinney Moller Institute and Department of Chemical Engineering, Biotechnology and Environmental Technology, Faculty of Engineering, University of Southern Denmark, Odense, Denmark

Ole Zoffmann Andersen Interdisciplinary Nanoscience Center (iNANO), Aarhus University, Aarhus C, Denmark

Goran Bajic Department of Molecular Biology and Genetics, Aarhus University, Aarhus C, Denmark

Christiane Beer Department of Public Health, Aarhus University, Aarhus C, Denmark

Brandon Buckway Department of Pharmaceutics and Pharmaceutical Chemistry, University of Utah, Salt Lake City, UT, USA

Utah Center for Nanomedicine, Nano Institute of Utah, University of Utah, Salt Lake City, UT, USA

Huntsman Cancer Institute, University of Utah, Salt Lake City, UT, USA

Raymond Chiasson Laboratory of Nanomedicine and Biomaterials Brigham and Women's Hospital, Harvard Medical School, Boston, MA, USA

Ayelet David Department of Clinical Biochemistry and Pharmacology, Faculty of Health Sciences, Ben-Gurion University of the Negev, Beer-Sheva, Israel

Omid C. Farokhzad Laboratory of Nanomedicine and Biomaterials, Brigham and Women's Hospital, Harvard Medical School, Boston, MA, USA

Morten Foss Interdisciplinary Nanoscience Center (iNANO), Aarhus University, Aarhus C, Denmark

Hamidreza Ghandehari Department of Pharmaceutics and Pharmaceutical Chemistry, University of Utah, Salt Lake City, UT, USA

Utah Center for Nanomedicine, Nano Institute of Utah, University of Utah, Salt Lake City, UT, USA

Huntsman Cancer Institute, University of Utah, Salt Lake City, UT, USA

Department of Bioengineering, University of Utah, Salt Lake City, UT, USA

Biana Godin Department of Nanomedicine, Houston Methodist Research Institute, Houston, TX, USA

Adi Golani-Armon Department of Clinical Biochemistry and Pharmacology, Faculty of Health Sciences, Ben-Gurion University of the Negev, Beer-Sheva, Israel

Moaraj Hasan Laboratory of Nanomedicine and Biomaterials, Brigham and Women's Hospital, Harvard Medical School, Boston, MA, USA

Yi-Ping Ho Department of Molecular Biology and Genetics, Interdisciplinary Nanoscience Center (iNANO), Aarhus University, Aarhus C, Denmark

Kenneth A. Howard Department of Molecular Biology and Genetics, Interdisciplinary Nanoscience Center (iNANO), Aarhus University, Aarhus C, Denmark

Morten Leth Jepsen Department of Molecular Biology and Genetics, Interdisciplinary Nanoscience Center (iNANO), Aarhus University, Aarhus C, Denmark

Nazila Kamaly Laboratory of Nanomedicine and Biomaterials, Brigham and Women's Hospital, Harvard Medical School, Boston, MA, USA

Department of Micro- and Nanotechnology, Technical University of Denmark, Lyngby, Denmark

Birgitta R. Knudsen Department of Molecular Biology and Genetics, Interdisciplinary Nanoscience Center (iNANO), Aarhus University, Aarhus C, Denmark

Emil Laust Kristoffersen Department of Molecular Biology and Genetics, Interdisciplinary Nanoscience Center (iNANO), Aarhus University, Aarhus C, Denmark

Dalit Landesman-Milo Laboratory of NanoMedicine, Department of Cell Research and Immunology, George S. Wise Faculty of Life Sciences, Tel Aviv University, Tel Aviv, Israel

Department of Materials Sciences and Engineering, The Iby and Aladar Fleischman Faculty of Engineering, Tel Aviv University, Tel Aviv, Israel

Center for Nanoscience and Nanotechnology, Tel Aviv University, Tel Aviv, Israel

S. Moein Moghimi School of Medicine, Pharmacy and Health, Durham University, Stockton-on-Tees, TS17 6BH, UK

Ryosuke Ogaki Interdisciplinary Nanoscience Center (iNANO), Aarhus University, Aarhus C, Denmark

Dan Peer Department of Cell Research & Immunology and Department of Material Science and Engineering Laboratory of Precision NanoMedicine, Tel-Aviv University, Tel Aviv, Israel

Shahd Qassem Laboratory of NanoMedicine, Department of Cell Research and Immunology, Tel Aviv University, Tel Aviv, Israel

Department of Materials Sciences and Engineering, The Iby and Aladar Fleischman Faculty of Engineering, Tel Aviv University, Tel Aviv, Israel

Center for Nanoscience and Nanotechnology, Tel Aviv University, Tel Aviv, Israel

Janus Asbjørn Schatz-Jakobsen Department of Molecular Biology and Genetics, Aarhus University, Aarhus C, Denmark

Dmitri Simberg Department of Pharmaceutical Sciences, The Skaggs School of Pharmacy and Pharmaceutical Sciences, University of Colorado Denver, Aurora, CO, USA

David Stepensky Department of Clinical Biochemistry and Pharmacology, Faculty of Health Sciences, Ben-Gurion University of the Negev, Beer-Sheva, Israel

Huw D. Summers Centre for Nanohealth, College of Engineering, Swansea University, Swansea, UK

Kiana C. Trippler Department of Pharmaceutical Sciences, The Skaggs School of Pharmacy and Pharmaceutical Sciences, University of Colorado Denver, Aurora, CO, USA

Thomas Vorup-Jensen Department of Biomedicine, Biophysical Immunology Laboratory, Aarhus University, Aarhus C, Denmark

Matthew J. Ware Centre for Nanohealth, College of Engineering, Swansea University, Swansea, UK

Department of Nanomedicine, Houston Methodist Research Institute, Houston, TX, USA

Laure Yatime DIMNP – UMR5235, University of Montpellier, Montpellier, France

Editor's Biography

Kenneth A. Howard is an Associate Professor and Group Leader at the Department of Molecular Biology and Genetics at the Interdisciplinary Nanoscience Center (iNANO), Aarhus University, Denmark. His research and teaching activities are focused on delivery science, nanomedicine, and RNA interference. Kenneth Alan Howard received a PhD in pharmaceutical science from the University of Nottingham, UK, and has held postdoctoral positions at the CRC Institute for Cancer Studies, University of Birmingham, UK, and the School of Pharmacy, University of London. Dr. Howard is an active member of the Controlled Release Society serving on the Board of Scientific Advisors (chair), Nominations Committee, and organizer of CRS Educational workshops "RNA Interference Biology and Therapeutics" and "Albumin: The Next Generation Therapeutic."

Dan Peer is a translational scientist and Professor that leads an NIH-funded lab in the Faculty of Life Science and the Faculty of Engineering at Tel Aviv University (TAU). He is also the Director of the Focal Technology Area (FTA) on Nanomedicines for Personalized Theranostics, a National Nanotechnology Initiative, and the Director of the Leona M. and Harry B. Helmsley Nanotechnology Research Fund. Prof. Peer's work was among the first to demonstrate systemic delivery of RNAi molecules using targeted nanocarriers to the immune system, and he pioneered the use of RNA interference (RNAi) for in vivo validation of new drug targets within the immune system. He is an editor of several books in the field of nanomedicine, Editor of *Molecular and Cellular Therapies* (Springer), Editor of *Biology and Medicine in Nanotechnology* (IOP), and an Associate Editor of the *Journal of Controlled Release* (Elsevier), *Journal of Biomedical Nanotechnology*, and *Biochemistry*; he is on the Editorial Boards of the *Biomedical Microdevices* (Springer), *Cancer Letters* (Elsevier), *Nanomedicine: Nanotechnology, Biology and Medicine* (Elsevier), and *Bioconjugate Chemistry* (ACS). Prof. Peer is currently the President of the Israeli Chapter of the Controlled Release Society and a Member of the Israel Young Academy of Sciences and Humanities.

Thomas Vorup-Jensen is head of the Biophysical Immunology Laboratory in the Department of Biomedicine and a member of the iNANO at Aarhus University. He received a PhD in medicine from Aarhus University on topics involving the discovery, characterization, and recombinant manufacture of molecules of the complement system. More recently, he was awarded from the same institution the Doctor of Medical Science degree following a dissertation on polyvalent interactions between molecules of the immune system and their targets, which includes certain nanomedicines. Following his work as Research Fellow in Pathology with Harvard Medical School and later employment with Aarhus University, his research has focused on understanding the nanoscience of protein ligand recognition. In addition to his academic efforts, Prof. Vorup-Jensen and his laboratory actively collaborate with industry partners to translate scientific findings into new treatments of clinical unmet needs, in particular focusing on autoimmune and infectious diseases. Prof. Vorup-Jensen is on the editorial boards of *Frontiers in Pharmacology*, *Frontiers in Immunology*, and *Molecular and Cellular Therapies*. Prof. Vorup-Jensen was recently elected member of the Danish Academy of Natural Sciences.

Chapter 1
Nanomedicine: Working Towards Defining the Field

Kenneth A. Howard

Abstract An aging population and poor clinical solutions for many disease types has fuelled the rapid emergence of nanomedicine. Defining what constitutes nanomedicine is crucial to its recognition as a distinct field and directing future research focus towards clinical translation. This chapter sets out to define the field by addressing the important constituent key elements and describing major nanomedicine initiatives and programmes, and product landscape to understand the development process and current status.

Keywords Nanomedicine • Definition • Nanoscale • Molecular processes • Clinical translation • Nanomedicine products

1.1 Introduction

Poor clinical solutions for many disease types and rising healthcare cost due to an ageing population, has fuelled the rapid emergence of nanomedicine to meet challenges such as early and rapid diagnosis, effective targeted drug delivery and discovery, organ and tissue replacement and personalized medicines. Immense expectation is the motivation behind the establishment of global roadmaps in nanomedicine [1, 2] with high industrial and governmental investment. Dedicated nanomedicine journals and global societies [3–6] have been established over recent years that perpetuate the wave of activity and interest in the field with a rapid increase in published articles. The field is extremely broad and multidisciplinary exemplified by the commonly used definition "*The medical application of nanotechnology for diagnosis, treatment and management of human health*" that spans nanoscience, physics, material science, molecular biology, pharmaceutics, medicine and clinical translation. Whilst utilization of properties occurring at the

K.A. Howard (✉)
Interdisciplinary Nanoscience Center (iNANO), Department of Molecular Biology and Genetics, Aarhus University, Gustav Wieds Vej 14, 8000 Aarhus C, Denmark
e-mail: kenh@inano.au.dk

K.A. Howard et al. (eds.), *Nanomedicine*, Advances in Delivery Science and Technology, DOI 10.1007/978-1-4939-3634-2_1

nanoscale having unique medical effects, is central to nanomedicine, understanding and controlling cellular molecular processes and clinical translation are other key elements that need to be considered for inclusion. Defining the field is crucial for the establishment of nanomedicine as a *bona fida* field of its own rather than rebranding of traditional disciplines, a measure of its societal impact, and identifying current status and steps needed in shaping its future development. This chapter sets out to clearly define the field by describing major nanomedicine initiatives and programmes and addressing the important constituent key elements, and marketed examples to give a global perspective and the impact of nanomedicine.

1.2 Nanomedicine Initiatives and Approaches

This section will describe some major initiatives in Europe and the United States of America and their respective views on what constitutes "nanomedicine", in order to give an historical perspective and understand the process and current status, and importantly, work towards consensus and a uniform definition required for focused efforts and the future development of the field. Whilst these serve to exemplify the process and status, there are numerous global initiatives in Asia and elsewhere in the nanomedicine field.

1.2.1 European Nanomedicine Initiatives

The European Medical Research Council (EMRC) within the European Science Foundation (ESF), initiated a process in 2003 towards preparation of a Scientific Forward Look report on Nanomedicine. Academic and industrial experts met to define the field, discuss future societal impact, identify Europe's strengths and recommend funding priorities and national and pan-European infrastructures needed for a coordinated scientific activity. This process was facilitated through a series of specialized workshops held during 2004 culminating in a final consensus meeting in November 2004 to prepare the ESF-EMRC forward look report published in 2005 [7]. In this report the field of nanomedicine was defined as: "*the science and technology of diagnosing, treating and preventing disease and traumatic injury, of relieving pain, and of preserving and improving human health, using molecular tools and molecular knowledge of the human body*".

Nanostructures from one nanometre to hundreds of nanometres, included in a device or a biological environment, were highlighted as an important component with key sub-disciplines; analytical tools, nanoimaging, nanomaterials and nanodevices, novel therapeutics and drug delivery systems, and clinical, regulatory and toxicological issues highlighted.

In September 2005, a European Commission Nanotechnology for Health Vision Paper was published that recommended the establishment of a European Union

European Technology Platform on NanoMedicine and a Strategic Research Agenda to be focused on nanotechnology-based diagnostics imaging, targeted drug delivery and release, and regenerative medicine [8]. It was an industrial-led consortium including academic participants, with a focus on strengthening competitive scientific and industrial positioning in the area of nanomedicine and to improve EU citizen's quality of life and healthcare. In this paper nanomedicine was defined as: "*The application of Nanotechnology to Health. It exploits the improved and often novel physical, chemical, and biological properties of materials at the nanometric scale. NanoMedicine has potential impact on the prevention, early and reliable diagnosis and treatment of diseases.*"

In contrast to the aforementioned ESF report, unique properties at the nanoscale were included within the definition. Following on from the Vision Paper, *Nanomedicine nanotechnology for health European technology platform strategic research agenda (SRA) for nanomedicine* [9] was published in November 2006, with defined objectives directed to its member states and the commission with the main aim to provide information "for decision making processes for policy makers and funding agencies". Although the document stressed technology development was driven by healthcare needs, it stated "nanomedicine is a strategic issue for the sustainable competitiveness of Europe". Cardiovascular diseases, cancer, musculoskeletal disorders, neurodegenerative diseases and psychiatric conditions, diabetes and bacterial and viral infectious diseases were identified as target diseases based on mortality rates, prevalence, and potential impact of nanotechnology in their diagnosis and treatment. Specific key research areas were diagnostics and imaging, targeted delivery-multi-tasking medicines and regenerative medicine that may overlap e.g. theranostics that include both diagnostic and therapeutic components.

The aforementioned initiatives, in combination with academic and industrial expert workshops in 2009, led to a joint European Commission/ETP nanomedicine expert report published in October 2009 *Roadmaps in nanomedicine towards 2020* [1]. This report focused on nano-diagnostics, nanopharmaceuticals and regenerative medicine and technological timelines towards the treatment of diseases highlighted in the SRA 2006, and to provide a future research funding focus in EU programmes. The industry-led and application driven approach was preferred with emphasis for detailed specific recommendations. Focus was placed on translational research and commercial potential and the necessity for industrial/academic interaction for turning academic results into products directed towards better healthcare for patients. The report highlighted a need for improved knowledge and communication between academics, small medium enterprises and industry for successful implementation of the roadmap and a requirement to exploit industries strengths in product development pathway and regulatory issues. The European nanomedicine roadmap was commissioned by the European Commission with economic considerations a factor. The involvement of industry has led to an application and product-driven approach that is focused and defined.

The ETP in nanomedicine is the main player in providing advice and suggestions including identification of priority areas to the European Commission [10]. It provides strategic documents and recommendations from the nanomedicine community including a nanomedicine white paper, the new EU Framework Programme for Research and Innovation Horizon 2020 [11], and dissemination of knowledge, regulatory and IPR issues, standardisation, ethical, safety, environmental and toxicity concerns. The ETP have prepared an interactive European Nanomedicine Map of the private and academic nanomedicine community to facilitate coordination and collaboration in Europe [12]. A key component of both the ESF and ETP is the focus on European strengths, a strategy that may neglect other potential important areas.

Most European member states have designated nanomedicine calls in its research funding programmes independent from the EU funded projects and allows diversity in the focus and approach. For example, in the UK, the Engineering and Physical Sciences Research Council (EPSRC) had a priority area in therapeutics and nanomedicine in the 2013 Centres for Doctoral Training (CDTs) doctoral training call. In 2008, "nanotechnology to medicine and healthcare grand challenge programme" was awarded by the EPSRC as part of the cross council programme Nanoscience through Engineering to Application. The £15 million call invited proposals for large-scale, integrated projects focused on Nanotechnologies for the targeted delivery of therapeutic agents and Nanotechnologies for healthcare diagnostics. The call was structured to support 8–10, 3 year projects in a first phase 2009–2012, followed by extension to selected projects prioritized on potential to attract alternative external funding in reaching a clinical aim. Interestingly, as part of the same scheme public engagement and dialogue workshop sessions discussing key issues such as research priorities and public perception were organized. The public particularly supported early diagnostic devices that would allow lifestyle changes managed by the patient, and drug delivery systems for difficult to reach organs and cure serious conditions. The public, however, were opposed to treatment controlled by the device and not the healthcare worker or patient, for example, theranostics based on insertion of devices that control drug release based on chemical levels detected by that device.

A model comprising of a specific cluster of dedicated nanomedicine hubs has been established in Denmark in which three new Nanomedicine Research Centers were established in 2010 through $17.7 million funding by the Lundbeck Foundation for an initial period of 5 years [13]. The Lundbeck Foundation Nanomedicine Centre for Individualized Management of Tissue Damage and Regeneration (LUNA) (Aarhus University), Lundbeck Foundation Center for Biomembranes in Nanomedicine (Copenhagen University) (CBN) and, the Lundbeck Foundation Nanomedicine Research Center for Cancer Stem Cell Targeting Therapeutics (NanoCAN) (University of Southern Denmark) were the three centres. The three centres of excellence investigate novel strategies for exploiting nanoscaled biostructures in combating major human diseases, including cancer as well as neurological, infectious, cardiovascular and musculoskeletal disorders. The centres are highly interdisciplinary including nanotechnology, biotechnology, molecular biology, biomedicine, chemistry, and physics with a key element to combine basic and clinical research. Independent funding from the Danish Research Council facilitates a sustained nanomedicine drive in the country.

1.2.2 USA Nanomedicine Initiatives

The National Institutes of Health have a common fund programme in nanomedicine also referred to as the NIH roadmap [2]. It is a 10 year programme, initiated in 2005, based on "High Risk, High Reward" with two main goals (1) study nanoscale cellular structures and processes using tools to probe and manipulate (Phase 1) and, (2) utilize this information to understand and treat specific disease (Phase 2). The programme started with a national network of eight nanomedicine centers, reduced to four in 2010 based on the most translational potential. The programme first phase focused on basic science and the second phase to clinical translation. The programmes definition of nanomedicine "*refers to highly specific medical intervention at the molecular scale for curing disease or repairing damaged tissues, such as bone, muscle, or nerve. A nanometer is one-billionth of a meter, too small to be seen with a conventional lab microscope. It is at this size scale – about 100 nanometres or less – that biological molecules and structures operate in living cells.*" Emphasis is placed on size and relationship with nanoscale biological structures, using the strengths of NIH-funded researchers to investigate and understand cellular mechanisms towards research directed at characterisation of molecular cellular components and re-engineering those processes for medical treatments. Funded centers included, the nanomedicine center for nucleoprotein machines, center for protein folding machinery, nanomedicine center for mechanobiology directing the immune response, cellular control: synthetic signalling and motility systems. Funding will be discontinued from the common fund in 2015 and funded from other sources as more focus is directed towards specific disease types.

Other initiative include the National Cancer Institute $144.3 million (2005–2010) *alliance for nanotechnology in cancer* that started in 2004, comprising of nine centers of excellence focused more towards clinical applications than basic science to use nanotechnology to diagnose, treat and prevent cancer [14]. Research includes tools to identify new biological targets, agents for early disease diagnosis using in vitro assays and in vivo imaging tools, multi-functional targeted devices to deliver multiple therapeutics to cancer, tools for real-time assessment of therapeutic and surgical efficacy. Emphasis is on multidisciplinary research between diverse institutions and organizations, with an alliance framework that includes nine centers of cancer nanotechnology excellence focused on discovery and tool development of nanotechnology in clinical oncology; 12 cancer nanotechnology platform partnerships that support individual projects and use nanotechnology to address fundamental questions on cancer; and six cancer nanotechnology training centers educating and training researchers from different fields in the use of nanotechnology-based approaches in cancer. This is a two phase programme, with the first phase 2005–2010, to establish multidisciplinary teams in focused areas and established a Nanotechnology Characterisation Laboratory as a nanomaterials characterisation hub for standardization of materials to better enable technology transfer from academia to companies to the clinic [15]. The second phase, 2010–2015 aims to translate nanotechnology findings into the clinic.

1.3 Nanomedicine Key Elements

Utilization of properties at the nanoscale having unique medical effects is central to nanomedicine programmes and easy to appreciate its inclusion. Understanding and controlling cellular molecular processes and clinical translation are, however, other components that constitute the field. This section will address each in turn towards a definition encompassing all these key elements.

1.3.1 Nanoscale Properties That Have Unique Medical Effects

Harnessing novel physicochemical and biological properties exhibited by nanoscale materials to improve human health is perhaps the fundamental element of nanomedicine. As far back as 1959 the future societal impact of nanotechnology was predicted by Nobel Laureate Richard Feynman "*Many of the cells are very tiny……; they do all kinds of marvelous things – all on a very small scale……Consider the possibility that we too can make a thing very small which does what we want – that we can manufacture an object that maneuvers at that level*" [16]. The rapid emergence of nanoscience in the new millennium has provided a wealth of knowledge into the physicochemical properties and biological performance of nanoscale materials that can be used to utilise nanotechnology for medical applications (Fig. 1.1). Examples of unique physical properties occurring at the nanoscale include distinct light-scattering and optical properties exhibited by metal nanoparticles of different size, shape and compositions that can be utilised in diagnostic biosensors [17]. The magnetic to superparamagnetic shift found at the nanoscale can be used for molecular imaging applications [18]. Furthermore, localized exposure of superparamagnetic iron oxide nanoparticles (SPION) to an oscillating magnetic field induce extreme polarity fluctuations that results in an increase in core temperature that has been utilised for thermal-induced destruction of tumours [19]. The increased solubility occurring at the nanoscale has been utilised in Nanocrystals for dissolution of poorly soluble drugs widely used within the Pharmaceutical Industry [20]. The saturation solubility becomes a function of particle size below a critical size ~1000 nm where solubility increases with decrease in size.

Nanoparticles used as drug delivery nanocarriers exhibit unique biological properties that can be utilised to create therapeutics with higher efficacy and reduced toxicity in patients. The ability for nanocarriers to circulate and extravasate across endothelial barriers into tissue and enter cells by endocytosis is a size dependent process [21] (Fig. 1.1). Furthermore, the large surface area of nanoscale particles potentiates surface functionalisation characteristics such as targeting and stealth. Nanoscale particles consequently have been shown to increase drug payload to diseased sites [22].

The ability to control cellular interactions, proliferation and stem cell differentiation by nanotopography and biomolecule patterning has great potential in tissue

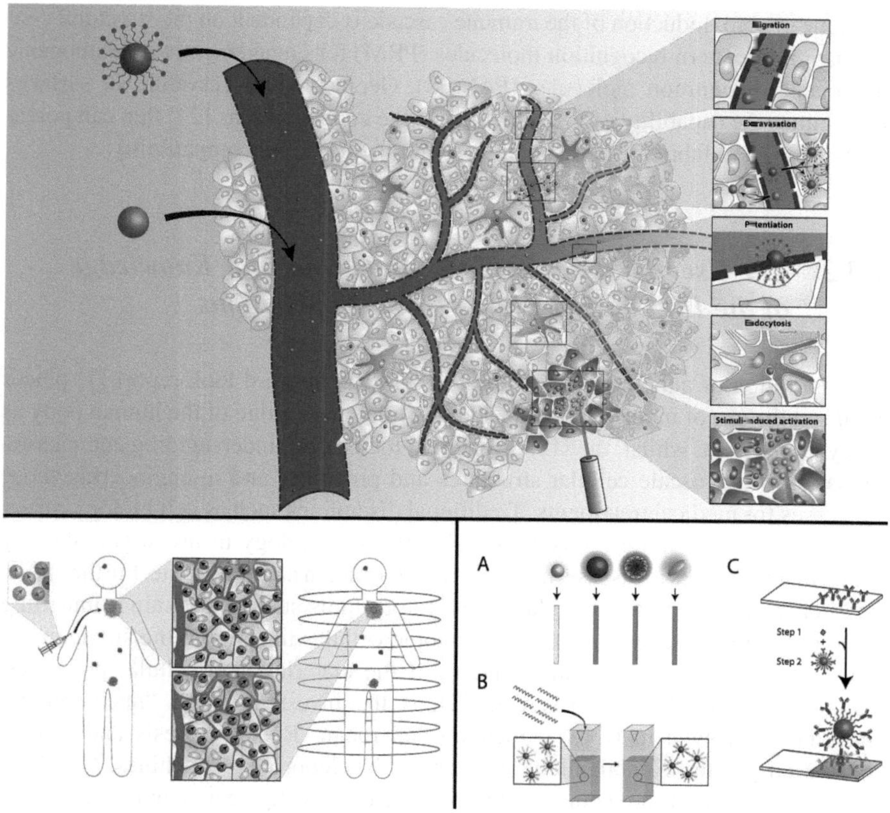

Fig. 1.1 Properties occurring at the nanoscale that exhibit unique medical effects. *Upper panel*: Nanoparticle-based drug delivery. Nanoscale particles can migrate within the bloodstream facilitated by stealth coatings and extravasate across endothelium into normal and diseased tissue. Surface engineered nanoparticles functionalised with targeting agents engage with cellular surface receptors potentiated by the nanoparticle high surface to volume ratio. Size-mediated endocytosis allows cellular entry and delivery of the drug cargo. Inorganic nanoparticles composed of gold or iron oxide can be used for thermal-induced disease death activated by thermal or magnetic stimuli, respectively. *Lower right panel*: In vitro diagnostics. The optical properties dependent on nanoparticle size, composition and shape (**a**) can be utilised for diagnostic purposes. Complementarity of specific oligonucleotide to oligonucleotide coated on particles induces aggregation and a detectable spectral shift (**b**). An antibody on a chip and a nanoparticle that target different epitopes on the same antigen can be used detect pathogenic antigens (**c**). *Lower left panel*: Molecular imaging. The shift to paramagnetism at the nanoscale where all the magnetic domains are aligned in the same direction can be used for improved disease contrast in magnetic resonance imaging. Graphic prepared by Morten Tobias Jarlstad Olesen, Aarhus University

engineering. Cells respond to nanoscale chemical patterns and topography cues provided by different protein formats such as protein folding or collagen banding found in the tissue. This can be mimicked by nanostructures functionalized onto implant surfaces with different biomolecule patterns [23]. The importance of interactions at the nano-bio interface is exemplified in recognition of materials by components of the immune system. Induction of the immune cascade is dependent on the structural conformation of pattern recognition molecules (PRM) for engagement with pathogenic associated recognition molecules (PAMPS). Geometry and curvature of surfaces determines the capability of PRM to adopt these conformations [24] that can potentially be used to fabricate nanomedicines with increased biocompatibility.

1.3.2 The Use of Molecular Tools and Molecular Knowledge of the Human Body for Medical Applications

The definition of nanomedicine in the ESF-EMRC forward look report [7] places focus on the use of molecular tools and molecular knowledge of the human body as a key component, whilst, the USA Roadmap focuses on understanding and characterisation of nanoscale cellular structures and processes and re-engineering those processes for medical treatments. Traditional disciplines such as cell biology, structural biology, proteomics, genomics and molecular biology to investigate the key molecular processes in disease, therefore, are relevant in nanomedicine. Furthermore, nano-characterisation and cellular investigatory tools such as scanning tunnelling microscopy (SPM) and real-time microscopic cellular imaging at the single molecule level allow a greater understanding of the materials and cellular processes. Examples within the NIH roadmap [2], is the investigation of "nucleoprotein machines" involved in DNA replication and repair, RNA synthesis and protein translation. At the Nanomedicine Center for Nucleoprotein Machines, GA Tech Research Corp. tools are being developed to better investigate and understand these fundamental processes, structure-function relationship such as assembly and disassembly, signaling and control mechanisms. The study of chaperone proteins, or "nanomachines" involved in folding proteins crucial to the cell homeostasis, and removal of damaging misfolded polypeptides towards identification of therapeutic or diagnostic targets is the focus at Center for Protein Folding Machinery at the Baylor college of Medicine.

The Nanomedicine Centre for Individualized Management of Tissue Damage and Regeneration (LUNA), Aarhus University, Denmark [25] combines basic with applied science using traditional and nano-approaches including X-ray diffraction, liquid- and solid-state NMR, magnetic resonance imaging (MRI), small-angle X-ray scattering, and cryo-electron microscopy and scanning tunnelling microscopy. The aim is to identify the central molecular processes involved in the inflammation and de/regenerative imbalance associated with certain cardiovascular and musculoskeletal diseases with focus on the role and regulation of pattern recognition molecules such as collectins and toll-like receptors. This information is to be used for new approaches in drug design, drug delivery, bioimaging, tissue regeneration towards therapeutics and early diagnostic tools. An example is the structure-based drug design for identification of complement blocking agents. X-ray diffraction has been used to study the atomic structure of C5 and its cleavage fragments C5a and C5b involved in innate immune-mediated autoimmune response. The aim is to identify novel blocking drugs, as an alternative to C5 antibody, with the possibility for specific inhibition of downstream cleavage fragments in autoimmunity without detriment to overall immune protection. The application of nanoparticle-based enabling technologies developed within a drug delivery component of the center can then be used for in vivo delivery, emphasising the link between the key elements of molecular knowledge and nanoscale properties. The functional importance to immunity of nm-scaled conformational changes in the large proteins of the immune system is a focus [26]. The capability to regulate ligand nanoscale spacing of carbohydrates such as mannose and N-acetyl glucosamine can be used to investigate by real-time monitoring using atomic force microscopy and small-angle X-ray scattering of the spacing of mannan-binding lectin (MBL) required for complement activation. This will allow insight into the conformational nature of the patterns recognized by MBL and allow design of antagonists to block MBL-mediated complement.

1.3.3 Nanoscience Merged with the Biomedical Sciences in a Clinical Setting

The application of nanotechnology is fundamental to nanomedicine and, thus, requires a nanoscience environment. The diverse disciplines ranging from nanoscience, physics, material science, molecular biology, pharmaceutics, medicine and clinical translation, however, necessitates interdisciplinary/translational research that are key to successful nanomedicine programmes. The environment is unlikely to be found in a single institution, and so calls for interdisciplinary programmes outlined in earlier section such as pan-European initiatives and NIH networks. The involvement of clinicians early in the programmes is needed to provide input and focus towards unmet clinical needs.

1.3.4 Nanomedicine: Redefined

This section has identified components of nanomedicine. The general definitions fall short of inclusion of all these key elements. A more precise, all inclusive definition could be "New scientific concepts through a greater understanding of human processes utilised for development of nano-based strategies in a clinical setting for prevention, diagnosis and treatment of diseases". Whilst this may not be appropriate for the public or press sound bites, it is important that scientists consider all elements and includes relevant participants in order to fulfil the potential of nanomedicine.

1.4 Nanomedicine Products

The necessity for a uniform definition of nanomedicine is crucial for an accurate assessment of the current status of nanomedicines on the market. A lack of consensus that includes differing size range criteria used, however, complicates clear identification of what constitutes a nanomedicine, exemplified by the different number and type described, and a lack of a definitive list. Furthermore, the inclusion of products approved before the "nanomedicine era", for examples, the liposomal Doxil® (FDA approval 1995) [27], and the PEGylated proteins Adagen® (FDA approval 1990) and Oncaspar® (FDA approval 1994) interferes with evaluating the impact of nanomedicine. An early review of the nanomedicine commercialization landscape was reported by Wagner et al. [28] based on science citation, medical and patent databases, and the business literature. Physical effects occurring at the nanoscale was the basis for inclusion such as iron oxide MRI contrast agents that exhibit paramagnetism, however, structures up to 1000 nm were also considered due to unique physiological behaviour. Products were categorized into drug delivery, biomaterials, in vivo imaging, in vitro diagnostics, active implants and, drugs and therapy. The predominant drug delivery category was highly represented by PEGylated proteins and liposomal-based systems. PEGylation that confers extended plasma half-life of proteins by Mw increase-mediated reduction in renal clearance can be viewed as an enabling technology, rather than a unique physical effect that occurs at the nanoscale. Similarly, the proposed ability for liposomes such as the liposomal doxorubicin formulation Doxil to extravasate across tumour endothelium is not dependent on transitions in physicochemical properties at the nanoscale. Both, however, are examples of changes in physiological interactions of particles of a certain size range. This has been highlighted by Etheridge et al. [29], and considerably expands the range of nanomedicines solely from a size range where quantum effects occur at ~100 nm occur to materials up to 500–100 nm e.g. cellular transcytosis or endothelial penetration which can occur, with particles of this size range. Nanocrystals formed by the micronization of drug to increase solubility has a size range from a few nm to microns and are widely used in the Pharmaceutical industry. This, however, reduces the distinction between micro and nano for categorization of a nanomedicine. Nanostructures to improve the mechanical strength of medical implants such as nanoparticle-based dental fillers and nanohydroxyapatite-based bone repair products are predominant. The therapeutic nanomedicine landscape continues to be dominated by liposomes and PEGylated proteins with the addition of Abraxane® by Celgene that is nanoparticle albumin bound paclitaxel (nab-paclitaxel) approved for cancer treatment [30]. It is proposed the nanoparticles disassemble into individual albumin drug-bound molecules after injection that results in high drug plasma levels and tumour accumulation attributed to the high metabolic uptake of albumin with a suggestion that SPARC expression in tumours is involved in an active albumin targeting process. There is, however, still a significant absence of targeted nanoparticles. Weissig et al. [31] focused on therapeutic nanomedicines commonly referred to as nanopharmaceutics in a more recent review. In this review nanopharmaceutical categories included liposomes, PEGylated proteins, polypeptides and aptamers, nanocrystals, protein-drug conjugates, polymer-based nanoformulations, metal-based nanoformulations such as imaging agents. A lack of a clear definition and designated data base is an issue, exemplified by the necessity to assign nanomedicines into categories of relevance ranging from "confirmed" to "questionable" in the Etheridge et al. review. There are no specific regulatory requirements for products containing nanoparticles that may reflect the current uncertainty of what constitutes a nanomedicine. The ability of nanoscale materials to penetrate tissue is of particular toxicology concerns for non-biodegradable metal nanoparticles rather than organic-based systems such as liposomes. The aforementioned initiative by the National Cancer Institute with the Nanotechnology Characterisation laboratory to standardize the characterisation of nanomaterials is needed for the clinical approval process as novel materials arise.

1.5 Conclusion

A clear definition of nanomedicine is required to identify it as a distinct field. Physical properties occurring at the nanoscale having unique medical effects would certainly serve this purpose but neglect other key elements. Inclusion of traditional disciplines such as cell biology, structural biology, proteomics and genomics, may first appear to limit its distinction, however, are relevant for understanding, controlling and re-engineering key molecular processes for disease treatment. Nanomedicine is a fledgling field trying to "find its feet"; knowledge of unique physical and biological properties of an expanding panel of new materials along with a greater understanding of molecular processes is part of its transformation into a distinct established field. The current product landscape reflects its early status, with inclusion of some materials above nanoscale. Whilst, a limited repertoire of products is expected in a field still emerging, it is important not to include materials that reduce the distinction between nano and micro and obstruct its unique field status and measuring the true impact of nanomedicine. Furthermore, it is important to resist "scientific rebranding" that can be damaging to its perception and acceptance as a unique field; "nanomedicine is more than a name". The recruitment of scientists, from various backgrounds and fields that "repurpose" respective technologies, towards development of nanomedicine, however, is a unique opportunity to take the field of nanomedicine forward. An example is the application of material scientist technologies for use in medical science. Defining the field is important to direct and focus research and education programmes, and standardisation of products. Key elements outlined in this chapter work towards consensus in the field required for focused efforts needed for realising the tremendous potential of nanomedicine.

References

1. ETPN Expert Report 2009—Roadmaps in Nanomedicine
2. National Institutes of Health. Office of Strategic Coordination—The Common Fund. http://commonfund.nih.gov/nanomedicine/index
3. European Foundation for Clinical Nanomedicine. http://www.clinam.org/
4. European Society for Nanomedicine. http://www.esnam.org/
5. American Society for Nanomedicine. http://amsocnanomed.org/

6. Japan Nanomedicine Society. http://www.nanobio.nagoya-u.ac.jp/nanomedicine/
7. European Science Foundation Forward Look on Nanomedicine (2005)
8. Vision Paper and Basis for a Strategic Research Agenda for Nanomedicine (2005)
9. European Technology Platform Strategic Research Agenda for Nanomedicine (2006)
10. European Technology in Nanomedicine. http://www.etp-nanomedicine.eu/public
11. European Technology Platform in Nanomedicine (2013). White paper to the horizon 2020 framework programme for research and innovation—recommendations from the nanomedicine community
12. http://www.etp-nanomedicine.eu/public/european-nanomedicine-map
13. Mollenhauer J, Stamou D, Flyvbjerg A, Wengel J, Gether U, Kjems J, Bjørnholm T, Besenbacher F (2010) David versus Goliath. Nanomed Nanotechnol Biol Med 6:504–509
14. National Cancer Institute Alliance for nanotechnology in cancer. http://nano.cancer.gov/
15. National Cancer Institute Nanotechnology Characterisation Laboratory. http://ncl.cancer.gov/
16. Feynman RP (1960) Engineering and Science, 23 (5), 22–36.
17. Rosi NL, Mirkin CA (2005) Nanostructures in biodiagnostics. Chem Rev 105(4):1547–1562
18. Larsen EKU, Nielsen T, Wittenborn T, Birkedal H, Vorup-Jensen T, Jakobsen MH, Østergaard L, Horsman MR, Besenbacher F, Howard KA, Kjems J (2009) Size-dependent accumulation of PEGylated silane coated magnetic iron oxide nanoparticles in murine tumours. ACS Nano 3(7):1947–1951
19. Rivera Gil P, Hühn D, del Mercato LL, Sasse D, Parak WJ (2010) Nanopharmacy: inorganic nanoscale devices as vectors and active compounds. Pharmacol Res 62:115–125
20. Junghanns J, Müller RH (2008) Nanocrystal technology, drug delivery and clinical applications. Int J Nanomedicine 3(3):295–309
21. Alexis F, Pridgen E, Molnar LK, Farokhzad OC (2008) Factors affecting the clearance and biodistribution of polymeric nanoparticles. Mol Pharm 5(4):505–515
22. Davis ME, Chen Z, Shin DM (2008) Nanoparticle therapeutics: an emerging treatment modality for cancer. Nat Rev Drug Discov 7:771–782
23. Dalby MJ, Gadegaard N, Tare R, Andar A, Riehle MO, Herzyk P, Wilkinson W, Oreffo OC (2007) The control of human mesenchymal cell differentiation using nanoscale symmetry and disorder. Nat Mater 6:997–1003
24. Pedersen MB, Zhou X, Larsen E, Sørensen U, Kjems J, Nygaard JV, Nyengaard JR, Meyer RL, Boesen T, Vorup-Jensen T (2010) Curvature of synthetic and natural surfaces is an important target feature in classical pathway complement activation. J Immunol 184:1931–1945
25. The Nanomedicine Centre for Individualized Management of Tissue Damage and Regeneration (LUNA), Aarhus University, Denmark. http://nanomedicine.au.dk/home/
26. Gjelstrup LC, Kaspersen JD, Behrens MA, Pedersen JS, Thiel S, Kingshott P, Oliveira CL, Thielens NM, Vorup-Jensen T (2012) The role of nanometer-scaled ligand patterns in polyvalent binding by large mannan-binding lectin oligomers. J Immunol 188(3):1292–1306.
27. Barenholz Y (2012) Doxil®—the first FDA-approved nano-drug: lessons learned. J Control Release 160(2):117–134
28. Wagner V, Dullaart A, Bock A, Zweck A (2006) The emerging nanomedicine landscape. Nat Biotechnol 24(10):1211–1217
29. Etheridge ML, Campbell SA, Erdman AG, Haynes CL, Wolf SM, McCullough J (2013) The big picture on nanomedicine: the state of investigational and approved nanomedicine products. Nanomed Nanotechnol Biol Med 9:1–14
30. Desai N, Trieu V, Yao Z, Louie L, Ci S, Yang A, Tao C, De T, Beals B, Dykes D, Noker P, Yao R, Labao E, Hawkins M, Soon-Shiong P (2006) Increased antitumor activity, intratumor paclitaxel concentrations, and endothelial cell transport of cremophor-free, albumin-bound paclitaxel, ABI-007, compared with cremophor-based paclitaxel. Clin Cancer Res 12(4):1317–1324
31. Weissig V, Pettinger TK, Murdock N (2014) Nanopharmaceuticals (part 1): products on the market. Int J Nanomedicine 9(1):4357–4373

Chapter 2
Complement Regulators and Inhibitors in Health and Disease: A Structural Perspective

Laure Yatime, Goran Bajic, Janus Asbjørn Schatz-Jakobsen, and Gregers Rom Andersen

Abstract The complement system is an important effector within the innate immune system as a defence against pathogens and maintaining homeostasis. Detection of pathogen- and damage-associated molecular patterns triggers the proteolytic cascade in complement. In healthy self-tissues effector proteins are tightly controlled by proteins acting as regulators of complement activation, and absence or malfunction of these regulators contribute to pathogenesis in a number of disease conditions in humans. Complement is highly relevant to nanomedicine due its role in adverse reactions on polymers and nanoparticle drug carriers, but also since complement hyperactivation contributes to pathogenesis in many disease conditions that are frequently addressed within nanomedicine. We review here the regulatory mechanisms that modulate complement activation and some of the most prominent cases linking complement dysregulation/deficiencies to pathogenesis as well as the strategies that have been considered for the development of therapeutic complement inhibitors and modulators to alleviate complement-mediated detrimental effects. In addition, this chapter summarizes the wealth of strategies adopted by pathogens to evade complement, such as inhibition of the proteolytic cascade, degradation of complement effector molecules and interference with transmembrane signaling by effectors, and highlights how structural and functional insight into their mode of function now provides leads for the development of novel complement therapeutics.

Keywords Innate immunity • Complement • Structural biology • Therapeutics

L. Yatime (✉)
DIMNP – UMR5235, University of Montpellier, Place Eugène Bataillon,
Bât. 24 cc107, 34095 Montpellier, Cedex 5, France
e-mail: laure.yatime@inserm.fr

G. Bajic • J.A. Schatz-Jakobsen • G.R. Andersen (✉)
Department of Molecular Biology and Genetics, Aarhus University,
Gustav Wieds Vej 10C, 8000 Aarhus C, Denmark
e-mail: gra@mb.au.dk

K.A. Howard et al. (eds.), *Nanomedicine*, Advances in Delivery
Science and Technology, DOI 10.1007/978-1-4939-3634-2_2

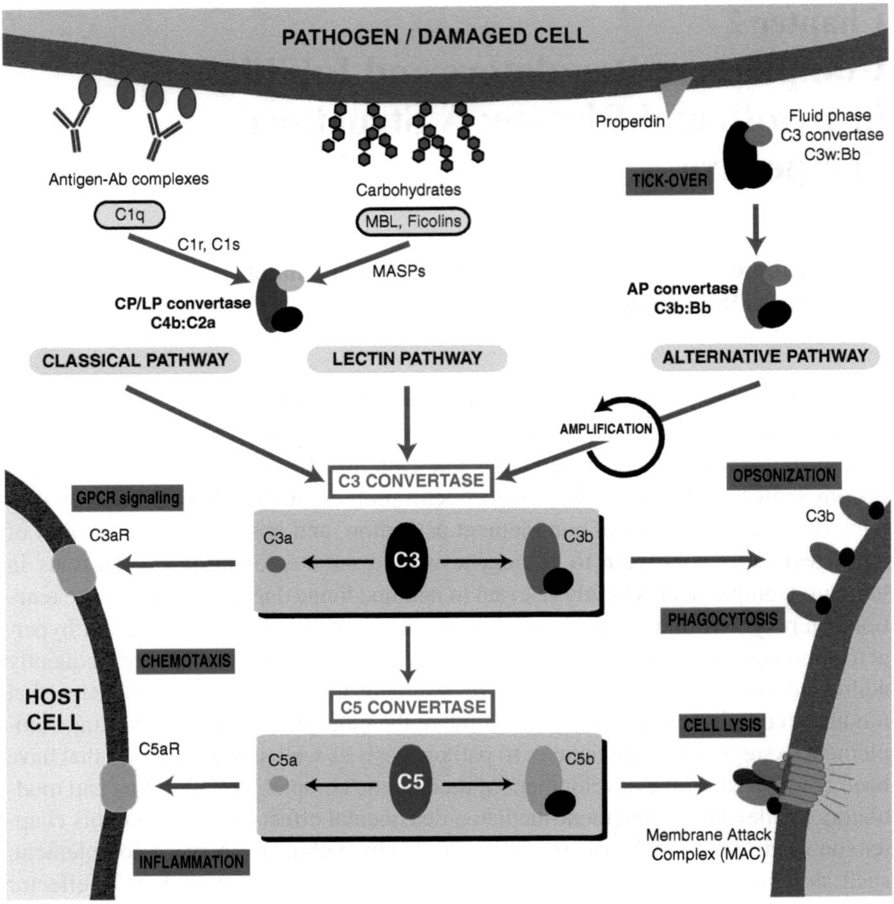

Fig. 2.1 General overview of the complement cascade emphasizing the initiation of the proteolytic cascade upon pattern recognition and the effector molecules acting on both host cells and pathogen/damaged cells

2.1 Introduction to Complement and Associated Diseases

2.1.1 Overview of the Complement Cascade

Complement is one of the major effectors of the innate immune system and is in the first line of defence against invading pathogens. Complement not only protects against infectious organisms, but also disposes of immune complexes, products of inflammatory injury and bridges the innate and the adaptive immunity [1–3]. Complement is a complex network of more than 50 circulating and membrane-bound proteins that can be activated through three different pathways: the classical (CP), lectin (LP) and alternative (AP) pathways (Fig. 2.1). The classical pathway is initiated upon recognition by the C1 complex of antibody (IgG or IgM):antigen complexes or pentraxins (CRP, PTX3 and SAP) binding directly to activators [4, 5]. The C1 complex consists of the pattern-recognition molecule C1q and the associated serine proteases C1r and C1s that are activated upon pattern recognition [6]. C1s then initiates the cascade by proteolytic cleavage of C4 into C4a and C4b. Through an internal reactive thioester C4b is covalently linked to the activator [7], and C2 subsequently joins to form the proconvertase C4bC2, which is cleaved, also by C1s, resulting in the appearance of the CP C3 convertase C4bC2a [8]. The lectin pathway is similar to the CP but differs in the molecular patterns activating it. The recognition is achieved by mannan-binding lectin (MBL), ficolins, and collectin-11 binding to glycan moieties of a variety of glycoproteins and glycolipids specific to pathogens (bacteria, viruses and fungi) and damaged self [9, 10]. The pattern recognition molecules are associated to serine proteases MASP-1 and MASP-2, and recent work has established that MASP-1 may autoactivate and then cleaves MASP-2 resulting in an activated MASP-2 which can cleave C4. Further processing of the proconvertase may then be carried out by either MASP-1 or MASP-2 [11, 12].

C3 convertases are proteolytic complexes able to cleave the central complement component C3 into C3a and C3b [13]. The anaphylatoxin C3a recruits immune cells to the site of infection and initiates an acute inflammatory response [14] whereas C3b is the major opsonin of the complement system. Like C4b, it covalently attaches to the activator through its thioester [15] thereby "tagging" foreign and altered-self objects leading to opsonization. This also leads to initiation of the alternative pathway as activator-bound C3b recruits and binds factor B which is then cleaved by factor D, yielding the AP C3 convertase C3bBb [16]. The C3 convertase in turn generates more C3b from C3 and in this way creates a powerful amplification loop that accounts for 80–90 % of the outcome in the terminal pathway (see below) when complement is activated through the CP [17] or the LP [18]. However, AP activation may also occur spontaneously upon hydrolysis (tick-over) of the thioester bond within C3 generating a water-reacted molecule, C3(H_2O) capable of forming the fluid-phase C3 convertase [19], which can then be stabilized on microbial surfaces and apopotic/necrotic host cells by properdin [20].

The C3 convertases may recruit a second molecule of C3b yielding the AP C5 convertase C3bBbC3b or the CP C5 convertase C4bC2aC3b [21]. These cleave C5 into C5a and C5b thereby initiating the terminal pathway (TP) [22]. The C5a anaphylatoxin signals through the G-protein coupled receptor, C5aR, and initiates inflammation (see below). C5b is devoid of the thioester and does not attach to activating surfaces but instead binds to C6, C7, C8 and multiple copies of C9, forming the (C5bC6C7C8)C9n complex known as the membrane attack complex (MAC) [23]. Active MAC is able to insert into membranes and form pores resulting in cell lysis. Although MAC is potentially a powerful weapon against invading pathogens, deficiencies in the TP proteins primarily leads to *meningococci* infections [24].

2.1.2 Complement Regulation

Obviously, uncontrolled C3b deposition causes inflammation and cytolysis. For this reason complement has to be tightly regulated on healthy tissues. Host cells express cell-surface and soluble regulators (Fig. 2.2). Low concentration/absence or mutations in these proteins are at the heart of pathogenesis when complement is involved. The serpin C1 esterase inhibitor (C1-INH) blocks the C1 complex and also the LP proteases MASP-1 and MASP-2 [25], but it is not specific to complement as it also targets plasmin, thrombin, factor Xa and kallikrein. In the AP, most regulators function at the level of the C3 convertases by stimulating their dissociation (decay acceleration activity) or by promoting the proteolytic degradation of C3b into iC3b by the serine protease factor I [26] (co-factor activity). iC3b cannot associate with factor B and is thus irreversibly unable to form the C3 convertase. iC3b may be further degraded into C3dg and finally C3d by factor I and plasmin [27].

Factor H (fH) is the major AP regulator and exhibits both decay and co-factor activity for both C3bBb and $C3(H_20)Bb$ C3 convertases [28]. A pathogen cell opsonized with C3b but not capable of stabilizing fH binding will bind fH weakly preventing inactivation of C3b, whereas a non-activating host cell presenting the appropriate glycosaminoglycans will associate efficiently with fH yielding protection through C3b degradation [29]. fH consists of 20 complement control protein (CCP) domains (Fig. 2.2a). X-ray crystallography and NMR have deciphered the three-dimensional structures of all fH CCP domains but CCPs 9, 14 and 17, while full length fH has been studied by solution scattering [30–34]. Most of the structure-function research has focused on CCPs 1–4 and 19–20 binding C3b, and CCPs 6–8 and 19–20 associating with self-surfaces [34–36].

There are six other proteins related to fH that bind to C3b or C3d, the fH-like protein 1 (a splice variant of fH) and five fH-related proteins (CFHR1–5). All these are also composed entirely of CCP domains with different degrees of sequence identity with fH (Fig. 2.2a). CFHR1 can associate into a homodimer as well as into a heterodimer with CFHR2 or CFHR5 [37]. CFHR1 inhibits the C5 convertase as well as MAC formation [38] by binding to C3b through its N-terminal homodimer-forming moiety, but CFHR1 lacks both co-factor and decay activity [39]. CFHR2 inhibits the formation of the C3 convertase [40] but does not compete with fH for the binding to C3b [40]. CFHR3 competes with fH for the binding to C3b but its role in complement control is not settled [41]. The activity of CFHR4 is also unclear, but it binds C3b and has cofactor activity. CHFR5 possesses both cofactor and decay activities and is also recruited to damaged self-surfaces [42].

C4b-binding protein (C4BP) is a large plasma-circulating glycoprotein. The major form of C4BP consists of seven identical α chains and one β chain, both consisting of CCP domains [43]. The chains are connected by disulfide bridges and the structure is spider-like with seven elongated subunits attached to a relatively small central body [44]. Each α chain contains a binding site for C4b. Once bound to C4BP, C4b serves as a substrate for factor I. C4BP can also act as cofactor in factor I-mediated proteolysis of C3b, although fH and CR1 (see below) are more efficient

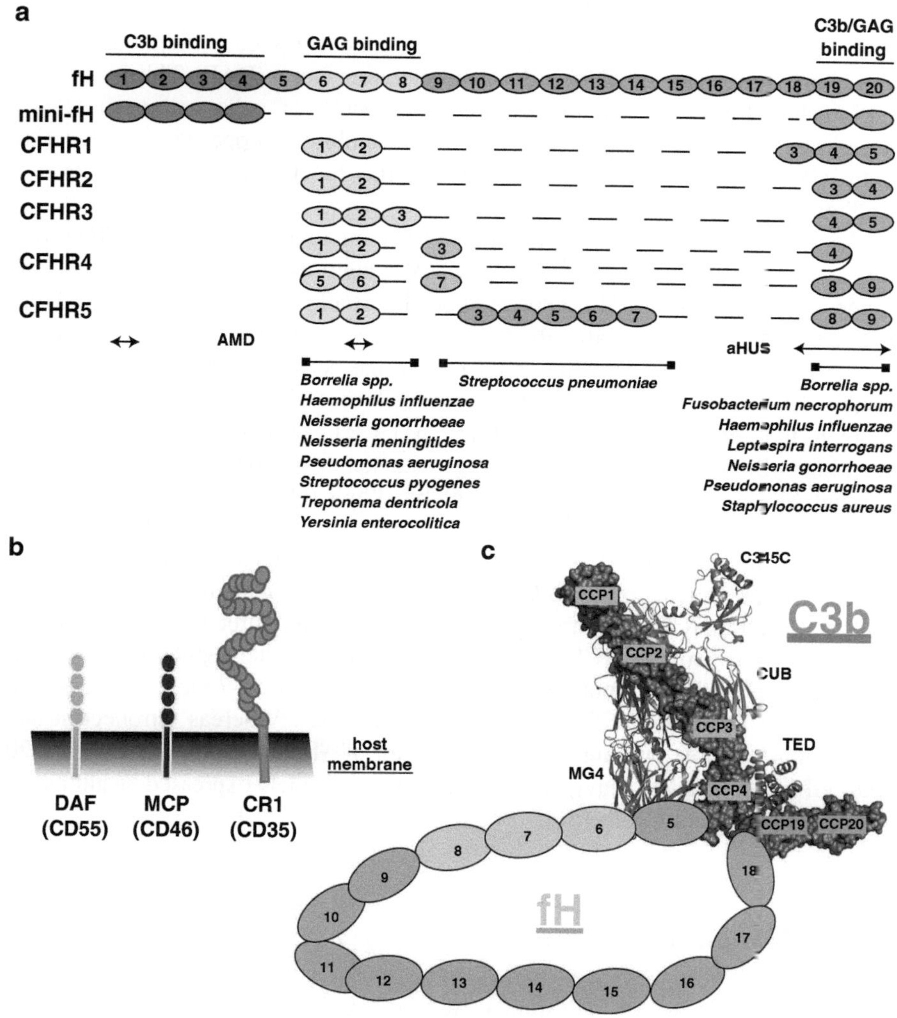

Fig. 2.2 Complement regulators built on complement control protein (CCP) modules. (**a**) Domain organization of human factor H. CCP domains involved in C3b binding are colored *red*, glycosaminoglycan (GAG) binding moieties *yellow*, CCPs binding C3d/GAG *blue*. Therapeutic molecule mini-fH is presented below. Factor H CCPs are conserved in CFHRs and are presented directly below. Regions mutated in AMD and aHUS are delimited with *arrows*. Bacterial species evading complement by binding fH are listed below with the respective fH binding regions. (**b**) Domain organization of the most prominent membrane-bound host regulators. (**c**) Structural model of fH binding to C3b [34, 36] (Color online)

cofactors for C3b. Humans have three CCP-based membrane-bound complement regulators as well (Fig. 2.2b). The first is CD46, also known as the membrane cofactor protein (MCP). It is a cofactor for factor I-mediated cleavage of C3b and C4b into iC3b and iC4b. CD46 is ubiquitously expressed on all nucleated cells, thus only erythrocytes lack CD46 [45]. The extracellular part of CD46 contains four CCP domains that harbor the C3b and C4b binding sites. The second membrane-bound complement regulator is CD55 or decay accelerating factor (DAF). This is a 70 kDa GPI-anchored glycoprotein expressed on a variety of cells and tissues [46]. Membrane-bound DAF exerts its complement-inhibitory properties by disrupting both AP and CP C3 and C5 convertases, and this function resides within its four CCP domains [47].

Another important complement regulator is complement receptor 1 (CR1). It is a type 1 transmembrane protein composed of numerous CCP domains (44 for the longest allelic variant). CR1 is expressed on almost all peripheral blood cells except platelets, NK- and T-cells [48, 49]. Apart from peripheral blood cells, CR1 is found in some tissues. It plays an important role in the germinal centers of the lymph nodes where it is found on follicular dendritic cells capturing complement-opsonized antigens that serve to stimulate B-cells [50, 51]. CR1 can bind to both C3b and C4b with high affinity and to iC3b and C3d(g) with somewhat lower affinity [52]. Both CP and AP C3 and C5 convertases are inhibited by CR1 via its decay-accelerating activity. It also serves as cofactor for factor I-mediated degradation of C3b and C4b.

Three proteins not based on CCP domains function as MAC assembly inhibitors. Vitronectin (S-protein) binds to C5b-7 through the C5b-7 membrane-binding site and the resulting SC5b-7 complex associates with C8 and three molecules of C9 to form the 1 MDa soluble SC5b-9 complex [53, 54]. Another regulator is clusterin (also called SP-40,40 or apoliporotein J) which impedes C5b-7 membrane association and the addition of C9 to C5b-8 and C5b-9 [55]. Whereas vitronectrin and clusterin are soluble proteins, the third inhibitor of MAC assembly, CD59, is a 20 kDa, GPI-anchored and heavily glycosylated protein widely expressed on almost all tissues and circulating cells [56].

2.1.3 *Complement-related Diseases*

Uncontrolled and excessive complement activation leads to tissue damage and pathogenesis. The molecular details of how complement proteins contribute to a wide variety of disease conditions and how control of complement may be re-established has recently been extensively reviewed [57–59]. Here we will present some prominent conditions in which animal models, studies of individuals with mutations in complement proteins, and clinical usage of a C5 antibody suggest that therapeutic control of complement is clinically relevant.

The most frequent cause of blindness (50 %) affecting elderly persons is age-related macular degeneration (AMD) with more than 30 million people affected worldwide. AMD causes significant changes in the retinal anatomy and deposition of drusen. This can develop into either neovascular/wet AMD with invasion of blood vessels into the retina or dry AMD with constriction of blood vessels, photoreceptor degeneration, and geographic atrophy [60]. The dry form is currently untreatable whereas wet AMD is treated with intravital injections of VEGF antibody.

AMD is strongly correlated with insufficient control of the AP caused by mutations in fH, C3, and fI but also in other complement proteins [61–64]. The best characterized mutant, fH Tyr402His, accounts for 50 % of the heritability of AMD, and causes the altered fH to bind more weakly to host cell glycans as the tyrosine is located in the ligand binding site of fH [32]. Since fH has AP convertase decay and co-factor activity for fI in C3b degradation, host cells are less protected than in the presence of normal fH. The same effect is caused by two C3 mutations for which fH co-factor activity is reduced [65], and fI variants which are less expressed and secreted [66]. Further supporting the link between AMD and high convertase activity, mutations in fB causing weaker association with C3b provide protection against AMD [67]. Despite these connections between AMD and complement, clinical trials with complement inhibitors targeting either the AP (C3 and factor D) or the TP (C5) have so far been disappointing, and so far no complement based therapeutic for AMD has reached phase III clinical trials [60].

Sepsis is a systemic inflammatory condition established by infectious agents, which leads to an excessive immune response resulting in host damage [68]. A 'cytokine storm' occurs and together with intravascular coagulation triggers multi-organ failure. The two C5a receptors C5aR and C5L2 have been implicated in the pathogenesis of sepsis by contributing to the cytokine storm and suppression of the oxidative burst in neutrophils, thereby, hampering the elimination of the infectious agent [69]. In a rat cecal ligation and puncture (CLP) sepsis model, C5a antibodies effectively decreased bacteremia, increased survival, restored H_2O_2 release by blood neutrophils and reduced coagulation [70, 71]. Neutrophil function was also restored in a mouse CLP model upon treatment with a cyclic peptide C5aR antagonist [72]. Furthermore, in a CLP model of sepsis simultaneous blockade of both C5aR and C5L2 with antibodies or a C5aR/C5L2 antagonist (C5a-based antagonist (see below)) increased survival compared to inhibition of the receptors one at a time [73].

Ischemia–reperfusion (I/R) injuries are caused by a reduction of blood flow to tissues and organs followed by re-establishment of the blood flow during which there is an accumulation of leukocytes in the vascular epithelium, upregulation of vascular pro-inflammatory molecules and reactive oxygen species [74]. Complement activation has been observed in many different organs undergoing I/R, including the gastrointestinal system, brain, lung, and kidneys [75]. I/R damage occurs during kidney transplantation, and locally-produced complement proteins are the important mediators of damage in this case [76, 77]. The molecular mechanism of complement-mediated damage during I/R appears to be intricate and tissue-dependent. In renal I/R, the TP plays a central role since both inhibition of C5b-9 assembly in C6 deficient mice and inhibition of C5aR with a small molecule antagonist reduce I/R damage [78, 79]. Whereas MBL-deficient mice are protected [80], absence of C4 does not provide protection [78], suggesting that C3 cleavage bypasses the C4b-based CP C3 convertase. In mouse models of myocardial and gastro-intestinal I/R injury both knockout of MASP-2 and MASP-2 inhibition with antibodies conferred protection and again C3 deposition was found to be independent of C4 [81]. Once C3b has been generated, further amplification takes place through the AP explaining the benefits in a mouse ischemic stroke model of fB absence, but with no protection afforded by C6 knockout [82]. In a mouse model of myocardiac I/R injury absence of MBL and MASP-2 [81, 83] attenuates injury, and interestingly, when given at pharmacological doses, the naturally occurring MAP-1 (Map44) protein competing with MASP-1/2 for binding to MBL and ficolins also inhibits I/R injury in addition to inhibiting thrombosis [84].

Rheumatoid arthritis (RA) is a chronic autoimmune inflammatory condition conferring synovial joint damage [85] to which complement TP has been acknowledged as an important contributor [86]. Aggregation of self-antigen-antibody complexes leads to the formation of complement-activating immune complexes in the synovial tissue [87]. Several complement proteins are involved in RA pathogenesis, but C5 is likely to play a pivotal role in complement-mediated tissue damage within RA [86]. Studies on C5-deficient murine models with type II collagen-induced arthritis (CIA), suggest a role of C5 in the pathogenesis [88], and administration of anti-C5 antibodies in C5-sufficient murine models prevents the onset of CIA and significantly reduces the severity of the disease during active CIA [89]. Signaling of C5a through C5aR and C5L2 is believed to play a significant role in the pathogenesis of RA [90–92] and vaccination with a C5a-fusion protein reduced arthritis severity and incidence in a mouse model [93]. However, somewhat disappointing considering these results from animal models, administration of the C5aR antagonist PMX53 failed to reduce synovial inflammation in RA patients [94].

In humans the terminal pathway and MAC formation is an important contributor to pathophysiology in paroxysmal nocturnal hemoglobinuria (PNH), paroxysmal cold hemoglobinuria [95], and atypical hemolytic urelytic syndrome (aHUS) [96]. PNH is caused by a *PIGA* gene mutation in haemopoietic stem cells resulting in deficiency of all GPI-anchored proteins on progeny cells [97–99] including CD59 and DAF. In PNH patients up to 90 % of erythrocytes are lysed by MAC assembly [100] resulting in anemia, intravascular hemolysis, and thrombosis. PNH is routinely treated by administration of a humanized monoclonal antibody (Eculizumab), preventing C5 cleavage and, thus, the assembly of MAC and the formation of C5a [101]. Eculizumab-treated patients are much more susceptible to neisserial infections and need to be immunized prior to treatment with the C5 antibody.

Asthma is a chronic airway disease characterized by inflammation of the upper respiratory tract, reversible airway obstruction, mucus cell hyperplasia and airway hyperresponsiveness (AHR) [102]. These outcomes are mediated by a T helper type 2 (Th2) polarized immune response, and the anaphylatoxins C3a and C5a have been implicated in both the sensitizing and the effector phases of the disease by regulating the adaptive immune response to allergens [103]. Through studies targeting C5aR by monoclonal antibodies or a C5a-based antagonist, it was suggested that C5a signaling mediates tolerance to aeroallergens by altering the ratio of immunogenic myeloid dendritic cells (mDCs) to tolerogenic plasmacytoid dendritic cells (pDCs), suppressing the Th2 immune response [104]. In contrast to C5a, C3aR knock-out mice develop less pronounced AHR when treated with ovalbumin (OVA) [105], and blocking of complement activation with the recombinant soluble form of the rodent-specific complement regulator Crry decreases airway inflammation in already sensitized mice, by a decrease in both pulmonary eosinophils and immunogenic

Th2 cytokine levels in bronchoalveolar lavage fluid (BALF) [106]. Increased levels of both C3a and C5a were found in BALF from asthmatic patients when compared to normal individuals [107], suggesting that both anaphylatoxins function during the effector phase [106, 108].

2.2 Inhibitors Targeting Complement

2.2.1 Inhibitors Targeting the Convertases

Convertases are obvious targets for complement inhibition, although therapeutics targeting these will in most complementrelated disease conditions have to be given systemically, thus, requiring high doses. Convertases targeting has been exploited by pathogens to evade complement and in research and drug development for controlling complement activation at either C3 or C5 cleavage stage [109]. To understand in detail the mechanism of known convertase inhibitors and how to develop new ones, it is helpful to investigate the structure of convertases, their substrates and the complexes these molecules form with their inhibitors (Fig. 2.3). Considerable structural information has been generated concerning the AP C3 convertase. Structures of the proconvertase C3bB by EM and crystallography revealed how fB, in a MIDAS-Mg^{2+} dependent manner, associates with the C-terminal C345c domain of C3b, that is flexibly attached to the remaining relatively rigid part of C3b [110, 111]. The proconvertase can exist in two states, open and closed, differing by a rotation of the fB SP domain, and only in the open conformation is the scissile bond region exposed and accessible to fD, which binds primarily through a fB exosite located 25 Å from the scissile bond [110]. The relevance of this exosite is evidenced by the ability of the anti-factor D mAb, developed for localized complement inhibition in the eye by Genentech, to prevent fD binding to this exosite by steric hindrance [112].

Once the activation of the convertase has taken place it dissociates in minutes, but the *S. aureus* protein SCIN (see below) binds tightly to the AP C3 convertase and prevents it from binding C3. A dimeric form of the C3bBb-SCIN complex in which two SCIN molecules bridge two C3bBb complexes were crystallized and revealed the basic architecture of C3bBb. The only contact here is between the C3b C345c domain and the Bb von Willebrand factor type A (vWA) domain, whereas, the SP domain is extending away from C3b [113]. Deeper insight into substrate recognition by the convertases was obtained with the structure of C5 in complex with the C3b homolog cobra venom factor (CVF). C5 and CVF interact in a head-to-head manner with the long axis of the two molecules aligned and with two separated points of contact. The largest of these is formed between the MG4 and MG5 domains from both proteins. The remaining intermolecular contacts are formed between the C5 MG7 domain and the CVF MG6 and MG7 domains [114]. Compared to the structure of both unbound C5 [115] and human C3 [116], CVF-bound C5 undergoes a significant conformational change that is necessary to establish the two-points interaction, and this conformation has also been captured by crystal

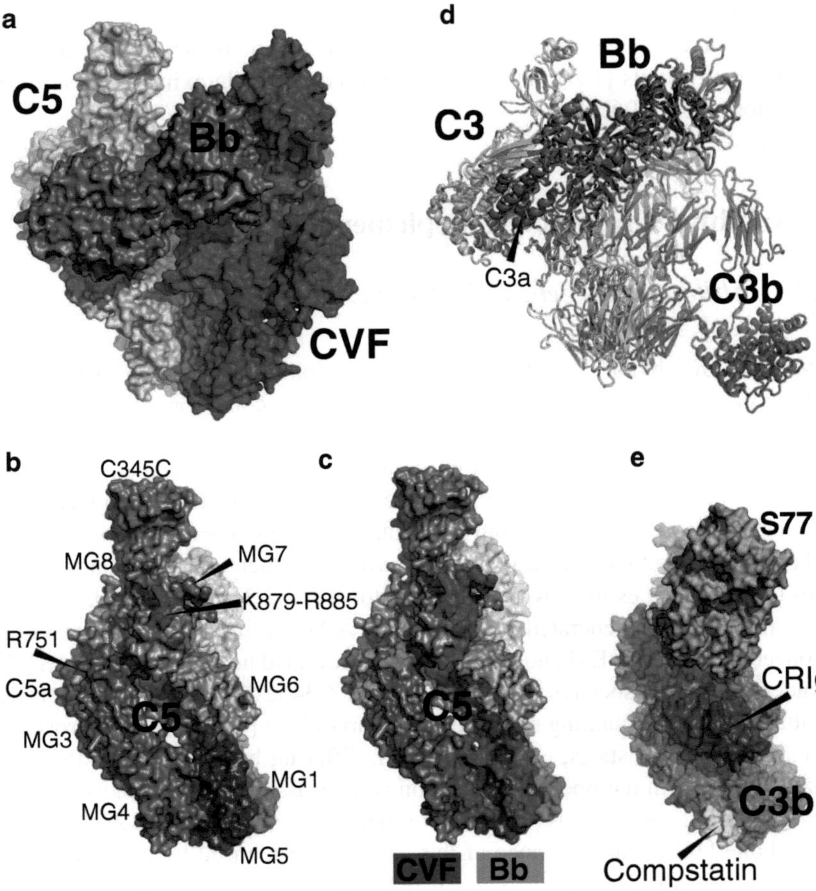

Fig. 2.3 A general model for convertase-substrate complexes explains the inhibition mechanism for convertase inhibitors. (**a**) The model of the CVFBb convertase bound to C5 [114]. (**b**) Complement C5 colored and labelled according to its domains as seen from CVF, the Eculizumab epitope K879-R885 is indicated. (**c**) As in *panel* **b**, but surface areas in contact with CVF and Bb in *panel* **a** are colored *magenta* and *green*, respectively. Notice the overlap between the Eculizumab epitope and the CVF-interacting area of the C5 MG7 domain explaining how the antibody prevents C5 cleavage. (**d**) A model of the C3 substrate bound to the AP convertase C3bBb. The C3a domain released is marked in *red*. (**e**) C3b within the AP C3 convertase as seen from the C3 substrate. The three convertase inhibitors S77, CRIg and Compstatin effectively prevent substrate recognition by steric hindrance (Color online)

packing for bovine C3 [117]. The combination of the SCIN-stabilized C3bBb structure and the C5-CVF structure led to the formulation of a general model for convertase-substrate interactions (Fig. 2.3). Since the catalytic subunit C2a/Bb is common for the CP/AP C3 and C5 convertases, it was suggested that the orientation of the substrates C3 and C5 with respect to the catalytic subunit is similar in the two types of convertases [114], although covalent or non-covalent association of either C3 convertase with C3b switches the specificity from C3 to C5 [118–120]. The additional C3b molecule lowers the K_m value for the C5 substrate by a factor of 100–1000 [121, 122], but whether this is through a direct interaction with C5 or induced conformational change in the C3 convertase remains open. The validity of this convertase-substrate model is emphasized by its ability to rationalize how some of the best characterized man-made and microbial complement inhibitors exert their effect. Prominent examples are the antibodies S77 and Eculizumab, the CRIg ectodomain, and the cyclic compstatin peptide, as described in the following.

The S77 mAb interferes with both AP convertases [123], which can be explained by its binding to C3b MG6 and MG7 domains (Fig. 2.3e), that are both predicted to recognize the substrate C3/C5 MG7 domain [114]. A similar inhibitor is the ectodomain of the CRIg complement receptor. This receptor is found on tissue resident macrophages and plays an important function in clearance of pathogens from circulation through interaction with C3b and iC3b on complement-opsonized activators [124]. The binding site on C3b has been mapped to the MG3, MG4, MG5, MG6 domains and the LNK region (Fig. 2.3e), and the ectodomain inhibits AP C3 and C5 convertase activity [125]. The Compstatin peptide is frequently used as a general complement inhibitor and through its binding to C3 it blocks both the CP and AP C3 convertases by interfering with binding of C3 to the convertases. Compstatin also binds to C3b, iC3b and C3c [126]. One disadvantage of compstatin is the high C3 plasma concentration (7 μM) and the rapid clearance of peptides, but new compstatin analogues with sub-nanomolar K_d have substantially increased half-lives [127]. The peptide binds in a groove between the MG4 and MG5 domains of C3 or its fragments [128] (Fig. 2.3e). As these domains in the substrate C3 are predicted to be recognized by the convertase, this explains the ability of compstatins to suppress C3 cleavage by either AP or CP convertases. Like compstatin, the C5 antibody Eculizumab hinders the binding of the substrate C5 to the convertase and, thereby, prevents its cleavage. However, Eculizumab binds to an epitope far from the convertase cleavage site in the MG7 domain centered on residues 879–885 [129, 130], which overlaps substantially with the area of C5 in contact with CVF in their complex [114] (Fig. 2.3b, c).

2.2.2 *Regulator-derived C3 and C5 Inhibitors*

Over the past years, the development of complement-targeted therapeutics has been inspired by natural complement regulators. The idea behind the rational design of such therapeutics is the use of naturally-occurring host proteins. Strategies involving complement targeting with systemically-administered inhibitors of such type have, however, a few drawbacks. Firstly, inhibiting complement systemically could have detrimental consequences in regard to the host immune defence. Secondly, due to the relatively low affinity of these molecules (μM range) the administered dose would need to be very high. Thus, a more focused strategy is needed. Indeed, direct targeting of the surfaces where complement activation occurs is a more intelligent and less costly approach. fH would be very attractive in therapeutics as it is a complement modulator that is a self-molecule, acts in the fluid-phase as well as on surfaces and does not interfere with complement defence against pathogens. Purified fH was shown to be able to control the C3 convertase in a mouse model of fH deficiency [131]. In vitro assays in the presence of aHUS-associated anti-fH autoantibodies showed that fH protected self-cells from complement. However, due to its size (155 kDa), presence of glycans and 40 disulfide bridges, recombinant production of full-length fH is challenging for routine administration to patients.

By the use of protein engineering it has been possible to develop mini-fH, a compact version of fH harbouring fH CCP 1–4 linked to CCP 19–20 (Fig. 2.2a). This molecule contains fH binding sites to C3b/iC3b/C3d, retains its decay acceleration and cofactor activity and, very importantly, has the ability of fH to recognize host surface-specific glycans. Interestingly, this molecule surpasses fH in both affinity towards C3 activation products and the ability to control AP activation [132, 133]. Mini-fH has potential as a therapeutic molecule in complement-mediated diseases such as aHUS, PNH or C3 glomerulopathies although its clinical properties remain to be investigated. Another inhibitor based on the fH ability to bind C3b is TT30 [134]. It is a chimeric protein developed by Alexion containing fH regulatory domains CCP 1–5 and CR2 CCP 1–4 responsible for binding to the C3b thioester domain. TT30 was extensively tested in the case of complement-mediated hemolysis and found to inhibit MAC formation resulting from AP but not CP or LP activation.

Part of the recent research in complement therapeutics has been focused on preventing complement activation on blood-exposed materials, such as implants. A great deal of research has been evolving around heparin as it has been expected to recruit fH and inhibit complement activation on the surface of biomaterials. As expected, coating the biomaterials with high densities of heparin resulted in inhibition of complement activation [135–137] but it was later found that this inhibition was fH-independent [138]. An alternative approach, employed recently, consists of coating the bioimplant surfaces with fH binding peptides but not interfering with its activity as complement regulator [139] but this approach remains to be tested in a clinical setting.

Since CR1 possesses cofactor activity for factor I-mediated proteolysis of C3b and C4b, its potential in therapeutics has been extensively studied. Soluble recombinant CR1 (sCR1) was studied in animal models of autoimmune and inflammatory disorders such as glomerulonephritis [140], myocardial infarction [141] and autoimmune thyroiditis [142]. In mice, administration of sCR1 resulted in resolution of inflammation and these encouraging results drove Avant Immunotherapeutics to develop sCR1 (TP10) for managing complement activation following coronary artery bypass graft surgery. TP10 was shown to be safe and well-tolerated in patients. However, its efficacy was far greater in male patients, and this formulation has been withdrawn from the market but its potential as a therapeutic agent has not been disputed. In fact, it has been successfully used in I/R injury when administered intravenously [143]. Nevertheless, since it is produced in Chinese hamster ovary cells as a 240 kDa glycoprotein [141, 144] and has to be injected systemically, it is an expensive therapeutic agent. An alternative approach has been used for the development of Microcept [145], a molecule readily produced in *E. coli* containing only the first 3 CCP domains of CR1, sufficient for inhibition of C3 and C5 activation by the AP and CP. Microcept contains a synthetic thiol-reactive myristoylated basic peptide tail attached through a C-terminal cysteine residue allowing its insertion into biological membranes and, thus, targeting and protecting the cells directly.

CD59 is an important complement regulator as evidenced by PNH patients, and the protein has been explored as a therapeutic agent since it only blocks MAC formation. Although CD59 has little effect as MAC inhibitor when administered systemically in a soluble form, if targeted to the sites of MAC formation, recombinant CD59 could be an effective complement modulator. Several strategies have explored generating CD59 chimeric proteins. One of them consisted of fusing CD59 to DAF and a GPI anchor, thus allowing its insertion into the membranes and its localization where C3 and C5 convertases are present [146]. Another strategy involved fusing CD59 to CR2 and DAF to CR2 [147]. These CR2 fusion proteins were efficient complement modulators in mouse models of lupus nephritis. In conclusion, fusing CD59 to a cell-surface targeting moiety may be an efficient therapeutic strategy in disease conditions where only MAC formation needs to be inhibited.

2.2.3 Inhibitors Targeting the Anaphylatoxin/Receptor Axis

The anaphylatoxins C3a and C5a produced by the complement proteolytic cascade and their associated receptors are important targets for the therapeutical treatment of inflammatory disorders. C3a and C5a mediate their pro-inflammatory effects by signaling through the G-coupled protein receptors C3aR and C5aR, respectively. This leads to chemotaxis, oxidative burst, production of pro-inflammatory molecules and activation of the adaptive immune system [148]. C3a is generally considered as a weaker pro-inflammatory inducer than C5a. To control the signaling exerted by both anaphylatoxins, carboxypeptidases cleave their C-terminal arginine resulting in C5a-desArg and C3a-desArg. C5a-desArg partially maintains C5aR binding and signaling activity whereas C3a-desArg is devoid of any signaling through C3aR [149, 150]. The 7TM receptor C5L2 binding C5a, C5a-desArg, C3a, and C3a-desArg, has long been considered as a decoy receptor, but there are now reports suggesting that it actively participates in orchestrating pro-inflammatory events [151]. Plasma levels of anaphylatoxins are elevated in various disease settings, and they can be used as biomarkers in a number of inflammatory disorders [148].

Targeting of the anaphylatoxin-receptor axis provides a selective way of down-regulating complement-mediated inflammation and is, therefore, an attractive therapeutic strategy since opsonization and MAC formation are preserved. Focus has so far been given to the C5a-C5aR axis, although C3aR antagonists are also in development as extensively reviewed recently [152]. In humans, C5a is a glycosylated protein that adopts either a compact four-helix bundle core with a flexible C-terminal extension [153] or a three-helix bundle for C5a-desArg [154]. The proposed binding interface for C5aR on C5a is hidden within the C5 molecule [115]. Thus targeting molecules can mostly be directed either towards the exposed C5a surface present on C5 or towards the C5aR binding surface only available on the released anaphylatoxin. The first strategy presents more risks since it can generate global C5 inhibitors, if large inhibitory molecules are employed impairing C5 cleavage and, thereby, the entire TP. This effect is in some clinical contexts undesirable, e.g. in the treatment of sepsis. Inhibitors directly interfering with C5a-C5aR interaction while preserving other C5 functions are, therefore, more appealing.

Several groups and companies have developed potent C5a monoclonal antibodies [155]. The use of C5a antibodies for preventing multiple organ failure and improving survival rate in sepsis has been documented since the 1980s [156]. Of interest among these, the anti-C5a mAb 137-26 directly binds to the C5a moiety on C5 without inhibiting C5 cleavage and subsequent MAC formation [157]. It is commercialized as TNX558 by Tanox/Genentech and has been in preclinical development for inflammatory diseases [158]. Another humanized C5a monoclonal antibody, CaCP29 (IFX-1), developed by InflaRx GmbH, has passed Phase I clinical trials in Germany for human sepsis [159]. Aptamer approaches have also been considered for the blockade of C5a function. NOXXON Pharma has developed a class of aptamers called Spiegelmers® [160] built on nucleotides containing L-ribose making them the mirror images of D-ribose containing RNA. As therapeutics, Spiegelmers are much more resistant to nuclease degradation compared to conventional aptamers giving high stability in the blood. C5a has been efficiently targeted by Spiegelmers® NOX-D19 and NOX-D20, which have shown promising results in reducing vascular injuries after transplantation and in attenuating organ damage during sepsis [151, 161]. NOX-D20 binds to both human and murine C5a with picomolar affinities but also human C5 with similar affinity although NOX-D20 does not prevent C5 cleavage. These mirror-image L-RNA aptamers, therefore, appear to be promising anti-C5a therapeutics. An alternative strategy involving immunization with a recombinant MBP-C5a fusion protein resulted in the production of neutralizing C5a antibodies [93]. A new study instead used a recombinant C5a molecule modified with unnatural amino acids. A single replacement was sufficient to induce the production of anti-C5a antibodies capable of blocking the C5a-C5aR interaction, leading to significant relief of the clinical symptoms in a mouse model of rheumatoid arthritis [162].

In relation to C5a-targeting molecules the anaphylatoxin receptors have also been a focus for inhibitor development. A large number of C5aR inhibitors have been developed. A well described cyclic peptide antagonist known as PMX53 or 3D53 targets C5aR and competes with C5a binding to human polymorphonuclear neutrophils (PMNs) with an IC_{50} value of 300 nM, but does not bind human C5L2 [148, 163]. The antagonistic activity of PMX53 was measured, showing an inhibition of myeloperoxidase release from PMNs with an IC_{50} of 20 nM [163]. NMR studies of PMX53 suggests a β-turn motif in the molecule [164]. Mitsubishi Pharmaceuticals Company developed the orally active small molecule C5aR antagonist W54011 with a K_i of 2.2 nM. It inhibits calcium mobilization in human neutrophils and C5a-induced neutropenia in gerbils, and shows no inhibition of

C5a-binding to C5L2 [165, 166]. Another small molecule, NDT9513727, specifically targeting C5aR, reduces the constitutive GTPγ[^{35}S] binding of human C5aR-coupled G-proteins, and, therefore, functions as an inverse agonist [167].

Proteins have also been selected to target C5aR. Developing human C5aR knock-in mice responding to endogenously produced C5a facilitated the selection of anti-C5aR antibodies showing promising results in preventing and reversing serum-induced inflammation in the knock-in mice. The most potent anti-C5aR antibodies bind to the second extracellular loop of the receptor [90], which was previously identified as important for balancing the activity of C5aR, since mutations in this region resulted in constitutively active receptors [168]. The C5a molecule has also been exploited as a scaffold to generate potent competitors of C5a receptors, and from a phage library an antagonist called ΔpIII-A8 was selected which reduced intestinal injury and lung vascular permeability and increased survival of (I/R) injured mice [169]. A shortened version ($A8^{\Delta71\text{-}73}$) targeting both C5aR and C5L2 was developed later and, interestingly, it was found that one particular residue determines agonism versus antagonism of A8 related proteins. In C5a, this amino acid is an aspartate and in $A8^{\Delta71\text{-}73}$ it is mutated to an arginine [170].

For C3aR inhibitor developments have mainly focused on peptide analogs of the C-terminal region of C3a. Hexapeptides mimicking the C-terminus of C3a were synthesized resulting in both agonists and antagonists. The most potent agonists shared an N-terminal phenylalanine, a tryptophan or leucine at the second position and the highly conserved C-terminal sequence Leu-Ala-Arg. Substituting the fourth leucine to the bulky cyclohexylalanine resulted in antagonists [171]. By NMR, one of the most potent agonistic peptides was found to adopt a *β*-turn motif similar to C5aR ligands [172]. As an alternative to peptides, small-molecule compounds targeting C3aR have also been discovered, with an example being the functional antagonist SB290157 with an IC_{50} of 200 nM [173]. Although this drug showed anti-inflammatory activity in a guinea pig model [174], a more recent paper found this molecule to have partial agonistic activity, a conflict that might be explained by differences in receptor density in the systems used [175]. Numerous other drugs including proteins, peptides and small molecules have been developed targeting the anaphylatoxin receptors and have been extensively reviewed elsewhere [148, 152, 176].

2.3 Bacterial Strategies for Immune Evasion: What Can Be Learnt from Them

Complement primary function resides in the host defence against pathogens, but many pathogens successfully evade complement [177]. One can learn from these evasion strategies on how to inhibit complement at various stages of the cascade and apply this knowledge to design inhibitors for therapeutic applications. Many pathogen inhibitors have also proven useful to gain structural insights into complement mechanistics by freezing complement proteins and their complexes in specific functional states with the ability of the *S. aureus* protein SCIN to stabilize the otherwise rapidly dissociating C3 convertase as the best example. We will review here some of the best characterized pathogen inhibitors and try to comprehend how their interaction with complement may inspire attempts to make potent, selective inhibitors of complement.

2.3.1 Pathogen Inhibitors Targeting C3

Staphylococci are quite versatile organisms with respect to complement evasion. Targets of choice for staphylococcal proteins are the C3 and C5 convertases [178] as already described for SCIN above. Three other potent complement inhibitors from *Staphylococcus aureus* - Efb, the closely related Ehp, and Sbi - are binders of C3 and its cleavage products, C3b and iC3b, through preferential interaction with the C3 thioester domain, with the C3b degradation product C3d as the minimal binding partner. Crystal structures of the Ehc:C3d, Efb-C:C3d and Sbi:C3d complexes revealed that their primary binding site on C3d is partially inaccessible within the intact C3b molecule [179–181] (Fig. 2.4a). As a consequence of their binding to this hindered site, Efb and Ehp induce an altered overall conformation of C3 and C3b. Displacement of the C3d thioester domain is relayed to the rest of the molecule, therefore preventing its further participation in the downstream events of the complement cascade, including formation of the AP convertases and covalent deposition on the pathogen surface. In the case of Sbi, a secondary binding site on C3d allows the inhibitory protein to form a covalent adduct with activated C3 forms (C3b and/or C3(H_2O)), thereby directly interfering with AP activation [181] (Fig. 2.4a). In addition, all three inhibitors compete with complement receptor 2 for the binding of C3d and, therefore, impede the stimulation of B cells mediated by this receptor [182].

2.3.2 Pathogen Inhibitors Targeting C5

Several pathogen inhibitors block the C5 convertase by interacting with C5. OmCI, a small protein from the soft tick *Ornithodoros moubata*, binds directly to C5 and inhibits its cleavage into C5a and C5b [183]. Structural studies of the OmCI:C5 complex suggested that OmCI binding to C5 fixes the C5 C345c domain and inhibits convertase binding to C5 by impairing the conformational flexibility of this domain [115]. A recombinant version of OmCI, rev576, has shown promising results in models of auto-immune neuromuscular diseases and sepsis [184–186]. The SSL7 protein from *Staphylococcus aureus* likewise binds C5 and prevents its proteolytic processing by the C5 convertase, thereby impairing MAC-mediated bacteriolysis and C5a release. Its mechanism of action was elucidated through structural analysis of the C5:SSL7 complex [187]. SSL7 binds to a surface patch quite distant from the C5 cleavage site suggesting a more complex mechanism than

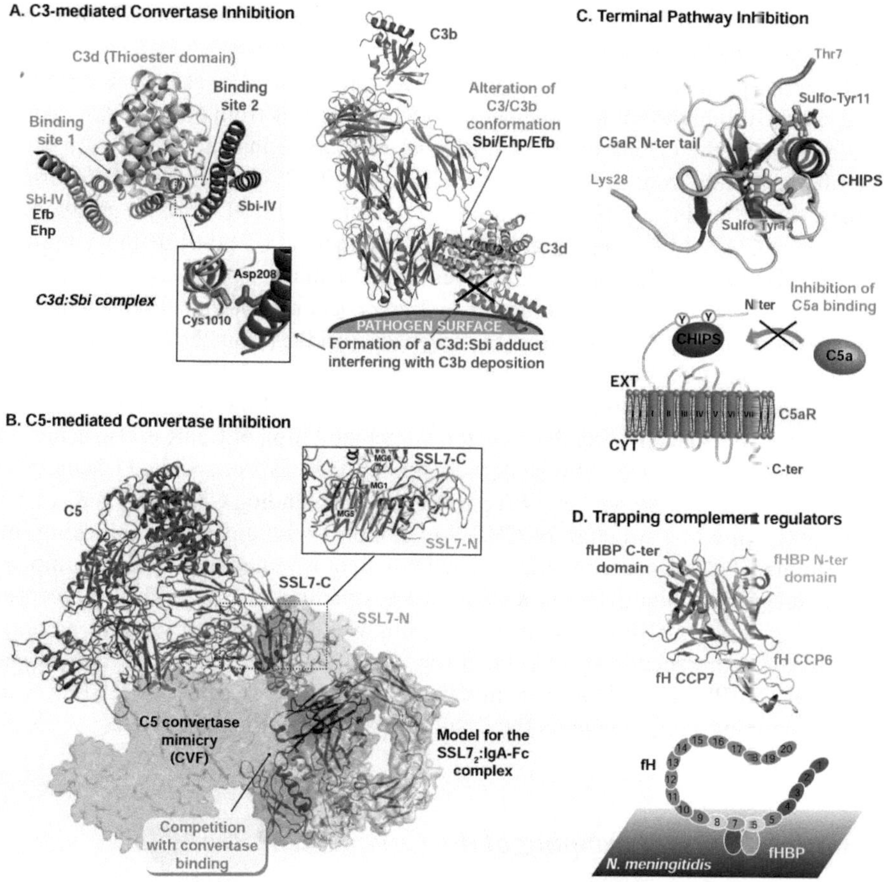

Fig. 2.4 Complement evasion by pathogens. (**a**) The two binding modes of staphylococcal protein Sbi onto C3d [181] and their docking onto C3b [210], revealing the structural model for their inhibition of the C3 convertase function. (**b**) Structural model for SSL7 inhibition of the C5 convertase by dual recruitment of C5 and IgA, based on the C5:SSL7 [187], C5:CVF [114] and SSL7:IgA-Fc [211] complex structures. (**c**) Structure of the inhibitor CHIPS in complex with a peptidic fragment of C5aR [193] and model for its inhibition of C5a binding by blocking the C5aR N-terminal docking site around the two sulfo-tyrosines. (**d**) Structure of the complex between fH CCPs 6–7 and fH-binding protein (fHBP) from *N. meningitidis* [199] and scheme for fH trapping on the pathogen surface

simple steric hindrance of the C5 cleavage site. This was later confirmed by the structure of the ternary complex between C5, SSL7 and CVF [114]. Together these studies explained how simultaneous recruitment of C5 by SSL7 C-terminal domain and IgA by SSL7 N-terminal domain prevented convertase recognition of C5 (Fig. 2.4b). Interestingly, the isolated C-terminal domain of SSL7 devoid of affinity for IgA, did not prevent C5 cleavage and permitted bacteriolysis while preserving a very low hemolytic activity on erythrocytes [187] suggesting that this domain may indicate a direction for development of a therapeutic agent targeting complement-dependent hemolytic diseases.

Bacterial proteins that directly target MAC formation have also been reported. Streptococcal Inhibitor of Complement (SIC) and variants from other *Streptococcus* strains bind to the C5b67 complex hindering membrane insertion [188]. An outer membrane protein from *Escherichia coli* K12 strain, TraT, also impairs MAC formation by direct binding to C5b6 preventing C7 recruitment [189]. *Borrelia burgdorferi* produces a CD59-like protein that blocks C9 polymerization [190]. Pathogen proteases are also potent effectors in complement evasion and C5a peptidases are found in various organisms. In particular, group A *streptococci* produce a cell-envelope proteinase (ScpA or ScpB) cleaving off the seven last residues from the C5a C-terminus, thereby shifting C5a from agonistic to antagonistic activity [190]. Proteinases from *Porphyromonas gingivalis* are also capable of shedding off the N-terminal region of C5aR [191], thus inactivating the receptor. Direct inhibition of C5aR is also achieved by pathogenic virulence factors. A well-studied case is the secreted chemotaxis inhibitory protein of *Staphylococcus aureus* (CHIPS) binding C5aR with a K_d of 1.1 nM [192]. An NMR structure of CHIPS in complex with a peptide mimicking the N-terminus of C5aR containing *O*-sulfated tyrosines revealed how CHIPS competitively antagonizes the interaction of the core structure of C5a with the receptor [193] (Fig. 2.4c). CHIPS or derivatives of this was suggested to serve as an anti-inflammatory therapeutic agent but this was questioned due to the high immunogenicity of CHIPS [194]. However, mutated or shortened versions of CHIPS with lower immunogenicity maintain the C5aR antagonizing effect [195, 196].

2.3.3 Inhibition/Trapping of the Complement Regulators

Bacteria and viruses have also developed approaches to hijack complement components to their own advantage with complement regulators fH, fHL-1 and C4BP being the preferred victims. The plethora of bacterial fH binders includes complement regulator-acquiring surface proteins (CRASP) and outer surface protein E (OspE) from *Borrelia burgdorferi*, M proteins, Fba, Scl1.6 and Hic from *Streptococci*, and fH binding protein (fHBP) from *Neisseria meningitidis*. CRASP-1 binds to CCP5–7 of fH as a dimer and clamps the bound CCP domains, thereby enhancing scavenging of the complement regulator on the pathogen cell surface [197]. OspE and neisserial FHBP, on the other hand, use a protein mimicry of the host carbohydrates to sequester fH by targeting either CCP19–20 [198] or CCP6–7 [199] (Fig. 2.4d). In both cases, the binding epitopes on fH are similar to the ones proposed for glycosaminoglycans binding on endothelial cells. Finally, certain viruses also possess complement regulators, such as the vaccinia virus complement-control protein (VCP) and the smallpox inhibitor of complement enzyme (SPICE) which both show cofactor activity for degradation of C3b [200]. http://www.nature.com/nrmicro/journal/v6/n2/full/nrmicro1824.html-B27

2.4 Discussion and Concluding Remarks

Although the concept of complement as a cornerstone in defence against microbial invasion and in homeostasis still holds true, there is also clearly a dark side to this highly complex machinery. The capacity of complement to recognize danger-associated molecular patterns from both pathogens and host cells and its interconnectivity with other branches of the innate and adaptive immune system and even the coagulation system [2], adds on to difficulty of maintaining tight control of the system. Evidently, any breach in its regulation may disturb the fine balance between protection and damage. The list of diseases that implicate complement as one of the causative elements keeps growing and encompasses a broad panel of pathologies ranging from inflammatory diseases to neurodegenerative disorders and cancers [59, 201]. Our knowledge of the intricate mechanisms at work during complement activation and complement crosstalk with parallel defence systems has gained considerable depth over the last decade, not least due to the increasing amount of available structural and biochemical data allowing a comprehension of the complement cascade in atomic details.

The success of Eculizumab [101] has encouraged many new initiatives in the field. A constantly growing number of molecules arising from both academic and industrial research efforts are now under development, with already promising preliminary results in clinical studies for several of them, as reviewed here. Many classes of molecules have been considered for these drug candidates, including antibodies, aptamers, small chemical compounds and recombinant versions of naturally occurring proteic inhibitors, e.g. based on bacterial inhibitors or host regulators. The constantly improved understanding of the mode of action of complement regulators and receptors has led to the conception of new generations of multimodular inhibitors incorporating functionalities of one or several complement regulators for efficient convertase decay and/or opsonin degradation [57]. Although many regulatory concepts have already been exploited, new ideas are emerging thanks to the availability of both functional and structural data allowing comprehending the mechanistics of complement activation at the molecular level. As one example, recent structures of C4 and its complex with MASP-2 revealed an important exosite interaction of MASP-2 with the C4 C345c domain and it could be demonstrated that a recombinant version of this domain functions as a CP/LP pathway inhibitor [202].

Evidently, targeting of complement at specific stages of the cascade is highly desirable to allow selective containment of the dysregulated pathway while preserving protective functions of the overall immune system. While targeting of central complement components such as the C3 convertase will lead to complete shutdown of the system, more refined approaches can be directed towards the initiation or the terminal steps. Interfering with the initial steps of complement activation may, for example, offer a way to specifically target one of the activation pathways while retaining normal complement defence functions through the other untargeted pathways—granted that the identified pathway is the major contributor to the pathological condition.

Targeting of the terminal steps of complement cascade has already been extensively exploited through the inhibition of C5a-mediated signaling [70, 152] and the blockade of C5 by Eculizumab [95, 101]. Nevertheless, improvement of the known strategies may still be needed as exemplified by Eculizumab. Although the antibody effectively prevents hemolytic activity by impairing MAC formation, it does not interfere with AP complement activation and subsequent opsonization of PNH cells, which thus are still preferentially marked for extravascular lysis [203]. Furthermore, the increased susceptibility to neisserial infections of the patients treated with Eculizumab still constitutes a disadvantage [204]. Another major concern for lifelong Eculizumab treatment for PNH and for some aHUS patients is the annual cost of €460,000 for an adult [130]. Despite these drawbacks, no efficient substitute has been produced yet and clinical trials for the use of Eculizumab in acute inflammatory disorders, such as antibody-mediated transplant graft rejection, are currently conducted [205]. In such settings where C5aR-mediated signaling is seen as a major contributor to the underlying inflammation, a C5a or C5aR antagonist might simply be a more promising drug candidate [203].

An emerging idea is that the inhibitor design has to be rethought for each particular disease [57]. During acute inflammation (e.g. in sepsis), large amounts of complement effectors (C3b, C4b, C5b, anaphylatoxins) will be produced. Thus, efficient inhibitors should have fast, high affinity binding capacities towards their target and a slow dissociation rate, to allow rapid and complete blocking of the complement cascade. In chronic inflammation, on the other hand, a milder modulation of complement may be sufficient to re-establish a proper balance between protective function and injury—complete shutdown of the system being avoidable in that case. Another trend within complement inhibitors is the specific cell/tissue targeting of these [206]. Local delivery at the site of injury is highly desirable if one wants to preserve systemic complement function. Efforts in that sense have already been made for example by fusing regulator-mimicking molecules to targeting modules in order to deliver engineered versions of these inhibitors to sites of complement activation, with the factor H-CR2 fusion protein TT30 as an excellent example [134, 207, 208]. Such targeting approaches may in the future be combined with nanoparticle-based delivery systems for delivery of anti-inflammatory drugs. More thoughts are to be put in this design strategy and organ- or tissue-specific delivering strategies used for other systems should be addressed as well in the complement therapeutics field. Finally, the control of complement activation may become an important issue for the future success of nanomedicine as surfaces considered as "foreign" to the complement system are introduced into the human body and often elicit activation of the system [209] and deeper structural knowledge of the mechanisms at play during complement activation will undoubtfully provide new tools to successfully overcome these challenges.

References

1. Carroll MC (2004) The complement system in regulation of adaptive immunity. Nat Immunol 5:981–986
2. Ricklin D, Hajishengallis G, Yang K, Lambris JD (2010) Complement: a key system for immune surveillance and homeostasis. Nat Immunol 11:785–797
3. Walport MJ (2001) Complement. First of two parts. N Engl J Med 344:1058–1066
4. Inforzato A, Doni A, Barajon I et al (2013) PTX3 as a paradigm for the interaction of pentraxins with the complement system. Semin Immunol 25:79–85
5. Kojouharova M, Reid K, Gadjeva M (2010) New insights into the molecular mechanisms of classical complement activation. Mol Immunol 47:2154–2160
6. Arlaud GJ, Gaboriaud C, Thielens NM et al (2001) Structural biology of C1: dissection of a complex molecular machinery. Immunol Rev 180:136–145
7. Law SK, Lichtenberg NA, Levine RP (1980) Covalent binding and hemolytic activity of complement proteins. Proc Natl Acad Sci U S A 77:7194–7198
8. Muller-Eberhard HJ, Polley MJ, Calcott MA (1967) Formation and functional significance of a molecular complex derived from the second and the fourth component of human complement. J Exp Med 125:359–380
9. Ma YJ, Skjoedt MO, Garred P (2013) Collectin-11/MASP complex formation triggers activation of the lectin complement pathway--the fifth lectin pathway initiation complex. J Innate Immun 5:242–250
10. Kjaer TR, Thiel S, Andersen GR (2013) Toward a structure-based comprehension of the lectin pathway of complement. Mol Immunol 56:413–422
11. Degn SE, Jensen L, Hansen AG et al (2012) Mannan-binding lectin-associated serine protease (MASP)-1 is crucial for lectin pathway activation in human serum, whereas neither MASP-1 nor MASP-3 is required for alternative pathway function. J Immunol 189:3957–3969
12. Heja D, Kocsis A, Dobo J et al (2012) Revised mechanism of complement lectin-pathway activation revealing the role of serine protease MASP-1 as the exclusive activator of MASP-2. Proc Natl Acad Sci U S A 109:10498–10503
13. Xu Y, Narayana SV, Volanakis JE (2001) Structural biology of the alternative pathway convertase. Immunol Rev 180:123–135
14. Elsner J, Oppermann M, Czech W, Kapp A (1994) C3a activates the respiratory burst in human polymorphonuclear neutrophilic leukocytes via pertussis toxin-sensitive G-proteins. Blood 83:3324–3331
15. Tack BF, Harrison RA, Janatova J, Thomas ML, Prahl JW (1980) Evidence for presence of an internal thiolester bond in third component of human complement. Proc Natl Acad Sci U S A 77:5764–5768
16. Lesavre PH, Muller-Eberhard HJ (1978) Mechanism of action of factor D of the alternative complement pathway. J Exp Med 148:1498–1509
17. Harboe M, Ulvund G, Vien L, Fung M, Mollnes TE (2004) The quantitative role of alternative pathway amplification in classical pathway induced terminal complement activation. Clin Exp Immunol 138:439–446
18. Harboe M, Garred P, Karlstrom E, Lindstad JK, Stahl GL, Mollnes TE (2009) The downstream effects of mannan-induced lectin complement pathway activation depend quantitatively on alternative pathway amplification. Mol Immunol 47:373–380
19. Pangburn MK, Schreiber RD, Muller-Eberhard HJ (1981) Formation of the initial C3 convertase of the alternative complement pathway. Acquisition of C3b-like activities by spontaneous hydrolysis of the putative thioester in native C3. J Exp Med 154:856–867
20. Kemper C, Atkinson JP, Hourcade DE (2010) Properdin: emerging roles of a pattern-recognition molecule. Annu Rev Immunol 28:131–155

21. Rawal N, Pangburn MK (2001) Structure/function of C5 convertases of complement. Int Immunopharmacol 1:415–422
22. Laursen NS, Magnani F, Gottfredsen RH, Petersen SV, Andersen GR (2012) Structure, function and control of complement C5 and its proteolytic fragments. Curr Mol Med 12:1083–1097
23. Tegla CA, Cudrici C, Patel S et al (2011) Membrane attack by complement: the assembly and biology of terminal complement complexes. Immunol Res 51:45–60
24. Degn SE, Jensenius JC, Thiel S (2011) Disease-causing mutations in genes of the complement system. Am J Hum Genet 88:689–705
25. Degn SE, Thiel S, Jensenius JC (2013) Recombinant expression of the autocatalytic complement protease MASP-1 is crucially dependent on co-expression with its inhibitor, C1 inhibitor. Protein Expr Purif 88:173–182
26. Liszewski MK, Farries TC, Lublin DM, Rooney IA, Atkinson JP (1996) Control of the complement system. Adv Immunol 61:201–283
27. Nilsson SC, Sim RB, Lea SM, Fremeaux-Bacchi V, Blom AM (2011) Complement factor I in health and disease. Mol Immunol 48:1611–1620
28. Pangburn MK, Schreiber RD, Müller-Eberhard HJ (1977) Human complement C3b inactivator: isolation, characterization, and demonstration of an absolute requirement for the serum protein beta1H for cleavage of C3b and C4b in solution. J Exp Med 146:257–270
29. Ferreira VP, Pangburn MK, Cortes C (2010) Complement control protein factor H: the good, the bad, and the inadequate. Mol Immunol 47:2187–2197
30. Jokiranta TS, Jaakola VP, Lehtinen MJ, Parepalo M, Meri S, Goldman A (2006) Structure of complement factor H carboxyl-terminus reveals molecular basis of atypical haemolytic uremic syndrome. EMBO J 25:1784–1794
31. Hocking HG, Herbert AP, Kavanagh D et al (2008) Structure of the N-terminal region of complement factor H and conformational implications of disease-linked sequence variations. J Biol Chem 283:9475–9487
32. Prosser BE, Johnson S, Roversi P et al (2007) Structural basis for complement factor H linked age-related macular degeneration. J Exp Med 204:2277–2283
33. Morgan HP, Mertens HD, Guariento M et al (2012) Structural analysis of the C-terminal region (modules 18–20) of complement regulator factor H (FH). PLoS One 7:e32187
34. Morgan HP, Schmidt CQ, Guariento M et al (2011) Structural basis for engagement by complement factor H of C3b on a self surface. Nat Struct Mol Biol 18:463–470
35. Kajander T, Lehtinen MJ, Hyvärinen S et al (2011) Dual interaction of factor H with C3d and glycosaminoglycans in host-nonhost discrimination by complement. Proc Natl Acad Sci U S A 108:2897–2902
36. Wu J, Wu YQ, Ricklin D, Janssen BJ, Lambris JD, Gros P (2009) Structure of complement fragment C3b-factor H and implications for host protection by complement regulators. Nat Immunol 10:728–733
37. Goicoechea de Jorge E, Caesar JJ, Malik TH et al (2013) Dimerization of complement factor H-related proteins modulates complement activation in vivo. Proc Natl Acad Sci U S A 110:4685–4690
38. Heinen S, Hartmann A, Lauer N et al (2009) Factor H-related protein 1 (CFHR-1) inhibits complement C5 convertase activity and terminal complex formation. Blood 114:2439–2447
39. Timmann C, Leippe M, Horstmann RD (1991) Two major serum components antigenically related to complement factor H are different glycosylation forms of a single protein with no factor H-like complement regulatory functions. J Immunol 146:1265–1270
40. Eberhardt HU, Buhlmann D, Hortschansky P et al (2013) Human factor H-related protein 2 (CFHR2) regulates complement activation. PLoS One 8:e78617
41. Fritsche LG, Lauer N, Hartmann A et al (2010) An imbalance of human complement regulatory proteins CFHR1, CFHR3 and factor H influences risk for age-related macular degeneration (AMD). Hum Mol Genet 19:4694–4704
42. McRae JL, Cowan PJ, Power DA et al (2001) Human factor H-related protein 5 (FHR-5). A new complement-associated protein. J Biol Chem 276:6747–6754

43. Blom AM (2002) Structural and functional studies of complement inhibitor C4b-binding protein. Biochem Soc Trans 30:978–982
44. Dahlback B, Smith CA, Muller-Eberhard HJ (1983) Visualization of human C4b-binding protein and its complexes with vitamin K-dependent protein S and complement protein C4b. Proc Natl Acad Sci U S A 80:3461–3465
45. Andrews PW, Knowles BB, Parkar M, Pym B, Stanley K, Goodfellow PN (1985) A human cell-surface antigen defined by a monoclonal antibody and controlled by a gene on human chromosome 1. Ann Hum Genet 49:31–39
46. Spiller OB, Hanna SM, Morgan BP (1999) Tissue distribution of the rat analogue of decay-accelerating factor. Immunology 97:374–384
47. Medof ME, Kinoshita T, Nussenzweig V (1984) Inhibition of complement activation on the surface of cells after incorporation of decay-accelerating factor (DAF) into their membranes. J Exp Med 160:1558–1578
48. Fearon DT (1980) Identification of the membrane glycoprotein that is the C3b receptor of the human erythrocyte, polymorphonuclear leukocyte, B lymphocyte, and monocyte. J Exp Med 152:20–30
49. Tedder TF, Fearon DT, Gartland GL, Cooper MD (1983) Expression of C3b receptors on human be cells and myelomonocytic cells but not natural killer cells. J Immunol 130:1668–1673
50. Reynes M, Aubert JP, Cohen JH et al (1985) Human follicular dendritic cells express CR1, CR2, and CR3 complement receptor antigens. J Immunol 135:2687–2694
51. Heesters BA, Chatterjee P, Kim YA et al (2013) Endocytosis and recycling of immune complexes by follicular dendritic cells enhances B cell antigen binding and activation. Immunity 38:1164–1175
52. Ross GD, Lambris JD, Cain JA, Newman SL (1982) Generation of three different fragments of bound C3 with purified factor I or serum: I. Requirements for factor H vs CR1 cofactor activity. J Immunol 129:2051–2060
53. Moskovich O, Fishelson Z (2007) Live cell imaging of outward and inward vesiculation induced by the complement c5b-9 complex. J Biol Chem 282:29977–29986
54. Podack ER, Kolb WP, Muller-Eberhard HJ (1978) The C5b-6 complex: formation, isolation, and inhibition of its activity by lipoprotein and the S-protein of human serum. J Immunol 120:1841–1848
55. Tschopp J, French LE (1994) Clusterin: modulation of complement function. Clin Exp Immunol 97(Suppl 2):11–14
56. Meri S, Waldmann H, Lachmann PJ (1991) Distribution of protectin (CD59), a complement membrane attack inhibitor, in normal human tissues. Lab Invest 65:532–537
57. Ricklin D, Lambris JD (2013) Progress and trends in complement therapeutics. Adv Exp Med Biol 735:1–22
58. Sturfelt G, Truedsson L (2012) Complement in the immunopathogenesis of rheumatic disease. Nat Rev Rheumatol 8:458–468
59. Wagner E, Frank MM (2010) Therapeutic potential of complement modulation. Nat Rev Drug Discov 9:43–56
60. Ambati J, Atkinson JP, Gelfand BD (2013) Immunology of age-related macular degeneration. Nat Rev Immunol 13:438–451
61. Zipfel PF, Lauer N, Skerka C (2010) The role of complement in AMD. Adv Exp Med Biol 703:9–24
62. Seddon JM, Yu Y, Miller EC et al (2013) Rare variants in CFI, C3 and C9 are associated with high risk of advanced age-related macular degeneration. Nat Genet 45:1366
63. Zhan X, Larson DE, Wang C et al (2013) Identification of a rare coding variant in complement 3 associated with age-related macular degeneration. Nat Genet 45:1375
64. Khandhadia S, Cipriani V, Yates JR, Lotery AJ (2012) Age-related macular degeneration and the complement system. Immunobiology 217:127–146
65. Heurich M, Martinez-Barricarte R, Francis NJ et al (2011) Common polymorphisms in C3, factor B, and factor H collaborate to determine systemic complement activity and disease risk. Proc Natl Acad Sci U S A 108:8761–8766

66. van de Ven JP, Nilsson SC, Tan PL et al (2013) A functional variant in the CFI gene confers a high risk of age-related macular degeneration. Nat Genet 45:813–817
67. Montes T, Tortajada A, Morgan BP, Rodriguez de Cordoba S, Harris CL (2009) Functional basis of protection against age-related macular degeneration conferred by a common polymorphism in complement factor B. Proc Natl Acad Sci U S A 106:4366–4371
68. Cohen J (2002) The immunopathogenesis of sepsis. Nature 420:885–891
69. Ward PA (2010) The harmful role of c5a on innate immunity in sepsis. J Innate Immun 2:439–445
70. Czermak BJ, Sarma V, Pierson CL et al (1999) Protective effects of C5a blockade in sepsis. Nat Med 5:788–792
71. Laudes IJ, Chu JC, Sikranth S et al (2002) Anti-c5a ameliorates coagulation/fibrinolytic protein changes in a rat model of sepsis. Am J Pathol 160:1867–1875
72. Huber-Lang MS, Sarma JV, McGuire SR et al (2001) Protective effects of anti-C5a peptide antibodies in experimental sepsis. FASEB J 15:568–570
73. Rittirsch D, Flierl MA, Nadeau BA et al (2008) Functional roles for C5a receptors in sepsis. Nat Med 14:551–557
74. Damman J, Daha MR, van Son WJ, Leuvenink HG, Ploeg RJ, Seelen MA (2011) Crosstalk between complement and Toll-like receptor activation in relation to donor brain death and renal ischemia-reperfusion injury. Am J Transplant 11:660–669
75. Gorsuch WB, Chrysanthou E, Schwaeble WJ, Stahl GL (2012) The complement system in ischemia-reperfusion injuries. Immunobiology 217:1026–1033
76. Vieyra MB, Heeger PS (2010) Novel aspects of complement in kidney injury. Kidney Int 77:495–499
77. Sacks SH, Zhou W (2012) The role of complement in the early immune response to transplantation. Nat Rev Immunol 12:431–442
78. Zhou W, Farrar CA, Abe K et al (2000) Predominant role for C5b-9 in renal ischemia/reperfusion injury. J Clin Invest 105:1363–1371
79. Arumugam TV, Shiels IA, Strachan AJ, Abbenante G, Fairlie DP, Taylor SM (2003) A small molecule C5a receptor antagonist protects kidneys from ischemia/reperfusion injury in rats. Kidney Int 63:134–142
80. Moller-Kristensen M, Wang W, Ruseva M et al (2005) Mannan-binding lectin recognizes structures on ischaemic reperfused mouse kidneys and is implicated in tissue injury. Scand J Immunol 61:426–434
81. Schwaeble WJ, Lynch NJ, Clark JE et al (2011) Targeting of mannan-binding lectin-associated serine protease-2 confers protection from myocardial and gastrointestinal ischemia/reperfusion injury. Proc Natl Acad Sci U S A 108:7523–7528
82. Elvington A, Atkinson C, Zhu H et al (2012) The alternative complement pathway propagates inflammation and injury in murine ischemic stroke. J Immunol 189:4640–4647
83. Walsh MC, Bourcier T, Takahashi K et al (2005) Mannose-binding lectin is a regulator of inflammation that accompanies myocardial ischemia and reperfusion injury. J Immunol 175:541–546
84. Pavlov VI, Skjoedt MO, Siow Tan Y, Rosbjerg A, Garred P, Stahl GL (2012) Endogenous and natural complement inhibitor attenuates myocardial injury and arterial thrombogenesis. Circulation 126:2227–2235
85. McInnes IB, O'Dell JR (2010) State-of-the-art: rheumatoid arthritis. Ann Rheum Dis 69:1898–1906
86. Okroj M, Heinegard D, Holmdahl R, Blom AM (2007) Rheumatoid arthritis and the complement system. Ann Med 39:517–530
87. Weissmann G (2004) Pathogenesis of rheumatoid arthritis. J Clin Rheumatol 10:S26–S31
88. Linton SM, Morgan BP (1999) Complement activation and inhibition in experimental models of arthritis. Mol Immunol 36:905–914
89. Wang Y, Rollins SA, Madri JA, Matis LA (1995) Anti-C5 monoclonal antibody therapy prevents collagen-induced arthritis and ameliorates established disease. Proc Natl Acad Sci U S A 92:8955–8959

90. Lee H, Zahra D, Vogelzang A et al (2006) Human C5aR knock-in mice facilitate the production and assessment of anti-inflammatory monoclonal antibodies. Nat Biotechnol 24:1279–1284
91. Grant EP, Picarella D, Burwell T et al (2002) Essential role for the C5a receptor in regulating the effector phase of synovial infiltration and joint destruction in experimental arthritis. J Exp Med 196:1461–1471
92. Woodruff TM, Strachan AJ, Dryburgh N et al (2002) Antiarthritic activity of an orally active C5a receptor antagonist against antigen-induced monarticular arthritis in the rat. Arthritis Rheum 46:2476–2485
93. Nandakumar KS, Jansson A, Xu B et al (2010) A recombinant vaccine effectively induces c5a-specific neutralizing antibodies and prevents arthritis. PLoS One 5:e13511
94. Vergunst CE, Gerlag DM, Dinant H et al (2007) Blocking the receptor for C5a in patients with rheumatoid arthritis does not reduce synovial inflammation. Rheumatology (Oxford) 46:1773–1778
95. Parker C (2009) Eculizumab for paroxysmal nocturnal haemoglobinuria. Lancet 373:759–767
96. Waters AM, Licht C (2011) aHUS caused by complement dysregulation: new therapies on the horizon. Pediatr Nephrol 26:41–57
97. Armstrong C, Schubert J, Ueda E et al (1992) Affected paroxysmal nocturnal hemoglobinuria T lymphocytes harbor a common defect in assembly of N-acetyl-D-glucosamine inositol phospholipid corresponding to that in class A Thy-1- murine lymphoma mutants. J Biol Chem 267:25347–25351
98. Hillmen P, Bessler M, Mason PJ, Watkins WM, Luzzatto L (1993) Specific defect in N-acetylglucosamine incorporation in the biosynthesis of the glycosylphosphatidylinositol anchor in cloned cell lines from patients with paroxysmal nocturnal hemoglobinuria. Proc Natl Acad Sci U S A 90:5272–5276
99. Takeda J, Miyata T, Kawagoe K et al (1993) Deficiency of the GPI anchor caused by a somatic mutation of the PIG-A gene in paroxysmal nocturnal hemoglobinuria. Cell 73:703–711
100. Wiedmer T, Hall SE, Ortel TL, Kane WH, Rosse WF, Sims PJ (1993) Complement-induced vesiculation and exposure of membrane prothrombinase sites in platelets of paroxysmal nocturnal hemoglobinuria. Blood 82:1192–1196
101. Rother RP, Rollins SA, Mojcik CF, Brodsky RA, Bell L (2007) Discovery and development of the complement inhibitor eculizumab for the treatment of paroxysmal nocturnal hemoglobinuria. Nat Biotechnol 25:1256–1264
102. Kudo M, Ishigatsubo Y, Aoki I (2013) Pathology of asthma. Front Microbiol 4:263
103. Zhang X, Kohl J (2010) A complex role for complement in allergic asthma. Expert Rev Clin Immunol 6:269–277
104. Kohl J, Baelder R, Lewkowich IP et al (2006) A regulatory role for the C5a anaphylatoxin in type 2 immunity in asthma. J Clin Invest 116:783–796
105. Humbles AA, Lu B, Nilsson CA et al (2000) A role for the C3a anaphylatoxin receptor in the effector phase of asthma. Nature 406:998–1001
106. Taube C, Rha YH, Takeda K et al (2003) Inhibition of complement activation decreases airway inflammation and hyperresponsiveness. Am J Respir Crit Care Med 168:1333–1341
107. Krug N, Tschernig T, Erpenbeck VJ, Hohlfeld JM, Kohl J (2001) Complement factors C3a and C5a are increased in bronchoalveolar lavage fluid after segmental allergen provocation in subjects with asthma. Am J Respir Crit Care Med 164:1841–1843
108. Abe M, Shibata K, Akatsu H et al (2001) Contribution of anaphylatoxin C5a to late airway responses after repeated exposure of antigen to allergic rats. J Immunol 167:4651–4660
109. Ricklin D (2012) Manipulating the mediator: modulation of the alternative complement pathway C3 convertase in health, disease and therapy. Immunobiology 217:1057–1066
110. Forneris F, Ricklin D, Wu J et al (2010) Structures of C3b in complex with factors B and D give insight into complement convertase formation. Science 330:1816–1820
111. Torreira E, Tortajada A, Montes T, Rodriguez de Cordoba S, Llorca O (2009) 3D structure of the C3bB complex provides insights into the activation and regulation of the complement alternative pathway convertase. Proc Natl Acad Sci U S A 106:882–887

112. Katschke KJ Jr, Wu P, Ganesan R et al (2012) Inhibiting alternative pathway complement activation by targeting the factor D exosite. J Biol Chem 287:12886–12892
113. Rooijakkers SH, Wu J, Ruyken M et al (2009) Structural and functional implications of the alternative complement pathway C3 convertase stabilized by a staphylococcal inhibitor. Nat Immunol 10:721–727
114. Laursen NS, Andersen KR, Braren I, Spillner E, Sottrup-Jensen L, Andersen GR (2011) Substrate recognition by complement convertases revealed in the C5-cobra venom factor complex. EMBO J 30:606–616
115. Fredslund F, Laursen NS, Roversi P et al (2008) Structure of and influence of a tick complement inhibitor on human complement component 5. Nat Immunol 9:753–760
116. Janssen BJ, Huizinga EG, Raaijmakers HC et al (2005) Structures of complement component C3 provide insights into the function and evolution of immunity. Nature 437:505–511
117. Fredslund F, Jenner L, Husted LB, Nyborg J, Andersen GR, Sottrup-Jensen L (2006) The structure of bovine complement component 3 reveals the basis for thioester function. J Mol Biol 361:115–127
118. Kinoshita T, Takata Y, Kozono H, Takeda J, Hong KS, Inoue K (1988) C5 convertase of the alternative complement pathway: covalent linkage between two C3b molecules within the trimolecular complex enzyme. J Immunol 141:3895–3901
119. Pangburn MK, Rawal N (2002) Structure and function of complement C5 convertase enzymes. Biochem Soc Trans 30:1006–1010
120. Takata Y, Kinoshita T, Kozono H et al (1987) Covalent association of C3b with C4b within C5 convertase of the classical complement pathway. J Exp Med 165:1494–1507
121. Rawal N, Pangburn M (2001) Formation of high-affinity C5 convertases of the alternative pathway of complement. J Immunol 166:2635–2642
122. Rawal N, Pangburn MK (2003) Formation of high affinity C5 convertase of the classical pathway of complement. J Biol Chem 278:38476–38483
123. Katschke KJ Jr, Stawicki S, Yin J et al (2009) Structural and functional analysis of a C3b-specific antibody that selectively inhibits the alternative pathway of complement. J Biol Chem 284:10473–10479
124. Helmy KY, Katschke KJ Jr, Gorgani NN et al (2006) CRIg: a macrophage complement receptor required for phagocytosis of circulating pathogens. Cell 124:915–927
125. Wiesmann C, Katschke KJ, Yin J et al (2006) Structure of C3b in complex with CRIg gives insights into regulation of complement activation. Nature 444:217–220
126. Ricklin D, Lambris JD (2008) Compstatin: a complement inhibitor on its way to clinical application. Adv Exp Med Biol 632:273–292
127. Qu H, Ricklin D, Bai H et al (2013) New analogs of the clinical complement inhibitor compstatin with subnanomolar affinity and enhanced pharmacokinetic properties. Immunobiology 218:496–505
128. Janssen BJ, Halff EF, Lambris JD, Gros P (2007) Structure of compstatin in complex with complement component C3c reveals a new mechanism of complement inhibition. J Biol Chem 282:29241–29247
129. Nishimura J, Yamamoto M, Hayashi S, et al (2012) A rare genetic polymorphism in C5 confers poor response to the anti-C5 monoclonal antibody eculizumab by nine Japanese patients with PNH. Abstract at 54th ASH annual meeting and exposition, Atlanta
130. Zuber J, Fakhouri F, Roumenina LT, Loirat C, Fremeaux-Bacchi V (2012) Use of eculizumab for atypical haemolytic uraemic syndrome and C3 glomerulopathies. Nat Rev Nephrol 8:643–657
131. Rohrer B, Long Q, Coughlin B et al (2009) A targeted inhibitor of the alternative complement pathway reduces angiogenesis in a mouse model of age-related macular degeneration. Invest Ophthalmol Vis Sci 50:3056–3064
132. Schmidt CQ, Bai H, Lin Z et al (2013) Rational engineering of a minimized immune inhibitor with unique triple-targeting properties. J Immunol 190:5712–5721
133. Hebecker M, Alba-Dominguez M, Roumenina LT et al (2013) An engineered construct combining complement regulatory and surface-recognition domains represents a minimal-size functional factor H. J Immunol 191:912–921

134. Fridkis-Hareli M, Storek M, Mazsaroff I et al (2011) Design and development of TT30, a novel C3d-targeted C3/C5 convertase inhibitor for treatment of human complement alternative pathway-mediated diseases. Blood 118:4705–4713
135. Mollnes TE, Brekke OL, Fung M et al (2002) Essential role of the C5a receptor in E coli-induced oxidative burst and phagocytosis revealed by a novel lepirudin-based human whole blood model of inflammation. Blood 100:1869–1877
136. Sperling C, Houska M, Brynda E, Streller U, Werner C (2006) In vitro hemocompatibility of albumin-heparin multilayer coatings on polyethersulfone prepared by the layer-by-layer technique. J Biomed Mater Res A 76:681–689
137. Kopp R, Bernsberg R, Kashefi A, Mottaghy K, Rossaint R, Kuhlen R (2005) Effect of hirudin versus heparin on hemocompatibility of blood contacting biomaterials: an in vitro study. Int J Artif Organs 28:1272–1277
138. Andersson J, Sanchez J, Ekdahl KN, Elgue G, Nilsson B, Larsson R (2003) Optimal heparin surface concentration and antithrombin binding capacity as evaluated with human non-anticoagulated blood in vitro. J Biomed Mater Res A 67:458–466
139. Wu YQ, Qu H, Sfyroera G et al (2011) Protection of nonself surfaces from complement attack by factor H-binding peptides: implications for therapeutic medicine. J Immunol 186:4269–4277
140. Couser WG, Johnson RJ, Young BA, Yeh CG, Toth CA, Rudolph AR (1995) The effects of soluble recombinant complement receptor 1 on complement-mediated experimental glomerulonephritis. J Am Soc Nephrol 5:1888–1894
141. Weisman HF, Bartow T, Leppo MK et al (1990) Soluble human complement receptor type 1: in vivo inhibitor of complement suppressing post-ischemic myocardial inflammation and necrosis. Science 249:146–151
142. Metcalfe RA, McIntosh RS, Morgan BP, Levin JL, Weetman AP (1996) The effect of soluble complement receptor 1 (sCR1) and human thyroid antibodies on the course of experimental autoimmune thyroiditis in rats. Autoimmunity 23:1–8
143. Rioux P (2001) TP-10 (AVANT immunotherapeutics). Curr Opin Investig Drugs 2:364–371
144. Weisman HF, Bartow T, Leppo MK et al (1990) Recombinant soluble CR1 suppressed complement activation, inflammation, and necrosis associated with reperfusion of ischemic myocardium. Trans Assoc Am Physicians 103:64–72
145. Dodd I, Mossakowska DE, Camilleri P et al (1995) overexpression in Escherichia coli, folding, purification, and characterization of the first three short consensus repeat modules of human complement receptor type 1. Protein Expr Purif 6:727–736
146. Fodor WL, Rollins SA, Bianco-Caron S et al (1995) Primate terminal complement inhibitor homologues of human CD59. Immunogenetics 41:51
147. Song H, He C, Knaak C, Guthridge JM, Holers VM, Tomlinson S (2003) Complement receptor 2-mediated targeting of complement inhibitors to sites of complement activation. J Clin Invest 111:1875–1885
148. Klos A, Wende E, Wareham KJ, Monk PN (2013) International union of pharmacology: LXXXVII. Complement peptide C5a, C4a, and C3a receptors. Pharmacol Rev 65:500–543
149. Bokisch VA, Muller-Eberhard HJ (1970) Anaphylatoxin inactivator of human plasma: its isolation and characterization as a carboxypeptidase. J Clin Invest 49:2427–2436
150. Cain SA, Monk PN (2002) The orphan receptor C5L2 has high affinity binding sites for complement fragments C5a and C5a des-Arg(74). J Biol Chem 277:7165–7169
151. Khan MA, Maasch C, Vater A et al (2013) Targeting complement component 5a promotes vascular integrity and limits airway remodeling. Proc Natl Acad Sci U S A 110:6061–6066
152. Woodruff TM, Nandakumar KS, Tedesco F (2011) Inhibiting the C5-C5a receptor axis. Mol Immunol 48:1631–1642
153. Zhang X, Boyar W, Toth MJ, Wennogle L, Gonnella NC (1997) Structural definition of the C5a C terminus by two-dimensional nuclear magnetic resonance spectroscopy. Proteins 28:261–267
154. Cook WJ, Galakatos N, Boyar WC, Walter RL, Ealick SE (2010) Structure of human desArg-C5a. Acta Crystallogr D Biol Crystallogr 66:190–197

155. Ward PA, Guo RF, Riedemann NC (2012) Manipulation of the complement system for benefit in sepsis. Crit Care Res Pract 2012:427607
156. Stevens JH, O'Hanley P, Shapiro JM et al (1986) Effects of anti-C5a antibodies on the adult respiratory distress syndrome in septic primates. J Clin Invest 77:1812–1816
157. Sprong T, Brandtzaeg P, Fung M et al (2003) Inhibition of C5a-induced inflammation with preserved C5b-9-mediated bactericidal activity in a human whole blood model of meningococcal sepsis. Blood 102:3702–3710
158. Sarma JV, Ward PA (2012) New developments in C5a receptor signaling. Cell Health Cytoskelet 4:73–82
159. Noris M, Mescia F, Remuzzi G (2012) STEC-HUS, atypical HUS and TTP are all diseases of complement activation. Nat Rev Nephrol 8:622–633
160. Vater A, Klussmann S (2003) Toward third-generation aptamers: spiegelmers and their therapeutic prospects. Curr Opin Drug Discov Devel 6:253–261
161. Hoehlig K, Maasch C, Shushakova N et al (2013) A novel C5a-neutralizing mirror-image (l-) aptamer prevents organ failure and improves survival in experimental sepsis. Mol Ther 21:2236
162. Kessel C, Nandakumar KS, Peters FB, Gauba V, Schultz PG, Holmdahl R (2013) A single functional group substitution in C5a breaks B and T cell tolerance and protects from experimental arthritis. Arthritis Rheum 66:610
163. Finch AM, Wong AK, Paczkowski NJ et al (1999) Low-molecular-weight peptidic and cyclic antagonists of the receptor for the complement factor C5a. J Med Chem 42:1965–1974
164. Zhang L, Mallik B, Morikis D (2008) Structural study of Ac-Phe-[Orn-Pro-dCha-Trp-Arg], a potent C5a receptor antagonist, by NMR. Biopolymers 90:803–815
165. Sumichika H, Sakata K, Sato N et al (2002) Identification of a potent and orally active nonpeptide C5a receptor antagonist. J Biol Chem 277:49403–49407
166. Scola AM, Higginbottom A, Partridge LJ et al (2007) The role of the N-terminal domain of the complement fragment receptor C5L2 in ligand binding. J Biol Chem 282:3664–3671
167. Brodbeck RM, Cortright DN, Kieltyka AP et al (2008) Identification and characterization of NDT 9513727 [N,N-bis(1,3-benzodioxol-5-ylmethyl)-1-butyl-2,4-diphenyl-1H-imidazole-5-methanamine], a novel, orally bioavailable C5a receptor inverse agonist. J Pharmacol Exp Ther 327:898–909
168. Klco JM, Wiegand CB, Narzinski K, Baranski TJ (2005) Essential role for the second extracellular loop in C5a receptor activation. Nat Struct Mol Biol 12:320–326
169. Heller T, Hennecke M, Baumann U et al (1999) Selection of a C5a receptor antagonist from phage libraries attenuating the inflammatory response in immune complex disease and ischemia/reperfusion injury. J Immunol 163:985–994
170. Otto M, Hawlisch H, Monk PN et al (2004) C5a mutants are potent antagonists of the C5a receptor (CD88) and of C5L2: position 69 is the locus that determines agonism or antagonism. J Biol Chem 279:142–151
171. Ember JA, Johansen NL, Hugli TE (1991) Designing synthetic superagonists of C3a anaphylatoxin. Biochemistry 30:3603–3612
172. Scully CC, Blakeney JS, Singh R et al (2010) Selective hexapeptide agonists and antagonists for human complement C3a receptor. J Med Chem 53:4938–4948
173. Ames RS, Lee D, Foley JJ et al (2001) Identification of a selective nonpeptide antagonist of the anaphylatoxin C3a receptor that demonstrates antiinflammatory activity in animal models. J Immunol 166:6341–6348
174. Allendorf DJ, Yan J, Ross GD et al (2005) C5a-mediated leukotriene B4-amplified neutrophil chemotaxis is essential in tumor immunotherapy facilitated by anti-tumor monoclonal antibody and beta-glucan. J Immunol 174:7050–7056
175. Mathieu MC, Sawyer N, Greig GM et al (2005) The C3a receptor antagonist SB 290157 has agonist activity. Immunol Lett 100:139–145
176. Monk PN, Scola AM, Madala P, Fairlie DP (2007) Function, structure and therapeutic potential of complement C5a receptors. Br J Pharmacol 152:429–448

177. Lambris JD, Ricklin D, Geisbrecht BV (2008) Complement evasion by human pathogens. Nat Rev Microbiol 6:132–142
178. Jongerius I, Kohl J, Pandey MK et al (2007) Staphylococcal complement evasion by various convertase-blocking molecules. J Exp Med 204:2461–2471
179. Hammel M, Sfyroera G, Ricklin D, Magotti P, Lambris JD, Geisbrecht BV (2007) A structural basis for complement inhibition by Staphylococcus aureus. Nat Immunol 8:430–437
180. Hammel M, Sfyroera G, Pyrpassopoulos S et al (2007) Characterization of Ehp, a secreted complement inhibitory protein from Staphylococcus aureus. J Biol Chem 282:30051–30061
181. Clark EA, Crennell S, Upadhyay A et al (2011) A structural basis for Staphylococcal complement subversion: X-ray structure of the complement-binding domain of Staphylococcus aureus protein Sbi in complex with ligand C3d. Mol Immunol 48:452–462
182. Ricklin D, Ricklin-Lichtsteiner SK, Markiewski MM, Geisbrecht BV, Lambris JD (2008) Cutting edge: members of the Staphylococcus aureus extracellular fibrinogen-binding protein family inhibit the interaction of C3d with complement receptor 2 J Immunol 181:7463–7467
183. Nunn MA, Sharma A, Paesen GC et al (2005) Complement inhibitor of C5 activation from the soft tick Ornithodoros moubata. J Immunol 174:2084–2091
184. Halstead SK, Humphreys PD, Zitman FM, Hamer J, Plomp JJ, Willison HJ (2008) C5 inhibitor rEV576 protects against neural injury in an in vitro mouse model of Miller Fisher syndrome. J Peripher Nerv Syst 13:228–235
185. Soltys J, Kusner LL, Young A et al (2009) Novel complement inhibitor limits severity of experimentally myasthenia gravis. Ann Neurol 65:67–75
186. Barratt-Due A, Thorgersen EB, Lindstad JK et al (2011) Ornithodoros moubata complement inhibitor is an equally effective C5 inhibitor in pigs and humans. J Immunol 187:4913–4919
187. Laursen NS, Gordon N, Hermans S et al (2010) Structural basis for inhibition of complement C5 by the SSL7 protein from Staphylococcus aureus. Proc Natl Acad Sci U S A 107:3681–3686
188. Fernie-King BA, Seilly DJ, Willers C, Wurzner R, Davies A, Lachmann PJ (2001) Streptococcal inhibitor of complement (SIC) inhibits the membrane attack complex by preventing uptake of C567 onto cell membranes. Immunology 103:390–398
189. Pramoonjago P, Kaneko M, Kinoshita T et al (1992) Role of TraT protein, an anticomplementary protein produced in Escherichia coli by R100 factor, in serum resistance. J Immunol 148:827–836
190. Pausa M, Pellis V, Cinco M et al (2003) Serum-resistant strains of Borrelia burgdorferi evade complement-mediated killing by expressing a CD59-like complement inhibitory molecule. J Immunol 170:3214–3222
191. Jagels MA, Ember JA, Travis J, Potempa J, Pike R, Hugli TE (1996) Cleavage of the human C5A receptor by proteinases derived from Porphyromonas gingivalis: cleavage of leukocyte C5a receptor. Adv Exp Med Biol 389:155–164
192. Postma B, Poppelier MJ, van Galen JC et al (2004) Chemotaxis inhibitory protein of Staphylococcus aureus binds specifically to the C5a and formylated peptide receptor. J Immunol 172:6994–7001
193. Ippel JH, de Haas CJ, Bunschoten A et al (2009) Structure of the tyrosine-sulfated C5a receptor N terminus in complex with chemotaxis inhibitory protein of Staphylococcus aureus. J Biol Chem 284:12363–12372
194. Wright AJ, Higginbottom A, Philippe D et al (2007) Characterisation of receptor binding by the chemotaxis inhibitory protein of Staphylococcus aureus and the effects of the host immune response. Mol Immunol 44:2507–2517
195. Gustafsson E, Rosen A, Barchan K et al (2010) Directed evolution of chemotaxis inhibitory protein of Staphylococcus aureus generates biologically functional variants with reduced interaction with human antibodies. Protein Eng Des Sel 23:91–101
196. Bunschoten A, Feitsma LJ, Kruijtzer JA, de Haas CJ, Liskamp RM, Kemmink J (2010) CHIPS binds to the phosphorylated N-terminus of the C5a-receptor. Bioorg Med Chem Lett 20:3338–3340

197. Caesar JJ, Wallich R, Kraiczy P, Zipfel PF, Lea SM (2013) Further structural insights into the binding of complement factor H by complement regulator-acquiring surface protein 1 (CspA) of Borrelia burgdorferi. Acta Crystallogr Sect F Struct Biol Cryst Commun 69:629–633
198. Bhattacharjee A, Oeemig JS, Kolodziejczyk R et al (2013) Structural basis for complement evasion by Lyme disease pathogen Borrelia burgdorferi. J Biol Chem 288:18685–18695
199. Schneider MC, Prosser BE, Caesar JJ et al (2009) Neisseria meningitidis recruits factor H using protein mimicry of host carbohydrates. Nature 458:890–893
200. Sfyroera G, Katragadda M, Morikis D, Isaacs SN, Lambris JD (2005) Electrostatic modeling predicts the activities of orthopoxvirus complement control proteins. J Immunol 174:2143–2151
201. Ricklin D, Lambris JD (2013) Complement in immune and inflammatory disorders: pathophysiological mechanisms. J Immunol 190:3831–3838
202. Kidmose RT, Laursen NS, Dobo J et al (2012) Structural basis for activation of the complement system by component C4 cleavage. Proc Natl Acad Sci U S A 109:15425–15430
203. Risitano AM, Perna F, Selleri C (2011) Achievements and limitations of complement inhibition by eculizumab in paroxysmal nocturnal hemoglobinuria: the role of complement component 3. Mini Rev Med Chem 11:528–535
204. Struijk GH, Bouts AH, Rijkers GT, Kuin EA, ten Berge IJ, Bemelman FJ (2013) Meningococcal sepsis complicating eculizumab treatment despite prior vaccination. Am J Transplant 13:819–820
205. Barnett ANR, Asgari E, Chowdhury P, Sacks SH, Dorling A, Mamode N (2013) The use of eculizumab in renal transplantation. Clin Transplant 27:E216–E229
206. Smith GP, Smith RA (2001) Membrane-targeted complement inhibitors. Mol Immunol 38:249–255
207. Fraser DA, Harris CL, Williams AS et al (2003) Generation of a recombinant, membrane-targeted form of the complement regulator CD59: activity in vitro and in vivo. J Biol Chem 278:48921–48927
208. Souza DG, Esser D, Bradford R, Vieira AT, Teixeira MM (2005) APT070 (Mirococept), a membrane-localised complement inhibitor, inhibits inflammatory responses that follow intestinal ischaemia and reperfusion injury. Br J Pharmacol 145:1027–1034
209. Moghimi SM, Andersen AJ, Ahmadvand D, Wibroe PP, Andresen TL, Hunter AC (2011) Material properties in complement activation. Adv Drug Deliv Rev 63:1000–1007
210. Janssen BJC, Christodoulidou A, McCarthy A, Lambris JD, Gros P (2006) Structure of C3b reveals conformational changes that underlie complement activity. Nature 444:213–216
211. Ramsland PA, Willoughby N, Trist HM et al (2007) Structural basis for evasion of IgA immunity by Staphylococcus aureus revealed in the complex of SSL7 with Fc of human IgA1. Proc Natl Acad Sci U S A 104:15051–15056

Chapter 3
The Art of Complement: Complement Sensing of Nanoparticles and Consequences

S. Moein Moghimi, Kiana C. Trippler, and Dmitri Simberg

Abstract The complement system is a complex network of plasma and membrane-associated proteins and represents one of the major effector mechanisms of the innate immune system. The function of complement in innate host defence is accomplished through highly efficient and tightly orchestrated opsonisation, lytic and inflammatory processes. Nanoparticle-based medicines may trigger complement and a number of consequences ensue from complement activation. These comprise both beneficial and adverse reactions, depending on the extent and severity of complement activation as well as microenvironmental factors. These concepts are briefly discussed in relation to therapeutic applications of nanoparticles and anti-cancer nanomedicines.

Keywords Adjuvanticity • Adverse reactions • Complement system • Immune responses • Nanomedicine • Nanoparticles • Opsonization • Phagocytes

3.1 Introduction

Synthetic nanoparticles and particulate drug carriers (e.g., liposomes, polymeric nanoparticles and nanocapsules) by virtue of their size, shape and surface characteristics (e.g., display of architecture with repetitive epitopes such as those arising from surface projected polymers or polyanionic/polycationic clusters) may resemble viruses or microorganisms. These 'pathogen-mimicking' properties make nanoparticles prone to interception by different components of the host defences following entry into the body [1–4]. Accordingly, the type and the extent of immune responses will depend on physicochemical characteristics of nanoparticles,

---

S.M. Moghimi (✉)
Professor and Chair in Pharmaceutics, Division of Pharmacy,
School of Medicine, Pharmacy and Health, Durham University, Queen's Campus,
Stockton-on-Tees TS17 6BH, UK
e-mail: moein.moghimi@gmail.com

K.C. Trippler • D. Simberg
Department of Pharmaceutical Sciences, The Skaggs School of Pharmacy and Pharmaceutical Sciences, University of Colorado Denver,
Anschutz Medical Campus, 12850 E. Montview Blvd., Aurora, CO 80045, USA

K.A. Howard et al. (eds.), *Nanomedicine*, Advances in Delivery Science and Technology, DOI 10.1007/978-1-4939-3634-2_3

administered dose, frequency and route of administration [2]. A key component of immunity that plays a central role in the sensing of particulate matters, their processing and elimination is the complement system [5].

3.2 The Complement System

The complement system, which consists of at least 30 soluble and membrane-bound proteins, is a key effector of both innate and cognate immunity [6]. Complement acts in a wide variety of host defence, inflammatory, homeostatic and immune reactions [6]. Complement recognizes danger signals primarily through pattern recognition. Indeed, many particulate drug carriers and nanomedicines are composed of polymeric components and other patterned nanostructures that makes them prone to complement sensing [5]. Complement activation proceeds through a series of enzymatic reactions that lead to the assembly of the lytic membrane attack complex (MAC) for target destruction [6]. There are three established pathways of complement activation: the classical, the lectin and the alternative pathways. Each pathway is triggered on binding of complement initiating molecules such as C1q, mannan-binding lectin (MBL), ficolins and properdin (either individually or in combination) to the target, but they converge to generate the same set of effector molecules [6]. For details of complement activation pathways and enzymatic cascades the reader is referred to excellent reviews elsewhere [5, 6]. Here, we limit our discussion to sensing of nanoparticles by complement initiation molecules and consequences of nanoparticle-mediated complement activation.

3.2.1 Complement Initiating Molecules in Nanoparticle-Mediated Complement Activation

The initiating molecule of the classical pathway is C1q, a pattern-recognition molecule composed of six identical subunits with cationic globular heads and long collagen-like tails [6, 7]. The classical pathway activation occurs when the globular head regions of C1q binds either directly to polyanionic surfaces (e.g., certain classes of negatively charged liposomes and polymeric nanoparticles such as polystyrene nanospheres) [5, 8, 9] or particulate materials that have become coated with immunoglobulins (e.g., IgM, IgG1, IgG2 and IgG3 classes) [10, 11] or C-reactive protein [12]. For instance, liposomes on binding to natural anti-phospholipid and anti-cholesterol antibodies, can trigger C1q-dependent classical pathway of the complement system [5, 8, 9], whereas lysophosphatidylcholine vesicles do so through initial binding of C-reactive protein [13].

MBL (also known as mannose-binding lectin, mannan- or mannose-binding protein) and ficolins are the initiating molecules of the lectin pathway and recognize nanometre scale oligosaccharide-based or acetyl-based molecular patterns on surfaces [5, 6, 14].

The role of lectin pathway in complement activation by nanoparticles displaying surface projected poly(ethylene oxide) has been demonstrated [15–17]. Indeed, surface projected poly(ethylene oxide) chains in close proximity may form dynamic 'pathogen-mimicking' clusters transiently resembling structural motifs of the D-mannose, which serves as a platform for MBL/ficolin docking. Also, in mouse sera MBL was found to bind to dextran-coated iron oxide nanoparticles [18], although the significance of this finding to complement activation and pharmacokinetic of diagnostic iron oxide nanocrystals in humans remains to be evaluated.

Initiation of the alternative pathway activation requires the presence of preformed C3b (arises from enzymatic cleavage of the third complement protein, C3), or C3(H_2O) (a meta-stable form of C3 arising from spontaneous hydrolysis of the thioester in C3, which has C3b-like properties) [5, 6]. It is the internal thioester bond in the α-chain of C3b, which undergoes a nucleophilic attack in the presence of a foreign surface structure rich in nucleophilic groups [6]. There are many examples of liposomes, oil-in-water emulsions, polymeric nanoparticles and metallic nanoparticles (e.g., gold nanoparticles and iron oxide nanoparticles, etc.) that activate complement through the alternative pathway [5]. For instance, a recent study has shown that nanoparticles grafted with a glucose-containing polymer trigger the alternative pathway of complement at a lower grafting density compared with nanoparticles grafted with a galactose-containing polymer [19]. However, no explanation was provided on the mechanisms of the initiation steps, but galactose polymer modified nanoparticles adsorbed more factor H (a negative regulator of complement) [6], than the glucose surface, providing a reason for its lower level of complement activation. In contrast to these observations, others have demonstrated a unique set of conformational changes related to a target adsorbed form of intact C3, for instance through transformation into C3(H_2O), leading to initiation of complement activation via the alternative pathway [19]. For example, zwitterionic dimyristoylphosphatidylcholine (DMPC) liposomes trigger the alternative pathway through contact activation of C3(H_2O) and not C3b binding [20]. However, these studies were performed with liposomes that were frozen inappropriately without the use of cryo-preserves. Accordingly, on thawing these vesicles may contain many membrane defects and altered morphologies that may explain the binding of C3(H_2O). In our hand, complement activation in human serum/plasma by freshly prepared DMPC unilamellar vesicles of 100 nm in size only proceeds through the classical pathway (Moghimi, unpublished data). Another mode by which the alternative pathway turnover can be enhanced is through sufficient surface deposition of properdin. Properdin is a highly cationic pattern-recognition protein that binds to certain polymers, nucleic acids, anionic clusters and apoptotic cells [15, 21, 22]. For instance, polymeric nanoparticles displaying surface projected poly(ethylene oxide) chains accelerate alternative pathway turnover through binding of both nascent C3b and properdin, but the latter is controlled by poly(ethylene oxide) surface density and conformation [15].

The literature surrounding the role of nanoparticle size in complement activation pathways and processes is somewhat confusing [2, 5, 23–27]. The reported discrepancies may be a reflection of misunderstanding of the effect of surface curvature in modulating the affinity and topological strain of complement initiating molecules such as IgM and C1q [25, 26, 28, 29].

3.3 Consequences of Nanoparticle-Mediated Complement Activation and Fixation

3.3.1 Opsonisation and Phagocytic Elimination

Complement activation and fixation remains a central component for efficient clearance and destruction of particulate invaders at the hand of phagocytic cells. The C3 molecule is central to opsonisation [6, 30]. It's first cleavage product, C3b, acts as an opsonin and becomes covalently bound to the activating nanoparticle surface and facilitates nanoparticle binding to phagocytes via complement receptor (CR) 1 [6, 30, 31]. C3b is further degraded to iC3b, C3c and C3dg, products that serve as ligands for other complement receptors on leukocytes [31]. For instance, iC3b is the primary ligand of CR3 (Mac-1, CD11b/CD18) and CR4 (CD11c/CD18, p150,95), but CR3 is the predominant receptor for phagocytic recognition and safer elimination of complement-opsonised particles and complexes [31].

Complement opsonisation and subsequent macrophage recognition of particulate drug carriers offers an unprecedented opportunity for intravenous delivery of therapeutic agents and immunomodulators to phagocytic cells in the liver (Kupffer cells) and the spleen (marginal zone and the red-pulp macrophages) [1–3]. In murine, Kupffer cells express a CR of the immunoglobulin superfamily (CRIg) that binds C3b- and iC3b-opsonized particles [32]. Depending on the species, complement opsonized particles may further interact with CRs of erythrocytes (CR1 in primates) [33] and platelets (CR1 in rats) [34, 35]. These modes of interaction may further modulate nanoparticle pharmacokinetics and biological fate.

Some studies have now confirmed that stealth nanoparticles such as PEGylated liposomes and polymeric nanospheres although fixing complement, still show considerable resistance to macrophage phagocytosis [36, 37]. This is due to the strong steric barrier of PEG on the nanoparticle surface, which interferes with the binding of surface localized C3b and iC3b to their corresponding phagocyte CRs.

3.3.2 Nanoparticle Integrity in the Blood

Complement activation can significantly affect the integrity of certain classes of drug carriers in the blood. For example, insertion of MAC into the liposomal bilayer may lead to substantial leakage of entrapped (aqueous) cargo [38], but this may be minimized or prevented through surface PEGylation [37]. Similarly, MAC may also affect the integrity of colloidal dispersions of cubosomes and hexosomes (liquid crystal phases of cubic and hexagonal structures in water [39]) which also act as drug carriers.

3.3.3 Adjuvanticity

Nanoparticles have been used as immune potentiators or adjuvants triggering elements of innate immunity that subsequently assist the generation of potent and persistent adaptive immune responses [40]. Most of these efforts are being directed to enhance the immunogenicity of subunit vaccines through both antigen protection and targeting to antigen-presenting cells as well as immunostimulation. The role of the complement system in some of these processes is rather intriguing. For instance, some complement activation products (e.g., C3d) can induce B lymphocyte activation [41]. Likewise, simultaneous binding of CR3 and CR2 to C3 fragments may modulate trafficking of complement-opsonised immune complexes and nanoparticles from macrophages to B lymphocytes and follicular dendritic cells in lymph nodes [42].

3.3.4 Nanoparticle-Mediated Infusion-Related Adverse Reactions

Acute allergic-like reactions with haemodynamic, respiratory, cardiovascular, cutaneous and gastrointestinal manifestations, which are not initiated or mediated by pre-existing IgE antibodies, have been reported to occur in approximately 45 % of individuals within a few minutes of infusion of nanomedicines (e.g., liposomal drugs such as Abelcet®, Ambisome®, DaunoXome®, Doxil®, Myocet® and Visudyne®; micelle-solubilized drugs such as Taxol®, Taxotere® and Vumon®) and diagnostic nanoparticles (e.g., dextran-coated iron oxide nanocrystals such as Feridex® and Combidex®) [43, 44]. In some isolated cases this has been fatal (e.g., Taxol®) [44]. Compelling evidence suggests that inadvertent activation of the complement system is an important factor in eliciting these reactions [43–46]. This is partly due to the liberation of potent complement bioactive products (e.g., C3a, C5a and C5b-9) with the ability to modulate the function of a variety of immune cells (e.g., monocytes, polymorphonuclear cells, platelets, mast cells) and vascular endothelial cells [43–46], and partly to cross-talk with Toll-like receptors (TLRs) [47]. The latter may include TLR2, 4 and 9. For instance, complement synergistically enhances TLR-induced production of proinflammatory cytokines such as TNF-α, IL-1β, and IL-6 through anaphylatoxin (C3a and C5a) signaling [48]. The pathways of the aforementioned TLRs converge with anaphylatoxin signaling at the level of mitogen-activated protein kinases ERK1/2 and JNK63 [48]. Some nanoparticles may directly interact with TLRs and TLR activation can also modulate expression of complement components or receptors [49].

3.3.5 *Nanoparticle-Mediated Tumour Growth*

A recent study in immunocompetent mice, as well as in C5 and C5a receptor knockout animals bearing a syngeneic tumour, has strongly indicated that intratumoral accumulation of complement activating long-circulating nanoparticles can accelerate tumour growth through C5a liberation [50]. Tumour growth was more significant with nanoparticles capable of directly enhancing the alternative pathway turnover of the complement system compared with nanoparticles that triggered complement predominantly through the lectin pathway. These observations are consistent with the findings that concentration of local C5a within the tumour microenvironment is a critical factor in determining tumour progression [51]. Indeed, C5a has been suggested to enhance tumour growth by promoting the recruitment of regulatory T cells (resulting in deregulation or suppression of $CD8^+$ cytotoxic T cell activity), immunosuppressive monocytes and alternatively activated macrophages into malignant tumours [51, 52]. Complement activation, therefore, is of serious concern for successful development of intravenous anti-cancer nanomedicine and other anti-cancer nanotechnology initiatives [52]. Possible therapeutic function of complement inhibitors (e.g., small molecule C5a receptor antagonists) in cancer treatment is an unexplored avenue, and, perhaps, a worthy strategy for exploration and better engineering of anti-cancer nanomedicines.

3.4 Conclusions

The interaction between medically relevant nanomaterials and the complement system is complex and regulated by inter-related factors comprising morphology, dimension, chemical makeup and surface characteristics. Understanding of nanomaterial properties that incite complement is a prerequisite for design and engineering of immunologically safer nanomedicine and biomedical devices. Many pathogenic microorganisms have deployed intriguing strategies that bypass complement activation [52, 53]. Translation of microbial strategies could provide effective means for design and engineering of 'complement-safe' as well as 'phagocyte-resistant' particulate drug carriers and biomaterials [52, 54]. However, it is imperative that such engineered nanoparticles must be structurally simple with attributes that will allow for production of affordable, viable (e.g., considering scaling up and Good Manufacturing Practice) and clinically acceptable pharmaceutical products.

Acknowledgement SMM acknowledges financial support by the Danish Agency for Science, Technology and Innovation, references 09-065736 (Det Strategiske Forskningsråd), and 12-126894 (Technology and Production). Financial support by Lundbeckfonden (reference R100-A9443) and the European Community's Seventh Framework Programme (FP7-NMP-2012-Large-6) under grant agreement No. 310337-2 CosmoPHOS CP-IP is also acknowledged.

References

1. Moghimi SM, Hunter AC, Murray JC (2005) Nanomedicine: current status and future prospects. FASEB J 19:311–320
2. Moghimi SM, Hunter AC, Andresen TL (2012) Factors controlling nanoparticle pharmacokinetics: an integrated approach and perspective. Annu Rev Pharmacol Toxicol 52:481–503
3. Moghimi SM, Parhamifar L, Ahmadvand D et al (2012) Particulate systems for targeting of macrophages: basic and therapeutic concepts. J Innate Immun 4:509–528
4. Moghimi SM, Farhangrazi S (2014) Nanoparticle in medicine: nanoparticle engineering for macrophage targeting and nanoparticles that avoid macrophage recognition. In: Boraschi D, Duschl A (eds) Nanoparticles and the Immune System: Safety and Effects. Elsevier, San Diego, pp 77–89
5. Moghimi SM, Andersen A, Ahmadvand D et al (2011) Material properties in complement activation. Adv Drug Deliv Rev 63:1000–1007
6. Ricklin D, Hajishengallis G, Yang K et al (2010) Complement—a key system for immune surveillance and homeostasis. Nat Immunol 11:785–797
7. Fust G, Medgyesi GA, Rajnavolgyi E et al (1978) Possible mechanisms of the first step of the classical complement activation: binding and activation of C1. Immunology 35:873–884
8. Moghimi SM, Hunter AC (2001) Recognition by macrophages and liver cells of opsonized phospholipid vesicles and phospholipid headgroups. Pharm Res 18:1–8
9. Moghimi SM, Hamad I (2008) Liposome-mediated triggering of complement cascade. J Liposome Res 18:195–209
10. Kishore U, Gupta SK, Perikoulis MV et al (2003) Modular organization of the carboxy-terminal, globular head region of human C1q A, B and C chains. J Immunol 171:812–820
11. Gaboriaud C, Thielens NM, Gregory LA et al (2004) Structure and activation of the C1 complex of complement: unravelling the puzzle. Trends Immunol 25:368–373
12. McGrath FD, Brouwer MC, Arlaud GJ et al (2006) Evidence that complement protein C1q interacts with C-reactive protein through its globular head region. J Immunol 176:2950–2957
13. Volanakis JE, Wirtz KWA (1979) Interaction of C-reactive protein with artificial phosphatidylcholine bilayers. Nature 281:155–157
14. Heja D, Kocsis A, Dobo J et al (2012) Revised mechanism of complement lectin-pathway activation revealing the role of serine protease MASP1 as the exclusive activator of MASP-2. Proc Natl Acad Sci U S A 109:10498–10503
15. Hamad I, Al-Hanbali O, Hunter AC et al (2010) Distinct polymer architecture mediates switching of complement activation pathways at nanosphere-serum interface: implications for stealth nanoparticles engineering. ACS Nano 4:6629–6638
16. Andersen AJ, Robinson JT, Dai H et al (2013) Single-walled carbon nanotubes surface control of complement sensing and activation. ACS Nano 7:1108–1119
17. Hamad I, Hunter AC, Moghimi SM (2013) Complement activation by Pluronic 127 gel and micelles: suppression of copolymer-mediated complement activation by elevated serum levels of HDL, LDL, and apolipoproteins A-I and B-100. J Control Release 170:167–174
18. Simberg D, Park JH, Karmali PP et al (2009) Differential proteomics analysis of the surface heterogeneity of dextran iron oxide nanoparticles and the implications for their in vivo clearance. Biomaterials 30:3926–3933
19. Yu K, Lai BFL, Foley JH et al (2014) Modulation of complement activation and amplification on nanoparticle surfaces by glycopolymer conformation and chemistry. ACS Nano 8:7687–7703
20. Klapper Y, Hamad OA, Teramura Y et al (2014) Mediation of a non-proteolytic activation of complement component C3 by phospholipid vesicles. Biomaterials 35:3688–3696
21. Schwaeble WJ, Reid KB (1999) Does properdin crosslink the cellular and the humoral immune responses. Immunol Today 20:17–21
22. Kemper C, Mitchell LM, Zhang L et al (2008) The complement protein properdin binds apoptotic T cells and promotes complement activation and phagocytosis. Proc Natl Acad Sci U S A 105:9023–9028

23. Bradley DV, Wong K, Serrano K et al (1994) Liposome-complement interactions in rat serum: implications for liposome survival studies. Biochim Biophys Acta 1191:43–51
24. Lundqvist M, Stigler J, Elia G et al (2008) Nanoparticle size and surface properties determine the protein corona with possible implications for biological impacts. Proc Natl Acad Sci U S A 105:14265–14270
25. Pedersen MB, Zhou X, Larsen EKU et al (2010) Curvature of synthetic and natural surfaces is an important target feature in classical pathway complement activation. J Immunol 184:1931–1945
26. Wibroe PP, Moghimi SM (2012) Complement sensing of nanoparticles and nanomedicines. In: Hepel M, Zhong CJ (eds) Functional nanoparticles for bioanalysis, nanomedicine and bioelectronic devices, vol 2, ACS Symposium Series. American Chemical Society, Washington, DC, pp 365–382
27. Tenzer S, Docter D, Kuharev J et al (2013) Rapid formation of plasma protein corona critically affects nanoparticle pathophysiology. Nat Nanotechnol 8:772–781
28. Ling WL, Biro A, Bally I et al (2011) Proteins of the innate immune system crystallize on carbon nanotubes but are not activated. ACS Nano 5:730–737
29. Pondman KM, Sobik M, Nayak A et al (2014) Complement activation by carbon nanotubes and its influence on the phagocytosis and cytokine response by macrophages. Nanomedicine 10:1287–1299
30. Crrol MV, Sim RB (2011) Complement in health and disease. Adv Drug Deliv Rev 63:965–975
31. van Lookeren Campagne M, Wiesmann C, Brown EJ (2007) Macrophage complement receptors and pathogen clearance. Cell Microbiol 9:2095–2102
32. Helmy KY, Katschke KJ Jr, Gorgani NN et al (2006) CRIg: a macrophage complement receptor required for phagocytosis of circulating pathogens. Cell 124:915–927
33. Cornacoff JB, Hebert LA, Smead WL et al (1983) Primate erythrocyte-immune complex-bearing mechanism. J Clin Invest 71:236–247
34. Reinish LW, Bally MB, Loughrey HC et al (1988) Interaction of liposomes and platelets. Thromb Haemost 60:518–523
35. Loughrey HC, Bally MB, Reinish LW et al (1990) The binding of phosphatidylglycerol liposomes to rat platelets is mediated by complement. Thromb Haemost 64:172–176
36. Gbadamosi JK, Hunter AC, Moghimi SM (2002) PEGylation of microspheres generates a heterogeneous population of particles with differential surface characteristics and biological performance. FEBS Lett 532:338–344
37. Moghimi SM, Hamad I, Andresen TL et al (2006) Methylation of the phosphate oxygen moiety of phospholipid-methoxypoly(ethylene glycol) conjugate prevents PEGylated liposome-mediated complement activation and anaphylatoxin production. FASEB J 20:2591–2593
38. Haxby JA, Gotze O, Muller-Eberhard HJ et al (1969) Release of trapped marker from liposomes by the action of purified complement components. Proc Natl Acad Sci U S A 64:290–295
39. Yaghmur A, Laggner P, Almgren M et al (2008) Self-assembly in monoelaidin aqueous dispersions: direct vesicles to cubosomes transition. PLoS One 3:e3747. doi:10.1371/journal.pone.0003747
40. Moghimi SM (2009) The innate immune responses, adjuvants and delivery systems. In: Jorgensen L, Nielsen HM (eds) Delivery Technologies for Biopharmaceuticals. Peptides, Proteins, Nucleic Acids and Vaccines. Wiley, Chichester, pp 113–127
41. Dempsey PW, Allison ME, Akkaraju S et al (1996) C3d of complement as a molecular adjuvant: bridging innate and acquired immunity. Science 271:348–350
42. Bajic G, Yatime L, Sim RB et al (2013) Structural insight on the recognition of surface-bound opsonins by the integrin I domain of complement receptor 3. Proc Natl Acad Sci U S A 110:16426–16431
43. Moghimi SM, Andersen AJ, Hashemi SH et al (2010) Complement activation cascade triggered by PEG-PL engineered nanomedicines and carbon nanotubes: the challenges ahead. J Control Release 146:175–181

44. Szebeni J (2014) Complement activation-related pseudoallergy: a stress reaction in blood triggered by nanomedicines and biological. Mol Immunol 61:163–173
45. Moghimi SM, Wibroe PP, Helvig SY et al (2012) Genomic perspectives in inter-individual adverse responses following nanomedicine administration: the way forward. Adv Drug Deliv Rev 64:1385–1393
46. Moghimi SM, Farhangrazi ZS (2013) Nanomedicine and complement paradigm. Nanomedicine 9:458–460
47. Hajishengallis G, Lambris JD (2010) Crosstalk pathways between Toll-like receptors and the complement system. Trends Immunol 31:154–163
48. Zhang X, Kimura Y, Fang C et al (2007) Regulation of Toll-like receptor-mediated inflammatory response by complement in vivo. Blood 110:228–236
49. Kaczorowski DJ, Afrazi A, Scott MJ et al (2010) Pivotal advance: The pattern recognition receptor ligands lipopolysaccharide and polyinosine-polycytidylic acid stimulate factor B synthesis by the macrophage through distinct but overlapping mechanisms. J Leukoc Biol 88:609–618
50. Moghimi SM (2014) Cancer nanomedicines and the complement system activation paradigm: anaphylaxis and tumour growth. J Control Release 190:556–562
51. Moghimi SM, Farhangrazi ZS (2014) Just so stories: random acts of anti-cancer nanomedicine performance. Nanomedicine 10:1661–1666
52. Moghimi SM, Hunter AC, Murray JC (2001) Long-circulating and target-specific nanoparticles; theory to practice. Pharmacol Rev 53:283–318
53. Schneider MC, Prosser BE, Caesar JJE et al (2009) *Neisseria meningitidis* recruits factor H using protein mimicry of host carbohydrates. Nature 458:890–893
54. Wu YQ, Qu HC, Sfyroera G et al (2011) Protection of nonself surfaces from complement attach by factor H-binding peptides: implications for therapeutic medicine. J Immunol 186:4269–4277

Chapter 4
The Nanoscience of Polyvalent Binding by Proteins in the Immune Response

Thomas Vorup-Jensen

Abstract Recent research has demonstrated that the successful use of nanometer-scaled material, such as nanoparticles, as medicines is often challenged by the host immune system. Mechanisms of the innate immunity seem to provide a swift response to administration of particulate nanomedicines, which may clear or in other way incapacitate the function of these drugs. To rationalize, why and how, the innate immune system especially interacts with nanomedicines, this chapter points to the prominent role of polyvalent interactions by large, immunoactive proteins with the surfaces of nanoparticles. From addressing the thermodynamics and ultra-structural properties of these interactions, it is proposed that the nm-scaled ligand presentation and symmetry on such surfaces is a determinant in the binding of these proteins. Better control over nanomedicine ultrastructure is consequently likely to provide important ways of regulating the interactions, wanted or unwanted, with the innate immune system.

Keywords Polyvalent interactions • Avidity entropy • Innate immunity • Pattern recognition • Complement • Mannan-binding lectin

4.1 Introduction

Any new field such as nanomedicine suffers a challenge to establish reasonably rigid borders to permit a definition what is inside and outside of its realm. In this process, it easy to embody parts of other scientific areas by essentially a change of nomenclature, which will not, however, add to the scientific insight previously established. Following more than century of investigations, one might suspect that the classic topic of protein structure in consequence cannot rightfully be claimed to

T. Vorup-Jensen, M.Sc., Ph.D., D.M.Sc. (✉)
Department of Biomedicine, Biophysical Immunology Laboratory,
Aarhus University, Aarhus C, Denmark
e-mail: vorup-jensen@biomed.au.dk

K.A. Howard et al. (eds.), *Nanomedicine*, Advances in Delivery Science and Technology, DOI 10.1007/978-1-4939-3634-2_4

be part of nanoscience or nanomedicine. Nevertheless, it is one of the purposes of the present chapter to argue that, in particular, the connection between structure and function of proteins in many ways can be treated as a *bona fide* nanoscience. This serves both as a better understanding of how proteins interacts with their ligands and what methodologies or theoretical approaches are suitable for analyzing protein function. With the prominent role of proteins in human health, development of diseases, and, more recently, as therapeutics, it seems wholly appropriate to consider the nanoscience of proteins as an important part of nanomedicine. In choosing the immune system as a particular focus, the effort is greatly supported by the vast insight now available on the structure of the proteins associated with functions of the immune system.

4.1.1 The Nanoscience of Protein Chemistry and Structure

Proteins in the human body are polymers made from a selection of 20 of zwitterionic amino acids, all composed of sulfur, carbon, oxygen, nitrogen, or hydrogen atoms. The chemical structure of the amino acids involves a central carbon atom (C_α) bonded to a hydrogen atom, a carboxyl group, and an amine group as well as in most cases a side chain. The cellular synthesis catalyzes the formation of the peptidic bond between a carboxyl groups and an amine group, this way creating the peptidic "back bone" of proteins. In a classic report by Pauling [1], the length of the peptide bond was determined to 0.272 nm. This simple structural information is a determinant to why proteins are part of the nanoscience since many proteins contain hundreds of residues. As just one example, the abundant plasma protein human serum albumin contains 585 residues and, hence, a contour length reaching 585×0.272 nm ~ 160 nm. In its normally folded state albumin takes approximately a diameter of 8 nm whereas the molecule under certain circumstances can be found in an extended, but still folded, form with a length of 22 nm [2]. Taken together, in quantitative terms these numbers certainly classify albumin as a type of nanocarrier and, consequently, the engineering of medicines to enable such carriage as an important part of nanomedicine [3].

To rationalize the three-dimensional properties of proteins, it is useful to introduce the hierarchal ordering of their structure [4]. The primary structure represents the sequence of amino acid residues. Through hydrogen bonds formed between –NH and –CO groups of in the peptide back bone, proteins may form (alpha) helical or (beta) sheet structure, these elements being referred to as the secondary structure. The spatial organization of alpha helices and beta sheets within a globular domain make up the tertiary structure. Finally, the quaternary structure describes the spatial organization of the multiple subunits. In some cases, the protein structure is formed by contacts between multiple protein chains. A striking example is the collagens. Here, the primary structure contains a repeated motif of glycine followed by any selection of two residues (typically referred to as a Gly-X-Y motif) enables three chains to form a helical structure. Covalent bonds between cysteine residues are another important part of the architecture of large proteins, which permit formation of stable complexes through intermolecular bonds.

The proteinous molecules encountered in the body are, of course, highly heterogeneous with regard to size. For that reason, it is helpful to consider a more narrow group of proteins, which will be discussed further below in terms of polyvalency and interactions with nm-scaled materials. A detailed enquiry into the size of plasma proteins forming a part of the complement system has been made earlier [5]. The complement system is part of innate immunity largely acting to defend the body by catalyzing the deposition of complement proteins on, for instance, viral or bacterial surface. Thereby, these agents are targeted for uptake by phagcocytic leukocytes, e.g., macrophages, or directly inactivated through insertion of pores in their membranes in addition to multiple other antimicrobial functions of the complement system. The hydrodynamic radii (R_G) of these proteins range from 4 to 12 nm with some of the smaller proteins being the part of the pore-forming agents while, strikingly, proteins initiating the deposition of complement are larger with R_G at ~12 nm and cross sectional diameter of ~30–35 nm.

These latter molecular species are easily comparable with engineered nanoparticles (NPs) as can be outlined in at least three ways.

First, both large molecules of the complement system and NPs up to about a diameter of 100 nm will not sediment without the aid of high-speed centrifugation, typically exceeding 100,000 × *g*. Indeed, the major forces affecting their colloidal stability are the physicochemical properties of their environment, e.g., ionic strength, viscosity, and availability of electric charge interactions. This also implies that hydrophobic or hydrophilic properties of their surfaces are determinants in both colloidal stability and ability to interaction with host tissue in vivo. Second, their sizes are a critical component in their tissue distribution inside the body of experimental animals or humans. This includes the penetration of physiological barriers such the endothelium of blood vessels, the blood-brain barrier, mucosal epithelial surfaces, or retention in certain tissues, e.g., zones of inflammation or malignant tissue. Third, the dimensions of NP and proteins both enable and limit the types of interaction these can form with other molecules or tissue presenting ligands. In the case of polyvalent interactions, an appropriate distance between ligands for a polyvalent molecule is a critical determinant in the binding as discussed further in Sect. 4.2.2.

4.1.2 Affinity, Avidity, and Polyvalent Interactions

The function of proteins can be quantified in a number of ways depending on what precise functional properties are supported by that protein. A classic example is the rate of catalysis of certain chemical reaction, e.g., the cleavage of nucleotide phosphates or protein chains. Another example is the binding strength in interactions between proteins and their ligands. This section will explore, in more detail, this topic as some recent reports have linked binding strength to nm-scaled properties of proteins. Furthermore, the binding strength is also an important parameter when using proteins as drugs or when considering their interactions with drugs.

For quantification of the binding strength between molecules, it is necessary to address what type of reaction is under consideration. A simple scheme is the 1:1 binding reaction, where

$$R + L \overset{K}{\leftrightarrow} R \cdot L. \tag{4.1}$$

The constant K describes the concentrations at chemical equilibrium of reactants R and L as well as the formed complex $R \cdot L$. In this case, the association constant is defined by

$$K_A = \frac{[R \cdot L]}{[R] \cdot [L]}. \tag{4.2}$$

However, usually in molecular biology the dissociation constant, $K_D = 1/K_A$, is the preferred measure of binding strength. This preference is elegantly explained from considering the Langmuir Equation [6]

$$\theta = \frac{c}{c + K_D}, \tag{4.3}$$

where θ is the fractional saturation of binding sites at the concentration c of species binding these binding sites with a dissociation constant K_D at chemical equilibrium. At this condition, the formula implies at $c = K_D$, 50 % of all binding sites would be occupied. In other words, the K_D value is a practical measure of what concentration would be required to reach a significant fractional saturation in chemical binding experiments. Equation (4.3) becomes especially useful in situations when one of the reactants is in vast stoichiometric excess over the other reactant, i.e., $[L] \gg [R]$. For the reaction scheme in Eq. (4.1), it follows that [L] can then be approximated by the total concentration of L, $[L]_0$. Obviously, both technical and theoretical aspects limit the number of analyses where it is practicable to "flood" the chemical reaction with one of the reactants. A very useful alternative are methodologies generating a steady-state equilibrium by fixing the concentration by continuous supply of one of the reactants, for instance as performed with the popular surface plasmon resonance (SPR) sensors supplied with the BIAcore instruments [7]. The "steady-state scenario" could be thought of as mimicking the encounter between a microbe and blood proteins of the human immune system, which are subject to continuous synthesis and, hence, often remains at a essentially constant concentration. In this case Eq. (4.3) also applies. With the fixation of one reactant concentration it is also simple to describe the time course of the formation of the complex $R{\cdot}L$ by the differential equation

$$\frac{d[R\cdot L]}{dt} = k_a [R]_0 \cdot ([L]_0 - [R\cdot L]) - k_d \cdot [R\cdot L], \tag{4.4}$$

where k_a and k_d are the association and dissociation rates, respectively. As discussed further below, the integration of Eq. (4.4) permits determination of these rates from time-resolved data on the binding between R and L.

Many molecules in the immune system permit a polyvalent binding, typically due to an oligomeric structure with identical repeated binding domains. Important examples present in blood are the immunoglobulins (Ig), certain soluble lectins, and other lectins expressed in the cell membrane of leukocytes. This repeated presentation of binding domains enables a surprisingly stable and strong interaction with ligand presenting surfaces, where the binding strength grows *quasi* exponentially with the number of interactions. The resulting "functional affinity" [8], also referred to as avidity, has been studied for almost 100 years [9–11]. Specifically, several attempts were made to relate the equilibrium constant for the monovalent interaction between R and L, as defined in Eq. (4.1), to the equilibrium constant, K_{Avidity}, for the polyvalent interaction

$$R_m + L_n \overset{K_{Avidity}}{\leftrightarrow} R_m \cdot L_n(i). \tag{4.5}$$

The indices m and n are here meant to indicate the number binding sites in the receptor and ligand, respectively, while i denotes the actual number of bonds formed between the two species.

A satisfying thermodynamical description of polyvalent interaction was provided by Kitov amd Bundle [12]. A starting point is made by dividing the polyvalent binding into a sequentially ordered series of binding events [12, 13]. The initial step is an intermolecular interaction, with a Gibbs free energy of $\Delta G_{\text{inter}}^0$, followed by multiple intramolecular binding step for simplicity assumed to contribute same Gibbs free energies, $\Delta G_{\text{intra}}^0$ (Fig. 4.1). To further characterize these interactions, the chemical binding is deconstructed into macroscopically and microscopically different interaction. Macroscopically different interaction differ in the number of bonds formed, i.e., i, while microscopically different interactions share the number of bonds formed, but differ with regard to exact interactions sites joined in R and L (Fig. 4.2). In the context of the present writings, it should perhaps be noted that the distinction between macro and microscopically different interactions are purely a nomenclature to enable distinction. The differences between these types of complex formation either within the class or between the classes are all on a nm-scale when dealing with biological macromolecules. This permits writing of the chemical binding reaction for polyvalent interaction as:

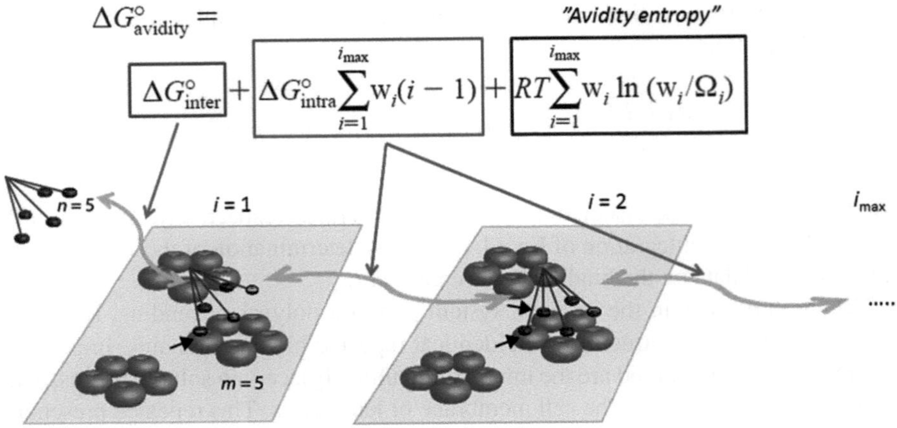

Fig. 4.1 Sequential binding model for polyvalent interactions according to Kitov and Bundle [12]. A polyvalent molecule with five ligand binding domains ($n=5$) and a polyvalent ligand with five binding sites ($m=5$) forms an initial, intermolecular contact ($i=1$), characterized by a Gibbs free energy of $\Delta G_{\text{inter}}^{0}$. All subsequent, intramolecular contacts are characterized by the second term of the equation, which includes the Gibbs free energy of $\Delta G_{\text{intra}}^{0}$ and summed to the number of maximal bond formation, i_{max}. Each bound state (i) is weighed by its occupancy, w_i. A third term constitutes the contribution by the avidity entropy. This involves the degeneracy factor, Ω_i, which may be calculated as shown in Fig. 4.3 under assumption of certain simple geometries of the polyvalent binding molecule and ligand

$$R + L \overset{K_{Avidity}}{\leftrightarrow} R \cdot L(1) + R \cdot L(2) \ldots R \cdot L(i), \tag{4.6}$$

i.e., where the product of the interaction between R and L is an ensemble of macroscopically different interactions. It is straight forward to suggest that the number of bonds formed is the critical component in regulating the binding strength of the interaction and hence the K_{Avidity} can also be written as a sum of equilibrium constants for the macroscopic identical interactions, i.e., those complexes with same i (Fig. 4.2).

$$K_{Avidity} = \frac{\sum [RL(i)]}{[R][L]} = \frac{[R_{bound}]}{[R][L]} = \frac{[R]_0 - [R]}{[R][L]} = \sum_{i_{max}}^{1} K_{RL(i)} \tag{4.7}$$

Obviously, the number of bonds formed between these species, i, cannot exceed $min(m,n)$.

Even though the microscopically different interactions have identical equilibrium constants, and hence are energetically identical, they nevertheless form a crucial contribution to the binding strength of polyvalent interactions through a phenomenon known as avidity entropy. This entropic contribution to the binding

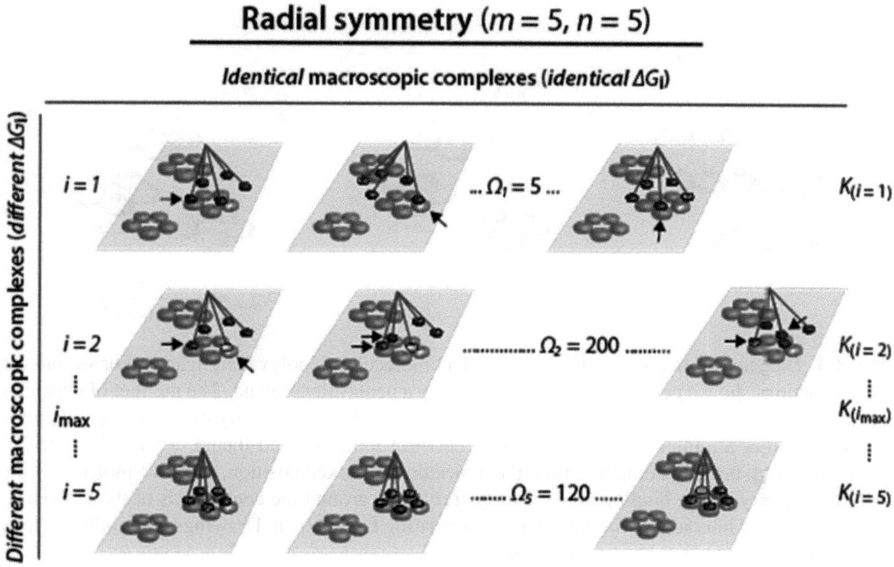

Fig. 4.2 Macroscopic and microscopic differences in complex formation. A molecule with radial symmetry, i.e., free rotation of the "arms" around the connecting point, with five binding sites binds a ligand with five binding sites. Complexes with identical macroscopic bonding (identical i) but with microscopic differences (the exact position of the molecule-ligand bonds) are compared with complexes with differences in the macroscopic bonding (different i). For each number of bond formation ($i = 1$–5), the number of possible macroscopically identical complexes (Ω_i) are calculated according to the formula shown in Fig. 4.3. According to equation for the Gibbs free energy in polyvalent interactions derived by Kitov and Bundle [12] shown in Fig. 4.1, the Gibbs free energy is identical for complexes with the same number of bond formation. This enables the assignment of a single equilibrium constant, K_i, for each bound state, i

increases with the number of ways a molecule can bind a ligand. The exact number of such possible bonding depends on the valency of R and L (i.e., m and n), the number of bonds formed, i, and geometry of the interaction (Fig. 4.3). This makes polyvalent interactions uniquely sensitive to the geometry of the binding molecules and the presentation of ligands, for instance as found on a microbial surface. The implications of this observation are explored further below.

4.1.3 *Quantification of Protein-Ligand Binding Kinetics of Polyvalent Interactions*

The possibilities for determining equilibrium constants for the binding of large bio-macromolecule, such as proteins, to their ligands are supported by a large and increasing number of methodologies. Among the routinely used techniques are

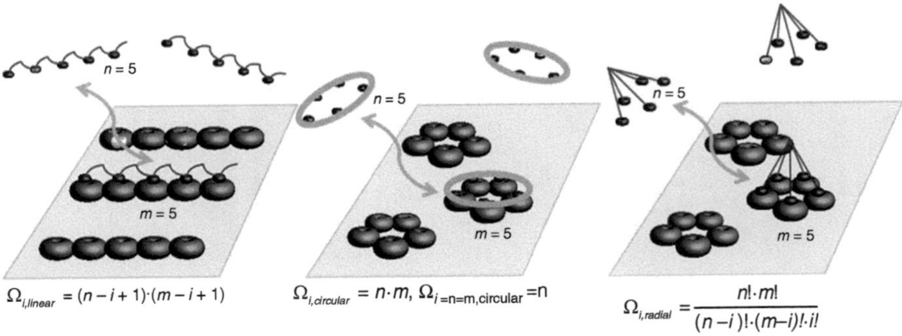

Fig. 4.3 Degeneracy in the binding by polyvalent molecules to polyvalent ligands. For simplicity all interactions show a pentavalent molecule binding a pentavalent ligand. The number of degenerate states (macroscopically identical complexes) depends on the geometry of the interaction, here shown as either a mutually linear interaction between molecule and ligand, a mutually circular interaction with binding domains within the molecule in a locked position, or with a molecule with radial symmetry where binding domains may free rotate around the central axis of the molecule. For each type of interaction, the formula for calculating Ω_i is shown. Drawings are modified from Kitov and Bundle [12] as shown earlier in Ref. [10]

various isothermal calorimetry, analytical ultracentrifugation and, as a more recent addition, microscale thermophoresis. It is beyond the scope of the present chapter to go into detail on how these methods are capable of affinity measurements. However, none of these techniques enable measurements of the underlying binding kinetics. Indeed, only two techniques routinely used are capable of measuring binding kinetics of large biomacromolecules, namely biosensors based on the principle of SPR or the quartz crystal microbalance (QCM). Excellent reviews describe the precise functioning of these sensors [7, 14, 15]. In both cases, a ligand is immobilized on the sensor surface followed by injection of a binding molecule, typically as provided in flow stream keeping the concentration of the binding molecule constant. The sensors will provide real-time data on the binding of the injected molecules to the sensor surface. In the case of SPR instruments, the binding is measured as a change in response level, *S*. *S* is proportional to the formation of the complex $R\cdot L$ in Eq. (4.1).

Binding kinetics are estimated from both the association phase, when the binding molecule is supplied to the surface, and the dissociation phase, when injection of the binding molecule is ceased and the release of the binding molecule from the surface can be followed. For the simple 1:1 interaction in Eq. (4.1), integration of Eq. (4.4) describes the expected deposition of binding molecule on the ligand-coated sensor surface

$$S\left(k_d, K_D, c, t\right) = \frac{c\cdot S_{Max}}{c+K_D}\cdot\left(1-e^{-\frac{k_d}{K_D}\cdot(c+K_D)\cdot(t-t_0)}\right), \tag{4.8}$$

where S_{Max} is the signal at binding saturation, t is the time since start of the experiment, and t_0 the time point for injection of the binding molecule. During the dissociation phase, S is described by a simple exponential function:

$$S(k_d, K_D, c, t) = \left(\frac{c \cdot S_{Max}}{c + K_D} \cdot \left(1 - e^{-\frac{k_d}{K_D} \cdot (c + K_D) \cdot (t_c - t_0)} \right) \right) \cdot \left(e^{-k_d \cdot (t - t_c)} \right), \qquad (4.9)$$

where t_c is time point for the end of the association phase with a duration of $t_c - t_0$. Fitting of these equations to experimental data will return K_D and k_d. k_a may then be calculated from $K_D = k_a / k_d$.

While often used, the simple 1:1 binding scheme can be compromised by experimental aspects not always appreciated in the studies using these methodologies. One such aspect is the valency of the binding molecule. Experimental studies have, in particular been made on immunoglobulins such as IgG. The IgG molecule is bivalent. Nevertheless, the binding of IgG to surfaces with multiple ligands (formally referred to as epitopes) has in multiple cases been studied by applying the simple 1:1 binding scheme. Karlsson et al. solved the equations to better describe bivalent binding [16]. However, until recently, no description was available for the more general problem of how to quantify the binding kinetics of polyvalent molecules. A solution to this problem has been proposed [10] by combining the experimental observation of Gjelstrup et al. [17] with the theoretical insights offered by Kitov and Bundle [12].

Gjelstrup et al. [17] investigated experimentally the binding of the human plasma protein mannan-binding lectin (MBL) to carbohydrate ligands using SPR. While the 1:1 binding scheme failed to fit the binding isotherms from the SPR biosensor, this was achieved when applying a more recently described algorithm for resolving heterogeneous binding. In the context of SPR instruments or similar devices such as the QCM, heterogeneous interactions are observed when the injected binding molecules interact with the immobilized ligands with multiple association and dissociation rates [18]. This can happen when, for instance, the ligand is randomly coupled to the surface, thereby, at least in principle, partially damming the binding site within the ligand. Other examples involve binding molecules capable of forming multiple interactions with one ligand molecule. Here, a recently studied case is certain types of adhesion molecules in the cell membrane of leukocytes, namely integrins [19–21]. Some of these receptors appear to be able to bind the –COOH groups in the side chains of aspartate and glutamate [20]. Since such side chains are abundant in most proteins, each protein will provide multiple binding sites. Even though there is probably little other contact with other parts of the ligand protein, when binding involves especially the short side chain of aspartate there could be a contribution by the nearby residues, at least in a way which would limit the accessibility of the aspartate –COOH [21]. In this way, an ensemble of largely similar, yet, with regard to the detailed binding kinetics, heterogeneous, interactions are provided by single ligand proteins with multiple acidic side chains. A solution to the analysis of such heterogeneous interactions has been contributed by Schuck and his colleagues [18, 22]. In the algorithm by Svitel et al. [18], the distribution of interactions, $P(k_d, K_D)$, is formulated such that the integral $P(k_d^*, K_D^*)dk_d dK_D$ is the fractional abundance of interactions with a dissociation rate between k_d^* and $k_d^* + dk_d$ and a dissociation constant between K_D^* and $K_D^* + dK_D$. The time-course of analyte binding in the association and dissociation phase, now a sum of multiple interactions, is then

$$S_{Tot}(c,t) = \int_{K_{D,\max}}^{K_{D,\min}} \int_{k_{d,\max}}^{k_{d,\min}} S(k_d, K_D, c, t) P(k_d, K_D) dk_d dK_D, \tag{4.10}$$

where the function S is defined by Eqs. (4.8) and (4.9) during the injection and dissociation phases, respectively. A robust and physically meaningful determination of the function P is a critical part of this analysis.

In considering the application of this strategy for resolving the binding kinetics of polyvalent interactions, a starting point is the observation that K_{Avidity} may be written as a simple sum of equilibrium constants for each macroscopically distinct complex (Eq. 4.7) [12]. This suggests that polyvalent interactions share an intrinsic property with the heterogeneous interactions analyzed by Svitel et al. [18], namely that they are an ensemble of energetically, and hence binding kinetically, distinct interactions. In general, this algorithm has been used to resolve heterogeneity in the immobilized ligand. Nevertheless, a pointed out by Svitel et al. [18], with some assumptions mentioned below the algorithm can also handle binding heterogeneity as a consequence of a heterogeneous composition of the injected analyte. It is helpful to combine the nomenclature of Eqs. (4.7) and (4.10) by noting at binding equilibrium that

$$\sum_i [RL(i)] = \sum_i [R_{bound}] \cdot P(k_{d,i}, K_{D,i}) \propto S_{Tot}, \tag{4.11}$$

where $k_{d,i}$ and $K_{D,i}$ are dissociation rate and constant for the $RL(i)$ complex. P is again the fractional abundance of the ith class of interaction such that $\sum P = 1$. Since $[R_{bound}] \propto S(k_{d,i}, K_{D,i}, c, t)$ Eq. (4.11) is similar to Eq. (4.10) with regard to the calculation of S_{Tot} except that the classes of interactions are assumed to be discrete while Eq. (4.10) represents a continuous distribution of binding kinetics. Consequently, it would be expected that Eq. (4.10) is an acceptable approximation of the binding kinetics of polyvalent molecules. Importantly, this approximation rests on the assumption of non-competing interactions between the components of the analyte. Considering the strong binding by highly polyvalent molecules compared to molecules with lesser capacity for polyvalent interactions it is actually difficult to rule out some level of competition. Also, although reports are now emerging on binding data for polyvalent interactions resolved by the use of Svitel et al.'s algorithm [17, 23], a more theoretical approach to the topic is needed to clarify the results from the analysis. Specifically, since a precise estimation of the P_is depends on a precise estimation of S_{Tot} while non-competition between binding species is required, it seems unavoidable that S_{Tot} must remain determined from extrapolation.

4.2 Examples of Polyvalent Binding by Proteins in the Immune System

The above section briefly outlined how large proteins of the immune system may be envisaged as part of nanoscience and how the binding may be detected and quantified. In this section, two aspects of polyvalent interactions are explored. One is the link to conformational change in the surface-bound protein, which is critically linked to the polyvalent interaction as well as the surface topology. The other involves how the binding strength is regulated by the polyvalent interaction, here especially taking into account the geometry of ligand presentation.

4.2.1 Conformational Change and Polyvalent Interactions

Conformational changes in protein structure is probably an over-used explanation of functional regulation [24]. In reality, conformational changes are far from trivial to record or just, in any sense, to document. One problem is the time scale, which typically does not permit the use of biosensors such as those based on SPR for detecting conformational changes when protein are, or become, surface bound. The past several decades of research clearly seems to suggest that conformational changes in large proteins are best investigated through means of characterizing protein structure, most notably through the use of electron microscopy (EM), X-ray crystallography, and small-angle X-ray scattering (SAXS). Below, conformational changes studied in the immunoglobulin IgM and mannan-binding lectin (MBL) are provided to highlight the link between these changes and polyvalency.

IgM is part of the immunoglobulin family of plasma proteins, which also include IgG and IgA and IgE. The Igs serve an important role in the immune system. Their cellular sources are selected through complex processes of maturation and genomic recombination to form Ig protein products that bind certain structures (epitopes) present on, e.g., microbial threats to the body while avoiding binding to the body's own molecules. The cellular maturation process eventually provide Igs with high affinity and specificity for the target structure as well as a compartment of memory cells, which can easily be enrolled if the body is subsequently challenged with a previously met microbial challenge. The "immunological memory" is one hall mark of so-called adaptive immunity, since past exposures of the immune system to non-self antigens will shape the abundance and reactivities of both immune cells and antibodies. However, while the formation of antibodies in this sense is classic part of adaptive immunity, recent research has identified especially IgM antibodies in mice, which seems to be generated and provide immunity in a ways, which are closer to the innate immune system [25].

Structural evidence suggests that IgM is capable of undergoing a significant conformational change when binding to epitope-presenting surfaces. In a classic study using EM, Feinstein and Munn reported a staple-like conformation of the molecule as imaged when bound to a bacterial flagellum from *Salmonella* that likely constituted an antigen for the IgM molecule [26]. By contrast, analysis with SAXS suggested a planar conformation of the molecule in solution, i.e., in the unbound state [27]. The dimensions of planar IgM compared to the staple-like conformation indicated that the C1 complex would only form efficient contact with the stable-like conformation. This elegantly rationalized how complement activation through the classic pathway is regulated by conformational changes in surface bound IgM while avoiding activation in the unbound, soluble state. More recently, investigations by Czajkowsky and Shao using AFM prompted the suggestion that such complement activation does not involve conformational change in the IgM molecule [28]. In their model, IgM-mediated complement activation is driven by surface binding in a planar or at least stabilized conformation as a consequence of the surface attachment. This will act to restrict structural fluctuations of the molecule compared to the soluble molecule such that strong binding by the C1 complex is enabled. Unfortunately, unlike the study by Feinstein and Munn [26], no direct observation of epitope-bound IgM was reported by Czajkowsky and Shao. Furthermore, the ability of Perkins et al. to obtain a full-size envelope from the SAXS studies [27] suggests a level of rigidity in the soluble molecule, which would be inconsistent with a highly fluctuating structure. Finally, functional data supported by structural calculations presented by Pedersen et al. [29] also support the idea that appropriate conformational change in IgM is critical for complement activation. In these studies, human sera were incubated with iron oxide nanoparticles (NPs) coated with dextran. Antibodies to dextran are commonly found in human sera [30] although the levels (usually expressed as a titer) differ. It was possible to show that complement activation by these particles correlated with the IgM titer to dextran suggesting that this protein was particularly important in the activation [29]. A striking finding was that the complement activation differed between particles depending on the size of these particles. Particles presenting the dextran antigen and with a diameter of 100–250 nm activated complement while smaller (50 nm) or larger (600 nm) particles failed in this activation [29]. From the work by Perkins et al. [31], it was possible to relate these experimental findings to the expected conformation of IgM on these particle surfaces. To take the staple-like conformation in a way permitting C1 binding, the bound IgM molecule must bind the Fab fragments in an angle of approximately 68° relative to the plane of the Fc regions. The resulting angle of the surface-bound molecule is probably influenced by several factors, for instance the precise location of the contact between the Fab and its epitope as well as the orientation of the epitopes on the surface. From geometrical considerations it is possible to show, however, that a major factor in regulating the structure of the surface-bound molecule is the curvature of the particle surface, most notably when the curvature are comparable to the dimensions of the molecule (Fig. 4.4) [29]. Flat or moderately curved surfaces will not be able to deflect the planar conformation of IgM in solution into the staple-like conformation. By contrast, more curved surface, e.g., as found on particles with diameters in the range of the cross-sectional diameter of the

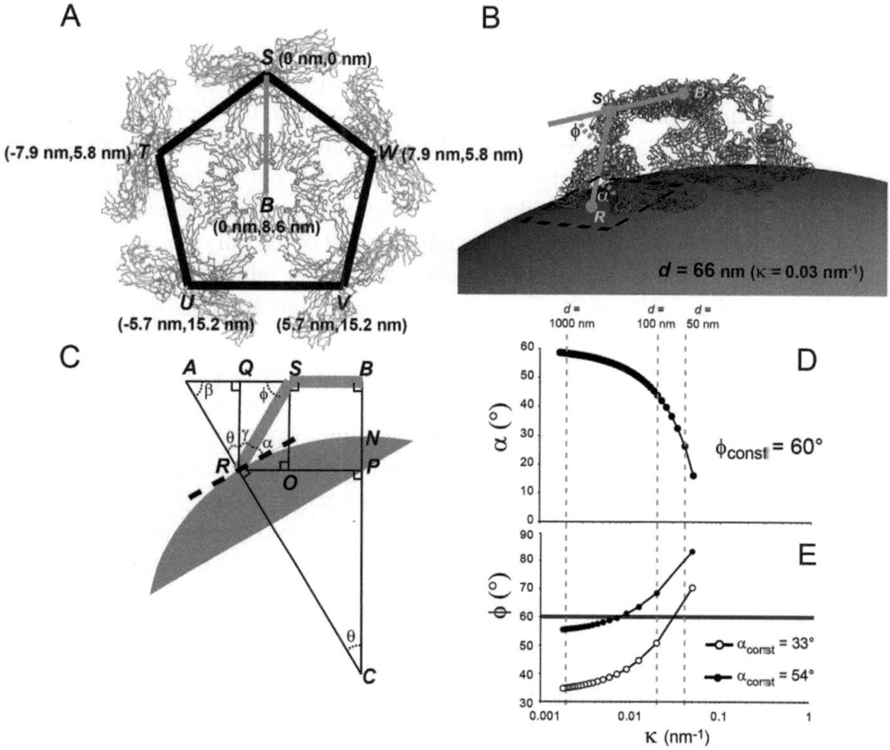

Fig. 4.4 Model of the structural requirements for the binding of IgM in the staple-like conformation [26, 62] to epitopes exposed on the surfaces of nanoparticles with differing curvatures. (**a**, **b**) A pseudo atomic model of the IgM pentamer in the staple-like conformation was made from the planar solution structure of IgM reported by Perkins et al. [27]. The regions corresponding to the five Fab_2' were rotated 60° downwards with respect to Fc_5/J chain disc. As shown in Panel A the centroid (***B***) of the Fc_5 disc, including the J chain indicated in green, was determined from considering the distances in two dimensions between equivalent residues (indicated with ***S***, ***T***, ***U***, ***V***, and ***W***, respectively) located in the Cμ2 domains, i.e., the hinge between the Cμ1 domain and the Fc_5/J chain disc. From these calculations the distance ***SB*** was 8.6 nm. In *panel* **b** the IgM molecule in the staple-like conformation is shown on the surface of particle with a diameter of 66 nm, i.e., with a curvature ($\kappa=0.03$ nm^{-1}); the variable loops are indicated in blue. The point ***R*** on the particle surface was placed equidistant between the variable loops of a Fab_2' on a line through the Cμ1 domains as indicated in the panel. The distance ***SR*** equals 10.6 nm. (**c**) Two-dimensional representation of geometric relationship between angles and intramolecular distances of the IgM molecule (sketched in *grey*). The angle ϕ measures bending of the Cμ1 and variable domains relative to the Fc_5 disk while α measures the angle between the line ***SR*** and the tangent to the circle with a radius ***CR*** (i.e. the particle surface) at the point ***R***. (**d**, **e**) Relationship between the angles α, ϕ, and the curvature of the particle κ. With ϕ at a fixed angle (60°) there is an analytical solution of Eq. (4.6) for α as a function of κ: $\alpha = \arcsin\left(\dfrac{SR \cdot \cos\varphi_{const} + SB}{CR}\right) - \varphi_{const}$. Based on this function α was plotted as a function of κ in *panel* **d**. In *panel* **e** values of ϕ was plotted as a function of κ with α at a fixed value of either 33° or 54°. For each plot in *panels* **d** and **e** the shortest distance from centroid in the Fc_5/J chain disc to the particle surface (the distance ***BN*** in *panel* **c**) is indicated for the minimum and maximum values of κ. Figure and legend were first published in Pedersen et al. [29] Copyright 2010. The American Association of Immunologists, Inc.

IgM molecule itself, would significantly deflect the planar structure of IgM when this molecule binds such as surface. Finally, some surfaces may be so steeply curved that they may not accommodate IgM binding, at least not in a way accommodating full engagement of all Fabs on the surface. Taken together, these three settings may well explain why certain sizes of particles will activate the complement system through IgM polyvalent binding while other sizes will not or prove considerably less efficient. It is evident from the proposed geometry of the activation that the sensitivity toward surface curvature is wholly a result of the polyvalency of the IgM binding. These findings encourage the expectation that other types of polyvalent protein interactions with surface-bound ligands may show a similar dependence on surface curvature.

A second example of conformational changes as part of complement protein binding involves the MBL molecule. MBL is a well-established part of the innate immune system and deficiency is associated with susceptibility to infections although rarely in a way that is life threatening [32]. Effector mechanism of innate immunity are not subject to immunological memory although the cross-talk between innate and adaptive immunity probably facilitates development of memory in the adaptive immune system through activation of the innate immune system [33]. The proteins of the innate immunity are encoded by germ-line genes, which remain unaltered in somatic cells. The gene products have been selected through the eons of biological evolution to provide protection against infectious agents. In many cases the protein architecture permits recognition of certain structures, or, as famously coined by Charles Janeway, "patterns", present on these microbial agents but not on host cells [34]. MBL is an excellent example of such a protein. Through the lectin domains MBL binds carbohydrates, notably *N*-acetyl glucosamine or mannose, but also other types. The MBL proteins chains consist of a cystein-rich N-terminal segment, enabling covalent inter chain binding, a collagen-like region which form homo trimers of MBL polypeptide chains, and finally a Ca^{2+}-dependent carbohydrate recognition domain (CRD) binding ligands [32]. The collagenous trimerization of polypeptide chains creates a structural unit (here designated MBL_3). These units are able to form larger oligomers, typically with 3–8 units ($3–8 \times MBL_3$) as judged from biochemical characterization and AFM imaging [17, 35]. The MBL oligomers are associated with proteases (MBL-associated serine proteases, or MASP), which permit activation of the complement system once the MBL oligomers are bound to target surfaces. The biochemical mechanism of distinguishing "self" from "non-self" is discussed further below. At this point, the studies on the MBL ultrastructure are helpful to outline the challenges in documenting conformational changes in large proteins. The MBL ultrastructure is actually similar to both IgM and C1q in the sense that these proteins are approximately 40 nm in cross sectional diameter, contains a pseudo rotational symmetry, and possesses multiple binding domains [35–37]. For this reason, it is tempting to suggest that complement activation by MBL involves a conformational change when the molecule attaches to a target surface as also proposed for IgM. To make a study on this topic Dong et al. characterized the structure of MBL in solution by SAXS [37]. The ultrastructure of $3 \times MBL_3$ oligomer is moderately bend and highly ramified with an internal maximum distance not exceeding 32 nm and a cross sectional diameter of 37 nm.

The CRD_3 domains are flexibly attached to the collagenous region [37, 38]. To provide ultrastructural data on the ligand-bound state, AFM imaging was made on the same preparation of MBL applied to a surface with carbohydrate ligand [37]. In this case, the cross sectional diameter was found to be roughly 60 nm on average when liquid-state imaging was performed. Higher resolution was obtained by dry-state imaging, but under these conditions the molecular structure deteriorated. Taken together, these measurements would nevertheless suggest that some kind of unbending or stretching (or both) of the molecule occurs in the surface-bound state.

Are conformational changes in MBL then critical for complement activation? Degn et al. noted recently [39] that a simple accumulation of MBL/MASP complexes would probably permit autoactivation of the MASPs, and, hence, activation of the complement cascade, as judged from what is known about enzymology of these proteins. In this model, the ligand binding activity of MBL simply acts as a conveyor of MASPs to the target surfaces eventually raising the enzyme concentration beyond a level critical for enzymatic activation. In principle, no conformational change in MBL is required per se for MASP activation. On the other hand, few, if any, of the surfaces, which were routinely used for testing complement activation by the MBL/MASP complexes, are flat compared to the size of the MBL oligomers. *Staphylococcus aureus* was used as complement-activating agent in studies suggesting catalytic activation of MASP by juxtaposition of complexes carrying these proteases [40]. *S. aureus* synthesizes its cell wall in such a way that the surface presents ridges organized in concentric circles [41]. The ridge height seems to vary considerably, with amplitudes from ~1 to 5 nm depending on the age of the material. Other structural features of the cell wall involve central depressions, where novel cell wall is synthesized. In a radius of ~30 nm from such depressions, the ridge height reaches 30 nm. When MBL binds through multiple points of contact to a surface wrinkled in these ways, it seems likely to involve some sort of structural change in the bound MBL compared to the soluble state. This suggestion lends support from the consideration made above on IgM, namely that polyvalent interactions—unlike their monovalent equivalents—are sensitive to surface topography of the bound surface. The study by Dong et al. [37] seems at least to suggest that such changes in MBL can be accommodated by the molecule. To settle the specific role of conformational changes in MBL as part of complement activation will consequently remain a challenge in a foreseeable future.

4.2.2 Avidity, Polyvalency, and Immune Target Recognition by Soluble Molecules

One of the most striking consequences of polyvalency is the increase in binding strength of a polyvalent molecule compared to its monovalent source. The polyvalent nature of the interaction is doing more to target recognition, however, than merely increasing binding strength or serving as anchor points to convey conformational changes as can be appreciated from as few cases already introduced.

In the case of MBL, it was suggested that the CRD trimer permit a critical polyvalent interactions with polymeric carbohydrates [42, 43]. The trimerization creates a distance between the CRD ligand binding sites of roughly 5 nm. Binding of these multiple CRDs to carbohydrate chains consequently requires branching of the glycan or a glycan of some length, longer than those found on mammalian cells but typical of fungal cell wall carbohydrates. This forms a biochemical principle for pattern recognition by this molecule although the exact target organisms recognized by such principles can be disputed [10, 44]. Binding data provided by Gjelstrup et al. [17] seems to suggest that the major source of avidity is the oligomerization of structural MBL_3 units while the CRD_3 unit only binds weakly to glycosylated targets. In any case, it is possible to rationalize this type of pattern recognition as "topological patterns", where the actual structural presentation of ligands will enable or discharge polyvalent interactions and hence MBL binding [10]. From the evolutionary selection of MBL to form an efficient tool to bind such patterns, it may be speculated that the ultrastructure of this, and other, similar pattern recognition molecules is imprinted by the size and topology of their binding targets, i.e., the patterns found on microbes. In this sense the topological patterns are of a fundamentally different structure than most of those recognized by other pattern recognition molecules, namely Toll-like receptors (TLR). The critical part for recognition by these molecules is not thought to involve any major ultrastructural presentation of ligands from microbial organisms such as parts of the cell wall (peptidoglycan) or flagella (flagellin) [10, 45, 46]. For instance, the surface recognition involved in TLR5 binding of flagellin, a protein component of bacterial flagella, would cover ~13 nm^2 in the flagellin [46]. By contrast, MBL pattern recognition may cover ligands presented over an area of as much as ~3000 nm^2 [37]. As judged from the work by Feinstein on Munn on IgM [26] something similar would probably be the case for this large molecule as well. In this context avidity entropy, mentioned in Sect. 4.1.2, is a particularly interesting phenomenon. Microbial organisms are made "bottom-up" implying that repeated biomacromolecular elements are found in microbial surfaces with certain symmetries in their presentation. In particular, work on nanomicrobiology by Yves Dufrêne and colleagues [47] as well as the work cited above on the cell wall structure of *S. aureus* [41] have unraveled a previously unappreciated ultrastructure of microbial cell surfaces. It seems certainly possible that an important part of topological pattern recognition would involve symmetries of ligand presentation on these cell surfaces and hence link the binding strength to aspects of avidity entropy.

4.3 Nanomedicines and Polyvalent Proteins of the Immune System

Almost three decades of immunological research has highlighted the importance of the innate immune system in acting as a first-line of defense against invasive microorganisms. As briefly discussed above, innate immunity distinguishes "friend from foe" by the ability of certain germ-line encoded proteins to bind patterns exposed by non-self surfaces or molecular constituents. Whilst there is plenty of evidence that such a system is required to protect the body from infections with pathogenic microbes, it is also now clear that the innate immune system also constitutes a formidable barrier to nearly any kind of drug delivery involving particulate materials. Below, the connection is made between the nanoscience of polyvalent interactions and the challenges in nanotechology to develop efficient drug delivery vehicles.

4.3.1 How Nanomedicines Resemble Targets for the Innate Immune System

It is a truism that nanoscience is a science of the importance of size. In nanomedicine, this is extended to include biological materials, where nearly any phenomenon involving biomacromolecules, in principle, is a variant nanoscience. In this line, with the molecular basis of the pathologies of diseases these could be argued both to be rationalized in terms of nanoscience as well as be treated by agents capable of acting on a nm scale. This notion was famously captured by the physicists Richard Feynman's vivid foresight in 1959 of a future, where the patient could "swallow the doctor" to enable this nano-sized medical practitioner to directly interfere with undue molecular mechanisms at a similar size scale [48]. The best examples of medical therapies related to this vision are probably the remotely similar application of NPs in medicine. Below, a few of examples are identified and put in a context of why they challenge, in particular, the innate immune system.

If one follows the European Union's definition of NPs as "with one or more dimensions of the order of 100 nm or less", several infectious agents, notably viruses are also on this list. Unsurprisingly, this has prompted the idea that the sheer size of viruses is their strongest resemblance to NPs and that this property somehow is in a major way responsible for NP recognition by the immune system [5, 10, 49]. However, arguably the mechanistic insight to support this end has been lacking. Indeed, at least two other properties are likely to be important, particularly in the context of the recognition of NP by polyvalent proteins of the innate immune system.

First, although it is possible to make certain types of NPs top-down by grinding and subsequent appropriate liquid chromatography (e.g., [50]), in more general such material is manufactured by precipitation or aggregation reactions (e.g., [51]). These "bottom-up" strategies basically assemble the particles from smaller metallic or organic polymer compounds, in some cases producing a semi-crystalline material. While this material is not necessarily highly ordered, the limited types of constituents are likely to create some degree of repeated structures in and on the surface of the resulting particles. An important example of this type is ultra small paramagnetic iron oxide particles (USPIO), which amongst other applications are used as contrast reagent in magnetic resonance imaging. These NPs are usually modified with carbohydrates (dextran) [52]. This is not unlike viral capsids, which also are rigid, semi crystalline particles, in many cases with diameters of ~100 nm [53]. Obviously, viral particles are also assembled in host cells "bottom up" from their macromolecular constituents. In this case, a high degree of ordered presentation of these constituents is the results [53]. Also, the viral capsids may be covered with a lipid envelope as well as integrate glycosylated proteins, as for instance known from human immunodeficiency virus, further strengthening the similarity to USPIO formulations. Second, the attempt to create biocompatible NPs has prompted a choice of materials for making the particles or their surface coating from biological sources. While this in many cases would ensure degradability of the material and avoidance of toxic responses from hepatic, pulmonary or renal accumulation, it does not prevent immune responses. It has become increasingly clear biologically-derived materials create a significant problem from immune recognition as "biomolecules" of apparently innocuous origin [54], e.g. chitosan derived from the shells of shrimp, triggers an instantaneous immune response from the innate immune system [55, 56]. Chitosan is made from deacetylating chitin, thereby, creating highly positively charged carbohydrate polymers of repeating glucosamine residues, however, with varying degrees of deacetylation leaving behind unmodified *N*-acetyl glucosamine residues as well. Since this latter constituent is an efficient ligand for, e.g., MBL and induces chitosan triggers antibody production [56], it not surprising that chitosan NPs trigger complement activation, nor that complement activation is likely to inflict serious challenges in keeping the particles functional or even in circulation. One of the reasons that the immune system to this extent attacks chitosan is that a similar material is generated by the enzymatic breakdown of fungal cell walls, which may contain chitin [57]. In brief, the choice of chitosan will make the particles effectively mimic surfaces of a known treat to the human body, i.e., fungi. Presumably, the innate immune system has been subject to selection for enabling a response to these organisms and the biological origin of the material is, thus, of no avail in avoiding immune responses. A similar point can be raised for dextran. Dextran-like carbohydrates are made by gram-negative organisms [58], which, in turn, may cause formation of host antibodies to this material. Human blood abounds with high level of dextran antibodies [30], which would also effectuate the clearance of the dextran-coated particles. Some success of engineering so-called stealth particles was achieved with synthetic polyethylene glycol layers [51]. Nevertheless, also in this case certain MBL-like molecules of the immune system, namely the ficolins, may be capable of triggering a response through complement activation [59]. These examples clearly advocate a more refined strategy, if these materials, particularly biomaterials, are to be used as drug delivery vehicles.

4.3.2 Surface Ultrastructure, Polyvalent Interactions and the Functionality of NPs

The point has been made above that many nanoparticulate systems by mode of synthesis and choice of materials are targets for molecules of the innate immune system as extensively reviewed elsewhere [54, 60, 61]. The view in the present text is that a considerable part of the challenges induced by the immunoreactivity reflects the polyvalent interactions often supported by the immune system. Of course, this begs the question if our current insight on how polyvalent interactions are made can be used to improve NP drug functionality.

The work by Pedersen et al. [29] highlight the point that size, or specifically curvature, of NPs is important when conformationally labile molecules such as IgM bind their surfaces. Whilst it is not surprising that certain particles are too small to accommodate this binding and will escape triggering IgM immune functions, the contrary finding that large particles with a lower curvature can do so, seems to offer new avenues of drug design. At this point it is, of course, difficult to determine if appropriate particle size will enable concomitant escape from multiple types of immune recognition such as phagocytic capture. At least Pedersen et al.'s work offers a rational to explain why the size of NPs, i.e., particles with a size close to the dimensions of the large biomolecules of the immune system, is a critical parameter. Indeed, when wrongly chosen, this will impede the function of the NPs with respect to clearance or other incapacitation of function by the immune system. Polyvalent interactions seems here particular important consider as they are highly sensitive to the topology of the surface.

A more surprising aspect relates to the thermodynamic aspects of polyvalent interactions, especially the sensitivity to the symmetry of ligand presentation via avidity entropy. From a simple experimental enquiry, Gjelstrup et al. [17] reported that ligand density may dramatically regulate polyvalent binding of MBL to surfaces coated with variable amounts of coated ligand [10]. It seems likely that avidity entropy plays a role in the binding of these surfaces, but arguably a more carefully engineered surface would be required to better analyze this point. In this context, it is important to remember that the schematic representation and associated calculations of degeneracy of ligand binding (Fig. 4.3) is a theoretical approach, which may not capture the full complexity of the actual ultrastructural properties of, for instance, microbial surfaces as discussed above [10, 12]. Nevertheless, it is clear that any complete or distorted symmetries of ligands on such surfaces, or the engineered surfaces of NPs, are likely to be an important factor in polyvalent interaction. While the current state-of-art in designing NPs mostly involves processes that only permits a partial control over the surface ultrastructure, it would be envisaged, from the considerations made above, that such control would enable better control over what molecules would bind the NP surfaces. Not only would this address the important aspect of avoiding particle clearance of the particles, it could also contribute novel means drug targeting, e.g., in drugs targeting immune cells or functions of the immune system by facilitating the design of antagonists. This potential was actually experimentally demonstrated by Kitov and Bundle by use of dendrimers targeting IgM [12].

4.4 Concluding Remarks

This chapter wants to enable the Reader to understand basic principles of polyvalent binding by large, immuno active proteins to ligands, mainly presented on surfaces of microbes or NP look-alikes. The few examples of such mechanisms are far from exhaustive, nor scientifically well understood. In consequence, the Reader is encouraged to more broadly consider in what sense and where polyvalent interaction are part of nanoscience and nanotechnology.

At this stage, it seems safe to conclude that the topic of large proteins and their interactions with NPs and other nanostructured materials is treatable as part of nanoscience with significant implications for nanomedicine. It is an interesting question if our insight on the structure, or simply the size, of proteins of the immune system can benefit the design of, for instance, NPs, which are better tolerated and with better performance in the body. At least two aspects come to mind from the current presentation. First, the chemical nature of polyvalent interactions, specifically with regard to the entropy-driven part of the binding, seems to advocate against highly symmetric systems. Second, the conformational influence of molecules bound or otherwise adsorbed onto surface can be critical for the outcome of such adsorption, in particular when the molecules in question are immunoactive. While adsorption affects conformational change in the finer protein structure, this chapter points to changes at the ultrastructural level—at the nm scale—as also potentially important. This is at least technically significant as nanoscience now offers multiple approaches to studying the ultrastructural changes in molecular architecture. In this sense, an outlook on future studies on the interaction between the immune system and nanostructured materials would include a vast increase in our knowledge about protein structure influenced by nanostructured materials. It appears as a fair expectation that such insights will benefit the design of efficient nanomedicines that are capable of better escaping unwanted activation of the immune system than current methodologies.

Acknowledgements This paper summarizes ideas presented earlier as part of a public defense on September 5th 2014 for the degree of Doctor Medical Science awarded to TV-J. I would like to thank the official opponents on that occasion, Prof. Søren Buus, University of Copenhagen, Dr. A. Christy Hunter, Manchester Pharmacy School, and Prof. Kristian Stengaard-Pedersen, Aarhus University, for their helpful discussion and contributions to developing my thoughts on this topic. Our original papers were made with generous financial support from The Danish Multiple Sclerosis Association, The Danish Rheumatism Association, The Carlsberg Foundation, The Novo Nordisk Foundation, The Lundbeck Foundation, The LEO Pharma Research Foundation, The Danish Council for Independent Research|Medical Sciences, and The Danish Council for Independent Research| Natural Sciences.

References

1. Pauling L, Corey RB, Branson HR (1951) The structure of proteins: two hydrogen-bonded helical configurations of the polypeptide chain. Proc Natl Acad Sci U S A 37(4):205–211
2. Leggio C, Galantini L, Pavel NV (2008) About the albumin structure in solution: cigar expanded form versus heart Normal shape. Phys Chem Chem Phys 10(45):6741–6750. doi:10.1039/b808938h
3. Howard KA (2015) Albumin: the next-generation delivery technology. Ther Deliv 6(3):265–268. doi:10.4155/tde.14.124
4. Linderstrøm-Lang KU (1952) Proteins and enzymes. In: Lane medical lectures, vol 6. Stanford University Publications, University Series, Medical Sciences. Stanford University Press, CA. https://books.google.dk/books?id=JUisAAAAIAAJ&printsec=frontcover&source=gbs_ge_summary_r&cad=#v=onepage&q&f=false
5. Vorup-Jensen T, Boesen T (2011) Protein ultrastructure and the nanoscience of complement activation. Adv Drug Deliv Rev 63(12):1008–1019. doi:10.1016/j.addr.2011.05.023, S0169-409X(11)00145-1 [pii]
6. Langmuir I (1916) The constitution and fundamental properties of solids and liquids: Part I. Solids. J Am Chem Soc 38:2221–2295
7. Schuck P (1997) Use of surface plasmon resonance to probe the equilibrium and dynamic aspects of interactions between biological macromolecules. Annu Rev Biophys Biomol Struct 26:541–566. doi:10.1146/annurev.biophys.26.1.541
8. Karush F (1976) Multivalent binding and functional affinity. Contemp Top Mol Immunol 5:217–228
9. Burnet FM, Keogh EV, Lusk D (1937) The immunological reactions of filterable viruses. Aust J Exp Biol Med Sci 15:226–368
10. Vorup-Jensen T (2012) On the roles of polyvalent binding in immune recognition: perspectives in the nanoscience of immunology and the immune response to nanomedicines. Adv Drug Deliv Rev 64(15):1759–1781. doi:10.1016/j.addr.2012.06.003
11. Mammen M, Choi SK, Whitesides GM (1998) Polyvalent interactions in biological systems: implications for design and use of multivalent ligands and inhibitors. Angew Chem Int Ed 37(20):2755–2794
12. Kitov PI, Bundle DR (2003) On the nature of the multivalency effect: a thermodynamic model. J Am Chem Soc 125(52):16271–16284. doi:10.1021/ja038223n
13. Jencks WP (1981) On the attribution and additivity of binding energies. Proc Natl Acad Sci U S A 78(7):4046–4050
14. Jason-Moller L, Murphy M, Bruno J (2006) Overview of Biacore systems and their applications. Curr Protoc Protein Sci. Editorial board, John E. Coligan et al. Chapter 19:Unit 19.13. doi:10.1002/0471140864.ps1913s45
15. Hunter AC (2009) Application of the quartz crystal microbalance to nanomedicine. J Biomed Nanotechnol 5(6):669–675
16. Karlsson R, Mo JA, Holmdahl R (1995) Binding of autoreactive mouse anti-type II collagen antibodies derived from the primary and the secondary immune response investigated with the biosensor technique. J Immunol Methods 188(1):63–71
17. Gjelstrup LC, Kaspersen JD, Behrens MA, Pedersen JS, Thiel S, Kingshott P, Oliveira CL, Thielens NM, Vorup-Jensen T (2012) The role of nanometer-scaled ligand patterns in polyvalent binding by large mannan-binding lectin oligomers. J Immunol 188(3):1292–1306. doi:10.4049/jimmunol.1103012, jimmunol.1103012 [pii]
18. Svitel J, Balbo A, Mariuzza RA, Gonzales NR, Schuck P (2003) Combined affinity and rate constant distributions of ligand populations from experimental surface binding kinetics and equilibria. Biophys J 84(6):4062–4077. doi:10.1016/S0006-3495(03)75132-7, S0006-3495(03)75132-7 [pii]
19. Vorup-Jensen T (2012) Surface plasmon resonance biosensing in studies of the binding between beta integrin I domains and their ligands. Methods Mol Biol 757:55–71. doi:10.1007/978-1-61779-166-6_5

20. Vorup-Jensen T, Carman CV, Shimaoka M, Schuck P, Svitel J, Springer TA (2005) Exposure of acidic residues as a danger signal for recognition of fibrinogen and other macromolecules by integrin alphaXbeta2. Proc Natl Acad Sci U S A 102(5):1614–1619. doi:10.1073/pnas.0409057102, 0409057102 [pii]
21. Bajic G, Yatime L, Sim RB, Vorup-Jensen T, Andersen GR (2013) Structural insight on the recognition of surface-bound opsonins by the integrin I domain of complement receptor 3. Proc Natl Acad Sci U S A 110(41):16426–16431. doi:10.1073/pnas.1311261110
22. Gorshkova II, Svitel J, Razjouyan F, Schuck P (2008) Bayesian analysis of heterogeneity in the distribution of binding properties of immobilized surface sites. Langmuir 24(20):11577–11586. doi:10.1021/la801186w
23. Kapinos LE, Schoch RL, Wagner RS, Schleicher KD, Lim RY (2014) Karyopherin-centric control of nuclear pores based on molecular occupancy and kinetic analysis of multivalent binding with FG nucleoporins. Biophys J 106(8):1751–1762. doi:10.1016/j.bpj.2014.02.021
24. Davis SJ, van der Merwe PA (2006) The kinetic-segregation model: TCR triggering and beyond. Nat Immunol 7(8):803–809. doi:10.1038/ni1369, ni1369 [pii]
25. Capolunghi F, Rosado MM, Sinibaldi M, Aranburu A, Carsetti R (2013) Why do we need IgM memory B cells? Immunol Lett 152(2):114–120. doi:10.1016/j.imlet.2013.04.007
26. Feinstein A, Munn EA (1969) Conformation of the free and antigen-bound IgM antibody molecules. Nature 224(5226):1307–1309
27. Perkins SJ, Nealis AS, Sutton BJ, Feinstein A (1991) Solution structure of human and mouse immunoglobulin M by synchrotron X-ray scattering and molecular graphics modelling. A possible mechanism for complement activation. J Mol Biol 221(4):1345–1366. doi:10.1016/0022-2836(91)90937-2
28. Czajkowsky DM, Shao Z (2009) The human IgM pentamer is a mushroom-shaped molecule with a flexural bias. Proc Natl Acad Sci U S A 106(35):14960–14965. doi:10.1073/pnas.0903805106, 0903805106 [pii]
29. Pedersen MB, Zhou X, Larsen EK, Sorensen US, Kjems J, Nygaard JV, Nyengaard JR, Meyer RL, Boesen T, Vorup-Jensen T (2010) Curvature of synthetic and natural surfaces is an important target feature in classical pathway complement activation. J Immunol 184(4):1931–1945. doi:10.4049/jimmunol.0902214, jimmunol.0902214 [pii]
30. Foote JB, Mahmoud TI, Vale AM, Kearney JF (2012) Long-term maintenance of polysaccharide-specific antibodies by IgM-secreting cells. J Immunol 188(1):57–67. doi:10.4049/jimmunol.1100783, jimmunol.1100783 [pii]
31. Perkins SJ, Nealis AS, Sim RB (1991) Oligomeric domain structure of human complement factor H by X-ray and neutron solution scattering. Biochemistry 30(11):2847–2857
32. Holmskov U, Thiel S, Jensenius JC (2003) Collections and ficolins: humoral lectins of the innate immune defense. Annu Rev Immunol 21:547–578. doi:10.1146/annurev.immunol.21.120601.140954, 120601.140954 [pii]
33. Heesters BA, Myers RC, Carroll MC (2014) Follicular dendritic cells: dynamic antigen libraries. Nat Rev Immunol 14(7):495–504. doi:10.1038/nri3689
34. Janeway CA Jr (1989) Approaching the asymptote? Evolution and revolution in immunology. Cold Spring Harb Symp Quant Biol 54(Pt 1):1–13
35. Jensenius H, Klein DC, van Hecke M, Oosterkamp TH, Schmidt T, Jensenius JC (2009) Mannan-binding lectin: structure, oligomerization, and flexibility studied by atomic force microscopy. J Mol Biol 391(1):246–259. doi:10.1016/j.jmb.2009.05.083, S0022-2836(09)00677-9 [pii]
36. Lu JH, Thiel S, Wiedemann H, Timpl R, Reid KB (1990) Binding of the pentamer/hexamer forms of mannan-binding protein to zymosan activates the proenzyme C1r2C1s2 complex, of the classical pathway of complement, without involvement of C1q. J Immunol 144(6):2287–2294
37. Dong M, Xu S, Oliveira CL, Pedersen JS, Thiel S, Besenbacher F, Vorup-Jensen T (2007) Conformational changes in mannan-binding lectin bound to ligand surfaces. J Immunol 178(5):3016–3022. doi:10.4049/jimmunol.178.5.3016

38. Gjelstrup LC, Boesen T, Kragstrup TW, Jorgensen A, Klein NJ, Thiel S, Deleuran BW, Vorup-Jensen T (2010) Shedding of large functionally active CD11/CD18 Integrin complexes from leukocyte membranes during synovial inflammation distinguishes three types of arthritis through differential epitope exposure. J Immunol 185(7):4154–4168. doi:10.4049/jimmunol.1000952, jimmunol.1000952 [pii]
39. Degn SE, Thiel S (2013) Humoral pattern recognition and the complement system. Scand J Immunol 78(2):181–193. doi:10.1111/sji.12070
40. Degn SE, Kjaer TR, Kidmose RT, Jensen L, Hansen AG, Tekin M, Jensenius JC, Andersen GR, Thiel S (2014) Complement activation by ligand-driven juxtaposition of discrete pattern recognition complexes. Proc Natl Acad Sci U S A 111(37):13445–13450. doi:10.1073/pnas.1406849111
41. Touhami A, Jericho MH, Beveridge TJ (2004) Atomic force microscopy of cell growth and division in Staphylococcus aureus. J Bacteriol 186(11):3286–3295. doi:10.1128/JB.186.11.3286-3295.2004
42. Weis WI, Drickamer K (1994) Trimeric structure of a C-type mannose-binding protein. Structure 2(12):1227–1240
43. Sheriff S, Chang CY, Ezekowitz RA (1994) Human mannose-binding protein carbohydrate recognition domain trimerizes through a triple alpha-helical coiled-coil. Nat Struct Biol 1(11):789–794
44. Hoffmann JA, Kafatos FC, Janeway CA, Ezekowitz RA (1999) Phylogenetic perspectives in innate immunity. Science 284(5418):1313–1318
45. Park BS, Song DH, Kim HM, Choi BS, Lee H, Lee JO (2009) The structural basis of lipopolysaccharide recognition by the TLR4-MD-2 complex. Nature 458(7242):1191–1195. doi:10.1038/nature07830, nature07830 [pii]
46. Yoon SI, Kurnasov O, Natarajan V, Hong M, Gudkov AV, Osterman AL, Wilson IA (2012) Structural basis of TLR5-flagellin recognition and signaling. Science 335(6070):859–864. doi:10.1126/science.1215584, 335/6070/859 [pii]
47. Dufrene YF (2014) Atomic force microscopy in microbiology: new structural and functional insights into the microbial cell surface. MBio 5(4):e01363–e01314. doi:10.1128/mBio.01363-14
48. Feynman R (1960) There's plenty of room at the bottom. Eng Sci 23(23):22–36
49. Vorup-Jensen T (2012) Wrong resemblance? Role of the immune system in biocompatibility of nanostructured materials. In: Peer D (ed) Handbook of harnessing biomaterials in nanomedicine. Pan Stanford Publishing, Singapore, pp 283–301
50. Guerreiro AR, Chianella I, Piletska E, Whitcombe MJ, Piletsky SA (2009) Selection of imprinted nanoparticles by affinity chromatography. Biosens Bioelectron 24(8):2740–2743. doi:10.1016/j.bios.2009.01.013
51. Larsen EK, Nielsen T, Wittenborn T, Birkedal H, Vorup-Jensen T, Jakobsen MH, Ostergaard L, Horsman MR, Besenbacher F, Howard KA, Kjems J (2009) Size-dependent accumulation of PEGylated silane-coated magnetic iron oxide nanoparticles in murine tumors. ACS Nano 3(7):1947–1951. doi:10.1021/nn900330m
52. Bernd H, De Kerviler E, Gaillard S, Bonnemain B (2009) Safety and tolerability of ultrasmall superparamagnetic iron oxide contrast agent: comprehensive analysis of a clinical development program. Invest Radiol 44(6):336–342. doi:10.1097/RLI.0b013e3181a0068b
53. Klein JS, Bjorkman PJ (2010) Few and far between: how HIV may be evading antibody avidity. PLoS Pathog 6(5):e1000908. doi:10.1371/journal.ppat.1000908
54. Moghimi SM, Andersen AJ, Ahmadvand D, Wibroe PP, Andresen TL, Hunter AC (2011) Material properties in complement activation. Adv Drug Deliv Rev 63(12):1000–1007. doi:10.1016/j.addr.2011.06.002, S0169-409X(11)00147-5 [pii]
55. Liu Y, Yin Y, Wang L, Zhang W, Chen X, Yang X, Xu J, Ma G (2013) Engineering biomaterial-associated complement activation to improve vaccine efficacy. Biomacromolecules 14(9):3321–3328. doi:10.1021/bm400930k
56. Tokura S, Tamura H, Azuma I (1999) Immunological aspects of chitin and chitin derivatives administered to animals. EXS 87:279–292

57. Barreto-Bergter E, Figueiredo RT (2014) Fungal glycans and the innate immune recognition. Front Cell Infect Microbiol 4:145. doi:10.3389/fcimb.2014.00145
58. Robyt JF, Yoon SH, Mukerjea R (2008) Dextransucrase and the mechanism for dextran biosynthesis. Carbohydr Res 343(18):3039–3048. doi:10.1016/j.carres.2008.09.012
59. Andersen AJ, Robinson JT, Dai H, Hunter AC, Andresen TL, Moghimi SM (2013) Single-walled carbon nanotube surface control of complement recognition and activation. ACS Nano 7(2):1108–1119. doi:10.1021/nn3055175
60. Moghimi SM, Hunter AC, Andresen TL (2012) Factors controlling nanoparticle pharmacokinetics: an integrated analysis and perspective. Annu Rev Pharmacol Toxicol 52:481–503. doi:10.1146/annurev-pharmtox-010611-134623
61. Szebeni J, Muggia F, Gabizon A, Barenholz Y (2011) Activation of complement by therapeutic liposomes and other lipid excipient-based therapeutic products: prediction and prevention. Adv Drug Deliv Rev 63(12):1020–1030. doi:10.1016/j.addr.2011.06.017, S0169-409X(11)00194-3 [pii]
62. Roux KH (1999) Immunoglobulin structure and function as revealed by electron microscopy. Int Arch Allergy Immunol 120(2):85–99. doi:10.1159/000024226

Chapter 5
Microfluidics-based Single Cell Analytical Platforms for Characterization of Cancer

Emil Laust Kristoffersen, Morten Leth Jepsen, Birgitta R. Knudsen, and Yi-Ping Ho

Abstract Individual cancer cells in a tumor are very diverse both genetically and functionally. Moreover, in many cancers, whether a tumor flourishes or dies after a given treatment depends on a small fraction of cells in the tumor, for instance, the cancer stem cells, rather than the bulk population. Traditionally, scientists only obtain averaged information from the entire population of cells in a bulk tumor, but it is critical to develop an effective and user-friendly platform capable of interrogating single cells in order to understand and treat the disease better. This chapter reviews recent progress of microfluidics-based tools for single cell analysis that are relevant for cancer characterization, as well as how nanotechnology may advance the analysis with improved signal responses. We hope that this general introduction may catalyze the adoption of these advanced single-cell analysis approaches for cancer studies.

Keywords Nanosensors • Circulating tumor cells • Genomic analysis • Enzymatic activities

5.1 The Importance of Single Cell Analyses for Cancer Characterization

Cells are defined as the basic functional unit of living organisms. They sense environmental stimuli, respond and adapt to the environmental changes chemically or physically in order to form and maintain the distinct tissues in complex biological organisms. Therefore, even cells of identical genetic identities can display a wide disparity of physical responses (such as cell morphologies) or chemical responses (such as RNA or protein expression level regulation). These asynchronous responses

E.L. Kristoffersen • M.L. Jepsen • B.R. Knudsen • Y.-P. Ho (✉)
Department of Molecular Biology and Genetics, Interdisciplinary Nanoscience Center (iNANO), Aarhus University, Gustav Wieds Vej 14, 8000 Aarhus C, Denmark
e-mail: megan.ypho@inano.au.dk

K.A. Howard et al. (eds.), *Nanomedicine*, Advances in Delivery Science and Technology, DOI 10.1007/978-1-4939-3634-2_5

to the microenvironments, often due to random fluctuations, or noise in gene expression [1–3], make precise characterization of cells in a population difficult, especially when techniques with single cell sensitivity are not available. Hence, currently our understanding of cells are in many aspects limited by the inherent ensemble average obtained by analyzing bulk populations of cells, which does not necessary precisely represent the individual behavior of each cell within a population.

While many population-level studies using bulk assays are presented, it remains ambiguous whether the population data may faithfully reflect the dynamic response of individual cells. In fact, cell-to-cell heterogeneity has been evidenced in many examples. For instance, the transcription events in mammalian cells are subject to random fluctuations and lead to large variation in messenger RNA (mRNA) copy numbers [4, 5]. Moreover, cell heterogeneity is observed to impact cell-fate decisions in mouse multipotent progenitor cells [6]. In particular, cancers are typically characterized by a very high level of cell-to-cell variation in regard to both morphological, physiological and genetic features [1, 7, 8]. The obvious reason for such heterogeneity is that cancers arise as a consequence of multiple and successive mutations in the genome of the cell leading to the generation of many different subpopulations of cells constituting the tumor. According to the cancer stem cell theory, which was presented a few decades ago, cancer initiates by the accumulation of mutations in stem cells [9]. This leads to the transformation of the stem cells into cancer stem cells, with self-renewal potential and the capability to differentiate into other types of cancer cells. Due to these characteristics, isolation and characterization of individual cells within a population of cancer cells (e.g. a tumor) are envisioned of particular value for understanding the origin and progression of this disease.

Especially as a predictive tool for individualized cancer treatment, single cell characterization may prove of tremendous value. Recent studies suggest that the drugs developed by measuring therapeutic responses from traditional bulk analysis, tend to suppress the growth of bulk tumor cells but not as effective in eliminating cancer stem cells (CSCs), where the relapse of tumor may occur from [10]. The different drug responses between bulk tumor and CSCs further demonstrate that the fidelity of bulk analysis is insufficient for the evaluation of antitumor treatments. Furthermore, a potential innate consequence of the heterogenic nature of tumor cells is great fluctuations in the amount/activity of cellular drug targets or important determinants of drug efficiency, such as growth rate or metabolic activity.

For example, the heterogeneity of prostate cancer poses a significant challenge to effective targeted therapy [11, 12], such as those involved introduction of a homologous double-stranded RNA (dsRNA), or RNA interference (RNAi) [13], simply because the individual cancerous cells respond very differently to the same treatment. Survival of only a few cancer cells after initial systemic treatment may result in recurrent tumors, which may be even more aggressive or chemoresistant than the initial tumor. Therefore, the identification of these rare cells may be imperial for tailoring treatment to the individual patient in the best possible way.

Finally, when it comes to detection of cancers for diagnostic purposes, the capture and registration of single tumor cells from the bloodstream may prove a powerful tool. Indeed, in combination with protocols for fast and reliable single-cell characteristics of captured cells, such techniques may pave the road for future diagnosis integrated with individually tailored treatment.

Previous development of microfluidics has shown great success in biomedical applications, but mainly centered on the analysis of targeting biomolecules from minute amount of samples, typically nanoliters to hundreds of picoliters. More recently, the exploitation of microfluidics for cell manipulation and analysis, taking advantages of the matching length scale of microfluidics and the size of individual cells (tens of micrometers) has increased. Of particular importance is the detection and analysis of diseased cells, which are typically rare among the whole cell population, difficult to detect and retrieve, and yet critical for accurate analysis of the biological processes. In this chapter, we intend to review the existing development of microfluidics-based tools for single cell analysis that are relevant for cancer characterization, as well as how nanotechnology may advance the analysis with improved signal responses. We hope that this chapter will give the audience a general introduction and catalyze the adoption of these advanced single-cell analysis approaches [2, 4, 14] for cancer studies.

5.2 Enrichment of Circulating Tumor Cells

5.2.1 What Are Circulating Tumor Cells?

Metastasis, the spread of cancer cells from the primary site to other organs in the body, represents the major cause of cancer-related patient death. Previous evidence suggests that the tumor cells are shed from the primary tumor at an early stage of metastasis development, break through the vascular wall, travel via the peripheral blood to sites distant from the primary tumor, and form a secondary tumor [15]. Cells escaping from the primary tumor, are called circulating tumor cells (CTCs). The significant role of CTCs in the metastatic spread of tumor has rendered them valuable biomarkers for both detection of the onset of cancer metastasis and clinical evaluation of treatment outcome. Moreover, detecting CTCs as a cancer marker is advantageous in the clinics because it makes noninvasive detection possible through capturing CTCs in a liquid biopsy, such as a blood sample. Recent evidence on how the CTCs may reflect the molecular features of the primary tumor cells further displays the importance of CTCs in cancer biology [15, 16]. For instance, the presence of mesenchymal markers on CTCs envisages more accurate prognosis than the expression of cytokeratins alone, implying that the currently used assays based on epithelial antigens may overlook the most aggressive subpopulation. However, CTCs are extremely rare and appear in very low concentrations down to one per millions of normal blood cells. Therefore, their detection remains a great challenge in cancer characterization [15, 17].

5.2.2 State-of-Art

CTC isolation is typically evaluated by many factors, including the capture efficiency (i.e. 100 % capture efficiency being isolation of all of the CTCs in the liquid biopsy, therefore, allowing identification of the cancer occurrence), isolation purity (i.e. 100 % isolation purity being isolation of only the CTCs, with no other cell types), isolation speed and the required sample volume. A large panel of macro-scale sorting techniques have been previously reported for CTC enrichment, such as immunomagnetic beads separation [18], laser scanning cytometry [19], fiber-optic array scanning technology (FAST) [20], as detailed in previously published reviews [18, 21]. In general, these approaches utilize the differences between CTCs and normal hematologic blood cells in physical (size, density, electric charges, deformability) or biochemical (surface protein expression, invasion capacity) properties, as illustrated in Fig. 5.1. For example, separation without labeling through the physical properties of the cells is adopted in the isolation by size of epithelial tumor cells (ISET) [22–24]. Dielectrophoretic field-flow fractionation (DEP-FFF) utilizes membrane resistance in combination with size to sort different responses to dielectrophoresis. Biochemical separation relies on immunological procedures using antibodies against tumor-associated antigens and common leukocytes antigens. For instance, the CellSearch® and the Ariol® select CTCs by utilizing magnetic beads coated with antibodies against genes that are highly expressed in CTCs [18, 21, 25]. Subsequently, the antigen-antibody complex is separated from the liquid phase *via* exposure to a magnetic field. However, many of the currently available approaches remain relatively ineffective in isolation efficiency with CTC identification in 50–90 % of patients, while 5–10 mL of sample volume is typically required [21]. Therefore, the search for sensitive, specific, and economical analytical techniques continues.

5.2.3 Microfluidics Based CTC Enrichment

Microfluidics based CTC enrichment has garnered considerable attention, due to the matching length scale of microfluidic channels to the cell sizes. Secondly, the micron-sized geometric features used in microfluidics greatly reduce the sample consumption. To date, microfluidics-based CTC enrichment has shown great promise by identification of CTCs in close to 99 % of patients [26], while requiring very minute sample size of 10 μl [27].

The very early demonstration of microfluidics based CTC enrichment is the so-called “CTC-chip” developed by Toner’s group [26]. The CTC-chip consists of an array of microposts coated with anti-EpCAM antibodies, as shown in Fig. 5.2, where the positive selection is implemented by the antibodies against the epithelial cellular adhesion molecule (EpCAM), relevant to epithelial growth and differentiation. Over expression of EpCAM has been observed in many human carcinomas including prostate, colon and rectum, breast, lung, esophagus, and pancreas.

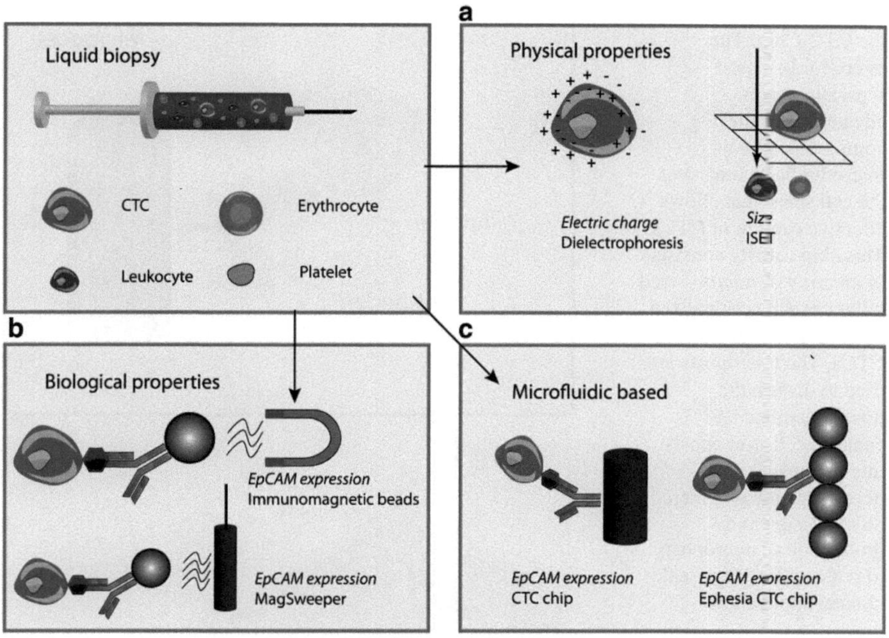

Fig. 5.1 Strategies to isolate circulating tumor cells (CTCs): Isolation of CTCs from normal hematologic blood cells, such as leukocyte, lymphocyte and erythrocyte, typically relies on dissimilar physical or biochemical properties of CTCs. (**a**) Separation by physical properties normally does not involve additional labeling, such as isolation by size of epithelial tumor cells (ISET), or the dielectrophoretic field-flow fractionation (DEP-FFF) that isolates cells by their responses to dielectrophoresis, which is determined by the size and membrane resistance. (**b**) Separation by biological properties: This category of separation usually involves immunological procedures which involve antibodies against tumor-associated antigens and common leukocytes antigens. The CTCs-specific antibodies are typically bound to micron-sized magnetic beads, which allow a separation when applying a magnetic field. (**c**) Microfluidics-based separation: Typical form of microfluidics-based CTC enrichment consists of microposts functionalized with antibodies specific to the surface-markers of tumor cells, such as epithelial cellular adhesion molecule (EpCAM). The microposts may be constructed by a solid structure, or by a pile of magnetic beads as recently demonstrated in the Ephesia CTC chip

In contrast, hematologic cells do not express EpCAM [28, 29]. Hence, EpCAM appears as an effective cancer biomarker and an appropriate target molecules for CTCs enrichment in a liquid biopsy. Combining immunolabeling with controlled laminar flow conditions in the microfluidic chip, the CTC-chip has shown to successfully identify CTCs in peripheral blood from 99 % of patients carrying metastatic lung, prostate, pancreatic, breast, and colon cancer, or more precisely, 115 identifications out of 116 investigated samples [26]. The capture efficiency, as defined previously, is improved by two essential parameters operated in microfluidics: (a) Low flow speed: permitting cells to interact with the microposts for extended duration and, thus, increasing the likelihood of cells sticking to the posts and (b) Low shear stress: enabling the cells to flow through the channel with minimum physical

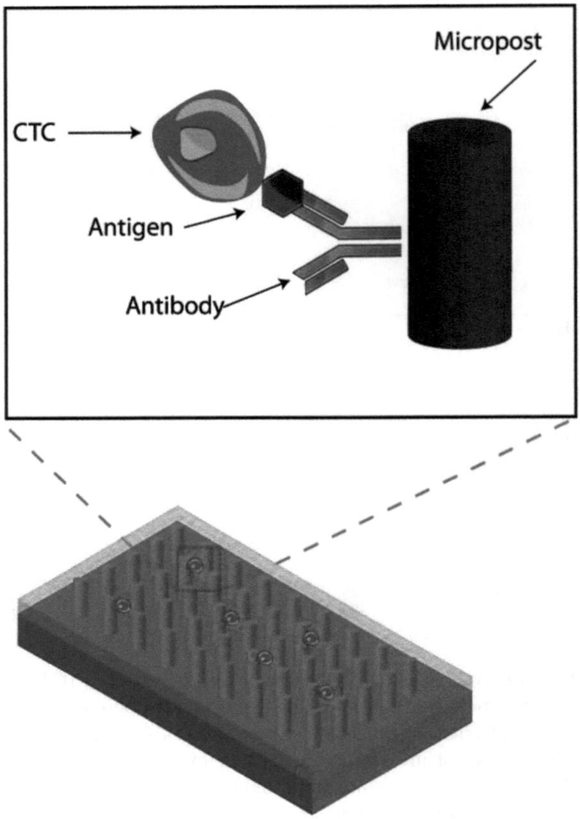

Fig. 5.2 Typical design of the CTC-Chip: The microfluidic-based separation takes advantages of the comparable size of microfluidic channel and the cell sizes, that allows effective capture of CTCs. This chip usually consists of an array of micron-sized pillar posts functionalized by antibodies specific to CTCs. The microposts are used to disturb the flow-streamline and enhance cell-microposts interactions. Current development has centered on the design and production of microposts to scale up for large-scale clinical applications

distress. The CTC chip appears very gentle to the cells with a shear force of less than 0.4 dyn/cm^2. The low shear supplied in the CTC-chip even allows capture of T-24 cells, which have a relatively low expression of EpCAMs. Other than EpCAM, current characterization of CTCs is typically through immunostaining of the cells with markers such as KLK3 (prostate specific antigen) for prostate cancer or TTF-1 (thyroid transcription factor-1) for lung adenocarcinoma. Furthermore, to accelerate the isolation speed, Stoot et al. later demonstrated an automated system for prostate cancers with enrichment and quantitative analysis of positive CTCs for prostate specific antigen (PSA) [30].

As an alternative to the microposts based CTC-chip, an innovative geometric improvement termed the Herringbone-chip (HB-chip) [31], has been demonstrated to prime the enrichment of CTC. The chevrons, or the herringbones, as depicted in Fig. 5.3, function by disrupting the laminar flow, which would then enhance the mixing between streamlines and encourage the collisions of cells and the antibody-coated herringbone structures. The HB-chip has shown a 26.3 % improvement in capture efficiency compared to the previously discussed CTC-chip, along with significantly higher purity of the captured CTCs. Moreover, the use of transparent and chemically stable materials allows imaging of the captured cells with standard staining or assays such as Fluorescent In Situ Hybridization (FISH).

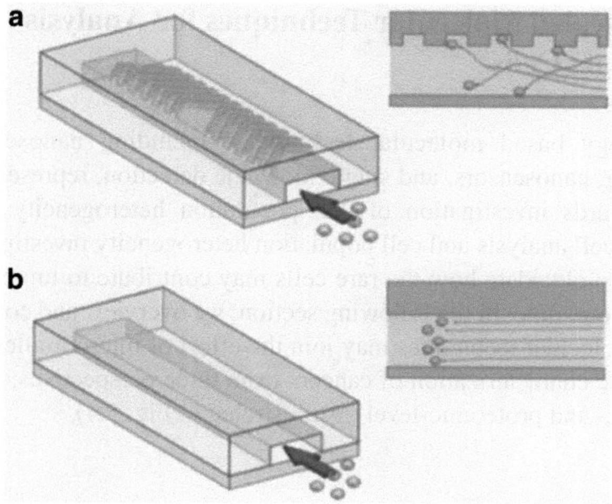

Fig. 5.3 The Herringbone (HB)-Chip: Conventional design of CTC-Chip relies on laminar flow, which limits the interactions of target cells with surfaces, and the complex micropost structure is also challenging to scale up for high-throughput production. The HB-chip represents an alternative strategy, which takes advantages of the design of herringbones, hence the name. (**a**) The surface ridges help to break up streamlines, maximizing collisions between target cells and the antibody-coated walls. (**b**) As a comparison, there is a lack of mixing under low Reynolds number regime in traditional flat-walled design. Figure adapted from Ref. 31 with permission

Saliba et al. have recently developed a system, named Ephesia, in which they combine superparamagnetic beads with microfluidic technology [27]. In this platform, superparamagnetic beads are pre-functionalized with antibodies. When introduced into the chip, upon application of an external magnetic field, the beads would stack up due to dipole-dipole interactions, forming microposts out of stacks of beads. To reduce production cost and technical complexity, the magnetic pattern is generated through microcontact printing with water-based ferrofluid, or "magnetic ink". Major benefits of this method, compared to the conventional design of CTC-chip, are the greatly reduced production cost and possible batch-functionalization of the magnetic beads. It is also of particular note that the proposed self-assembly process offers an aspect ratio beyond the most sophisticated nanofabrication techniques. In summary, the microfluidic technologies have emerged as an attractive micro-scale CTC isolation system. The unique features of fluidic mechanics at the micro- and nano-scale, such as channel dimensions, flow operated at the laminar flow regime, and surface area to volume ratio, have enabled improved capture efficiency and isolation purity of CTC. However, perhaps similar to the challenges for other microfluidics-based applications, it is necessary to validate the reproducibility and robustness of technology with extensive clinical relevant testing. Further, it is expected that the reduced technical complexity, together with low cost of production and testing, would encourage the widespread adoption of microfluidics-based CTC enrichment in the clinics. Audiences interested in the latest development of microfluidics-based CTC detection are referred to recently published reviews [32–34].

5.3 Nanoscaled Molecular Techniques for Analysis of Single Cells

Nanotechnology based molecular techniques, including nanoscale molecular manufacturing, nanosensors, and single molecule detection, represent a significant evolution towards investigation of cell population heterogeneity. As described above, single cell analysis and cell population heterogeneity investigation is necessary in order to elucidate how the rare cells may contribute to tumor development and treatment outcome. In the following section, we overview and comment on how nanoscaled molecular techniques may join the effort of microfluidics in single-cell analysis for the characterization of cancers from three perspectives: The genomic-, transcriptomic- and proteomic-level (as illustrated in Fig. 5.4).

5.3.1 Single Cell Genome Analysis for Cancer Characterization

Cancers are typically caused by errors, or mutations, induced in the genome, which subsequently prompt the cells to malfunction, or grow uncontrollably. Such mutations can either be subtle genomic alterations (such as single base mutations) or gross

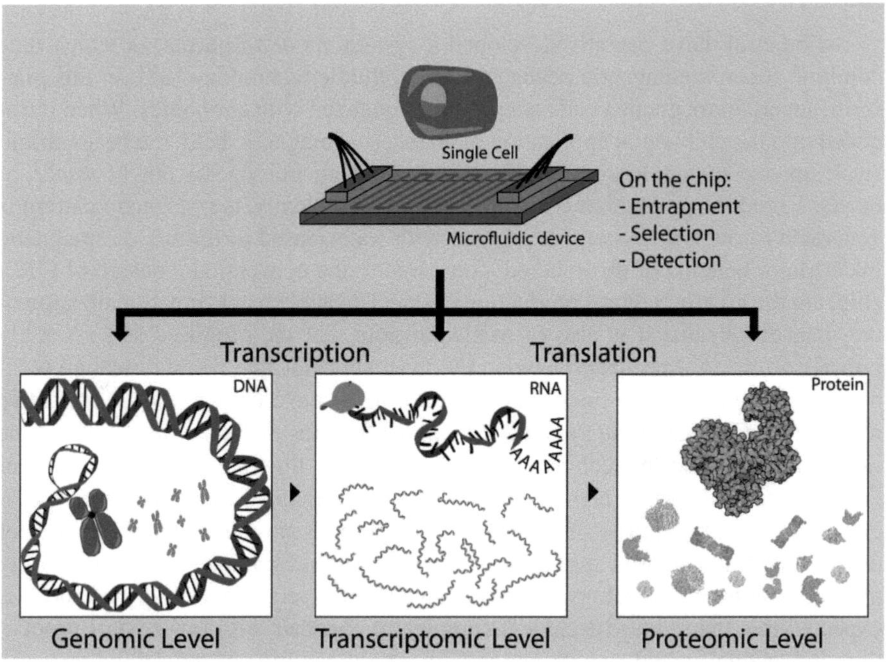

Fig. 5.4 Overview of how microfluidic systems may assist in single cell analysis: High-throughput cell-based screens can benefit considerably from the unique liquid-handling capabilities offered by microfluidic systems

genomic alterations (such as deletions, translocations, insertions, rearrangement or even loss or gain of entire chromosomes) [35]. As the disease progresses, the evolutional accumulation of cancer causing mutations result in a high degree of genetic diversity between or among cancer cells [36–38]. Furthermore, it is becoming established that CSCs, which account for only a minor part of the bulk tumor, exist in a variety of cancers, such as colon, ovarian, and small cell lung cancers [39]. CSCs are known to have a distinct gene expression phenotype, rendering them highly resistant to chemotherapeutic treatments [39–42].

Identification of genomic alterations at the single cell level are routinely accomplished today by standard cancer diagnostic techniques such as chromosome staining and FISH [43–45], or FISH based on nanoparticles-labeled probes [46]. These refined techniques are based on microscopic readouts, and can be used to identify large genomic alterations in single cells. In some cases FISH can even be used to detect Single Nucleotide Polymorphisms (SNPs) if the site and polymorphism is known [47]. In recent years, genomic sequencing has been adapted for single human cell genome analysis [48, 49]. This late arrival of DNA-sequencing as a tool for single cell analysis reflects the preceding barriers: (1) sufficiently sensitive sequencing techniques for single cell analysis have emerged not until the last decade, (2) the data obtained by sequencing is often very comprehensive and the analysis of which requires well-established bioinformatics tools and/or statistics as well as appropriate references. With this said, a combination with microfluidics might provide a unique means for isolation and/or enrichment of specific populations of cells, for example the CSCs.

As a pioneer in the full genome sequencing of single human cells using single cell based approaches, Wigler's group combined flow-sorting, Whole Genome Amplification (WGA) and next-generation sequencing for the investigation of single breast cancer cells [49]. More than 400 cells were sequenced with around 6 % genome coverage of each cell. Due to the suboptimal genome coverage (~6 %), sequencing data was clustered in 54 kilo base (kb) sized "bins", to obtain proper statistical significance. These bins were then used to determine the copy numbers of genomic areas mapped to the healthy genome, which enabled the analysis of the evolutionary history of the cancer. Another example of genome sequencing of single cells is provided by Frumkin and colleagues [48] however, this was done without the use of any microfluidics or flow based techniques. In this study, cancer lineage relations were investigated by cutting out single tumor cells from tissue sections of mouse lymphoma by microdissection. Around 50 single cells were genotyped using Sanger sequencing and data was used to produce a "lineage tree" for the analyzed cells populating the tumor.

These studies provide a demonstration of the need of single cell sequencing in characterization of cancer, along with pros and cons of the utilized approaches. For example, microdissection on tissue samples allows the selection of specific cells. However, microdissection holds a general limitation in relation to number of analyzed cells: ~50 in Frumkin's study compared to >400 in the study by Navin et al. using a flow based single cell analysis. Moving forward, microfluidics based approaches are expected to ease the handling of sample and enhance the enrichment efficiency prior to the genome analysis. Moreover, the automation offered by microfluidics also makes the experiments less prone towards human errors, such as the biased selection of cells by manually picking, and contamination of sample by human handling.

5.3.2 Single Cell Transcription Analysis for Cancer Characterization

The genetic information held by the genome is transcribed into RNA either as coding RNA such as messenger RNA (mRNA), or as non-coding RNAs such as transfer RNA (tRNA), ribosomal RNA (rRNA) and the regulatory RNAs (miRNA, siRNA and other types) [50]. The set of all the RNA molecules is henceforth termed transcriptome. The noncoding RNAs play direct roles in the cellular functions in e.g. constituting the core elements of protein synthesis, regulating gene expression or protection against viral infection. In contrast, the coding RNA, such as mRNAs in most cases are categorized as mediator molecules without direct cellular effect of their own. Rather, mRNAs convey the genetic information from DNA to the ribosome, where they are then translated into a polymer of amino acids, or a protein, as stated in the central dogma of molecular biology. The transcriptome in a cell, regardless of their functions, is extensively regulated by the cellular microenvironments and the request of the cells. Therefore, probing the transcriptome would provide important information about the cell. But why is "single cell" transcription analysis necessary? It has been reported that the cells encode a subtle set of analogue parameters to modulate the responses to environmental stimuli, e.g. Tay et al. have observed a heterogeneous activation of mouse fibroblast (3T3) cells in response to the signaling molecule tumor-necrosis factor (TNF)-α [51]. In this scenario, bulk measurements provide relatively limited biologically relevant information in the development of cancers, especially how individual cells respond to the changes of environment differently. Moreover, considerable cell-to-cell variation with regard to transcription is an inherent feature of cancers [52, 53]. Therefore, methods for analyzing the transcriptomic level of single cells is of critical need to bring forth new and insightful information about cancer development and the potential treatment against the diseases.

In the clinical settings, RNA based methods, such as quantitative polymerase chain reaction (qPCR), are widely used in cancer diagnosis and prognosis [54–56], where the expression level of specific genes is analyzed mainly in bulk set-ups. Hitherto, transcription analyses in the single cell manner have not yet found its way to the clinic. Nevertheless, microfluidics based single cell studies at the transcriptomic level, made possible by the commercialized system supplied by Fluidigm, have been widely adopted for scientific purposes by combining quantitative Reverse Transcriptase Polynucleotide Chain Reaction (qRT-PCR) and microfluidics in cancer characterization [57–63]. The commercialized single cell qRT-PCR platform is based on the prototype reported by Quake's group (Fig. 5.5) [64]. The basic principle is to trap single cells sorted by Fluorescence-Activated Cell Sorting (FACS) in

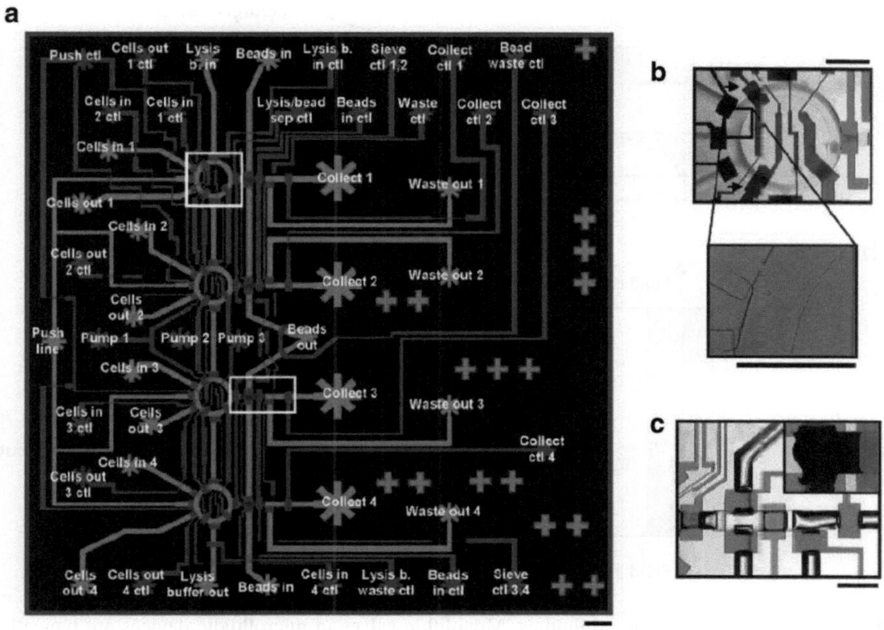

Fig. 5.5 Single-cell mRNA isolation and cDNA synthesis: (**a**) Overview of the device, which implements five steps, including cell capture, cell lysis, mRNA purification, cDNA synthesis, cDNA purification, in one integrated device. Flow channels and control channels are depicted in green and blue, respectively. The red unrounded (rectangular profile) flow channels are where designed for affinity column construction. White boxed regions are zoomed in (**b**) and (**c**), individually. (**b**) The lysis ring and an NIH/3T3 cell captured in the ring. (**c**) The affinity column construction area and a stacked column. Scale bars are 400 μm. Figure adapted from Ref. 62 with permission (Color online)

confined wells, and perform qRT-PCR reactions in the individual wells. The tiny nanoliter wells are generated by a fluidic circuit composed of a fluidic layer and a valve layer. These are separated by a thin elastomeric rubber member, as shown in Fig. 5.6. When pressurized, gas is applied to the valve layer, the membrane deflects and interrupts the flow in the fluidic layer [65]. As a result, trapped single cells may be analyzed within the individualized well without cross-contamination. Using this platform, Dalerba and colleagues have shown that human colon cancer tissues contain distinct cell populations of which the transcriptional identities mirror those of the different cellular lineages of the normal colon tissue [57]. In their work, more than 230 genes were evaluated in 336 single cells of three different colon epithelial cells and cancer cell lineages. Other reports have used a similar approach to investigate transcription in breast cancers [58, 59] and leukemia [60]. As an alternative to the commercial qRT-PCR platform, White and colleagues have developed a fully integrated microfluidic qRT-PCR device that implements all steps, including cell entrapment, lysis, reverse transcription and qPCR analysis in one single device, which produces a throughput of 300 cells per run [66]. The device is capable of

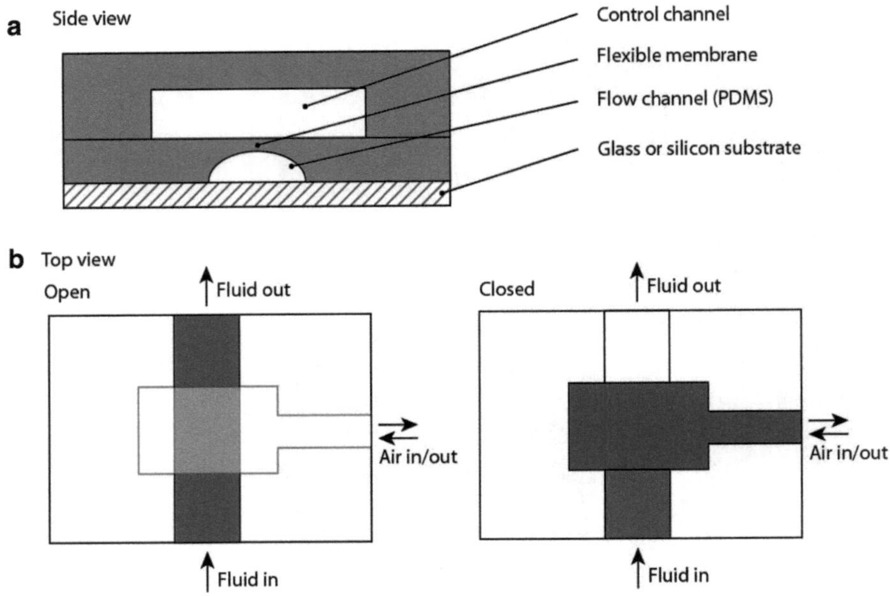

Fig. 5.6 Valve control in microfluidics: Monolithic valves in microfluidic devices are typically produced by soft-lithography techniques using polydimethylsiloxane (PDMS). Shown here is a typical two-layer PDMS "push-down" microfluidic valve. (**a**) *Side view*: A thin elastomeric membrane is placed in between the flow and control channels. (**b**) *Top view*: When the control channel is pressurized, the thin membrane would be "pushed" downward and close the flow in the flow channel. The flow channel is typically positioned orthogonal to the control channel

measuring RNA levels, as well as performing single nucleotide variant detection of single cells in a high throughput manner.

In contrast to the techniques mentioned above, capable of handling many cells simultaneously but only allowing a limited number of RNAs per cell (up to 96 RNAs) to be analyzed, Ramsköld et al. have recently presented a non-microfluidic based whole transcriptomic analysis of 12 single human CTCs using mRNA-sequencing [67]. The mRNA-sequencing study was performed on fewer cells compared to the microfluidic based qRT-PCR studies (12 cells in Ramsköld et al. [67] compared to >500 by qRT-PCR in Dalerba et al. [57]). In future studies, incorporation of microfluidics with mRNA-sequencing is expected to improve the throughput of transcriptional analyses of single cells.

5.3.3 *Single Cell Protein Analysis for Cancer Characterization*

The human genome codes for ~25,000 genes that are transcribed and translated into proteins [68]. The process of mRNA splicing makes the number of different proteins expressed by the cell many fold higher than the number of genes [69].

These proteins, which include active enzymes, are the main determinants for the cellular processes. Many currently used drugs in cancer therapy function by targeting specific enzymatic reactions in the cell [70–73]. Consistently, changes in the activity of the target enzyme may result in elevated chemo-resistance. Since the activity of enzymes are often modified on the post-translational level, such changes may not always be recognized at the genomic or transcriptional level [74], posing a need for analysis at the post-translational protein level.

Various methods have been applied for analysis of proteins at the single cell level, including mass spectrometry, enzyme-linked immunosorbent assay (ELISA), enzymatic detection or optical approaches [75–80]. Such analyses can involve measurement of the "expressed protein amounts" or "protein activity". For example, Qihui et al. have proposed a microfluidic device that can isolate 0–5 cells in a 2 nL volume chamber and assay up to 11 different proteins per chamber on the chip using immunostaining procedures [78]. Regarding the throughput, the microfluidic device is able to isolate and assay 100 single cells per chip.

The strategy of detecting protein amount by immunohistochemistry is well established and antibodies can be designed to recognize almost all proteins and even specific features of a protein such as posttranslational modifications (e.g. phosphorylation and other modifications) [81–84]. However, in most cases it is the function of the proteins, such as the catalytic activity for enzymes, and not the amount per se, that determines the effect of the given protein on cellular conditions, such as health, drug response etc. From that perspective, our group has developed an array of DNA nanosensors that measure cancer relevant enzymatic activities, such as human topoisomerase I (hTopI) [85–87], topoisomerase II (hTopII) [88], tyrosyl-DNA phosphodiesterase 1 (Tdp1) [89, 90], which are emerging targets for anticancer therapy. The major working principle, termed rolling circle enhanced enzyme activity detection (REEAD), utilizes the catalytic activity of the target enzyme to generate an intrinsic amplification. Take the detection of hTopI activity for example (Fig. 5.7a), the DNA nanosensors are designed as a linear DNA substrate which fold upon itself, forming a dumbbell shape. hTopI is able to recognize the substrate, cleave from the 3′ end and religate the 5′ hydroxyl group, turning the linear DNA into a circularized one. Subsequent process is designed to differentiate the circles from the linear pieces by isothermal rolling circle amplification (RCA). As a result, the measured fluorescence represents the individual cleavage-religation event generated by active hTopI. Perhaps the best example demonstrating the potential of combining the microfluidics with nanotechnology for single cell analysis, water-in-oil droplets generated by a flow-focusing type of droplet generator is introduced to encapsulate individual cells along with the above-mentioned DNA nanosensors. The droplets provide a confined environment for the serial of biochemical reactions, enabling the enzymatic activities to be observed at the single cell level.

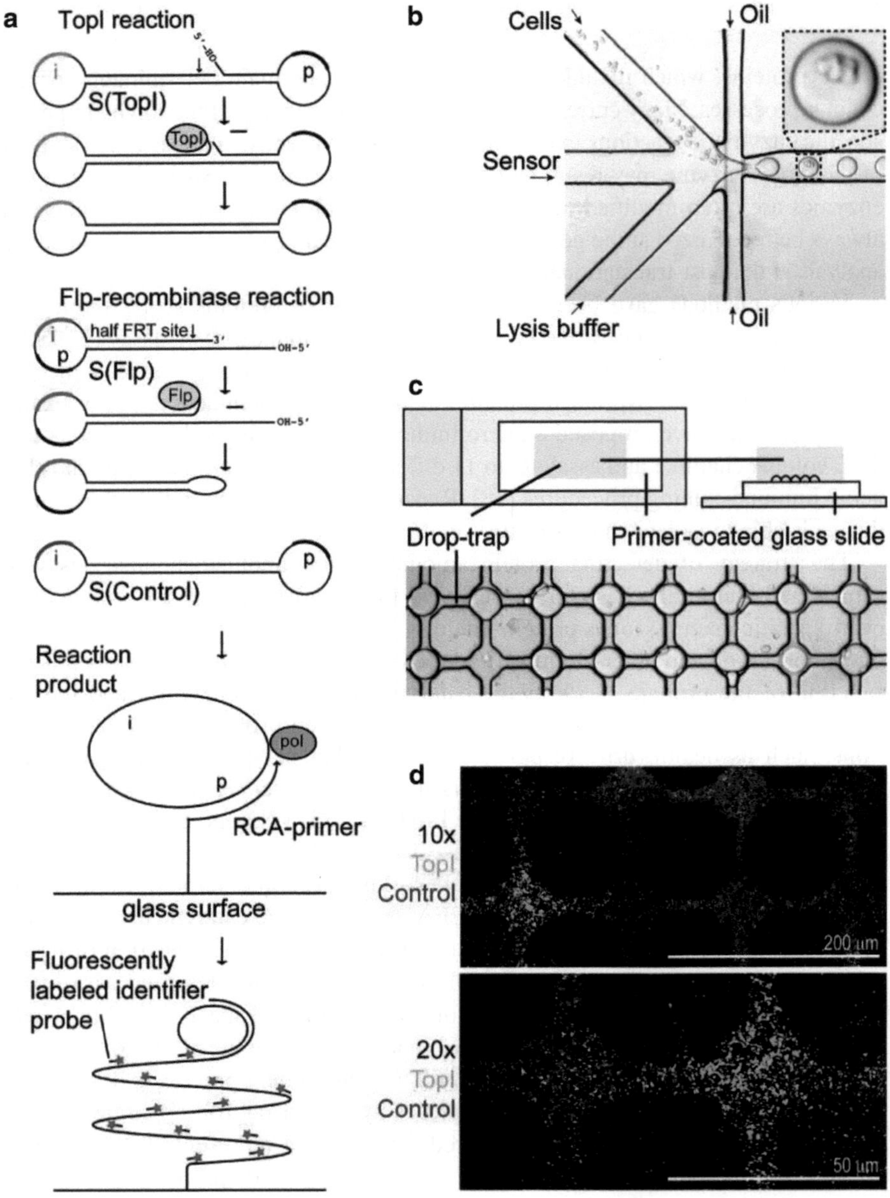

Fig. 5.7 Combining DNA nanosensors and microfluidics for single cell based analysis of enzymatic activities: (**a**) DNA substrates S(TopI) and S(Flp) are oligonucleotides that target cleavage-ligation by human topoisomerase I (TopI) and Flp, respectively. The detection, termed rolling circle enhanced enzyme activity detection (REEAD), initiates by recognition of enzymes to the substrates, which results in a circularized product. Subsequently, the circles allow a solid-support rolling circle amplification (RCA), which generate ~10^3 tandem repeat of amplified products. The results are visualized by florescence microscopy at the single-molecule level by hybridization of fluorescently labeled probes. (**b**) The droplet microfluidics is introduced to encapsulate individual Fig. 5.7 (continued) cells along with DNA substrates and lysis buffer, in picoliters of water-in-oil droplets. (**c**) The droplets containing circularized DNA are then confined in a drop-trap on a primer-coated glass slide, where the RCA takes place. (**d**) The observed enzymatic activities from single cells are visualized as fluorescence signals: *green* (TopI) and *blue* (control). Figure adapted from Ref. 78 with permission (Color online)

5.4 Conclusion

Microfluidics-based approaches have been promoted for many biochemical applications, such as drug screening [91], nucleic acid amplification [92], and analysis of chemical reactions [93]. Therefore, it becomes a natural extension to take advantage of microfluidics for single cells interrogation. The device can be tailored to exploit physical and/or biological differences for isolation of particular cell types from a population of cells, such as enrichment of CTCs, and separation of CSCs from non-stem cells. The small dimensions of microfluidic devices have also enabled many unique features, such as gentle capture of live or rare cells [34], which avoids possible interferences for the subsequent analysis. Microfluidics also presents an opportunity to integrate many functions, such as isolation, biochemical reaction and detection, in a single device. For example, recent evidence has shown a high degree of heterogeneity even within a population of CTCs [94]. Therefore, combining CTC enrichment and molecular analysis at the single cell level is clinically important for the characterization of rare cell phenotypes, including CTCs in various stages during the cancer progression or perhaps study of how CSCs behave differently compared to non-stem cells. Furthermore, the isolated cells can be directed to next-stage analysis on-chip (e.g., genomic, transcriptomic or proteomic analysis), or on-chip cell culture as part of the analysis. Such integration will speed up the cancer characterization process while eliminating the intermediate sample transfer procedures typically required in macro-scale approaches.

Furthermore, advancement in cellular, microscopic, or nanoscaled molecular techniques is pivotal to the single cell analysis. The development of reliable biomarkers shall closely follow, if not precede, the emergence of microfluidics. For this purpose, immunostaining remains the prevailing technique for cellular or protein recognition. Recent progress in nanotechnology has joined the league by providing immunomodulatory agents engineered with nanostructure materials, including metallic nanoparticles, quantum dots, or nanotubes, to either improve the efficiency immunorecognition or enhance the detection sensitivity [95]. On the other hand, nanoparticles possessing unique photophysical properties, such as semiconductor nanocrystals and noble metal nanoclusters, also serve as unique fluorescence analogs in illuminating various forms of biological analytes through different signal transduction pathways. Taken together, the latest development of both nanotechnology and microfluidics is expected to encourage new excitement in the field of single cell analysis.

Despite the large variety of available approaches targeting and analyzing single cells, we are still in the infancy to uncover the implications of cellular heterogeneity. Many challenges have yet to come. One of the foremost, existing available approaches for single cell analysis is still expensive and technically complex, in particular those require high precision pumps or delicate valving control, which prevent the adaptation in the clinical studies. Future development of microfluidics-based single cell analysis is expected to reduce the production cost, sample consumption and to avoid human operational error by integration of sample preparation and subsequent data acquisition onto one single device. However, tradeoff of microfluidics remains on the analysis speed, especially when it comes to large sample volume. The capability of parallel processing, when made possible, will hopefully accelerate the adoption of microfluidics for large-scale single cell analysis. Accessible single cell analysis, however, requires collective efforts including further engineering optimization of microfluidic systems and suitable nanoscaled molecular analysis approaches on the chip. The new possibility to reveal the characteristics of rare cells is expected to ultimately lead to clinical implications in the combat of cancers.

References

1. Bertucci F, Birnbaum D (2008) Reasons for breast cancer heterogeneity. J Biol 7:6
2. Brouzes E et al (2009) Droplet microfluidic technology for single-cell high-throughput screening. Proc Natl Acad Sci U S A 106:14195–14200
3. Cai L, Friedman N, Xie XS (2006) Stochastic protein expression in individual cells at the single molecule level. Nature 440:358–362
4. Levsky JM, Shenoy SM, Pezo RC, Singer RH (2002) Single-cell gene expression profiling. Science 297:836–840
5. Raj A, Peskin CS, Tranchina D, Vargas DY, Tyagi S (2006) Stochastic mRNA synthesis in mammalian cells. PLoS Biol 4:e309
6. Chang HH, Hemberg M, Barahona M, Ingber DE, Huang S (2008) Transcriptome-wide noise controls lineage choice in mammalian progenitor cells. Nature 453:544–547
7. Nwosu V, Carpten J, Trent JM, Sheridan R (2001) Heterogeneity of genetic alterations in prostate cancer: evidence of the complex nature of the disease. Hum Mol Genet 10:2313–2318
8. Shackleton M, Quintana E, Fearon ER, Morrison SJ (2009) Heterogeneity in cancer: cancer stem cells versus clonal evolution. Cell 138:822–829
9. Williams JL (2012) Cancer stem cells. Clin Lab Sci 25:50–57
10. Merlos-Suárez A et al (2011) The intestinal stem cell signature identifies colorectal cancer stem cells and predicts disease relapse. Cell Stem Cell 8:511–524
11. Kumar-Sinha C, Tomlins SA, Chinnaiyan AM (2008) Recurrent gene fusions in prostate cancer. Nat Rev Cancer 8:497–511
12. Mackinnon AC, Yan BC, Joseph LJ, Al-Ahmadie HA (2009) Molecular biology underlying the clinical heterogeneity of prostate cancer: an update. Arch Pathol Lab Med 133:1033–1040
13. Devi GR (2006) siRNA-based approaches in cancer therapy. Cancer Gene Ther 13:819–829
14. Spiller DG, Wood CD, Rand DA, White MRH (2010) Measurement of single-cell dynamics. Nature 465:736–745
15. Alix-Panabières C, Schwarzenbach H, Pantel K (2012) Circulating tumor cells and circulating tumor DNA. Annu Rev Med 63:199–215
16. Pantel K, Brakenhoff RH (2004) Dissecting the metastatic cascade. Nat Rev Cancer 4:448–456
17. Yu M, Stott S, Toner M, Maheswaran S, Haber DA (2011) Circulating tumor cells: approaches to isolation and characterization. J Cell Biol 192:373–382
18. Pantel K, Alix-Panabières C, Riethdorf S (2009) Cancer micrometastases. Nat Rev Clin Oncol 6:339–351
19. Rolle A et al (2005) Increase in number of circulating disseminated epithelial cells after surgery for non-small cell lung cancer monitored by MAINTRAC(R) is a predictor for relapse: a preliminary report. World J Surg Oncol 3:18
20. Krivacic RT et al (2004) A rare-cell detector for cancer. Proc Natl Acad Sci U S A 101:10501–10504

21. Bednarz-Knoll N, Alix-Panabières C, Pantel K (2011) Clinical relevance and biology of circulating tumor cells. Breast Cancer Res 13:228
22. Pinzani P et al (2006) Isolation by size of epithelial tumor cells in peripheral blood of patients with breast cancer: correlation with real-time reverse transcriptase-polymerase chain reaction results and feasibility of molecular analysis by laser microdissection. Hum Pathol 37:711–718
23. Hofman VJ et al (2011) Cytopathologic detection of circulating tumor cells using the isolation by size of epithelial tumor cell method: promises and pitfalls. Am J Clin Pathol 135:146–156
24. Vona G et al (2000) Isolation by size of epithelial tumor cells: a new method for the immunomorphological and molecular characterization of circulatingtumor cells. Am J Pathol 156:57–63
25. Deng G et al (2008) Enrichment with anti-cytokeratin alone or combined with anti-EpCAM antibodies significantly increases the sensitivity for circulating tumor cell detection in metastatic breast cancer patients. Breast Cancer Res 10:R69
26. Nagrath S et al (2007) Isolation of rare circulating tumour cells in cancer patients by microchip technology. Nature 450:1235–1239
27. Saliba A-E et al (2010) Microfluidic sorting and multimodal typing of cancer cells in self-assembled magnetic arrays. Proc Natl Acad Sci U S A 107:14524–14529
28. Went P, Lugli A, Meier S, Bundi M (2004) Frequent EpCam protein expression in human carcinomas. Hum Pathol 35:122–128. doi:10.1016/S0046-8177(03)00502-1
29. Balzar M, Winter MJ, de Boer CJ, Litvinov SV (1999) The biology of the 17-1A antigen (Ep-CAM). J Mol Med (Berl) 77:699–712
30. Stott SL et al (2010) Isolation and characterization of circulating tumor cells from patients with localized and metastatic prostate cancer. Sci Transl Med 2:25ra23
31. Stott S, Hsu C, Tsukrov D (2010) Isolation of circulating tumor cells using a microvortex-generating herringbone-chip. Proc Natl Acad Sci U S A 107:18392–18397
32. Dong Y et al (2013) Microfluidics and circulating tumor cells. J Mol Diagn 15:149–157
33. Hyun K-A, Jung H-I (2013) Microfluidic devices for the isolation of circulating rare cells: a focus on affinity-based, dielectrophoresis, and hydrophoresis. Electrophoresis 34:1028–1041
34. Li P, Stratton ZS, Dao M, Ritz J, Huang TJ (2013) Probing circulating tumor cells in microfluidics. Lab Chip 13:602–609
35. Stratton MR, Campbell PJ, Futreal PA (2009) The cancer genome. Nature 458:719–724
36. Park SY, Gönen M, Kim HJ, Michor F, Polyak K (2010) Cellular and genetic diversity in the progression of in situ human breast carcinomas to an invasive phenotype. J Clin Invest 120:636–644
37. Torres L et al (2007) Intratumor genomic heterogeneity in breast cancer with clonal divergence between primary carcinomas and lymph node metastases. Breast Cancer Res Treat 102:143–155
38. Farabegoli F et al (2001) Clone heterogeneity in diploid and aneuploid breast carcinomas as detected by FISH. Cytometry 46:50–56
39. Jordan CT, Guzman ML, Noble M (2006) Cancer stem cells. N Engl J Med 355 1253–1261
40. Larzabal L et al (2013) Differential effects of drugs targeting cancer stem cell (CSC) and non-CSC populations on lung primary tumors and metastasis. PLoS One 8:e79798
41. Kuan W-C, Horák D, Plichta Z, Lee W-C (2014) Immunocapture of CD133-positive cells from human cancer cell lines by using monodisperse magnetic poly(glycidyl methacrylate) microspheres containing amino groups. Mater Sci Eng C Mater Biol Appl 34C:193–200
42. Ahmad A, Li Y, Bao B, Kong D, Sarkar FH (2013) Epigenetic regulation of miRNA-cancer stem cells nexus by nutraceuticals. Mol Nutr Food Res 58:79–86. doi:10.1002/mnfr.201300528
43. Nahi H, Sutlu T, Jansson M, Alici E, Gahrton G (2011) Clinical impact of chromosomal aberrations in multiple myeloma. J Intern Med 269:137–147
44. Neben K et al (2013) Progression in smoldering myeloma is independently determined by the chromosomal abnormalities del(17p), t(4;14), gain 1q, hyperdiploidy, and tumor load. J Clin Oncol 31:4325–4332
45. Sawyer JR (2011) The prognostic significance of cytogenetics and molecular profiling in multiple myeloma. Cancer Genet 204:3–12
46. Wu S-M et al (2006) Quantum-dot-labeled DNA probes for fluorescence in situ hybridization (FISH) in the microorganism Escherichia coli. Chemphyschem 7:1062–1067

47. Iacobucci I, Lonetti A, Papayannidis C, Martinelli G (2013) Use of single nucleotide polymorphism array technology to improve the identification of chromosomal lesions in leukemia. Curr Cancer Drug Targets 13:791–810
48. Frumkin D et al (2008) Cell lineage analysis of a mouse tumor. Cancer Res 68:5924–5931
49. Navin N et al (2011) Tumour evolution inferred by single-cell sequencing. Nature 472:90–94
50. Alberts B et al (2007) Molecular biology of the cell, 5th edn. http://www.google.dk/books?hl=da&lr=&id=DjMmAgAAQBAJ&pgis=1
51. Tay S et al (2010) Single-cell NF-kappaB dynamics reveal digital activation and analogue information processing. Nature 466:267–271
52. Narsinh KH et al (2011) Brief report: Single cell transcriptional profiling reveals heterogeneity of human induced pluripotent stem cells. J Clin Invest 121:1217–1221
53. Diercks A, Kostner H, Ozinsky A (2009) Resolving cell population heterogeneity: real-time PCR for simultaneous multiplexed gene detection in multiple single-cell samples. PLoS One 4:e6326
54. Kantarjian HM et al (2009) Significance of increasing levels of minimal residual disease in patients with Philadelphia chromosome-positive chronic myelogenous leukemia in complete cytogenetic response. J Clin Oncol 27:3659–3663
55. Graziano F et al (2011) Genetic activation of the MET pathway and prognosis of patients with high-risk, radically resected gastric cancer. J Clin Oncol 29:4789–4795
56. Hoshimoto S et al (2012) Association between circulating tumor cells and prognosis in patients with stage III melanoma with sentinel lymph node metastasis in a phase III international multicenter trial. J Clin Oncol 30:3819–3826
57. Dalerba P et al (2011) Single-cell dissection of transcriptional heterogeneity in human colon tumors. Nat Biotechnol 29:1120–1127
58. Diehn M, Cho R, Lobo N, Kalisky T (2009) Association of reactive oxygen species levels and radioresistance in cancer stem cells. Nature 458:780–783
59. Spike BT et al (2012) A mammary stem cell population identified and characterized in late embryogenesis reveals similarities to human breast cancer. Cell Stem Cell 10:183–197
60. Kikushige Y et al (2011) Self-renewing hematopoietic stem cell is the primary target in pathogenesis of human chronic lymphocytic leukemia. Cancer Cell 20:246–259
61. Lambolez B, Audinat E, Bochet P, Crépel F, Rossier J (1992) AMPA receptor subunits expressed by single Purkinje cells. Neuron 9:247–258
62. Bengtsson M, Ståhlberg A, Rorsman P, Kubista M (2005) Gene expression profiling in single cells from the pancreatic islets of Langerhans reveals lognormal distribution of mRNA levels. Genome Res 15:1388–1392
63. Tang F et al (2006) 220-plex microRNA expression profile of a single cell. Nat Protoc 1:1154–1159
64. Marcus JS, Anderson WF, Quake SR (2006) Microfluidic single-cell mRNA isolation and analysis. Anal Chem 78:3084–3089
65. Melin J, Quake SR (2007) Microfluidic large-scale integration: the evolution of design rules for biological automation. Annu Rev Biophys Biomol Struct 36:213–231
66. White AK et al (2011) High-throughput microfluidic single-cell RT-qPCR. Proc Natl Acad Sci U S A 108:13999–14004
67. Ramsköld D, Luo S, Wang Y, Li R (2012) Full-length mRNA-Seq from single-cell levels of RNA and individual circulating tumor cells. Nat Biotechnol 30:777–782
68. Venter JC et al (2001) The sequence of the human genome. Science 291:1304–1351
69. Han J, Xiong J, Wang D, Fu X-D (2011) Pre-mRNA splicing: where and when in the nucleus. Trends Cell Biol 21:336–343
70. Hamada S, Masamune A, Shimosegawa T (2013) Novel therapeutic strategies targeting tumor-stromal interactions in pancreatic cancer. Front Physiol 4:331
71. Kümler I, Brünner N, Stenvang J, Balslev E, Nielsen DL (2013) A systematic review on topoisomerase 1 inhibition in the treatment of metastatic breast cancer. Breast Cancer Res Treat 138:347–358
72. Burden D, Osheroff N (1998) Mechanism of action of eukaryotic topoisomerase II and drugs targeted to the enzyme. Biochim Biophys Acta 1400:139–154

73. Fortune J, Osheroff N (2000) Topoisomerase II as a target for anticancer drugs: when enzymes stop being nice. Prog Nucleic Acid Res Mol Biol 64:221–253
74. Poletto M et al (2012) Acetylation on critical lysine residues of Apurinic/apyrimidinic endonuclease 1 (APE1) in triple negative breast cancers. Biochem Biophys Res Commun 424:34–39
75. Fosbrink M, Aye-Han N-N, Cheong R, Levchenko A, Zhang J (2010) Visualization of JNK activity dynamics with a genetically encoded fluorescent biosensor. Proc Natl Acad Sci U S A 107:5459–5464
76. Huang B et al (2007) Counting low-copy number proteins in a single cell. Science 315:81–84
77. Mellors JS, Jorabchi K, Smith LM, Ramsey JM (2010) Integrated microfluidic device for automated single cell analysis using electrophoretic separation and electrospray ionization mass spectrometry. Anal Chem 82:967–973
78. Shi Q et al (2012) Single-cell proteomic chip for profiling intracellular signaling pathways in single tumor cells. Proc Natl Acad Sci U S A 109:419–424
79. Sun J et al (2010) A microfluidic platform for systems pathology: multiparameter single-cell signaling measurements of clinical brain tumor specimens. Cancer Res 70:6128–6138
80. Stougaard M, Juul S, Andersen FF, Knudsen BR (2011) Strategies for highly sensitive biomarker detection by Rolling Circle Amplification of signals from nucleic acid composed sensors. Integr Biol (Camb) 3:982–992
81. Perez-Hernandez D et al (2013) The intracellular interactome of tetraspanin-enriched microdomains reveals their function as sorting machineries toward exosomes. J Biol Chem 288:11649–11661
82. Varjosalo M et al (2013) Interlaboratory reproducibility of large-scale human protein-complex analysis by standardized AP-MS. Nat Methods 10:307–314
83. Chen Y et al (2013) Bcl2-associated athanogene 3 interactome analysis reveals a new role in modulating proteasome activity. Mol Cell Proteomics 12:2804–2819
84. Davies CC, Chakraborty A, Diefenbacher ME, Skehel M, Behrens A (2013) Arginine methylation of the c-Jun coactivator RACO-1 is required for c-Jun/AP-1 activation. EMBO J 32:1556–1567
85. Juul S, Ho Y, Stougaard M (2011) Microfluidics-mediated isothermal detection of enzyme activity at the single molecule level. Conf Proc IEEE Eng Med Biol Soc 2011:3258–3261, http://ieeexplore.ieee.org/xpls/abs_all.jsp?arnumber=6090885
86. Andersen FF et al (2009) Multiplexed detection of site specific recombinase and DNA topoisomerase activities at the single molecule level. ACS Nano 3:4043–4054
87. Marcussen LB et al (2013) DNA-based sensor for real-time measurement of the enzymatic activity of human topoisomerase I. Sensors (Basel) 13:4017–4028
88. Kristoffersen EL, Givskov A, Jørgensen LA, Andersen AH, Stougaard M, Jensen PW. Ho Y-P, Knudsen BR "Topoisomerase II enzymatic activity detection using self-assembled small catenated DNA circles," in preparation
89. Jensen PW et al (2013) Real-time detection of TDP1 activity using a fluorophore-quencher coupled DNA-biosensor. Biosens Bioelectron 48C:230–237
90. Jakobsen A-K, Stougaard M (2015) Combining a nanosensor and ELISA for measurement of Tyrosyl-DNA Phosphodiesterase 1 (TDP1) activity and protein amount in cell and tissue extract. Nano Life 05:1541001
91. Weltin A et al (2013) Cell culture monitoring for drug screening and cancer research: a transparent, microfluidic, multi-sensor microsystem. Lab Chip 14:138–146
92. Selck DA, Karymov MA, Sun B, Ismagilov RF (2013) Increased robustness of single-molecule counting with microfluidics, digital isothermal amplification, and a mobile phone versus real-time kinetic measurements. Anal Chem 85:11129–11136
93. Demello AJ (2006) Control and detection of chemical reactions in microfluidic systems. Nature 442:394–402
94. Van de Stolpe A, Pantel K, Sleijfer S, Terstappen LW, den Toonder JMJ (2011) Circulating tumor cell isolation and diagnostics: toward routine clinical use. Cancer Res 71:5955–5960
95. Smith DM, Simon JK, Baker JR (2013) Applications of nanotechnology for immunology. Nat Rev Immunol 13:592–605

Chapter 6
Nanotheranostics and In-Vivo Imaging

Brandon Buckway and Hamidreza Ghandehari

Abstract Advances in imaging and nanotechnology have provided the opportunity for simultaneous delivery and diagnosis. Modalities such as positron emission tomography (PET), single photon emission computerized tomography (SPECT), magnetic resonance imaging (MRI) and optical imaging have allowed researches to visualize nano-sized drug delivery vehicles which carry payloads in order to coordinate disease treatment. This important tool can be termed "Nanotheranostics." This chapter describes the potential utility of the combined approach. The importance of selecting the correct components for a particular disease will also be discussed allowing for researchers to design effective delivery systems in order to accelerate the development in this field.

Keywords Theranostics • Polymeric drug delivery • Image-guided delivery • Personalized medicine • Nanomedicine

---

B. Buckway
Department of Pharmaceutics and Pharmaceutical Chemistry, University of Utah, Salt Lake City, UT, USA

Utah Center for Nanomedicine, Nano Institute of Utah, University of Utah, 36 South Wasatch Dr., Salt Lake City, UT 84112, USA

Huntsman Cancer Institute, University of Utah, Salt Lake City, UT, USA

H. Ghandehari (✉)
Department of Pharmaceutics and Pharmaceutical Chemistry, University of Utah, Salt Lake City, UT, USA

Utah Center for Nanomedicine, Nano Institute of Utah, University of Utah, 36 South Wasatch Dr., Salt Lake City, UT 84112, USA

Huntsman Cancer Institute, University of Utah, Salt Lake City, UT, USA

Department of Bioengineering, University of Utah, Salt Lake City, UT, USA
e-mail: hamid.ghandehari@pharm.utah.edu

K.A. Howard et al. (eds.), *Nanomedicine*, Advances in Delivery Science and Technology, DOI 10.1007/978-1-4939-3634-2_6

6.1 Introduction

One of the main challenges to successful treatment of disease is the difficulty in site specific delivery. Historically, patients have been administered treatments in a "one-size-fits-all" approach [1]. For example, many cancer patients are given a cocktail of chemotherapeutic drugs and are dosed based on the maximum tolerance levels of these drugs and the fact that these levels have worked previously in other cases. This approach often leads to toxic effects which can severely harm the patient. Although there have been many strategies developed both pre-clinically and clinically for improved delivery to diseased tissue, there has been relatively few successful or approved technologies by the regulatory agencies [2].

Targeted therapies, such as monoclonal antibodies have been developed which take advantage of the high expression of related proteins to selectively treat the disease and avoid off target toxicity. However, response to these therapies has been limited. This can primarily be attributed to a lack of improvement in efficacy in the cancer population based on the wide variety of tumor characteristics present in, not only an individual patient, but in the population as a whole. Targeted therapies today rely on a specific phenotype of a tumor and that phenotype may be different from tumor-to-tumor, thus, leaving some anti-tumor response in one tissue while lack of treatment in another. This leads to non-efficient patient care and disease recurrence in many patients. There remains a need to develop diagnostics which can predict targeted therapeutic efficacy in a patient by assessing whether the patient's disease is expressing the specified target.

Theranostics, combine both diagnostic and therapeutic modalities and can be used for personalized therapy [3]. This includes the use of diagnostic tests including genetic testing, histology and/or imaging which can estimate potential response, predict safety and monitor progress of a specific therapy. One of the earliest examples of this type of approach involves the use of an excised biopsy tissue being evaluated for HER2 expression using the HercepTest® and other HER2 assays from the tumor of a breast cancer patient which gives prognostic information when treating with Herceptin® [4]. The patient can then be qualified or disqualified (stratifying patients) for the therapy based on the outcome of this diagnostic test, thus assuring that the patient is receiving a treatment that has a much greater chance of efficacy. However, diagnostics such as HER2 assays for breast cancer have their limitations. They rely on invasive tissue biopsies, sometimes leading to false negative results simply because of the limited sampling of the tissue. The use of modern imaging technologies can more accurately assess the tumors for prognostic factors because the whole diseased tissue is visualized in its natural environment.

Advances in modern imaging technologies have provided clinicians with the ability to actively assess in real-time the status of disease. High-resolution techniques such as X-ray computerized tomography (CT) and magnetic resonance imaging (MRI) can enable anatomical placement of tumors along with size and shape that can be progressively measured during treatment to monitor response. Through the advanced molecular imaging techniques used in nuclear medicine modalities such as single photon emission computerized tomography (SPECT) and positron emission tomography (PET) imaging, clinicians can also understand the underlying processes effecting tumor treatment response, including metabolic state, hypoxia or proliferation that are occurring in each tumor within an individual [5, 6]. Each of these modalities can be combined to give a more accurate and precise treatment plan for patients, leading to better overall treatment. Personalized medicine can be further accomplished by combining these imaging technologies with advanced drug delivery approaches, thus, providing a more accurate and precise method for a theranostic system.

Among the many potential drug delivery vehicles for use in theranostics one promising area is the use of nano-sized biocompatible nanomaterials. Nanomaterials can be functionalized with drugs, targeting moieties and imaging agents to provide multifunctional carriers for assessing both real-time localization of delivery to the tumor via imaging modalities. One example of a nano-sized material that has been demonstrated in the clinic is monoclonal antibodies. Zevalin®, a radioimmunotherapy based on a radiolabelled monoclonal antibody, ibritumomab tiuxetan, recognizes CD20 receptors highly expressed on B-cells. This approved therapy for treatment of lymphoma has an imaging version and a therapeutic version. The patients are administered a low dose of gamma emitting ^{111}In-ibritumomab tiuxetan to confirm normal biodistribution observed by SPECT imaging. If no abnormal biodistribution is observed, the patient is administered a high dose of beta emitting ^{90}Y-ibritumomab tiuxetan for an effective radiotherapy. This example demonstrates the potential selection and qualification of a patient for a given therapy. Many other synthetic biocompatible nanomaterials can also be designed in a similar approach. Because of their macromolecular nature they can be tuned for long-term circulation and site specific delivery by targeting to various diseased tissues.

In this chapter, the historical aspects of imaging and theranostics, including several approaches in which nanomaterials can be utilized, will be discussed. The various imaging modalities will be presented along with each modality's potential for use with nanomaterial based theranostics. The basis for selecting appropriate nanomaterial carriers for delivery of both an imaging agent and a therapeutic also will be described along with their specific advantages and disadvantages. Therapeutic choices are also of particular importance when considering a theranostic system and, therefore, will be presented with technologies for site specific activation. Finally, the future directions and summary of the current status of theranostic nanomaterials for imaging will be discussed.

6.2 Imaging and Theranostics

6.2.1 Historical Advances in Imaging Technology

The beginning of imaging in the modern sense started in nuclear medicine with researchers using hand-held Geiger Counters to measure biodistribution of radioactivity without obtaining an actual image [7]. The first device for full body imaging was known as the rectilinear scanner which was fitted with a photographic component for recording [8]. This led to the development of the gamma-camera for scintigraphic imaging of gamma-emitting radionuclides that were administered to living subjects. Various advances in gamma-cameras improved the quality and utility of the images through the development of 3D reconstructions of images using SPECT imaging. Further advances led to the development of images obtained from positron emitting isotopes also known as PET imaging beginning in the early 60s which provide even more sensitivity and higher resolution images of these radioisotopes. MRI was initially developed in the 70s and the use of paramagnetic contrast agents to visualize molecules distributing throughout the body has become one of the most useful tools in imaging [9]. Researchers have continued to develop other methods of detecting materials that are administered to subjects non-invasively through the use of fluorescence emission in optical imaging [10–12] and also sound transmission using ultrasound techniques [13–16]. Every one of these advances has opened a large field of research for the development of materials which can be safely administered to a patient and provide accurate real-time assessment of disease status or diagnostics.

Diagnostics have been utilized for treatment planning for many years. Diagnostic testing provides genomic, proteomic and anatomical information related to a disease state. These tests come from a wide range of sources, including tissue biopsies and blood tests for in vitro diagnostics or even more sophisticated minimally invasive methods such as imaging. Diagnostic testing utilizes these results to assay the disease status in order to take advantage of weak points in the disease that can be treated by a particular drug or therapeutic intervention. In vitro diagnostics for receptor expression have been found to be useful in qualifying patients for various targeted therapies. However, these assays are invasive and suspect to false positive or false negative results due to sampling error. A sample may or may not be reflective of the entire diseased tissue's expression profile. Also some may be too small and not detected within the patient and, therefore, tissue sampling for a diagnostic test cannot be performed in this case. Therefore, diagnostic methods involving imaging are extremely advantageous because they accurately assess the whole tissue for target expression.

6.2.2 Theranostics

Combining diagnostics with therapeutics in order to increase safety and/or efficacy was first described by Funkhouser and coworkers in 2002 [17] as theranostics. Successful development of theranostics has the potential of improving the treatment of diseases such as cancer, and can potentially lower healthcare costs and streamline regulatory approval of targeted therapies [18]. This concept of combining treatment with a diagnostic is gaining the attention of pharmaceutical companies. Many companies focused on development of therapeutics are able to partner with diagnostic companies to accelerate the drug development process by improving clinical trial outcomes. This is done by selecting patients based on potential therapeutic success on a "companion diagnostic." Early examples of this were Herceptin® and the HercepTest® [4]. The FDA simultaneously approved both Genentech's Herceptin® and Dako's HercepTest® for treatment and diagnosis of Stage IV breast cancer.

6.2.3 *Modern Diagnostics and Use in Theranostics*

Modern advances in medical imaging have brought theranostics to a whole new level of potential possibilities. Molecular imaging methods such as MRI, X-ray computed tomography (CT), ultrasound, SPECT and PET imaging are prevalent in many medical centers today and can accurately monitor disease status [7, 19, 20]. Theranostics are being accelerated in development with these imaging modalities. Imaging provides noninvasive and real-time information for diagnosis of a patient. MRI and CT are used for anatomical information and can be combined with other nuclear medicine modalities (SPECT and PET) to provide quantification of radiolabeled targeted probes and determine their localization within the patient. Another advantage to image guided therapies is the potential to use platforms which include both diagnostic and therapeutic on one material. Real-time assessment of both target and therapeutic can be measured simultaneously. As discussed earlier Zevalin® [21] is a clear clinical example of how theranostics can improve the treatment of not only cancer but other diseases as well. It improves the safety of radioimmunotherapy because the imaging reduces the risk of more radiation exposure to patients who would not see any benefit from the therapeutic version due to abnormal biodistribution. This strategy was also helpful for regulatory approval because it qualifies a patient for therapy and increases the overall chance for efficacious treatment of lymphoma without risking the patient to unnecessary dangerous ionizing radiation treatment due to lack of targeting efficiency.

6.2.4 *Theranostic Methods*

The methods that can be applied to nano-based materials for theranostics can be summarized in three main approaches (Fig. 6.1). Although many more approaches could potentially be imagined [22], these three main approaches basically encompass the methods utilized with nanomaterials with imaging and therapeutics. The first method utilizes the basic concept of monitoring deposition of a single nanomaterial containing both the imaging agent and the therapeutic within the entire subject. Pharmacokinetic and biodistribution information then provides a basis on which one can determine if enough of the therapeutic component reaches the intended site. If the therapeutic dose required at the site is known, the efficacy of the administered nanotherapeutic may be predicted. If the amount of carrier is insufficient, the subject can receive an increased dose on subsequent administrations. The second method utilizes imaging of the known diseased site for carrier accumulation and non-invasive triggering using

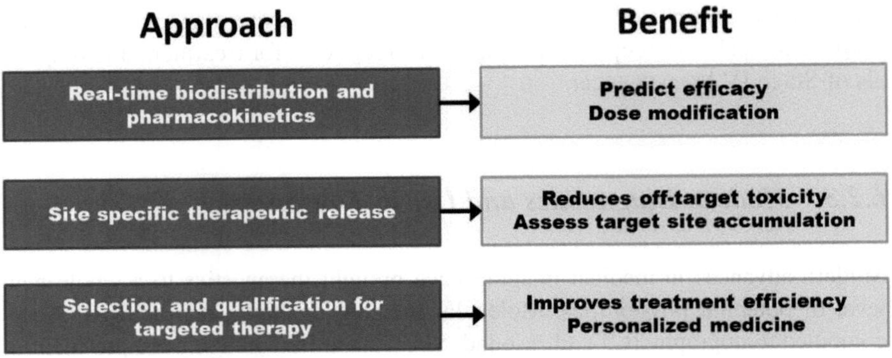

Fig. 6.1 Theranostic approaches for image based nanomaterials and their clinical benefit [22]

light, sound or heat to activate or release the therapeutic component. As long as the trigger is administered only in the required site, off-target toxicity is greatly diminished. If designed appropriately, therapeutic release can also be visualized by an imaging modality. The third approach for theranostics using nanomaterials with imaging is selection and qualification for targeted therapy. Targeted therapy relies on disease status which can change not only from patient-to-patient but from tumor-to-tumor and at different stages or times of the disease. An imaging version of the nano-carrier without the therapeutic can be administered before therapeutic versions are considered, to observe if the targeted receptor is adequately expressed. Thus reducing cost and optimizing time by allowing patients to be treated only with materials of which the desired therapeutic will be optimally effective. If the imaging results also demonstrate off-target accumulation that may compromise the subject's health, then the subject can also be eliminated from the current treatment and receive alternative treatments with better safety. These three methods describe the basic ideas and advantages behind the use of theranostics nanomaterials in imaging.

Imaging theranostics have great potential for improving the treatment of diseases. Many strategies can be utilized with these systems. However, there are several challenges for the development and successful use of these constructs when treating a disease. A wide range of imaging modalities, carriers, targeting strategies and therapeutics have been studied and currently only a select few have overcome the hurdles required for clinical use. The following sections will focus on the design criteria and components of theranostic nanomaterials for imaging.

6.3 Selecting the Appropriate Imaging Modality

Imaging modalities are becoming more sensitive, accurate and precise at detecting tracers in the body. Availability of sophisticated imaging devices is increasing; therefore, there is a growing need to rapidly develop probes for use in these different imaging modalities, especially in the area of theranostics. Personalized medicine is

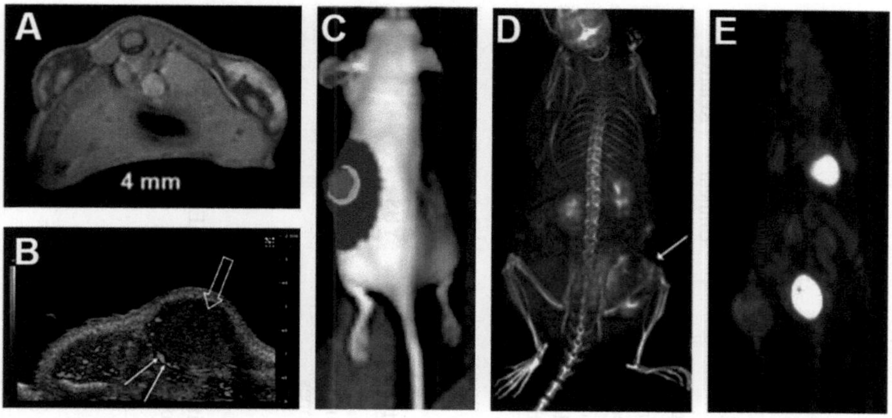

Fig. 6.2 Representative preclinical cancer images from common imaging modalities. *Arrow* denotes tumor. (**a**) Contrast enhanced MRI detailing the enhanced localization and tumor penetration of a gadolinium polymer conjugate treated with hyperthermia [23]. (**b**) Microbubble contrast imaging in a tumor model representing angiogenesis in the periphery of the tumor (*green*) [24]. (**c**) Representative luminescence imaging (optical imaging) of gene transfection events using silk-elastinlike polymer gene carrier [25]. (**d**) SPECT/CT image of ^{111}In-labeled HPMA copolymer localization in a tumor model [26]. (**e**) Representative PET image of tumor bearing mouse model injected with a proliferation biomarker (Courtesy of Jeffrey T. Yapp, Ph.D. Center for Quantitative Cancer Imaging at the Huntsman Cancer Institute, University of Utah) (Color online)

the ultimate goal of imaging in the clinic and its full potential is yet to be realized. This section will present the different strengths and weaknesses associated with various imaging modalities. Many of the modalities can be combined in such a way as to compensate for lack of sensitivity or accuracy in detection of different probes. The development of nanoparticles which have multiple methods of detection for multimodality imaging have been investigated providing information that is not possible with the use of one alone. Figure 6.2 is an overview of some of the available modalities for imaging. Table 6.1 shows a brief summary comparison of the different modalities. A more in-depth discussion is provided for each modality in the following subsections.

6.3.1 Magnetic Resonance Imaging

MRI is primarily used in the clinic for anatomical imaging. MRI signals are produced from changes in magnetic orientation from radiofrequency pulses of aligned protons in a strong magnetic field [28]. Signals from MRI are measured based on two typical responses, T1 and T2 relaxation demonstrated in Fig. 6.3. Typical use of MRI measures proton T1 or T2 relaxation signals in the body for high resolution images containing soft-tissue anatomical information. Differences in each proton's environment influence the T1 or T2 relaxation rate which in turn causes the variations that can be

Table 6.1 Summary comparison of the main imaging modalities [7, 27]

Modality	Temporal resolution	Spatial resolution	Quantitation	Depth of penetration	Probe sensitivity	Imaging agents	Clinical use	Relative cost
X-ray computed tomography (CT)	Minutes	0.5–1 mm	No	Limitless	Not determined	Iodine, barium suspensions	YES	$$
Magnetic resonance imaging (MRI)	Minutes-hours	~1 mm	Semi-quantitative	Limitless	10^{-3}–10^{-5} M	Gadolinum, iron oxide nanoparticles, ^{19}F-fluorine	YES	$$$
Fluorescence molecular tomography (FMT)	Seconds-minutes	1–3 mm	Quantitative	<2 cm	10^{-9}–10^{-12} M	NIR fluorochromes	NO	$
Ultrasound (US)	Seconds-minutes	0.01–2 mm	No	mm-cm	~10^{-12} M	Microbubbles	YES	$
Single photon emission computed tomography (SPECT)	Minutes	8–10 mm	Semi-quantitative	Limitless	10^{-10}–10^{-11} M	Gamma emitting isotopes	YES	$$
Positron emission tomography (PET)	Seconds-minutes	5–7 mm	Quantitative	Limitless	10^{-11}–10^{-12} M	Positron emitting isotopes	YES	$$$

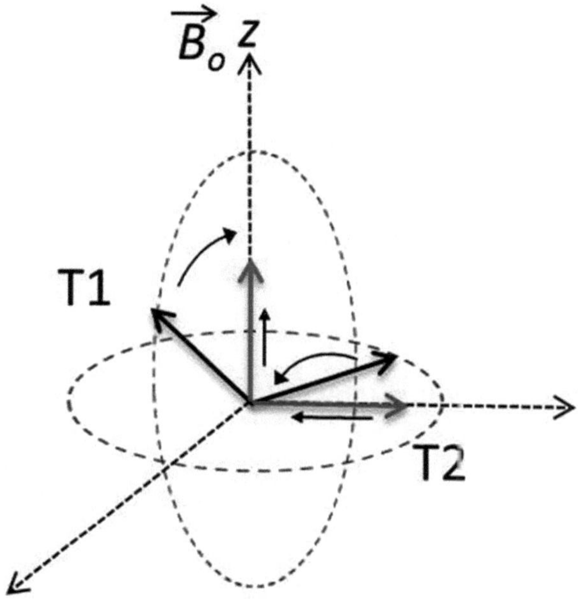

Fig. 6.3 Diagram of T1 and T2 relaxation in MRI. T1 relaxation is the rate at which the magnetic vector realigns or net alignment (*red arrow*) increases on z-axis which aligns with the external magnetic field after a 90° radiofrequency pulse. T2 relaxation is the rate at which the magnetic vector disappears or net x–y vector (*blue*) decreases in the x–y plane after a 90° radiofrequency pulse [29] (Color online)

converted into an image. Different materials used as MRI contrast agents also have various effects on protons in the body with these two relaxation signals [7, 28]. Gadolinium has been used as a contrast agent for T1 relaxation in image based theranostic systems [30]. An additional method for MRI imaging is the detection of other paramagnetic elements such as ^{19}F called magnetic resonance spectroscopy [31]. ^{19}F has been incorporated into many carriers [32–35]. One example is the system developed by Porsch et al. containing ^{19}F conjugated to amphiphilic polymers [36]. These polymers forming a micelle structure were subsequently loaded with doxorubicin for a theranostic drug delivery approach that exhibited MRI suitable signal-to-noise ratio (SNR) in phantoms [36]. However, one of the main challenges for MRI contrast agents is the lack of sensitivity of the MRI scanner. Large amounts of contrast are needed in order to produce a signal distinguishable from background. This makes MRI very difficult to use quantitatively. Although some techniques are in development to improve quantitative capacity of MRI, the physical nature of sensitivity is a large hurdle to overcome. Nonetheless, MRI is a valuable tool for anatomical imaging that can be used in conjunction with other imaging modalities which are more quantitative.

6.3.2 X-Ray Computed Tomography

X-ray computed tomography or CT is another method in which anatomical information is provided. CT uses X-ray projections that interact with high electron dense materials through a subject in multiple planes [7]. These planar images can then be reconstructed based on computer algorithms which produce a high-resolution image of the body.

Resolution depends on the electron density of the material and, therefore, primarily returns high-resolution images of hard-tissues. Contrast agents are based on heavier atoms such as iodine or barium. Iodine contrast agents are commonly used for angiography studies but require large frequent doses for a sustained signal output. The need for highly dense materials for contrast is the main limitation for use of CT with image-guided delivery. Therefore, CT is utilized in a multimodality approach which provides 3D anatomical information in tandem with other imaging information modalities. Risk due to ionizing radiation from prolonged CT exposure is of some concern. However, CT has made a substantial impact in detection and measurement of tumor sizes [37].

6.3.3 Optical Imaging

Optical imaging is based on light emitting probes which can be detected by a camera [10, 12]. Optical imaging has been successfully used for many years for the detection of molecular processes in in vitro assays. However, in vivo the challenge becomes increasingly difficult when trying to penetrate tissues which rapidly attenuate light signals. The attenuation of light prevents absolute quantification and limits resolution [38]. A modern technique called fluorescence molecular tomography (FMT) can measure signals from the visible to near-infrared spectrum (500–900 nm) of fluorescent probes in multiple orientations and use mathematical models which predict attenuation in the subject to produce three-dimensional images of probe localization [19, 39]. FMT is currently available for small animal imaging research, but clinical translation is yet to be viable. The tissue penetration in a human subject is more difficult. Therefore, FMT and other optical imaging techniques remain at the preclinical level. However, one of the advantages to using fluorescent probes in the development of theranostics is the potential to visualize drug release from a nano-construct. Recent developments in dyes which are conjugated to nanoparticles can be activated once released by targeted mechanisms from the nanoparticle surface [40]. Fluorescent dyes when in proximity (i.e., conjugated to the surface of a nanoparticle) lead to a strong fluorescence quenching due to fluorescence resonance energy transfer (FRET). Once the dyes are released and quenching is stopped, the signal will be visible and the resulting signal can be related to a mechanistic process [40]. Other imaging modalities are not capable of having signals that are activated in relation to cellular responses in vivo such as SPECT or PET imaging. Radionuclide signals are constitutively active and cannot be suppressed unlike fluorophores. One example of such a system provides information of cathepsin B protease activity [41]. This is performed by a FRET designed peptide sequence conjugated with two terminal fluorophores. When the peptide sequence is cleaved, FRET interactions cease and the signal representing protease activity can be visualized via FMT imaging. A similar type of system could potentially be used to visualize drug release by containing enzymatically degradable linkers conjugated with drug and FRET capable fluorophores. The released fluorophore would produce the optical signal that could be related to the release of drug from the linker.

6.3.4 Ultrasound

Ultrasound is probably the most cost effective and safe imaging modality available in the clinic today [7, 27]. Ultrasound utilizes high-intensity ultrasonic waves mechanically produced from a transducer. The sound waves then reflect or scatter from different tissues which can be detected by the transducer and converted into images. One advantage that ultrasound has is actual real-time imaging. Images are returned within seconds and, therefore, provide the highest temporal resolution available among the different modalities. Temporal resolution is a term describing the ability to distinguish between individual events [7]. Contrast agents for ultrasound are limited to gas bubbles which have specific properties that resonate in the 1–20 MHz frequency range, producing highly specific signals that can be recognized by the ultrasound device [19]. Ultrasound is limited to only microbubbles for visualizing probes and is, therefore, relatively limited in application to theranostics. It also suffers from a lack of penetration (Table 6.1) and requires contact of the device to the subject. Ultrasound resolution is highly dependent on the type of tissue and its depth and is, therefore, difficult to directly compare against other modalities. In the best case scenario, resolution is on the order of 10–100 μm [7]. Microbubbles are used as carriers for theranostic delivery. Further refinement in manufacturing is needed for clinical translation of this approach for theranostics.

6.3.5 Single Photon Emission Computerized Tomography

Nuclear medicine techniques such as single photon emission computerized tomography (SPECT) detect gamma emission from radioisotopes which are administered to a patient for purposes of diagnosis and treatment [42]. SPECT is a descendent of older gamma scintigraphic methods which were only capable of producing planar images that were not quantifiable and had very poor resolution. SPECT imaging takes modern advances in scanner and computer technology to obtain single γ-ray emissions using two to three gamma cameras that rotate around a patient who has been administered a gamma-emitting isotope tracer. Gamma emission is detected through thick collimators, plates of lead or tungsten, with small holes between the subject and gamma detector, that only allow photons emitted in the 90° direction of the gamma camera to interact with the detector. SPECT relies on reconstructing these multiple projections into a 3D image that gives precise localization of radioisotope biodistribution. Because only a select few gamma emissions are detected by the camera, the sensitivity is compromised to some degree due to lack of sufficient detector events [42]. Images are also dependent on the energy of the γ-rays emitted from the radioisotope. Lower energy radioisotopes have more attenuation than higher energy radioisotopes. Attenuation increases the scatter and noise detected from the camera, thus, compromising the resolution of the image. The quality and usefulness of images for image-guided approaches with SPECT are highly dependent on the type of radioisotope being utilized.

Table 6.2 Radioisotopes for SPECT imaging [19, 46, 47]

Radionuclides	$T_{1/2}$	γ-Ray energy
^{99m}Tc	6.01 h	140 keV
^{111}In	2.83 days	173, 247 keV
^{67}Ga	3.26 days	93, 184, 300, 393 keV
^{123}I	13.3 h	159 keV
^{131}I	8 days	365 keV
^{67}Cu	2.58 days	184.6 keV
^{201}Tl	3 days	69–81, 167 keV
^{133}Xe	5.2 days	81 keV

Radioisotopes for SPECT imaging must be selected based on the length of time needed to acquire information; otherwise, the half-life of the radioisotope must match the biological process that is being monitored. A list of clinically utilized radioisotopes for SPECT imaging with their emission properties are described in Table 6.2. Many isotopes can be utilized in SPECT for detection and the majority of them can be attached to nanoparticles via metal chelation, ionic interaction or covalent linkage. Transition metal based radioisotopes are easily conjugated to nanoparticles using stable bifunctional chelators of metals [43]. Some radioisotopes can be associated with a nanoparticle by ionic charge interactions [44]. Others, including halides, form stable covalent bonds in order to radiolabel a nanoparticle [45]. Many strategies exist to radiolabel a nanoparticle. Consideration for the method of labeling must not interfere with other functions (i.e., drug and targeting) for theranostics. One interesting strategy for SPECT imaging is the ability to image two different radionuclides with different energies at the same time. This may allow for imaging of distinct processes within a given construct. For example, a study was conducted using an ^{111}In labeled targeted peptide and ^{177}Lu labeled control nontargeting peptide injected into the same mouse in order to visualize differences in accumulation without inter animal variations [48]. The dual isotope approach could be applied in mixed micelles where the hydrophobic and hydrophilic components are labeled with different isotopes and could potentially give information as to the breakdown and ultimate fate of the two components.

SPECT is a promising modality for theranostics because it can provide accurate information on the relative location of carriers. However, there is the risk of ionizing radiation exposure that limits the amount of radioactive exposure a patient can receive. It also does not provide anatomical reference. This has been overcome by multimodal approaches such as combining images with both CT and MRI. The combined modalities can more accurately pin-point where the carriers are accumulating and give accurate information predicting both potential efficacy and safety of image guided constructs. The major challenge for SPECT imaging is obtaining accurate quantitative results. Detection events required for accurate quantitation are limited due to the use of collimators and a large range of gamma emission energies that are scattered and attenuated differently in tissue. Longer imaging times are also required for SPECT which can exaggerate motion effects leading to increased signal noise. There are methods for both attenuation and scatter correction which are being developed to make SPECT more

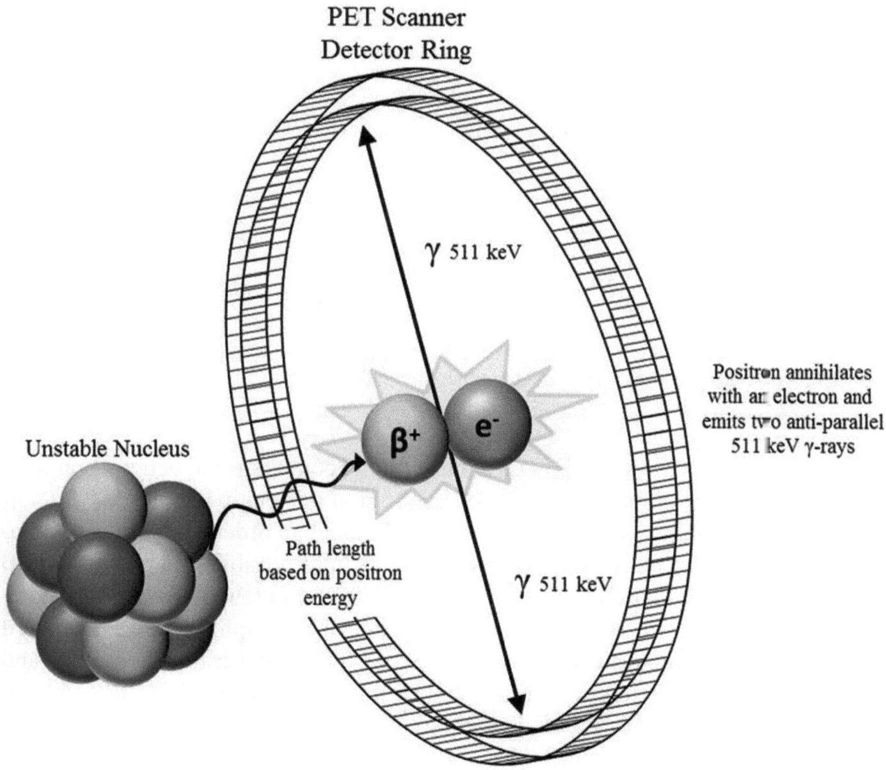

Fig. 6.4 Positron emission tomography [5]

quantitative. However, these methods and capacities vary greatly among scanners and groups. Ideally, further development and common adoption of correction techniques may eventually lead to more routine quantitative capability of SPECT.

6.3.6 Positron Emission Tomography

Positron emission tomography (PET) is another nuclear medicine technique which has made a major impact in the field of cancer treatment [5, 49]. PET detection is based on radionuclides emitting positrons which interact with nearby electrons and annihilate into two antiparallel gamma photons each with signature 511 keV energy. This allows for coincidence detection by a circular array of detectors that can trace back along the line of response to the origin of emission and produce high-resolution images of radioisotope probes. A basic diagram of PET is shown in Fig. 6.4. PET can detect radioisotopes down to the pico-molar (10^{-12}) range and has limitless depth penetration due to the high energy of its 511 keV gamma-rays [7].

Table 6.3 Physical characteristics of common positron emitting radioisotopes [5, 19, 50]

Radionuclides	$T_{1/2}$	Max β^+ energy (MeV)	Availability[a]
^{18}F	110 min	0.69	+++
^{11}C	20.4 min	0.96	+++
^{15}O	122.2 s	1.7	+++
^{13}N	9.97 min	1.20	+++
^{124}I	4.2 days	2.14	+
^{64}Cu	12.7 h	0.65	+
^{89}Zr	3.3 days	0.897	+
^{68}Ga	67.7 min	1.90	++
^{82}Rb	75 s	3.18	+

[a]Availability based on estimated cost and production facilities

In comparison with SPECT, PET has a much higher count rate and better resolution, thus, providing the ability for accurate quantitation of imaging agents. PET scanners are also combined with CT and other techniques in order to provide attenuation and scatter correction, thus, increasing its quantitative ability in comparison to SPECT. Due to the fact that only one gamma energy window is needed for detection around 511 keV, these correction techniques can be simplified when compared to SPECT. Radioisotopes used in the clinic for PET are generally short-lived and for the most part must be produced locally using a cyclotron. This increases the costs and availability of PET radionuclides. Like SPECT, ionizing radiation also limits the ability of patients to be continually administered radioisotopes for research studies. Table 6.3 lists some of the isotopes used in PET imaging with details of the characteristics and properties. PEt also does not provide anatomical reference; thus, almost all PET imaging devices are coupled with CT detectors for precise localization of signals with the patient. The CT images can also be used to correct for attenuation in the imaged subject for improved resolution and quantification.

Selection of radionuclides for PET imaging is important in that it can affect the resolution and relevance of the research study. The initial positron energy determines the path-length from the parent nuclide to the annihilation event. The higher the energy, the longer the path-length, thus, increasing image noise and reducing resolution [19]. Like SPECT, isotope half-lives must match the relevant biological process intended to be measured. This represents a significant challenge in imaging with PET, since most of the clinically available radionuclides have ½-lives on the order of minutes ($^{15}O \sim 2$ min ½-life) to a couple of hours ($^{18}F \sim 110$ min ½-life) which does not match the biological half-life of many macromolecular systems which circulate in the bloodstream for hours and days before being taken up by the target or eliminated. However, the majority of these short-lived isotopes are basic elements found in biology and, therefore, can be incorporated into drugs, sugars and other biological entities without influencing their structure.

A recent study using a positron emitting ^{64}Cu radionuclide conjugated to HPMA copolymers containing angiogenesis targeting peptides demonstrated measurable increased tumor localization by PET imaging in prostate tumor bearing mice [51]. Due to ^{64}Cu also having some beta-emission, the radionuclide can serve a dual purpose for imaging and therapy. This is an example of how radioisotopes can be utilized in theranostics. Clearly, more research and effort in this area need to be conducted to improve the availability of radionuclides. PET represents the future of nuclear medicine imaging and has already influenced the treatment of cancer profoundly [52]. However, costs and lack of radionuclide availability are detrimental for rapid development of theranostics based on PET imaging.

6.4 Choosing the Appropriate Nanomaterial Carrier

A large investment in research has been focused on the design and development of materials for targeted delivery. Each has their advantages and disadvantages for delivery of imaging and therapeutics. Some have had extensive clinical experience or testing. Recently, constructs based on nanomaterials have become a promising area of research [53–55]. The use of nano-scale constructs of 1–100 nm in size as therapeutic delivery systems has generated a promising venue for image-guided theranostics [56–58]. This property allows for multiple components to be incorporated on the surface or within these materials for targeted delivery of both imaging agents and therapeutics. They are also able to be tailored in ways as to interact with fenestrations, channels and surfaces in unique ways because of their small size [54]. Many types of carriers are synthesized in the nano-size range. Figure 6.5 displays some of the most investigated systems that have been utilized for image-guided delivery. The following are examples of nanoparticle based carriers that can be used as theranostics.

6.4.1 Water Soluble Polymers

Some examples of water soluble polymers that have been extensively studied are poly(ethylene glycol) (PEG), *N*-(2-hydroxypropyl)methacrylamide (HPMA) and poly-L-glutamic acid (PGA) [59–61]. A wide body of research has been conducted using these polymers. These polymers can have a linear or branched structure. PEG and HPMA copolymers are inherently non-degradable unless degradable sequences are included within the backbone of these copolymers [62–64]. However, PGA is naturally degradable in the body. The main advantage of polymer carriers is the ease to control size and afford multifunctionality [61]. Water soluble polymers, such as those described above, are biocompatible and have limited recognition by the immune system [65, 66]. This provides an advantage because they can circulate in the body for an extensive period of time, thus, increasing the likelihood of their ability to interact with the targeted tumor sites. PEG and HPMA copolymers also have properties which impart steric hindrance to degradation of the attached payload of drug and targeting moieties [67]. One advantage that HPMA copolymers have is the ability to

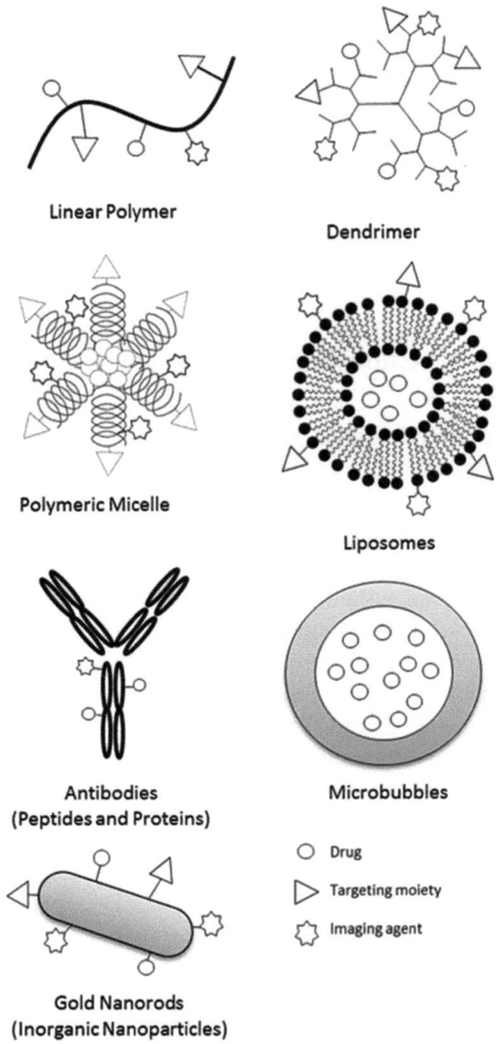

Fig. 6.5 Examples of carriers for nanotheranostics

incorporate multiple therapeutics and targeting agents into the side chains. Linear PEG has limitations in the amount of payload because it is limited to end group functionalization [60]. Some of the main disadvantages to traditional HPMA copolymers are nonbiodegradability and potential long-term exposure in the body leading to possible toxicity effects [68, 69]. PGA is biodegradable; however, this may not be ideal for a combined image-guided therapeutic because breakdown of the PGA backbone will eventually lead to imaging agent being separated from the carrier. HPMA copolymer doxorubicin drug conjugates were evaluated in clinical trials [70]. A matching HPMA copolymer was available with attached radioisotopes for imaging the biodistribution of the copolymers in patients [71]. The drug conjugate failed due to the lack of efficacy in some patients [72]. However, the trial may have had more success if they had used the imaging version of the HPMA copolymer for preselection of patients that would have been more susceptible to the HPMA copolymer-drug conjugate [22]. Recently, Yuan et al. designed integrin targeted HPMA copolymer-DOTA-^{64}Cu conjugates for PET imaging of angiogenesis [51]. These conjugates demonstrated increased uptake in tumors in-vivo. This approach could be applied to many HPMA copolymer drug conjugates for a combined imaging and drug delivery approach.

6.4.2 Polymeric Micelles

Micelles are constructed from a combination of hydrophobic and hydrophilic components or segments [73–75]. The structure is formed in aqueous solutions by self-assembly of a hydrophobic core and hydrophilic shell. Many chemotherapeutics are hydrophobic leading to solubility issues [73]. Hydrophobic drugs can be associated with the hydrophobic core and improve their solubility and protect them from metabolic enzymes in the bloodstream while associated with the nanoparticle [76]. Imaging agents and targeting groups can be associated with the hydrophilic components and allow for imaging of micelle biodistribution within the body [76, 77]. Another advantage that micelles have is their ability to have triggered therapeutic release based on pH change [76, 78]. Micelles however, have limited stability in the body and progressively breakdown to the initial components, especially when encountering biological milieu [54]. Therefore, imaging of the nanoparticle has limitations due to the eventual breakdown of the micelle. One example of a block copolymer micelle image-guided drug delivery system incorporated folate targeting, pH sensitive drug release of doxorubicin and Cy5.5 for optical imaging [79]. The polymeric conjugate was able to successfully image targeted conjugate localization and treatment in HeLa tumor cell bearing nude mice [79]. Another example of a promising micelle carrier was developed by Decato et al. [80]. This construct incorporates several perfluoro-*tert*-butyl groups as a fluorophilic tail in combination with hydrophilic PEG groups which self-assemble into stable spherical micelles. They demonstrated that these conjugates can potentially be detected by ^{19}F-MRI and in the future could be used for delivery of hydrophobic drugs and serve as potential imaging and drug delivery vehicles.

6.4.3 Dendrimers

Dendrimers are multibranched polymeric systems with a central initiator core [81, 82]. Dendrimer size can be precisely controlled by successive additions of layers of branched units. Dendrimers have exceptionally low polydispersity and in some cases can be unimolecular [83]. Because of their unimolecular structure regulatory hurdles to clinical approval can be significantly less challenging compared to other polydisperse polymers [84]. The branched structure allows for specific control of the amount and density of functional groups on the surface of the dendrimer. This can be utilized for surface decoration with various imaging agents, targeting moieties and drugs. With larger generations of dendrimers, a hollow core can be formed which can also be loaded with hydrophobic drugs. Several limitations, however, exist with dendrimers. Large dendrimers become difficult to synthesize, because the larger the dendrimers become, the more steric hindrance prevents chemical ligation of the branching units and surface modifications [85]. The dendrimers that have been extensively investigated for clinical development are poly(amido amine) or PAMAM dendrimers [82, 86]. Depending on surface charge and generation, PAMAM dendrimers can have toxicity based on the charge density of the surface when reacting with cells [87, 88]. One of the key aspects of PAMAM dendrimers is the alternating generations terminating with either primary amines or carboxyl groups. One limitation with PAMAM dendrimers is that their typical branched layered structure contains only one type of surface group for conjugation of targeting moieties, imaging agents and drugs. This limits the degree of control over how much of each component is incorporated unless different dendrons are constructed during conjugation. Several groups have begun strategically synthesizing "Janus" dendrimers which are able to have several different types of surface functional groups, thereby providing better surface conjugation control [89]. One particular example shows great promise as a platform for a theranostic [90]. Although the study only demonstrated the dendrimer with a near-infrared agent for optical imaging, the conjugate did not show any toxicity in presence of T98G human cells and the unique trifunctional surface groups showed potential for development as a theranostics treatment.

6.4.4 Liposomes

Liposomes are spherical lipid bilayer constructs with an aqueous core which can contain drugs or other therapeutic agents [91, 92]. Their lipid bilayer structure mimics the biological environment of a cell but can be made of many different materials with the majority using phospholipids. Liposomes are among the most studied drug delivery systems with many clinically approved formulations [93, 94]. They are primarily formulated to entrap hydrophilic drugs within their aqueous core or associate hydrophobic drugs into their lipid bilayer [91]. Due to their macromolecular nature, they can passively target tumors via the EPR effect [95]. Of particular mention is Doxil®, a liposome-based formulation of doxorubicin for treatment of advanced ovarian cancer [93]. Liposomes protect their encapsulated drugs from metabolism in the bloodstream and fuse with their biologically similar membranes of targeted cells. Once fused with cells, they release the payload into the cell for increased therapeutic efficacy. Phospholipids can be modified with different imaging agents and targeting moieties and be inserted into the lipid bilayer in order to make theranostic nanoparticle systems [92]. One particular challenge recognized by the FDA with liposomes is reproducibly controlling the size and polydispersity [96]. They are also subject to rapid degradation or reticuloendothelial system (RES) clearance from the body [97]. Methods have been developed to overcome some of the issues such as incorporation of PEG for "stealth" liposomes [98]. However, the instability of these systems in-vivo remains a particular challenge when considered for theranostics. Regardless, complex systems have been attempted to overcome these issues. For example, magnetic resonance high-intensity focused ultrasound (MR-HIFU) has been used with temperature sensitive liposomes loaded with doxorubicin [99]. MR imaging was used to guide the placement of localized hyperthermia within the tumor for triggered release of the liposomes in Vx2 rabbit models. This study shows the potential application of a liposomal formulation for image guided therapy.

6.4.5 Antibodies and Other Proteins

Nature has developed its own methods of recognizing diseased tissue which can be utilized for delivery of both drugs and imaging agents in a theranostic approach. Antibodies are of particular interest in delivery because of their biocompatibility and intrinsic ability to target antigens within the body, including those highly expressed in tumors. Many monoclonal antibodies (MoAbs) can be therapeutically active and have been approved, or are in, clinical trials for the treatment of cancer [100]. Therapeutics such as drugs and radionuclides conjugated to MoAbs have been shown to be effective for targeted delivery and image guided therapy [101]. One of the major concerns in using antibodies as image-guided therapies is their method of manufacture. MoAbs are produced from nonhuman sources (typically mice) and this can induce immune recognition within a patient by human anti-mouse antibodies (HAMA) that can either reduce the effectiveness of the therapeutic conjugate or cause severe life-threatening immune reactions [102]. However, advances in this technology have minimized mouse component of the antibodies, thus minimizing the potential for HAMA. Image guidance helps to select patients which may suffer from this effect as demonstrated in radioimmunotherapy [21]. Antibodies are not the only proteins that are used in therapeutic delivery. Albumin, a native protein in the blood-stream, serves as a macromolecular delivery system that is biocompatible and capable of being imaged. For example, a photosensitizer, chlorin e6, covalently linked to albumin formed nanoparticles that were used for image guided drug delivery [103]. Tumor localization was observed by optical imaging of the photosensitizer and then subsequently irradiated with light in the tumor for anticancer treatment in a mouse model.

6.4.6 Inorganic Nanoparticles

Nanotheranostics can also encompass an array of inorganic materials such as quantum dots, gold and iron oxide nanoparticles. Metal elements exhibit interesting properties when synthesized in the nano-size ranges. For example, gold nanoparticles can absorb light in the near infrared spectrum and produce heat for localized hyperthermia [104]. Quantum dots are semiconductor nanocrystals made of transition metals that have tunable emission spectra that are much stronger than organic dyes [105]. Iron-oxide nanoparticles can be detected using MRI. All of these materials can be decorated with targeting agents, imaging agents and therapeutics for image-guided therapeutic delivery in cancer [106]. One of the main concerns with these constructs is their toxicity. Typically, these constructs do not breakdown in the body and, therefore, are deposited in tissues for an extended period of time [106]. The nature of their toxicity is still being investigated, but the long-term effects of some of these materials raise many questions as to whether they can be used safely in the clinic.

6.5 Therapeutic Considerations for Nanotheranostics

The appropriate therapeutic type for a delivery system is an important part of designing an effective theranostic. In this section we will describe three major classifications of therapeutics that have been used in theranostics with considerations for their advantages or disadvantages that need to be considered before selecting one for any given system.

6.5.1 Chemotherapeutics

Chemotherapeutics are the main-stay of treatments for advanced stages of cancer. While surgical methods to remove known tumor sites are preferred, small invasive tumors are difficult to detect and remove. Therefore, they require small molecule agents to penetrate and destroy tumor sites which are not accessible. However, many small molecules are nonspecific in their cytotoxic action. Therefore, they cause a variety of side effects that reduce quality of life and endanger the patient. Most chemotherapeutics can be broken down into a few classes: alkaloids, antibiotics, platinates, antimetabolites, topoisomerase inhibitors, mitosis inhibitors and others [107]. In discussing chemotherapeutics for image-guided drug delivery it becomes necessary to discuss what has been performed in the past. Chemotherapeutic selection for drug delivery systems has been based primarily on clinically approved drugs which have limitations due to solubility, dose limiting toxicity, instability in the bloodstream or poor efficacy. For FDA approval, selecting drugs that are already approved for use is one less hurdle to cross and is, therefore, beneficial in image-guided therapeutics.

One of the many challenges for chemotherapeutic nanoparticle delivery is the incorporation methods and subsequent release. Some carriers entrap the chemotherapeutic cargo like liposomes, microbubbles and potentially micelles. The important part of image-guided drug delivery is that the nanoparticle drug complex must stay intact until it arrives at the targeted tumor site. Otherwise, the imaging of the carrier does not represent the actual drug localization. This is a challenge for nanoparticle formulations which have limited stability over time and ' leak" drug. Covalent linkages to nanoparticles can overcome this but methods must be developed in order to realize site-specific drug release. Linear carriers such as HPMA copolymers have utilized lysosomally degradable GFLG peptide sequences for drug release once the nanoparticle has been endocytosed into the cell [108]. Success of site-specific drug release has been variable using this approach. In the case of HPMA copolymers conjugated with docetaxel, rapid release was observed in cell culture media and, therefore, expected to observe similar behavior in vivo [109]. In another study, release kinetics was variable when HPMA copolymers were conjugated with both gemcitabine and doxorubicin via GFLG linkers [110]. Gemcitabine was rapidly released in the presence of cathepsin-B, a lysosomal enzyme, but docetaxel showed a very slow release even with the enzyme. These examples demonstrate the balance that is needed with site-specific drug release. Drugs must be stable enough before getting to the targeted site but have the ability to empty its payload rapidly for a maximally effective image-guided delivery.

6.5.2 Radiotherapeutics

Radioisotopes have been utilized for cancer treatment for a considerable amount of time. Some of the first clinically used radioisotopes were produced in the 1930s such as ^{131}I iodine [111]. Radiotherapy is primarily performed using isotopes which emit alpha and beta radiation. Both forms of radiation are efficient at forming radicals (usually with water) which exert damage to the DNA of cells and cause irreversible damage leading to cell death. However, the range of tissue in which these effects occur is different depending on the type and energy of particle [47]. Many of the radioisotopes in the clinic for radiotherapy have beta emission with a range of 50–5000 μm. More rare isotopes with alpha emission have a much shorter range of 40–90 μm. Radioisotopes with auger electrons have the shortest range of 0.01–1.0 μm. The greater the range the more risk to nontargeted tissues near tumors. However, for larger tumors, a larger range may be more advantageous. Selecting the proper isotope is a matter of ½-life and the range of emission that is ideal for the targeted tumor. Table 6.4 is a list of radioisotopes for potential use in radiotherapy. The key aspect required for successful radioisotopes is site-specific localization and ample radiation dose to cause sufficient irreversible DNA damage leading to tumor cell death. This is a challenge that requires tumor-targeted approaches which do not cause off-site accumulation. One of the first clinically available theranostic delivery systems is based on a radiotherapeutic approach. Iodine uptake is especially high in the thyroid and for more than 50 years, radiation oncologists have exploited this feature to treat thyroid cancer with ^{131}I-iodine [112]. Imaging can be performed using SPECT or simple gamma scintigraphy of ^{123}I, a gamma emitter, which gives information on the areas of accumulating iodine in the patient. Safety and efficacy concerns can be utilized to select the patient for the subsequent ^{131}I therapy. Other examples

Table 6.4 Isotopes for theranostic radiotherapy [47, 113]

Radionuclides	$T_{1/2}$	Type of emission	Average energy (keV)
^{131}I	8.0 day	β^-	181
^{90}Y	2.7 days	β^-	935
^{153}Sm	1.94 days	β^-	280
^{64}Cu	12.7 h	β^-	1670 (max)
^{67}Cu	2.58 days	β^-	141
^{211}At	7.2 h	α	5867
^{213}Bi	46 min	α	6000 (max)
^{67}Ga	3.26 days	Auger	0.04–9.5
^{125}I	60.5 days	Auger	27 (max)

were discussed early on in this chapter, specifically the radioimmunotherapies, Bexxar® and Zevalin® for lymphoma patients. Theranostic approaches have been extremely useful in the approval of these therapeutics in the clinic. Selecting the proper isotope is highly dependent on the ability to stably link the radionuclide to the nanoparticle platform. For efficacious and safe radiotherapy, loss of the radioisotope must not occur. Using nanoparticles to carry isotopes prevents them from accumulating in radiation sensitive areas such as bone marrow as long as the chelated radioisotope is stable in the bloodstream. The main advantage of using radioisotopes in nanoparticles is that their effect on cancer is independent of release, unlike chemotherapeutics. Nanoparticles in this case do not need to penetrate deep within the tumor to exert their anticancer effects. This advantage is profound because typically nanomedicines cannot penetrate deep within tumor tissues.

Several successful image guided radiotherapeutics have been developed and have shown promising results in treating cancer [113]. Wang et al. synthesized a multifunctional lipid-polymer hybrid nanoparticle that contained docetaxel and chelator for both ^{111}In and ^{90}Y radioisotopes [114]. The prostate tumor-targeting nanoparticle demonstrated superior antitumor efficacy of the combination chemotherapy and radiotherapy particle. Although the approach has promise, control over all components especially between loaded drug and ^{90}Y may prevent reproducible synthesis required for eventual translation.

6.5.3 *Nucleic Acid Delivery*

A promising area still in development for therapeutics is antisense oligonucleotides. This can include short strands of either DNA or RNA which are complementary to a chosen sequence. When introduced into the cell, they can knock-down expression of a disease-related gene. One such technology involves small interference RNA or siRNA which has the higher therapeutic index when compared to other gene modifying therapies [115]. Due to the siRNA charge, it cannot cross biological membranes and, therefore, polymers have been utilized to shield the charge, protect the nucleic acid in the bloodstream and transport them into the cytosol of tumor cells. Optical imaging offers an interesting opportunity with FRET-like systems based on siRNA delivery. Cationic quantum dots were conjugated with poly(ethylene imine) (PEI) in order to complex with siRNA that had been labeled with a fluorochrome [116]. Once complexed, the fluorescence signal was quenched and upon siRNA release, an increase in fluorescence was observed. This system could potentially be utilized to observe release and essentially be correlated with efficacy within tumor cells. However, systems such as these are in early-development. For example, Medarova et al. demonstrated the utility of using MRI and optical imaging to observe the nanoparticle uptake and subsequent knockdown of GFP expressing tumors by complexed siRNA [117].

6.6 Methods of Release or Activation

The therapeutic component of a theranostic system may or may not require the need to be released from the carrier in order to exhibit its function in the target tissue. For example, doxorubicin requires uptake into the cell and subcellular localization to the nucleus in order to intercalate into the DNA to inhibit biosynthesis of the cell [118]. In this case the carrier requires some type of release mechanism that promotes delivery to the nucleus of the cell to be most effective. On the other hand, therapeutic radioisotopes do not necessarily require detachment from the carrier because the ionizing radiation that causes cell death can penetrate cellular and subcellular walls depending on the amount of energy in the emitted particle [119]. Therefore, careful consideration must be given in choosing a proper mechanism of release for a therapeutic in order to exert its potential action against the diseased tissue of interest depending on where the therapeutic must exert its action. This section will describe some of the technologies that have been developed for site specific release or activation of a therapeutic component on a nano-carrier.

6.6.1 *Photo-activation*

One exogenous source of energy that can be used to activate or release drug within a diseased tissue is the use of light. Some of the first examples of light triggered therapy used photosensitizers mainly for photodynamic therapy (PDT). A photosensitizer is a light sensitive molecule that when activated at the specific wavelength of light creates reactive oxygen species which can interact in a destructive manner within cells [120–122]. PDT has been used for treatment of skin disorders, cancer and other various diseases. One benefit of photosensitizers is that their fluorescence emission can be observed by optical imaging, therefore, they can both provide detection and therapy. The potential for PDT is attractive, however, limitations in tissue penetration of light limit this therapy to superficial tumors or through endoscopically accessible organs such as the lung [123]. There is potential to use small light emitting optical fiber probes through guided needles into deeper tissues such as the pancreas but this remains challenging and invasive [124]. Multimodality imaging using PET or SPECT is also a possibility using nanoparticles conjugated with photosensitizers, thus, identifying accurately the location and depth of required light penetration for PDT [125]. However, PDT is generally limited to superficial and local treatment, thus, limited against metastatic cancer [126]. Despite this, PDT can be used for cancers which are not restricted by the limitations of this approach. Several examples of photosensitizers have been conjugated to nano-carriers for use in a theranostic approach [103, 127–132]. In a recent example, a multi-block based polymeric micelle containing folate for tumor targeting and a mitochondrial targeted porphyrin based photosensitizer was examined in-vitro and demonstrated enhanced uptake in HeLa cells and subcellular localization to the mitochondria [131]. Site specific toxicity relies on the lack of cytotoxicity of the construct until the appropriate wavelength of light interacts with the photosensitizer. The polymeric micelle mentioned above accomplished this goal of having cytotoxicity only after treatment with light.

6.6.2 Thermal-activation

Another potential exogenous source that can be used to release or activate a therapeutic drug delivery is the use of heat. Hyperthermia has been used for treatment of various diseases [133–135] and could potentially be used in conjunction with a nanotheranostic for targeted release or activation. However, it must be able to be administered only in the diseased tissue in order for it to be utilized for site specific delivery. One such method of localized delivery is the use of gold nanorods for plasmonic photothermal therapy (PPT) in cancer. PPT utilizes the surface plasmon resonance of gold nanorods to produce heat when activated at the appropriate laser light wavelength [136, 137]. Gold nanorods, depending on the size, can passively accumulate into tumors and, therefore, when treated with light in a controlled manner efficiently heat the surrounding tissue. However, because light is utilized, penetration for deep tissue administration remains a challenge. Another method that may be closer to clinical translation is high-intensity focused ultrasound (HIFU). HIFU utilizes ultrasonic waves to generate heat within biological tissues [138]. It can also be combined with MRI imaging with gadolinium thermal sensitive liposome conjugates for precise imaging of the nanoparticle localization within the tumor and subsequent drug release [139–141]. One example of thermal activation for a drug release involved the use of a temperature sensitive HPMA mono/dilactate grafted liposome which was tuned to the desired temperatures for release [142]. This was accomplished by varying the content of grafted copolymer and demonstrated efficient drug release in-vitro using HIFU. Hyperthermic delivery based on HIFU represents a promising future for targeted delivery of nanoparticles for therapeutics and imaging.

6.6.3 *Acoustic-activation*

Another method for site specific drug release that can be controlled locally at the disease site is the use of ultrasound with microbubbles. Microbubbles for therapeutic delivery are a special case of vesicular structure which encompasses gas and can be designed for image-guided therapeutic delivery [14]. Small gas bubbles, typically perfluorocarbons, are imaged via ultrasound based on their difference in echogenicity than liquid media [15]. In most cases, microbubbles are stabilized by a lipid bilayer which surrounds the gas bubble. The surrounding liposome can be loaded with drugs and targeting agents for tumor delivery. With increased sonic waves these liposomal bubbles can burst, thus, releasing their contents. This can be focused in the areas where the microbubbles are accumulating, including tumors. Imaging and drug release is, therefore, controlled by ultrasound, and, thus, is a promising theranostic. One such example is the stabilization of perfluorocarbon nanodroplets using block copolymers containing paclitaxel [143]. Due to the high ^{19}F-fluorine content, ^{19}F-MR spectroscopy in conjunction with ultrasound was utilized to determine precise anatomical location of the nanoconstructs. Complete tumor regression was observed in a pilot study with tumor focusing ultrasound in tumor bearing mice. The promise of ultrasound-mediated delivery using microbubbles is being investigated in several clinical trials [13]. However, challenges remain in reproducibly manufacturing these constructs, especially when complicated by associating targeting agents and drugs into their structure. Their size distribution also remains a challenge. Microbubbles span a large range in sizes affecting their distribution in the body. Strategies are needed to refine microbubble manufacturing process for more reproducible formulations, smaller size and less polydispersity. Stabilization of the liposomal structure is also important to ensure that drug leakage does not occur. Polymersomes made of block copolymers can be designed to encapsulate drug in the hydrophobic shell and form a stable liposomal structure for gas for detection with ultrasound. Specific delivery to the site can be accomplished by rupturing the shell structure with cavitation forces using the ultrasound device in a similar fashion as lipid counterpart. Polymersomes have been developed for this purpose, and the surface tuned to have functional groups which can be utilized to attach targeting moieties [144], thus, enhancing site specific delivery. The combined approach of targeting and site specific delivery is a promising direction in tumor treatment.

6.7 Challenges for Development of Image Based Theranostics

The previous sections have presented the different aspects, advantages, disadvantages and requirements for image-based theranostics. In order to overcome some of the challenges and limitations of certain imaging modalities, researchers have designed materials that can be visualized by more than one way. For example, Chakravarty et al. developed ^{69}Ge-superparamagnetic iron oxide nanoparticles for both MRI and

PET imaging [145]. The advantage of this system is increased sensitivity provided by the PET isotope component but high resolution of the MRI iron oxide contrast component for accurate localization within the targeted area. One must also consider for clinical translation the regulatory hurdles that need to be overcome for manufacturing of nanomaterial delivery systems with imaging agent, therapeutic and targeting.

Perhaps one of the more challenging hurdles to overcome is the fact that no true nanotheranostic system has been successfully tested in the clinic with the exception of monoclonal antibodies. This may be due to the fact that many questions about safety still need to be answered regarding the long-term effects of nanoparticles within the body. Many of the nanoparticles with the intent of being used as a theranostics agent have been developed. The literature is full of nanoparticle systems designed with both therapeutic and diagnostic components which have been well characterized and evaluated in-vitro. However, what is missing is well-designed studies using appropriate animal models that truly evaluate how advantageous these systems could be in the clinic. More studies are needed that demonstrate how nanotheranostics actually can predict efficacy or correlate imaging results with actual efficacy. Several researchers have designed and synthesized nanoparticle conjugates with evidence of efficacy but study designs often lack the main goal which is to determine whether imaging of the nanoparticle localization to the target does or does not correlate with efficacy. Steps toward translation will accelerate, if researchers utilize appropriate animal models that can mimic various stages of disease and utilize the imaging information from the nanotheranostic localization in the subject to correlate with the corresponding therapeutic.

Generally, the simpler the design is, the better the approach. Good manufacturing practices (GMP) required for FDA approval require that each component of the formulation have complete characterization at the full range of possibilities within the construct or formulation. With multiple components having multiple cross-interactions possible, the regulatory hurdle for just one image-based theranostic is significant. Much is still needed to address how each component of these nanotheranostics will be deposited in the human body and what long-term effects might occur. Thus, a simpler approach is more likely going to be logistically and economically feasible. Selecting components especially drugs with ample clinical experience will also lessen the burden because previous experience can yield much information in regards to potential problems. The FDA has recently begun to address these challenges by characterizing these materials using criteria provided by the National Nanotechnology Initiative [146]. The Center for Drug Evaluation and Research within the FDA has recognized that this technology will need specific guidance in regards to their eventual approval but have also acknowledged that they will be treated as any other diagnostic or therapeutic.

6.8 Conclusion

In summary nanotheranostics are in their earliest stages of development. With the exception of MoAbs, no nanotheranostic system has been approved for use in the treatment of diseases. As the industry evolves it is clear that imaging combined with therapeutic development with nano-sized carriers will be promising in personalized medicine. Questions, however, will still need to be answered regarding their overall safety. Many advances with nanomaterial design and imaging technologies have taken the field closer to realization. However, much is still to be done to make these systems widely applicable in the field of medicine.

References

1. Sumer B, Gao J (2008) Theranostic nanomedicine for cancer. Nanomedicine (Lond) 3:137–140
2. Eifler AC, Thaxton CS (2011) Nanoparticle therapeutics: FDA approval, clinical trials, regulatory pathways, and case study. Methods Mol Biol 726:325–338
3. Kelkar SS, Reineke TM (2011) Theranostics: combining imaging and therapy. Bioconjug Chem 22:1879–1903
4. Zieba A, Grannas K, Soderberg O, Gullberg M, Nilsson M, Landegren U (2012) Molecular tools for companion diagnostics. N Biotechnol 29:634–640
5. Bailey DL (2005) Positron emission tomography: basic sciences. Springer, New York
6. Chowdhury FU, Scarsbrook AF (2008) The role of hybrid SPECT-CT in oncology: current and emerging clinical applications. Clin Radiol 63:241–251
7. James ML, Gambhir SS (2012) A molecular imaging primer: modalities, imaging agents, and applications. Physiol Rev 92:897–965
8. Allen HC, Libby RL, Cassen B (1951) The scintillation counter in clinical studies of human thyroid physiology using ^{131}I. J Clin Endocrinol Metab 11:492–511
9. de Haën C (2001) Conception of the first magnetic resonance imaging contrast agents: a brief history. Top Magn Reson Imaging 12:221–230
10. Bremer C, Ntziachristos V, Weissleder R (2003) Optical-based molecular imaging: contrast agents and potential medical applications. Eur Radiol 13:231–243
11. Graves EE, Weissleder R, Ntziachristos V (2004) Fluorescence molecular imaging of small animal tumor models. Curr Mol Med 4:419–430
12. Ntziachristos V, Bremer C, Weissleder R (2003) Fluorescence imaging with near-infrared light: new technological advances that enable in vivo molecular imaging. Eur Radiol 13:195–208
13. Castle J, Butts M, Healey A, Kent K, Marino M, Feinstein SB (2013) Ultrasound-mediated targeted drug delivery: recent success and remaining challenges. Am J Physiol Heart Circ Physiol 304:H350–H357
14. Kiessling F, Fokong S, Koczera P, Lederle W, Lammers T (2012) Ultrasound microbubbles for molecular diagnosis, therapy, and theranostics. J Nucl Med 53:345–348
15. Liang HD, Blomley MJ (2003) The role of ultrasound in molecular imaging. Br J Radiol 76(Spec No 2):S140–S150
16. Zhao YZ, Du LN, Lu CT, Jin YG, Ge SP (2013) Potential and problems in ultrasound-responsive drug delivery systems. Int J Nanomedicine 8:1621–1633
17. Funkhouser J (2002) Reintroducing pharma: theranostic revolution. Curr Drug Discov 2
18. Blair ED, Stratton EK, Kaufmann M (2012) The economic value of companion diagnostics and stratified medicines. Expert Rev Mol Diagn 12:791–794
19. Weissleder R (2009) Molecular imaging: principles and practice. People's Medical Publishing House, Shelton, CT
20. Gambhir SS (2002) Molecular imaging of cancer with positron emission tomography. Nat Rev Cancer 2:683–693
21. Goldsmith SJ (2010) Radioimmunotherapy of lymphoma: Bexxar and Zevalin. Semin Nucl Med 40:122–135
22. Lammers T, Kiessling F, Hennink WE, Storm G (2010) Nanotheranostics and image-guided drug delivery: current concepts and future directions. Mol Pharm 7:1899–1912

23. Gormley AJ, Larson N, Banisadr A et al (2013) Plasmonic photothermal therapy increases the tumor mass penetration of HPMA copolymers. J Control Release 166:130–138
24. Bowden DJ, Barrett T (2011) Angiogenesis imaging in neoplasia. J Clin Imaging Sci 1:38
25. Cresce A, Dandu R, Burger A, Cappello J, Ghandehari H (2008) Characterization and real-time imaging of gene expression of adenovirus embedded silk-elastinlike protein polymer hydrogels. Mol Pharm 5:891–897
26. Buckway B, Frazier N, Gormley AJ, Ray A, Ghandehari H (2014) Gold nanorod-mediated hyperthermia enhances the efficacy of HPMA copolymer-90Y conjugates in treatment of prostate tumors. Nucl Med Biol 41:282–289
27. Rudin M, Weissleder R (2003) Molecular imaging in drug discovery and development. Nat Rev Drug Discov 2:123–131
28. Plewes DB, Kucharczyk W (2012) Physics of MRI: a primer. J Magn Reson Imaging 35:1038–1054
29. Brown MA, Semelka RC (2010) MRI: basic principles and applications, 4th edn. Wiley-Blackwell/John Wiley & Sons, Hoboken, NJ
30. Koh TS, Bisdas S, Koh DM, Thng CH (2011) Fundamentals of tracer kinetics for dynamic contrast-enhanced MRI. J Magn Reson Imaging 34:1262–1276
31. Chen J, Lanza GM, Wickline SA (2010) Quantitative magnetic resonance fluorine imaging: today and tomorrow. Wiley Interdiscip Rev Nanomed Nanobiotechnol 2:431–440
32. Zhang H, Zhang L, Myerson J et al (2011) Quantifying the evolution of vascular barrier disruption in advanced atherosclerosis with semipermeant nanoparticle contrast agents. PLoS One 6:e26385
33. Kok MB, de Vries A, Abdurrachim D et al (2011) Quantitative (1)H MRI, (19)F MRI, and (19)F MRS of cell-internalized perfluorocarbon paramagnetic nanoparticles. Contrast Media Mol Imaging 6:19–27
34. Neubauer AM, Myerson J, Caruthers SD et al (2008) Gadolinium-modulated 19F signals from perfluorocarbon nanoparticles as a new strategy for molecular imaging. Magn Reson Med 60:1066–1072
35. Waters EA, Chen J, Allen JS, Zhang H, Lanza GM, Wickline SA (2008) Detection and quantification of angiogenesis in experimental valve disease with integrin-targeted nanoparticles and 19-fluorine MRI/MRS. J Cardiovasc Magn Reson 10:43
36. Porsch C, Zhang Y, Ostlund A et al (2013) In vitro evaluation of non-protein adsorbing breast cancer theranostics based on 19F-polymer containing nanoparticles. Part Part Syst Charact 30:381–390
37. Hricak H (2011) Oncologic imaging: a guiding hand of personalized cancer care. Radiology 259:633–640
38. Hielscher AH (2005) Optical tomographic imaging of small animals. Curr Opin Biotechnol 16:79–88
39. Ntziachristos V, Ripoll J, Wang LV, Weissleder R (2005) Looking and listening to light: the evolution of whole-body photonic imaging. Nat Biotechnol 23:313–320
40. Licha K, Olbrich C (2005) Optical imaging in drug discovery and diagnostic applications. Adv Drug Deliv Rev 57:1087–1108
41. Ntziachristos V, Tung CH, Bremer C, Weissleder R (2002) Fluorescence molecular tomography resolves protease activity in vivo. Nat Med 8:757–760
42. Prekeges J (2013) Nuclear medicine instrumentation, 2nd edn. Jones & Bartlett Learning, Burlington, MA
43. Brechbiel MW (2008) Bifunctional chelates for metal nuclides. Q J Nucl Med Mol Imaging 52:166–173
44. Khalil MM (2011) Basic sciences of nuclear medicine. Springer, Heidelberg
45. Seevers RH, Counsell RE (1982) Radioiodination techniques for small organic molecules. Chem Rev 82:575–590
46. Saha GB (2010) Fundamentals of nuclear pharmacy, 6th edn. Springer, New York
47. Srivastava SC (2012) Paving the way to personalized medicine: production of some promising theragnostic radionuclides at Brookhaven National Laboratory. Semin Nucl Med 42:151–163

48. Hijnen NM, de Vries A, Nicolay K, Grull H (2012) Dual-isotope 111In/177Lu SPECT imaging as a tool in molecular imaging tracer design. Contrast Media Mol Imaging 7:214–222
49. Alberini JL, Edeline V, Giraudet al et al (2011) Single photon emission tomography/computed tomography (SPET/CT) and positron emission tomography/computed tomography (PET/CT) to image cancer. J Surg Oncol 103:602–606
50. Liu Y, Welch MJ (2012) Nanoparticles labeled with positron emitting nuclides: advantages, methods, and applications. Bioconjug Chem 23:671–682
51. Yuan J, Zhang H, Kaur H, Oupicky D, Peng F (2012) Synthesis and characterization of theranostic poly(HPMA)-c(RGDyK)-DOTA-64Cu copolymer targeting tumor angiogenesis: tumor localization visualized by positron emission tomography. Mol Imaging 12:203–212
52. Fass L (2008) Imaging and cancer: a review. Mol Oncol 2:115–152
53. Prabhu P, Patravale V (2012) The upcoming field of theranostic nanomedicine: an overview. J Biomed Nanotechnol 8:859–882
54. Zhang XQ, Xu X, Bertrand N, Pridgen E, Swami A, Farokhzad OC (2012) Interactions of nanomaterials and biological systems: implications to personalized nanomedicine. Adv Drug Deliv Rev 64:1363–1384
55. Farokhzad OC, Langer R (2006) Nanomedicine: developing smarter therapeutic and diagnostic modalities. Adv Drug Deliv Rev 58:1456–1459
56. Zhang H (2012) Multifunctional nanomedicine platforms for cancer therapy. J Nanosci Nanotechnol 12:4012–4018
57. Liu Y, Miyoshi H, Nakamura M (2007) Nanomedicine for drug delivery and imaging: a promising avenue for cancer therapy and diagnosis using targeted functional nanoparticles. Int J Cancer 120:2527–2537
58. Lee PY, Wong KK (2011) Nanomedicine: a new frontier in cancer therapeutics. Curr Drug Deliv 8:245–253
59. Kopecek J, Kopeckova P, Minko T, Lu ZR, Peterson CM (2001) Water soluble polymers in tumor targeted delivery. J Control Release 74:147–158
60. Torchilin VP (2008) Multifunctional pharmaceutical nanocarriers. Springer, New York
61. Kopecek J, Kopeckova P (2010) HPMA copolymers: origins, early developments, present, and future. Adv Drug Deliv Rev 62:122–149
62. Luo K, Yang J, Kopečková P, Kopeček J (2011) Biodegradable multiblock poly[N-(2-hydroxypropyl)methacrylamide] via reversible addition-fragmentation chain transfer polymerization and click chemistry. Macromolecules 44:2481–2488
63. Pan H, Sima M, Miller SC, Kopečková P, Yang J, Kopeček J (2013) Efficiency of high molecular weight backbone degradable HPMA copolymer-prostaglandin E1 conjugate in promotion of bone formation in ovariectomized rats. Biomaterials 34:6528–6538
64. Pan H, Sima M, Yang J, Kopeček J (2013) Synthesis of long-circulating, backbone degradable HPMA copolymer-doxorubicin conjugates and evaluation of molecular-weight-dependent antitumor efficacy. Macromol Biosci 13:155–160
65. Rihova B, Kovar M (2010) Immunogenicity and immunomodulatory properties of HPMA-based polymers. Adv Drug Deliv Rev 62:184–191
66. Pasut G, Veronese FM (2009) PEG conjugates in clinical development or use as anticancer agents: an overview. Adv Drug Deliv Rev 61:1177–1188
67. Ulbrich K, Subr V (2010) Structural and chemical aspects of HPMA copolymers as drug carriers. Adv Drug Deliv Rev 62:150–166
68. Duncan R (2011) Polymer therapeutics as nanomedicines: new perspectives. Curr Opin Biotechnol 22:492–501
69. Webster R, Elliott V, Park BK, Walker D (2009) PEG and PEG conjugate toxicity: towards an understanding of toxicity of PEG and its relevance to pegylated biologicals. In: Veronese FM (ed) PEGylated protein drugs: basic science and clinical applications. Birkhauser, Basel, pp 127–146
70. Duncan R, Vicent MJ (2010) Do HPMA copolymer conjugates have a future as clinically useful nanomedicines? A critical overview of current status and future opportunities. Adv Drug Deliv Rev 62:272–282

71. Julyan PJ, Seymour LW, Ferry DR et al (1999) Preliminary clinical study of the distribution of HPMA copolymers bearing doxorubicin and galactosamine. J Control Release 57:281–290
72. Duncan R (2009) Development of HPMA copolymer-anticancer conjugates: clinical experience and lessons learnt. Adv Drug Deliv Rev 61:1131–1148
73. Gong J, Chen M, Zheng Y, Wang S, Wang Y (2012) Polymeric micelles drug delivery system in oncology. J Control Release 159:312–323
74. Li G, Liu J, Pang Y et al (2011) Polymeric micelles with water-insoluble drug as hydrophobic moiety for drug delivery. Biomacromolecules 12:2016–2026
75. Kedar U, Phutane P, Shidhaye S, Kadam V (2010) Advances in polymeric micelles for drug delivery and tumor targeting. Nanomedicine 6:714–729
76. Oerlemans C, Bult W, Bos M, Storm G, Nijsen JF, Hennink WE (2010) Polymeric micelles in anticancer therapy: targeting, imaging and triggered release. Pharm Res 27:2569–2589
77. Lee HJ, Ponta A, Bae Y (2010) Polymer nanoassemblies for cancer treatment and imaging. Ther Deliv 1:803–817
78. Liu Z, Zhang N (2012) pH-Sensitive polymeric micelles for programmable drug and gene delivery. Curr Pharm Des 18:3442–3451
79. Tsai HC, Chang WH, Lo CL et al (2010) Graft and diblock copolymer multifunctional micelles for cancer chemotherapy and imaging. Biomaterials 31:2293–2301
80. Decato S, Bemis T, Madsen E, Mecozzi S (2014) Synthesis and characterization of perfluoro-tert-butyl semifluorinated amphiphilic polymers and their potential application in hydrophobic drug delivery. Polym Chem 5:6461–6471
81. Fréchet JMJ, Tomalia DA (2001) Dendrimers and other dendritic polymers. Wiley, Chichester, NY
82. Tomalia DA, Baker H, Dewald J et al (1985) A new class of polymers: starburst-dendritic macromolecules. Polym J 17:117–132
83. Gillies ER, Frechet JM (2005) Dendrimers and dendritic polymers in drug delivery. Drug Discov Today 10:35–43
84. Zolnik BS, Sadrieh N (2009) Regulatory perspective on the importance of ADME assessment of nanoscale material containing drugs. Adv Drug Deliv Rev 61:422–427
85. Walter MV, Malkoch M (2012) Simplifying the synthesis of dendrimers: accelerated approaches. Chem Soc Rev 41:4593–4609
86. Yellepeddi VK, Kumar A, Palakurthi S (2009) Surface modified poly(amido)amine dendrimers as diverse nanomolecules for biomedical applications. Expert Opin Drug Deliv 6:835–850
87. Sadekar S, Ghandehari H (2012) Transepithelial transport and toxicity of PAMAM dendrimers: implications for oral drug delivery. Adv Drug Deliv Rev 64:571–588
88. Duncan R, Izzo L (2005) Dendrimer biocompatibility and toxicity. Adv Drug Deliv Rev 57:2215–2237
89. Caminade AM, Laurent R, Delavaux-Nicot B, Majoral JP (2012) "Janus" dendrimers: syntheses and properties. New J Chem 36:217–226
90. Ornelas C, Pennell R, Liebes LF, Weck M (2011) Construction of a well-defined multifunctional dendrimer for theranostics. Org Lett 13:976–979
91. Kumar P, Gulbake A, Jain SK (2012) Liposomes as vesicular nanocarriers: potential advancements in cancer chemotherapy. Crit Rev Ther Drug Carrier Syst 29:355–419
92. Al-Jamal WT, Kostarelos K (2011) Liposomes: from a clinically established drug delivery system to a nanoparticle platform for theranostic nanomedicine. Acc Chem Res 44:1094–1104
93. Barenholz Y (2012) Doxil(R)—the first FDA-approved nano-drug: lessons learned. J Control Release 160:117–134
94. Meyerhoff A (1999) U.S. Food and Drug Administration approval of Am Bisome (liposomal amphotericin B) for treatment of visceral leishmaniasis. Clin Infect Dis 28:42–48, discussion 49–51
95. Sawant RR, Torchilin VP (2012) Challenges in development of targeted liposomal therapeutics. AAPS J 14:303–315
96. FDA (2002) Guidance for industry: liposome drug products

97. Desai N (2012) Challenges in development of nanoparticle-based therapeutics. AAPS J 14:282–295
98. Immordino ML, Dosio F, Cattel L (2006) Stealth liposomes: review of the basic science, rationale, and clinical applications, existing and potential. Int J Nanomedicine 1:297–315
99. Ranjan A, Jacobs GC, Woods DL et al (2012) Image-guided drug delivery with magnetic resonance guided high intensity focused ultrasound and temperature sensitive liposomes in a rabbit Vx2 tumor model. J Control Release 158:487–494
100. Oldham RK, Dillman RO (2008) Monoclonal antibodies in cancer therapy: 25 years of progress. J Clin Oncol 26:1774–1777
101. Barbet J, Bardies M, Bourgeois M et al (2012) Radiolabeled antibodies for cancer imaging and therapy. Methods Mol Biol 907:681–697
102. Mirick GR, Bradt BM, Denardo SJ, Denardo GL (2004) A review of human anti-globulin antibody (HAGA, HAMA, HACA, HAHA) responses to monoclonal antibodies. Not four letter words. Q J Nucl Med Mol Imaging 48:251–257
103. Jeong H, Huh M, Lee SJ et al (2011) Photosensitizer-conjugated human serum albumin nanoparticles for effective photodynamic therapy. Theranostics 1:230–239
104. Kennedy LC, Bickford LR, Lewinski NA et al (2011) A new era for cancer treatment: gold-nanoparticle-mediated thermal therapies. Small 7:169–183
105. Fernandez-Fernandez A, Manchanda R, McGoron AJ (2011) Theranostic applications of nanomaterials in cancer: drug delivery, image-guided therapy, and multifunctional platforms. Appl Biochem Biotechnol 165:1628–1651
106. Huang HC, Barua S, Sharma G, Dey SK, Rege K (2011) Inorganic nanoparticles for cancer imaging and therapy. J Control Release 155:344–357
107. Espinosa E, Zamora P, Feliu J, Gonzalez Baron M (2003) Classification of anticancer drugs—a new system based on therapeutic targets. Cancer Treat Rev 29:515–523
108. Ulbrich K, Zacharieva EI, Obereigner B, Kopecek J (1980) Polymers containing enzymatically degradable bonds: V. Hydrophilic polymers degradable by papain. Biomaterials 1:199–204
109. Ray A, Larson N, Pike DB et al (2011) Comparison of active and passive targeting of docetaxel for prostate cancer therapy by HPMA copolymer-RGDfK conjugates. Mol Pharm 8:1090–1099
110. Lammers T, Subr V, Ulbrich K et al (2009) Simultaneous delivery of doxorubicin and gemcitabine to tumors in vivo using prototypic polymeric drug carriers. Biomaterials 30:3466–3475
111. Kramer-Marek G, Capala J (2012) The role of nuclear medicine in modern therapy of cancer. Tumour Biol 33:629–640
112. Griggs WS, Divgi C (2008) Radioiodine imaging and treatment in thyroid disorders. Neuroimaging Clin N Am 18:505–515, viii
113. Zhang L, Chen H, Wang L et al (2010) Delivery of therapeutic radioisotopes using nanoparticle platforms: potential benefit in systemic radiation therapy. Nanotechnol Sci Appl 3:159–170
114. Wang AZ, Yuet K, Zhang L et al (2010) ChemoRad nanoparticles: a novel multifunctional nanoparticle platform for targeted delivery of concurrent chemoradiation. Nanomedicine (Lond) 5:361–368
115. Whitehead KA, Langer R, Anderson DG (2009) Knocking down barriers: advances in siRNA delivery. Nat Rev Drug Discov 8:129–138
116. Lee H, Kim IK, Park TG (2010) Intracellular trafficking and unpacking of siRNA/quantum dot-PEI complexes modified with and without cell penetrating peptide: confocal and flow cytometric FRET analysis. Bioconjug Chem 21:289–295
117. Medarova Z, Pham W, Farrar C, Petkova V, Moore A (2007) In vivo imaging of siRNA delivery and silencing in tumors. Nat Med 13:372–377
118. Lee YC, Byfield JE (1976) Induction of DNA degradation in vivo by adriamycin. J Natl Cancer Inst 57:221–224
119. Choppin GR, Liljenzin J-O, Rydberg J, Ekberg C (2013) Radiochemistry and nuclear chemistry, 4th edn. Elsevier/Academic Press, Amsterdam/Boston
120. Garland MJ, Cassidy CM, Woolfson D, Donnelly RF (2009) Designing photosensitizers for photodynamic therapy: strategies, challenges and promising developments. Future Med Chem 1:667–691

121. Shirasu N, Nam SO, Kuroki M (2013) Tumor-targeted photodynamic therapy. Anticancer Res 33:2823–2831
122. Allison RR, Downie GH, Cuenca R, Hu XH, Childs CJ, Sibata CH (2004) Photosensitizers in clinical PDT. Photodiagnosis Photodyn Ther 1:27–42
123. Vergnon JM, Huber RM, Moghissi K (2006) Place of cryotherapy, brachytherapy and photodynamic therapy in therapeutic bronchoscopy of lung cancers. Eur Respir J 28:200–218
124. Bown SG, Rogowska AZ, Whitelaw DE et al (2002) Photodynamic therapy for cancer of the pancreas. Gut 50:549–557
125. Pandey SK, Gryshuk AL, Sajjad M et al (2005) Multimodality agents for tumor imaging (PET, fluorescence) and photodynamic therapy. A possible "see and treat" approach. J Med Chem 48:6286–6295
126. Capella MA, Capella LS (2003) A light in multidrug resistance: photodynamic treatment of multidrug-resistant tumors. J Biomed Sci 10:361–366
127. Oh IH, Min HS, Li L et al (2013) Cancer cell-specific photoactivity of pheophorbide a-glycol chitosan nanoparticles for photodynamic therapy in tumor-bearing mice. Biomaterials 34:6454–6463
128. Rong P, Yang K, Srivastan A et al (2014) Photosensitizer loaded nano-graphene for multimodality imaging guided tumor photodynamic therapy. Theranostics 4:229–239
129. Vaidya A, Sun Y, Ke T, Jeong EK, Lu ZR (2006) Contrast enhanced MRI-guided photodynamic therapy for site-specific cancer treatment. Magn Reson Med 56:761–767
130. Vaidya A, Sun Y, Feng Y, Emerson L, Jeong EK, Lu ZR (2008) Contrast-enhanced MRI-guided photodynamic cancer therapy with a pegylated bifunctional polymer conjugate. Pharm Res 25:2002–2011
131. Xu J, Zeng F, Wu H, Hu C, Wu S (2014) Enhanced photodynamic efficiency achieved via a dual-targeted strategy based on photosensitizer/micelle structure. Biomacromolecules 15:4249–4259
132. Yoon HY, Koo H, Choi KY et al (2012) Tumor-targeting hyaluronic acid nanoparticles for photodynamic imaging and therapy. Biomaterials 33:3980–3989
133. Marchosky JA, Welsh DM, Moran CJ (1990) Hyperthermia treatment of brain tumors. Mo Med 87:29–33
134. Matsumine A, Takegami K, Asanuma K et al (2011) A novel hyperthermia treatment for bone metastases using magnetic materials. Int J Clin Oncol 16:101–108
135. Zablow A, Shecterle LM, Dorian R et al (1997) Extracorporeal whole body hyperthermia treatment of HIV patients, a feasibility study. Int J Hyperthermia 13:577–586
136. Link S, El-Sayed MA (2000) Shape and size dependence of radiative, non-radiative and photothermal properties of gold nanocrystals. Int Rev Phys Chem 19:409–453
137. Gormley AJ, Greish K, Ray A, Robinson R, Gustafson JA, Ghandehari H (2011) Gold nanorod mediated plasmonic photothermal therapy: a tool to enhance macromolecular delivery. Int J Pharm 415:315–318
138. Lynn JG, Zwemer RL, Chick AJ (1942) The biological application of focused ultrasonic waves. Science 96:119–120
139. Grull H, Langereis S (2012) Hyperthermia-triggered drug delivery from temperature-sensitive liposomes using MRI-guided high intensity focused ultrasound. J Control Release 161:317–327
140. Yudina A, de Smet M, Lepetit-Coiffe M et al (2011) Ultrasound-mediated intracellular drug delivery using microbubbles and temperature-sensitive liposomes. J Control Release 155:442–448
141. de Smet M, Heijman E, Langereis S, Hijnen NM, Grull H (2011) Magnetic resonance imaging of high intensity focused ultrasound mediated drug delivery from temperature-sensitive liposomes: an in vivo proof-of-concept study. J Control Release 150:102–110
142. van Elk M, Deckers R, Oerlemans C et al (2014) Triggered release of doxorubicin from temperature-sensitive poly(N-(2-hydroxypropyl)-methacrylamide mono/dilactate) grafted liposomes. Biomacromolecules 15:1002–1009
143. Rapoport N, Nam KH, Gupta R et al (2011) Ultrasound-mediated tumor imaging and nanotherapy using drug loaded, block copolymer stabilized perfluorocarbon nanoemulsions. J Control Release 153:4–15

144. Zhou W, Meng F, Engbers GH, Feijen J (2006) Biodegradable polymersomes for targeted ultrasound imaging. J Control Release 116:e62–e64
145. Chakravarty R, Valdovinos HF, Chen F et al (2014) Intrinsically germanium-69-labeled iron oxide nanoparticles: synthesis and in-vivo dual-modality PET/MR imaging. Adv Mater 26:5119–5123
146. Choi HS, Frangioni JV (2010) Nanoparticles for biomedical imaging: fundamentals of clinical translation. Mol Imaging 9:291–310

Chapter 7
Targeting Cancer Using Nanocarriers

Dalit Landesman-Milo, Shahd Qassem, and Dan Peer

Abstract Nanotechnology is an emerging multidisciplinary field that offers unprecedented access to living cells of target (i.e. cancer cells) and promises the state of the art in cancer detection and treatment. Development of nanocarriers that target cancer for diagnostics and therapy draws upon principles in the field of chemistry, medicine, physics, biology, and engineering. Given the zealous activity in the field as demonstrated by over 7000 published journal articles on the topic and given the promise of recent clinical results, nanocarrier-based approaches are anticipated to soon have a profound impact on cancer medicine and as a consecuence on human health. The versatility in size, material, and targeting agents of nanocarriers permits potential targeting for individual cancer cells. This chapter addresses nanocarriers spanning liposomes to polymeric nanoparticles, inorganic nanoparticles,polymers conjugates and dendrimers. The targeting approaches include conjugation of molecules such as receptor-specific ligands, antibodies and aptamers to the surface of the carrier. Targeting cancer with nanocarriers represent the next important milestone that is already impacts the lives of millions around the world. One such example, is of DaunoXome (Liposomal Daunorubicin) for Acute myeloid Leukemia (AML) treatment that shows an increased intratumor and intracellular levels of the drug, while normal tissue toxicity, including cardiotoxicity, may be reduced.

Keywords RNAi • Nanoparticles • Targeting agents • mAbs

D. Landesman-Milo • S. Qassem
Laboratory of NanoMedicine, Department of Cell Research and Immunology, George S. Wise Faculty of Life Sciences, Tel Aviv University, Tel Aviv 69978, Israel

Department of Materials Sciences and Engineering, The Iby and Aladar Fleischman Faculty of Engineering, Tel Aviv University, Tel Aviv 69978, Israel

Center for Nanoscience and Nanotechnology, Tel Aviv University, Tel Aviv 69978, Israel

D. Peer (✉)
Department of Cell Research & Immunology and Department of Material Science and Engineering, Laboratory of Precision NanoMedicine, Tel Aviv University, Tel Aviv Israel
e-mail: peer@tauex.tau.ac.il

K.A. Howard et al. (eds.), *Nanomedicine*, Advances in Delivery Science and Technology, DOI 10.1007/978-1-4939-3634-2_7

7.1 Introduction

Current therapy for cancer involves the cyclical administration of one or more chemotherapeutic drugs, with or without, surgical intervention and radiation. The effectiveness of the treatment is directly related to the ability to selectively kill the cancer cells without damaging healthy cells. By administering bolus doses of cytotoxic drugs, adverse effects are commonly observed and patients may only experience a marginal improvement on long-term survival. Therefore, more effective cancer therapies are needed both in terms of new classes of drugs and in terms of developing "smart delivery vehicles" to minimize the adverse effects and to enhance the survival of patients with metastatic cancer.

One approach to target the delivery of drugs specifically to cancer cells is through directing cellular events at the nanometer scale [1–6]. Where current technologies,mainly based on anatomical and physiological alteration require hundreds of thousands of cells to detect the presence of a tumor, while nanotechnology approaches utilize metabolic and molecular unique features of cancer cells (e.g. Glucose fluorescent conjugates) that could radically lower this requirement, enabling much earlier diagnosis/treatment regimens. Through working on the nano scale it becomes possible to differentiate between healthy and cancerous cells, thus, significantly offering a wide therapeutic index with reduced side effects. One major clinical advantage of nanocarrier-based strategies over drugs that are directly linked to a targeting moiety is the specific delivery of large amounts of chemotherapeutic agents per recognition event.

Typically, nano-carrier based approaches include a carrier, a targeting moiety that is bound to the carrier via specific conjugation chemistry, and a drug (Fig. 7.1). Carriers may be composed of liposomes, polymeric nanoparticles, inorganic nanoparticles, polymers or dendrimers. Targeting moieties may include high affinity ligands, peptides, antibodies, and nucleic acids, and they may be conjugated to the carriers utilizing a variety of chemistries. Given the wide array of potential nanocarrier-based strategies for targeting cancer, this review will focus on drug delivery strategies for supplying a lethal chemotherapeutic dose to target cells and nucleic acid drug delivery system with a focus on siRNA delivery platforms.

7.2 Targeting

There are generally two approaches for nanocarrier-mediated systemic drug targeting from the site of administration (i.e. injection):

1. Passive targeting to tumor tissues—via enhanced permeability of tumor vasculature.
2. Active targeting at the cellular level—to cancer cells without affecting normal cells.

Passive targeting to the tumor tissues relies on 'leaky' microvascular or the enhanced permeation and retention (EPR) effect [7, 8], which permits selective

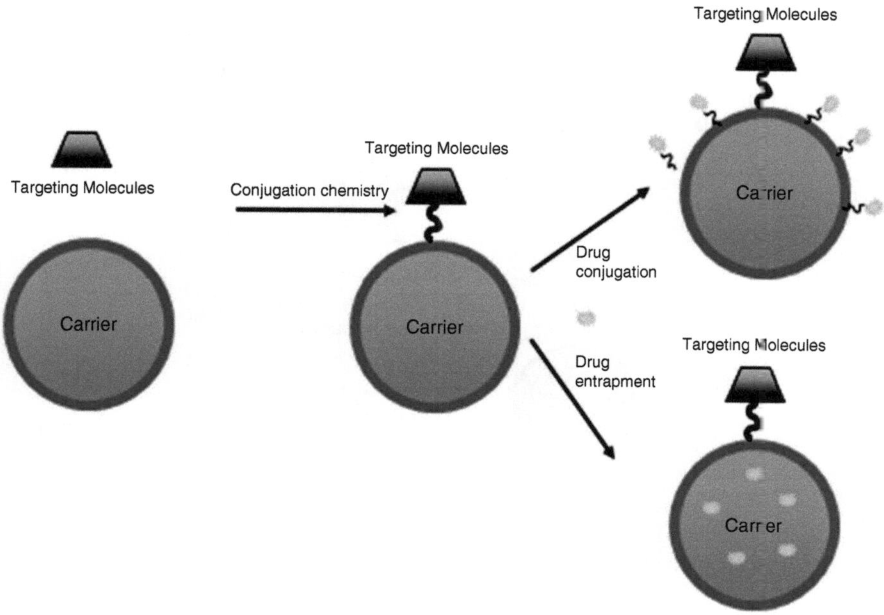

Fig. 7.1 Nanocarrier components for targeting cancer. Construction of the vehicle consists of a nano-scale carrier, a targeting moiety conjugate to the nanocarrier and the cargo (the desired drug)

permeation of nanoparticles into the desired tumor tissue. Active targeting to cancer cells is possible through promoting specific interactions with targeting ligands and surface receptors. An engineered nanoparticle should be able to overcome the physiological challenges of systemic circulation including tumor vascular biology and organs of mononuclear phagocytic system (MPS) whose function is to remove foreign material from the circulation [9].

Once at the target site, nanoparticles may release the drug in close proximity to cancer cells (though passive targeting), attached to the cancer cells, or within the cell following internalization (Fig. 7.2). Cell surface targets (often receptors), may recognize a variety of targeting agents such as ligands, antibodies, and nucleic acids.

7.3 Ligands

Targeting agents that may be broadly classified as ligands include proteins, peptides, vitamins and carbohydrates. One of the commonly used ligands for cancer targeting is the high affinity vitamin folic acid (folate) since folate receptors (FRs) are frequently overexpressed in a range of tumor cells [10, 11]. Folate is a small molecule (MW 441.40 Da) that strongly binds to FRs. Folate is water soluble and can be easily conjugated using its gamma-carboxyl group and, thus, it has been used as a targeting moiety in a wide array of drug delivery vehicles including liposomes, polymeric nanoparticles, linear polymers, and dendrimers to deliver drugs into cancer cells

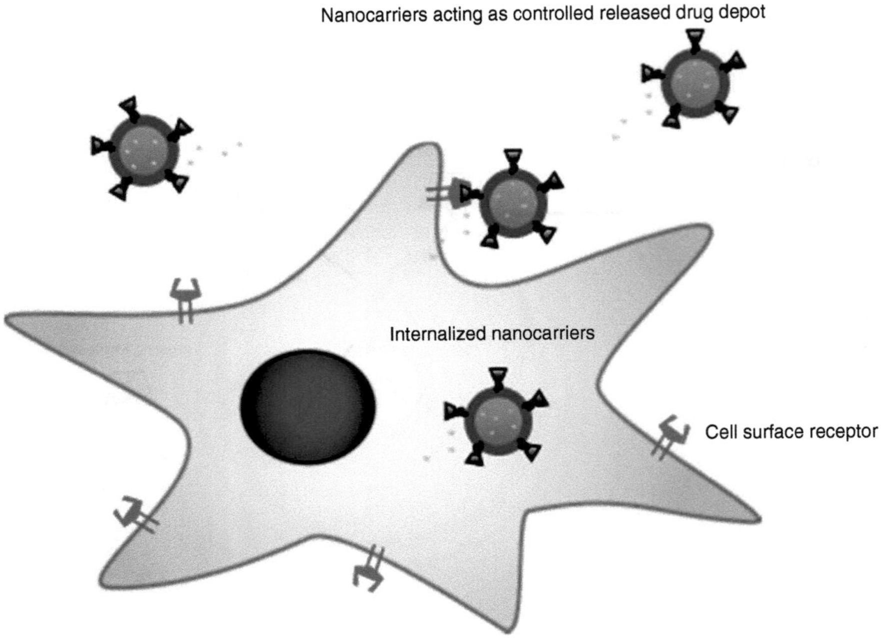

Fig. 7.2 Targeted nanocarriers-cell interactions. The Nanocarrier can release its content in close proximity of the target cells (passive targeting); attach to the membrane of the cell and act as an extracellular sustained release drug depot; or internalize into the cells

using FR-mediated endocytosis [11–16]. Another ligand with selectivity for cancer cells is transferrin (Tf) protein [17]. Tf receptors (TfR) are overexpressed on many tumor cells and coupling Tf directly to the liposomes, and other nanoparticles entrapping chemotherapies has resulted in improved intracellular delivery and therapeutic outcomes in cancer models [18–22]. Recently, Tf has also been used to facilitate siRNA delivery through transferrin receptors allowing a specific killing of tumor cells [23] and for effective brain tumors targeting [24]. In addition to cell surface antigens, extracellular matrices (ECM) such as heparin sulfate, chondroitin sulfate, and hyaluronan (also called Hyaluronic Acid, or HA), are usually responsible for the tumor's cells communication and loss of homeostatic control, and, thus, play an important role in the targeted drug delivery niche. HA is a naturally-occuring glycosaminoglycan that binds to CD44 and CD168 receptors. Both receptors are highly expressed on the surface of cancer cells. The potential of using ECM as targeting agents was demonstrated through intravenous injections of Paclitaxel loaded HA nanoparticles into tumor bearing mice, which reduced tumor size [25–28].

This effect is likely a result of competition of HA with its receptors on tumor cells. Although HA is ubiquitously expressed in normal tissues, normal cells bind it with low affinity, and seem to require activation to bind HA more effeciently [28–32]. This may help limit non-specific interactions thereby increasing the effectiveness of approaches that aim to utilize ECM components as targeting agents. Over the last 10 years, Chitin and Chitosan, amino sugar N-acetyl-glucosamines and other sugar residuse (e.g. mannose), biocompatable, biodegradable, low cost polysacharides and their derivatives have drawn attention as vehicles and targeting moieties for drug delivery [33]. Kawakami et al. have previously demonstrated an efficient mannose receptor-mediated gene transfer into macrophages [34]. In addition to proteins, peptides such as Arginine–Glycine–Aspartic acid (RGD) that target integrin $\alpha_v\beta_3$ on tumor cells' surface show an increased efficacy against different murine tumor models [35, 36]. Although RGD based peptide approaches are commonly employed, this peptide also binds to other integrins such as $\alpha_5\beta_1$ and $\alpha_4\beta_1$ and, thus, is not highly specific to cancer cells only. Alternatively, vasoactive intestinal peptide (VIP), which only binds to VIP receptors has been used to deliver paclitaxel to VIP receptors on the surface of breast cancer cells that resulted in inhibition of tumors in vivo [37–39].

Another example for targeting agent is TH10 peptide, which was isolated from a phage display library that targets neural/glial antigen 2 (NG2), a proteoglycan highly expressed in tumor-derived vascular pericytes lining the inside of blood vessels [40].

Interfering with pro-cancer and metastasis cellular triggering activities such as cell proliferation, chemotaxis and gene transcription by blocking a critical chemotactic axis interaction in tumor/metastasis progression is another targeting option. Blocking the chemokine receptor 4 (CXCR4) and its ligand (CXCL12) CXCR4/CXCL12 is an exapmle for this attitude [41].

7.4 Antibodies

One of the earliest cancer targeting is using a monoclonal antibody which was first described in 1981 by the Milstein group [42]. Over the past three decades, the feasibility of antibody based tissue targeting has been clinically demonstrated (reviewed in [43, 44]). Monoclonal antibodies have been used as biological therapeutics and as targeters. The first FDA approval for therapeutic monoclonal antibodies for the treatment of cancer came in 1997 where rituximab (Rituxan™) was approved for treating patients with a form of non-Hodgkin's lymphoma [45]. A year later, in 1998 Trastuzumab (Herceptin™), a humanized anti-HER2 (against EGFR) monoclonal antibody, has been approved for the treatment of breast cancer [46]. Over the past couple of decades more than two dozen of different monocolonal antibodies have already approved by the FDA to treat certain cancers [47]. Among them, the first angiogenesis inhibitor to treat colorectal cancer, Bevacizumab (Avastin™) an anti-VEGF mAb, was approved in 2004 [48–50]. Emerging approaches include delivery of a chemoimmunoconjugate containing anti-CD33 antibody and calicheamicin which is in clinical Phase III development for the treatment of acute myelogenous leukemia [51]. Recently, a targeting angiogenic agent was approved by the FDA—ramucirumab (Cyramza, Eli Lilly and Company), a mAb that binds to vascular endothelial growth factor receptor-2 for the treatment of advanced gastric cancer [52].

Despite the recent success of monoclonal antibodies as targeting moieties, this class of molecule continues to exhibit important suboptimal properties. Foremost, the biological production of monoclonal antibodies can be difficult and unpredictable. In addition, the performance of antibodies can vary from batch to batch, in particular when production is scaled up. However, despite these disadvantages, antibodies represent the most common and potentially effective strategy for targeting cancer. A relatively new targeting technology that addresses many of the shortcomings of antibody based approaches involves use of high affinity oligonucleotides called aptamers.

7.5 Aptamers

Aptamers are short single stranded DNA or RNA oligonucleotides that have been selected in vitro from a large number of random sequences (~10^{14}–10^{15}) and have a molecular weight (10–15 kDa), which is one order of magnitude lower than that of antibodies (~150 kDa) [53]. Aptamers are derived from an in vitro iterative protocol called systematic evolution of ligands by exponential enrichment (SELEX) (Fig. 7.3) that was first described independently by Tuerk and Gold [54] and by Ellington and Szostak [55] in 1990.

Since their original description, aptamers have been isolated to a wide variety of targets, including intracellular proteins, transmembrane proteins, soluble proteins,

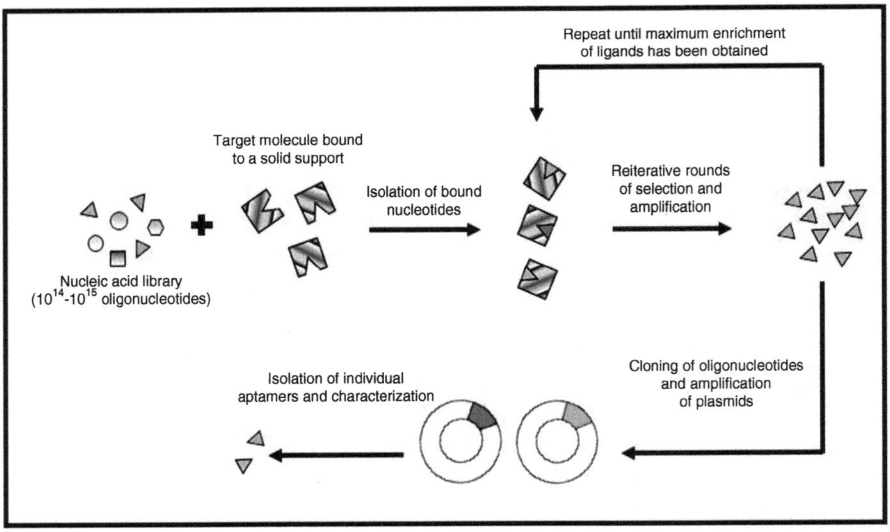

Fig. 7.3 Outline of the SELEX process. Target molecules are initially incubated with a combinatorial library of nucleic acids. Bound nucleotides are isolated followed by reiterative rounds of selection and amplification. This process is repeated with increased stringency with each round. Next, oligonucleotides are cloned and plasmids are amplified. Finally, through this enrichment process, a small number of tightly binding aptamers are obtained

carbohydrates, and small molecule drugs. Unlike antibodies, aptamers can be chemically synthesized at microgram or kilogram scales, and due to their small size and similarity to endogenous molecules, are believed to be non-immunogenic. It is possible to incorporate selection by intact biological entities such as cancer cells or tissues to identify an array of highly specific aptamers [56, 57]. Unlike antisense small interfering RNA compounds, which are single-stranded nucleic acids that affect the synthesis of a targeted protein by hybridizing to the mRNAs that encode it, aptamers may inhibit a protein's function through directly binding to it.

Due to their composition (nucleotides) and relative small size, aptamers may be degraded by nucleases or rapidly cleared from the blood. Unlike antibodies, aptamers require sheltering from degradation, which can be achieved through a variety of techniques [58–61]. A number of aptamers have been developed to bind specifically to receptors on cancer cells. For example, as guided delivery of nanoparticles for treatment of over expressing EGF receptor tumor cells [62] or as a selective delivery agent of chemotherapy based liposomes to prostate cancer cells [3, 63].

To date, there is one FDA approved aptamer based therapeutics called Macugen (Pegatanib) in treatment for age-related macular degeneration (AMD) and many more aptamers under clinical trials, one antitumor DNA based aptamer targeting nucleolin which is drawing attention accoridng to its clinical trials results is the AS1411 [64].

7.6 Incorporation of Drugs into Nanocarriers

Traditional drug administration such as intravenous injection or multiple daily oral dosing has proven to be suboptimal or unsuitable for a variety of medical conditions. Aside from requiring high doses to reach target tissues, these types of therapeutics typically produce peaks in drug concentration above the therapeutic range followed by a rapid decrease in concentration to a level below the therapeutic range. During the past five decades delivery strategies have evolved and methods are now available for achieving a constant or pulsatile release, within the therapeutic range, over long periods of time [65–69].

Chemotherapeutic anti-cancer drugs and their carrier may be joined via two major techniques. One way is to entrap drug molecules into delivery vehicles for controlled release of the drugs. Controlled release occurs when a natural or synthetic polymer is combined with a drug in such a way that the drug is encapsulated within the polymer system for subsequent release in a predetermined manner. Drug delivery vehicles that are designed as particles can range in size from <10 nm (dendrimers) and 100 nm (nanoparticles) to over 10 μm (microparticles) and can release the encapsulated drugs via surface or bulk erosion, diffusion, swelling followed by diffusion, or response to a stimulus (e.g. pH, temperature, ultrasound, electrical stimuli). The release of the active agent may be constant or cyclic over a long period, or may be triggered by the environment or other external events [70]. In general, controlled-release polymer systems can provide drug levels in the optimum range for an extended period of time than other drug delivery methods, thus, increasing the efficacy of the drug and maximizing patient compliance. The primary consideration of drug delivery is to achieve more effective therapies while eliminating the potential for both under- and overdosing. Other advantages of using controlled release delivery systems can include the maintenance of drug levels within a desired range, the need for fewer administrations, higher diffusion rates of the drug into the tumor due to a continuous-controlled release of it in the tumor proximity and optimal use of the drug in question. While these advantages can be significant, the potential toxicity of the materials used in these delivery systems may limit their translation into clinical practice in some cases [71].

Another approach is to conjugate drugs covalently to the bulk of drug carriers. carrier-drug conjugation usually inhibits metabolic activity until the conjugation bond is cleaved, resulting in activation of the drug at the target site [72]. The covalent conjugation between carriers and drugs has a number of advantages including: (1) Limiting the mechanism of drug release from the carriers and (2) reducing potential burst of drug as it is typically seen in entrapment approaches. A further detailed review on the polymer-drug conjugates and examples thereof can be found elsewhere [73].

7.7 Choosing a Carrier

In choosing an appropriate nanocarrier construct, one should consider the following criteria:

1. A carrier should be made from a material that is biocompatible, easily functionalizable, and well characterized.
2. A nanocarrier with targeting molecules should exhibit high uptake efficiency by the target cells unless they are required to accumulate on the cell surface and act as site specific drug depots.
3. The nanocarriers should be either soluble or colloidal in aqueous phase and exhibit an extended circulating half-life to increase the likelihood of their effectiveness.
4. The nanocarriers should have a low rate of aggregation and preferably a long shelf life.

A variety of potential nanocarrier compositions are available including liposomes, polymeric nanoparticles, inorganic nanoparticles, and dendrimers and these are reviewed below (Fig. 7.4).

7.8 Liposomes

Liposomes are the first nanocarriers described more than 50 years ago by Bangham [74–76], and the first that have been clinically approved by the FDA to carry chemotherapy (DaunoXome™ (50–80 nm) in 1996 [77], Doxil™ (100 nm) in 1997,

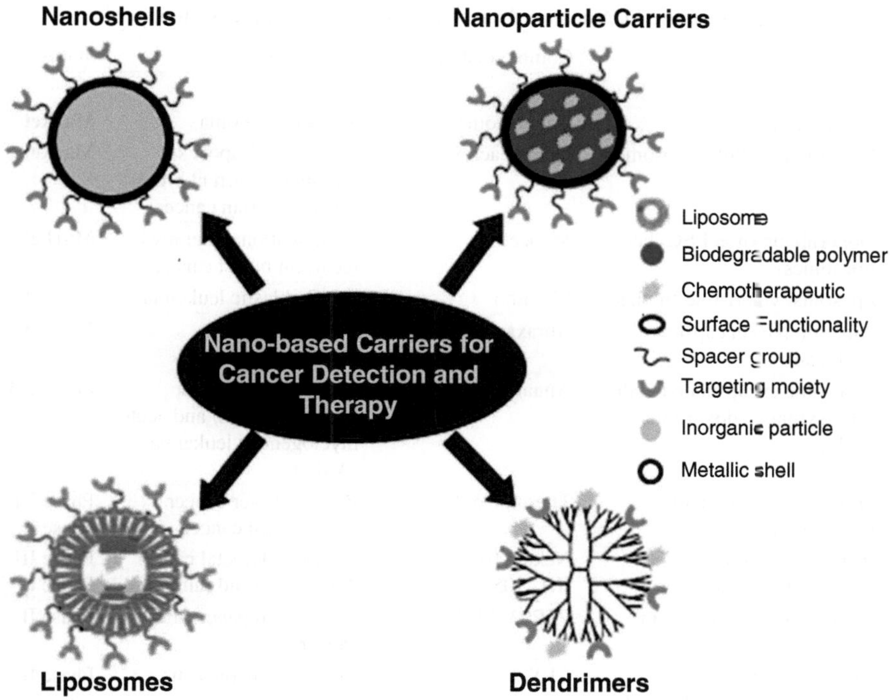

Fig. 7.4 The arsenal of nano-based carriers of targeting cancer. Liposomes (50–800 nm), polymeric nanoparticles (5–150 nm), dendrimers (2–10 nm) and nanoshells (5–100 nm)

Myocet™ (190 nm) in 1997 [78]) [79–81]. Liposomes are spherical, self-closed structures formed by one or several concentric lipid bilayers with an aqueous phase inside and between the lipid bilayers. The most common way to prepare liposomes is the lipid film method [82] where the lipids are dissolved in organic solvents and evaporated to dryness under reduced pressure. Then they are hydrated by a swelling solution with the desired drug to form a multilamellar vesicle (MLV). An extruder is used to downsize the MLV to a nano-scale, unilamellar vesicle (ULV) or to small ULV (SUV) having progressively smaller pore-size membranes, with several cycles per pore-size and the result is a uniform nano-scale population of liposomes with a narrow size distribution. There are today more than 20 different liposomal formulations in different stages of clinical trials [83] (Table 7.1). Among the active chemotherapeutic drugs entrapped in the liposomes are: daunorubicin, doxorubicin, cytarabine, vincristine, lurtotecan, platinum based compounds and more [84, 85]. Liposomes have many attractive biological properties including: biocompatibility, biodegradability, the ability to entrap both hydrophilic drugs (through their interwater compartment) and hydrophobic drugs (through the membrane), and to protect the drugs from the biological environment. Their size, charge and surface could be easily modified by adding more agents into the lipid membrane or by surface chemistry modifications.

To achieve improved selectivity, a targeting moiety is attached to the liposomal surface via a PEG spacer arm to reduce steric hindrance of binding to the target [86].

Table 7.1 Representative FDA approved nanomedicines or in clinical evaluation

Compound	Commercial name™	Indications	Status
Liposomes			
Daunorubicin	DaunoXome	Kaposi's sarcoma	**Market**
Doxorubicin (PEG-liposomes)	Doxil/Caelyx	Refractory Kaposi's sarcoma; recurrent breast cancer; ovarian cancer	**Market**
Doxorubicin (non PEGylated liposomes)	Myocet	Combinational therapy of recurrent breast cancer	**Market**
liposomal vincristine sulphate	Marqibo	lymphoblastic leukemia	**Market**
Albumin NPs encapsulate paclitaxel	Abraxane®	Breast cancer NSCLC	**Market**
Liposome encapsulated with semi-synthetic doxorubicin analogue	Annamycin	Acute lymphoblastic leukemia (ALL) and acute myelogenous leukemia (AML)	Phase I/II
Doxorubicin (thermally sensitive liposomes)	ThermoDox®	Primary Liver cancer RCW breast cancer	Phase III Phase II
Pegylated lipsomes encapsulated Irinotecan	Nektar-102 IHL-305	Breast, colorectal cancer Advanced solid tumors	Phase III Phase I
Nanoliposomal irinotecan	PEP02 (MM-398)	Gastric and pancreatic cancer	Phase II
Liposome entrapped irinotecan and HCl:floxuridine	CPX-1	Colorectal neoplasms	Phase II
Liposome enrapped paclitaxel	LEP-ETU	Ovarian, breast and lung cancers	Phase I
Liposome entrapped with cysplatin	Aroplatin	Colorectal cancer	PhaseI/II
Vinorelbine liposomes	Vinorelbine	Advanced solid tumors Hodgkins disease Non-Hodgkins lymphoma	Phase I
Liposome entapped with cisplatin	Lipoplatin	Pacreatic, head and neck, breast cancer	Phase III
Stealth liposomal cisplatin	STEALTH®	Head and neck, lung cancers	Phase III
Targeted liposomal cispaltin formulation (based on tumor triggered release mechanism)	LiPlaCis	Advanced or refractory solid tumors	Phase I
Vincristine	Onco TCS	Relapsed aggressive non-Hodgkin's lymphoma (NHL)	**Market**
Liposome entrapped cytarabine and daunorubicin	CPX-351	AML	Phase II
Mitoxantrone		Prostate cancer	Phase I/II
Liposomal SN-38 (DNA topoisomerase I inhibitor)	NK102	Colorectal cancer	Phase II
Liposomal Lurtotecan (Liposome entapped topoisomerase I inhibitor)	OSI-211	Ovarian cancer	Phase II/ III

(continued)

Table 7.1 (continued)

Compound	Commercial name™	Indications	Status
Cytarabine liposome	DepoCyt	Esophageal cancer Lymphomatous meningitis	**Market**
Plasmid DNA encoding HLA-B7	Allovectin-7®	Metastatic melanoma	Phase III
Polymeric micelles			
Cisplatin-incorporating polymeric micelles	NC-6004	Pancreatic cancer NSCLC	Phase III PhaseI/II
Paclitaxal encapsulated micelle	NK105	Breast cancer	Phase III
Liposome encapsulated with (S)-camptothecin (aerosolized)	9-Nitrocamptothecin liposome-L9NC	Lung, metastatic endometrial cancer	Phase II
Peptides/aptamers			
Peptide conjugate based on PLGA-LHRH	Lupron® (leuprolide)	Prostate cancer	**Market**
LHRH synthetic decapeptide	Goserelin (Zoladex)	Breast cancer	**Market**
Peg-L-Asparaginase	Oncaspar	Acute lymphoblastic leukemia	**Market**
Pegylated recombinant human granulocyte colony-stimulating factor (rhG-CSF)	HHPG-19K	Breast cancer	Phase II
Paclitaxel poly-L-glutamic acid conjugate (PG-TXL)	CT-2103 (XYOTAX)	Metastatic breast cancer Ovarian cancer	Phase I Phase III
Doxorubicin-N-(2-hydroxypropyl) methacrylamide HPMA copolymer	HPMA copolymer	Breast cancer	Phase II

Some examples of selective targeting include use of anti-CD19 long-circulating liposomes that deliver doxorubicin to B cell lymphoma [87] and anti-Her2-liposomes that showed effective targeting to breast cancer cells overexpressing erbB2 receptor [88, 89]. Folate-targeted-liposomes have been proposed also as vehicles for chemothrapy to improve theraputic activity [90].

Transferrin (Tf)-mediated liposome delivery is also an emerging strategy to target tumors overexpressing the transferring receptors. Some examples include Tf-coated liposomes entrapping doxorubicin and paclitaxel targeted to U-87brain glioma cells [24, 91] and to solid tumors [92].

Other approaches include the combination of fusogenic peptides, targeting moieties and pH sensitive lipids that will cause the liposome disruption in the endosome [93–98].

One of the potential challenges with drug containing liposomes is the inability to effectively reconstitute them after lyophilization and to maintain their dimensions. Successful attempts to address these challenges involve covalent immobilization of high molecular weight HA on unilamellar liposomes. This serves to cryoprotect against lyophilization [32, 99]. In addition, the HA coatings improves circulation time and enhances targeting to tumors expressing HA receptors (CD44 and RHAMM) [4, 100, 101]. We developed antibody coated HA-liposomes (Integrin beta 7 targeted) that can be lyophilized and stored as a dry powder which preserves the nano dimensions and the binding capacity of the immobilized mAb on its surface [102].

Chitosan based NPs hold several intrinstic adventages of biocompatability, biodegradablty and a prepared infrastructure for chemical reaction, made them a suitable vehicle for encapsulatiing a varitey of anti cancer drugs and surface readily modified for chemical cross link conjugation of different targeted moities [103], such as folic acid [104], poly lactic-co-glycolic acid (PLGA) [105], cell-targeting peptides (for example CP15) [106] or provascular agent bradykinin potentiating peptide (BPP) [107]. The main adnantage of these targeted NPs is longer circulation time, that enables significantly prolonged survival rates of tumor-bearing mice.

7.9 Polymeric Nanoparticles

One of the earliest reports of using nanoparticles for drug delivery was introduced in 1978 [108]. Since then, polymeric nanoparticles for clinical applications have been attempted using a wide variety of biodegradable polymers including: poly (lactic acid) (PLA) [109], poly (glycolic acid) (PGA), poly (lactic co-glycolic acid) (PLGA) [110], poly (orthoesters) [111], poly(caprolactone) [112], poly(butyl cyanoacrylate) [113], polyanhydrides [114], and poly-N-isopropylacrylamide [115]. Although many fabrication methods exist, polymeric nanoparticles are frequently made using an oil-in-water emulsion [116] which involves dissolving a polymer and drug in an organic solvent. Next, the organic phase is vigorously mixed with an aqueous phase, followed by removal of the organic solvent by evaporation which forces the polymer to precipitate as nanoparticles. The particles are then recovered by centrifugation and lyophilization. To prevent the aggregation of the nanoparticles after they are reconstituted, nanoparticles can be made from amphiphilic copolymers composed of two biocompatible blocks [65]. For example, copolymers containing both lipophilic (i.e. PLGA) and hydrophilic (i.e. PEG) polymers have been constructed where the PEG migrates to the surface of the nanoparticles which comes in contact with aqueous solution.

Although polymeric nanoparticles offer great potential for targeted drug delivery approaches, one key limitation is their polydispersity with respect to size which may increase non-specific uptake. Another class of nanoparticles made from inorganic materials represents an emerging class of nanocarriers that may be prepared with near monodispersity.

7.10 Inorganic Nanoparticles

One of the first uses of inorganic nanoparticles for cancer therapy was reported by Roy et al. in 2003 [117]. Aside from potential monodispersity, a unique and highly advantageous feature of inorganic nanoparticles is their ability to be non-invasively imaged by magnetic resonance imaging (MRI) and high resolution superconducting quantum interference device (SQUID) [118]. Additionally, Inorganic particles offer advantages including ease of size control,potential functionalization to introduce targeting molecules and drugs, and the ability to induce cell apoptosis via hyperthermia. Examples of cancer cells' targeting by inorganic nanoparticles include galactose-PEG modified gold nanoparticles [119] and superparamagnetic iron oxide nanoparticles conjugated with methotrexate (as a targeted drug) [120].

To demonstrate the potential of the dual imaging/therapuetic feature, immunotargeted nanoshells were used to detect and destroy breast carcinoma cells that overexpress HER2, a clinically relevant cancer biomarker [121]. 'Nanoshells' are composed of a core of silica and a metallic outer layer. Nanoshells have optical resonances that can be adjusted to absorb or scatter essentially anywhere in the electromagnetic spectrum, including the near infrared region (820 nm, 4 W/cm^2) where transmission of light through tissue is optimal. Absorbing nanoshells are suitable for hyperthermia-based therapeutics, while scattering nanoshells are desired for imaging applications. In addition to utilizing a passive approach where Nanoshells preferentially accumulate in tumors due to EPR, gold nanoshells can be readily conjugated to antibodies, peptides or aptamers for targeting tumor cells. Recently, a cancer therapy based on absorption of near infrared light by gold nanoshells leading to rapid localized heating was developed [121, 122]. Tissues heated above the thermal damage threshold displayed coagulation, cell shrinkage, and loss of nuclear staining, which are indicators of irreversible thermal damage, whereas control tissues appeared undamaged [122]. Specifically, exposure to near infrared light caused the tumors to increase temperature by approximately 40 °C, while controls without nanoshells was heated by less than 10 °C.

One of the challenges of those inorganic nanoparticles is the inability to achieve high drug loading. Other than utilizing thiol groups, the conjugation chemistries are limited.

7.11 Dendrimers

Dendrimers are synthetic monodispersed branched macromolecules that form a tree-like structure and their synthesis represents a relatively new field of polymer chemistry. Particularly, polyamidoamine (PAMAM) dendrimers, since first synthesized by Tomalia and co-workers using the divergent synthetic route [123], have shown great promise to be used in biomedical applications due to: (1) ease of functionalization (conjugation with biomolecules such as targeting molecules, imaging agents, and drugs), (2) high water solubility, (3) well-defined chemical structure, (4) biocompatibility, (5) low production of toxic metabolites due to the dominance of renal filtration as the main rout of clearance, and (6) the rate of clearance may be controlled by chemical modification [124]. Although cationic PAMAM dendrimers are biologically active and, thus, have been used as a non-viral gene delivery vectors [125], they lack specificity leading to substantial non-specific interactions [126]. Consequently, charge neutral dendrimers have been prepared by acetylation using acetic anhydride and play as a platform to which the multiplicity of their functional end groups can be covalently attached. Although dendrimers have not yet progressed to clinical trials, they show great promise for medical applications based on the ability to potentially synthesize high densities of accessible functional groups in a well-defined and standardized manner and based on their non-immungenic and minimally toxic properties [127]. A successful example of in vivo targeted delivery approach using dendrimers was demonstrated in 2005 where the Baker group used methotrexate conjugated to PAMAM dendrimer which led to a tenfold reduction in tumor size compared to the same molar concentration of free methotrexate [6]. PAMAM dendrimers have also been complexed with MRI detectable metal nanoparticles and in vivo biodistribution of the dendrimer nanocomposite was studied [128]. Other potential dendrimers for targeting cancer include polypeptide and polyester dendrimers [129]. Other examples of dendrimers are summarized in literature [130, 131]. Although promising, there are still many obstacle currently preventing progression to clinical trials and markets. Dendrimers are relatively expensive as compared to other nanoparticles and require many repetitive steps to be synthesized, posing a challenge for large-scale synthesis.

7.12 Polymer Conjugates

A versatile group composed from both natural and synthetic polymer conjugate molecules that can carry and efficiently deliver drugs via covalent or non-covalent carrier-drug bonds. The most important features of this group that make it a sutiable drug delivery agent, based on its triangular features of safety, quality and efficacy.

FDA approved delivery formulation is a conjugate of neocarzinostatin and poly(styrene-comaleic acid), SMANCS, that have shown a potential for a reduction of toxicity and immunogenicity in addition to an extended half-time [132].

Peptide conjugate formulation based on pegylation using covalet or non-covalent attachment of PEG are very common drug delivery approach for many anti cancer drugs. For example, *Oncaspar*, a pegylated version of asparaginase, a bacterial enzyme, use to treat Leukemia [133]. Neulasta, another pegylated based formulation of human G-CSF, use to treat cancer patients in order to reduce chemotherapy side effects [134].

PEGylation coupling polymer strategy has been proven as a sucsseful and powerfull delivery carrier especialy in the field of protein polymers conjugates [135]. The PEGylation backbone enhanes the instability and short half time besides improving theraputic effect of the free drug [136].

However there is no good without bad. Certain PEG conjugated formulations may cause toxicity side effects. Excessive toxicity signs were observed during phase II clinical trials in patients with advanced ovarian cancer treated with PEG-L-asparaginase (Oncaspar®): Pancreatitis, fatigue, neutropenia and hypoalbuminemia, weight loss, dehydration, decreased fibrinogen and hepersensitivity were identified and the trial was ended before completing [134, 137].

Other polymer conjugates which is under preclinical trials is N-(2-hydroxypropyl) methacrylamide (HPMA) copolymer [138].The first example of an anti-cancer agent coupled to HPMA is that of Duncan et al., that have demonstrated, although less effective than free daunomycin, an inhibitory activity of daunomycin-HPMA conjugates in vitro [139]. A more recent study by Satchi-Fainaro et al., have targeted angiogenesis with a conjugate of HPMA copolymer and caplostatin in order to overcome short circulation time in plasma and improve the therapeutic effect of the free drug [140].

Drug based polymer conjugates can also potenially improve cancer chemotherapy by preferentially target the tumor. Examples for chemotherapeutic drug conjugates wich are under clinical investigation are the (N-(2-hydroxypropyl) ethacrylamide) (HPMA) copolymer-Dox [141] for the treatment of breast, lung and colon cancers. Poliglumex (Xyotax) is a paclitaxel-poly-L-glutamic conjugate [142] and is in phase II clinical trials in ovarian cancer patients [143] and phase III for advanced non-small cell lung cancer patients [144].

In order to increase treatment effectiveness, a combination therapy for blocking both cancer cells and cancer stem cells based on HPMA copolymer PI3K/mTOR inhibitor and docetaxel was developed by Zhou et al. [47].

Main advantages of HPMA as drug carrier include increasing solubility, prolonging the circulation time, and improving pharmacokinetic/biodistribution profiles of small molecule drugs.

Trojan peptides known as cell-penetrating peptides (CPPs) composed from short cationic arginine-rich amino acid sequence caple to traverse cell membrane efficiently. CPPs can be conjugated to chemtherapy drugs and used as a non-receptor-mediated delivery platform [145, 146]. However, CPPs-based delivery platforms suffer from low specificity and a limited therapeutic effect.

Other proposed polymeric conjugates for delivery are: poly(ethyleneimine) (PEI), polyvinylpyrrolidone (PVP), polydimethylacrylamide (PDMA), polyacrylamide (PAM), polyvinyl alcohol (PVA), chitosan and dextran [135].

Polymers can be designed as passive drug delivery conjugates and its penetration to the cancer cells is mainly attributed by the EPR mechanism, or when including a targeting moiety to increase specificity to cancer cells microenvironment via an active cellular targeting strategy [98].

Using folic acid as a targeting agent which can be easlily conjugated to the polymer may facilate drug uptake by the cancer cells via receptor-mediated manner and improve theraputic effect as was demonstrated by pullulan conjugated doxorubicin delivery system [11].

Another tumor targeting conjugate is the arginine-glycine-aspartic acid (RGD) peptide, selectively target the celllular receptors for αvβ3 and αvβ5 integrins expreesed on tumor vasculature cells [147].

HA can be conjugated directly to chemotherapy drugs to target highly metastatic CD44-positive tumor [148, 149].

7.12.1 siRNA Conjugate Polymers

A novel treatment modality could utilize RNA interference (RNAi) for therapeutic intervention in variety of diseases including cancer. Small interfering RNAs (siRNA) are a class of short double-stranded RNA molecules that can knockdown the expression of genes that has the complementary nucleotide sequence, yet siRNA delivery is a major abstacle due to its limited biological stabilty and negative charge characteristics. Polymer conjugates use for improving siRNA stability in circulation. In order to inhibit the potential for lung metastasis, Bonnet et al. intravenously injected TSA-Luc mammary tumor-bearing mice with a stable complex of siRNA against survivin and cyclin B1, key cell cycle regulators, both involved in cell proliferation and survival processes. A strong inhibition of lung tumor metastases was demonstrated. An insignificant inhibition (<20 %) was observed with the classic siRNA-PEI complex [150]. No additive inhibitory effect was demonstrated when the two genes were simultaneously targeted. The therapeutic effect of Cis was increased (up to 90 % tumor inhibition) when injected after cell cycle blockage with the ssiRNA-PEI complex treatment, compared with the 40 % inhibition when cisplatin was injected alone. Mouse survival rate following co-treatment demonstrated a similar trend [150].

Shen et al. presented simultaneous inhibition of both metastasis and tumor cells by co-delivery of shRNA together with Ptx. The master regulator Twist mediated tumor metastasis by promoting epithelial to mesenchymal transition (EMT), which occurs in metastatic sites. In order to block EMT, Twist shRNA (TshRNA) and Ptx were both conjugated with pluronic P85 and polyethyleneimine (PEI) polymer to form the D-a-Tocopherylpolyethyleneglycol 1000 succinate complex. In the pulmonary metastasis mouse model, a significant synergistic inhibitory effect of both tumor growth and pulmonary metastasis formation was demonstrated. The IC_{50} of free Ptx was 63-fold higher in comparison with the TshRNA-Ptx complex. In addition, the complex had a prolonged circulation time and promoted increased accumulation of Ptx and TshRNA in lung and tumor tissues [151]. Finally, an elegant approach to conjugate siRNA to triantennary N-acetylgalactosamine (GalNAc) induce robust RNAi-mediated gene silencing in the liver, most probably due to an uptake mechanism mediated by the asialoglycoprotein receptor (ASGPR) on hepatocytes. This promising strategy is currently under clinical investigation [152]. It will be interesting to see if this strategy could also be applied to lung tumors and their metastases.

7.13 Conjugate Chemistry

Covalent conjugation has a number of advantages over physical adsorption. Specifically, (1) increased stability until the carrier reaches the targeting site, (2) control over the density of the targeting molecule on the nanocarrier surface, (3) linkers for improved availability. Conjugation of targeting moieties to nano-based carriers can be achieved through a variety of chemistries including simple succinimidyl ester chemistry at room temperature [63]. For example, through reactive NH_2 terminated aptamers with polymers containing carboxyl groups (COOH) in the presence of Ethyl-3-(3-dimethylaminopropyl)-carbodiimide (EDC) and N-Hydroxysuccinimide (NHS) under aqueous conditions, amide linkage is formed and a product yield of 90 % should be routinely achieved [153]. The conjugation methodology can be categorized into five major reaction types: reaction between activated carboxyl groups and amino groups, which yields amide bond; reaction between thiols, which yields disulfide bonds; reaction between maleimide derivatives and thiols, which yields thioether bonds; reaction between hydroxyl and carboxyl, which yields ester bonds; and reaction between *p*-nitrophenylcarbonyl- and amino-group, which yields carbamate (urethane) bonds (Fig. 7.5).

Chemotherapeutic drugs can be also covalently conjugated to drug carriers. A preferred linkage for drugs would be ester bond since ester groups can be immediately cleaved by cytosolic enzyme esterases after endocytosis and, thus, can exploit the prodrug conversion mechanism. Ester linked drug carriers can be achieved by the EDC conjugation chemistry using hydroxyl group containing polymers and carboxylic acids or NHS attached drugs such as methotrexate and Taxol [154, 155]. The readers are referred to excellent reviews on conjugation chemistry in [84, 156].

Characterization of the binding of antibodies or aptamers to the nanoparticles can be achieved through conjugation of a fluorescent probe such as Fluorescein isothiocyanate (FITC), AlexaFluors, rhodamine or any other fluorophors and examining the particles under a fluorescent microscope or with flow cytometry. A more precise way is conjugation of ^{111}In or ^{125}I directly to the targeting moiety (usually antibodies) prior to their conjugation to the particles and an accurate measurement of the radioactivity prior and post conjugation to the particle. This can give also an estimation of the number of mAb per particle with a known diameter [84, 156–158].

7.14 Conclusion and Future Prospective

The development of nano-based carriers for targeting cancer is currently a vibrant area of research which promises the state of the art in cancer diagnostics and treatment. Although numerous nano-carrier based strategies exist, choosing the appropriate nano-carrier is still not obvious and comparative studies, of which there are few, are difficult to interpret and a suitable system which demonstrates optimal characteristics remains elusive. Determining the optimal nano-carrier system is especially difficult given the numerous factors which effect biodistribution and targeting. Therefore, successful targeting strategies at present must be determined experimentally on case by case basis. In addition, systemic therapies employing nano-based carriers require methods for overcoming few factors, such as: (1) non-specific uptake of nanoparticles by the mononuclear phagocytic cells and non-targeted cells (2) lack of the EPR effect in some cancers (3) the presence of necrotic processes in malignant masses [159].

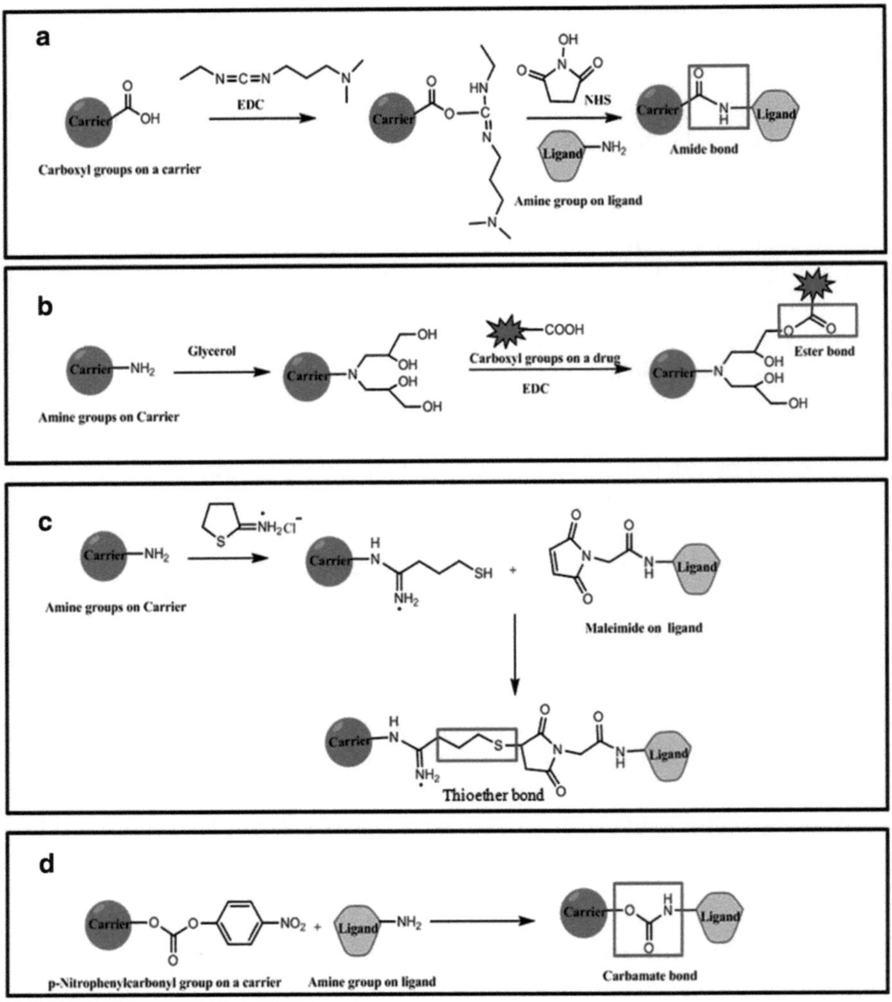

Fig. 7.5 Examples of common conjugation chemistry forming five major linkages. (**a**) Amide bond formation facilitated by EDC/NHS chemistry (two step reaction), (**b**) ester bond formation using glycerol as a linker (two step reaction), (**c**) thioether bond formation (two step reaction), and (**e**) formation of carbamate (urethane) linkage (two step reaction). Note that all the conjugation shown here, fall under the defenition of Click chemistry due to the simplicity of their conditions, and may be achieved under aqueous conditions at room temperature

From a succesful pharmaceutical delivery system prespective, there are few succesful liposomal formulations (13) and a fewer number of conjugats based drugs (3) on the market and above 300 different formulartions in different stages of clinical trials—representatives can be seen from Table 7.1.

As this field matures, there will undoubtedly be a great demand for simple scale-up methods and high throughput assays for quality assurance and this is certain to

Table 7.2 Exampls for targeting ligands and their receptors

Ligand	Receptor
Transferrin	Transferrin receptor [160]
Hyaluronan	CD44 [161]
Folic acid	Folate receptor
RGD peptide	Integrin αvβ3 [162]
Somastatin analogues	Somatostatin receptor

be an active area of development. With the fulfillment of these requirements, we are likely to enter an era where nano-carriers will play a fundamental role in the oncology field in the upcoming years (Table 7.2).

References

1. Fassas A, Anagnostopoulos A (2005) The use of liposomal daunorubicin (DaunoXome) in acute myeloid leukemia. Leuk Lymphoma 46(6):795–802
2. Ferrari M (2005) Cancer nanotechnology: opportunities and challenges. Nat Rev Cancer 5(3):161–171
3. Farokhzad OC et al (2006) Targeted nanoparticle-aptamer bioconjugates for cancer chemotherapy in vivo. Proc Natl Acad Sci U S A 103(16):6315–6320
4. Peer D, Margalit R (2004) Loading mitomycin C inside long circulating hyaluronan targeted nano-liposomes increases its antitumor activity in three mice tumor models. Int J Cancer 108(5):780–789
5. Duncan R (2006) Polymer conjugates as anticancer nanomedicines. Nat Rev Cancer 6(9):688–701
6. Kukowska-Latallo JF et al (2005) Nanoparticle targeting of anticancer drug improves therapeutic response in animal model of human epithelial cancer. Cancer Res 65(12):5317–5324
7. Maeda H (2001) The enhanced permeability and retention (EPR) effect in tumor vasculature: the key role of tumor-selective macromolecular drug targeting. Adv Enzyme Regul 41:189–207
8. Maeda H, Sawa T, Konno T (2001) Mechanism of tumor-targeted delivery of macromolecular drugs, including the EPR effect in solid tumor and clinical overview of the prototype polymeric drug SMANCS. J Control Release 74(1–3):47–61
9. Lucas AT, Madden AJ, Zamboni WC (2015) Formulation and physiologic factors affecting the pharmacology of carrier-mediated anticancer agents. Expert Opin Drug Metab Toxicol 11(9):1419–1433
10. Antony AC (1992) The biological chemistry of folate receptors. Blood 79(11):2807–2820
11. Scomparin A et al (2015) A comparative study of folate receptor-targeted doxorubicin delivery systems: dosing regimens and therapeutic index. J Control Release 208:106–120
12. Quintana A et al (2002) Design and function of a dendrimer-based therapeutic nanodevice targeted to tumor cells through the folate receptor. Pharm Res 19(9):1310–1316
13. Benns JM, Mahato RI, Kim SW (2002) Optimization of factors influencing the transfection efficiency of folate-PEG-folate-graft-polyethylenimine. J Control Release 79(1–3):255–269
14. Leamon CP, Low PS (1991) Delivery of macromolecules into living cells: a method that exploits folate receptor endocytosis. Proc Natl Acad Sci U S A 88(13):5572–5576
15. Lee RJ, Low PS (1994) Delivery of liposomes into cultured KB cells via folate receptor-mediated endocytosis. J Biol Chem 269(5):3198–3204
16. Scomparin A et al (2011) Novel folated and non-folated pullulan bioconjugates for anticancer drug delivery. Eur J Pharm Sci 42(5):547–558

17. Prost AC et al (1998) Differential transferrin receptor density in human colorectal cancer: a potential probe for diagnosis and therapy. Int J Oncol 13(4):871–875
18. Iinuma H et al (2002) Intracellular targeting therapy of cisplatin-encapsulated transferrin-polyethylene glycol liposome on peritoneal dissemination of gastric cancer. Int J Cancer 99(1):130–137
19. Ishida O et al (2001) Liposomes bearing polyethyleneglycol-coupled transferrin with intracellular targeting property to the solid tumors in vivo. Pharm Res 18(7):1042–1048
20. Gijsens A et al (2002) Targeting of the photocytotoxic compound AlPcS4 to Hela cells by transferrin conjugated PEG-liposomes. Int J Cancer 101(1):78–85
21. Kolhatkar R, Lote A, Khambati H (2011) Active tumor targeting of nanomaterials using folic acid, transferrin and integrin receptors. Curr Drug Discov Technol 8(3):197–206
22. Yu B et al (2010) Receptor-targeted nanocarriers for therapeutic delivery to cancer. Mol Membr Biol 27(7):286–298
23. Cinci M et al (2015) Targeted delivery of siRNA using transferrin-coupled lipoplexes specifically sensitizes CD71 high expressing malignant cells to antibody-mediated complement attack. Target Oncol 10(3):405–413
24. Kuang Y et al (2013) T7 peptide-functionalized nanoparticles utilizing RNA interference for glioma dual targeting. Int J Pharm 454(1):11–20
25. Bartolazzi A et al (1994) Interaction between CD44 and hyaluronate is directly implicated in the regulation of tumor development. J Exp Med 180(1):53–66
26. Stamenkovic I, Aruffo A (1994) Hyaluronic acid receptors. Methods Enzymol 245:195–216
27. Zeng C et al (1998) Inhibition of tumor growth in vivo by hyaluronan oligomers. Int J Cancer 77(3):396–401
28. Thomas RG et al (2015) Paclitaxel loaded hyaluronic acid nanoparticles for targeted cancer therapy: in vitro and in vivo analysis. Int J Biol Macromol 72:510–518
29. Lesley J, Hyman R (1998) CD44 structure and function. Front Biosci 3:D616–D630
30. Sneath RJ, Mangham DC (1998) The normal structure and function of CD44 and its role in neoplasia. Mol Pathol 51:191–200
31. Cohen ZR et al (2015) Localized RNAi therapeutics of chemoresistant grade IV glioma using hyaluronan-grafted lipid-based nanoparticles. ACS Nano 9(2):1581–1591
32. Mizrahy S et al (2014) Tumor targeting profiling of hyaluronan-coated lipid based-nanoparticles. Nanoscale 6(7):3742–3752
33. Narayanan D, Jayakumar R, Chennazhi KP (2014) Versatile carboxymethyl chitin and chitosan nanomaterials: a review. Wiley Interdiscip Rev Nanomed Nanobiotechnol 6(6):574–598
34. Kawakami S et al (2000) Mannose receptor-mediated gene transfer into macrophages using novel mannosylated cationic liposomes. Gene Ther 7(4):292–299
35. Shan D et al (2015) RGD-conjugated solid lipid nanoparticles inhibit adhesion and invasion of alphavbeta 3 integrin-overexpressing breast cancer cells. Drug Deliv Transl Res 5(1):15–26
36. Wang K et al (2014) Tumor penetrability and anti-angiogenesis using iRGD-mediated delivery of doxorubicin-polymer conjugates. Biomaterials 35(30):8735–8747
37. Dagar S et al (2003) VIP grafted sterically stabilized liposomes for targeted imaging of breast cancer: in vivo studies. J Control Release 91(1–2):123–133
38. Dagar S et al (2001) VIP receptors as molecular targets of breast cancer: implications for targeted imaging and drug delivery. J Control Release 74(1–3):129–134
39. Dagar A et al (2012) VIP-targeted cytotoxic nanomedicine for breast cancer. Drug Deliv Transl Res 2(6):454–462
40. Guan YY et al (2014) Selective eradication of tumor vascular pericytes by peptide-conjugated nanoparticles for antiangiogenic therapy of melanoma lung metastasis. Biomaterials 35(9):3060–3070
41. Mei L et al (2014) Enhanced antitumor and anti-metastasis efficiency via combined treatment with CXCR4 antagonist and liposomal doxorubicin. J Control Release 196:324–331
42. Warenius HM et al (1981) Attempted targeting of a monoclonal-antibody in a human-tumor xenograft system. Eur J Cancer Clin Oncol 17(9):1009–1015

43. von Mehren M, Adams GP, Weiner LM (2003) Monoclonal antibody therapy for cancer. Annu Rev Med 54:343–369
44. Weiner LM, Adams GP (2000) New approaches to antibody therapy. Oncogene 19(53):6144–6151
45. James JS, Dubs G (1997) FDA approves new kind of lymphoma treatment. Food and Drug Administration. AIDS Treat News (No 284):2–3
46. Albanell J, Baselga J (1999) Trastuzumab, a humanized anti-HER2 monoclonal antibody, for the treatment of breast cancer. Drugs Today (Barc) 35(12):931–946
47. Zhou Y et al (2015) Combination therapy of prostate cancer with HPMA copolymer conjugates containing PI3K/mTOR inhibitor and docetaxel. Eur J Pharm Biopharm 89:107–115
48. Ferrara N (2005) VEGF as a therapeutic target in cancer. Oncology 69(Suppl 3):11–16
49. Ferrara N, Hillan KJ, Novotny W (2005) Bevacizumab (Avastin), a humanized anti-VEGF monoclonal antibody for cancer therapy. Biochem Biophys Res Commun 333(2):328–335
50. Gerber HP, Ferrara N (2005) Pharmacology and pharmacodynamics of bevacizumab as monotherapy or in combination with cytotoxic therapy in preclinical studies. Cancer Res 65(3):671–680
51. Gibson AD (2002) Phase III trial of a humanized anti-CD33 antibody (HuM195) in patients with relapsed or refractory acute myeloid leukemia. Clin Lymphoma 3(1):18–19
52. Javle M, Smyth EC, Chau I (2014) Ramucirumab: successfully targeting angiogenesis in gastric cancer. Clin Cancer Res 20(23):5875–5881
53. White RR, Sullenger BA, Rusconi CP (2000) Developing aptamers into therapeutics. J Clin Invest 106(8):929–934
54. Tuerk C, Gold L (1990) Systematic evolution of ligands by exponential enrichment: RNA ligands to bacteriophage T4 DNA polymerase. Science 249(4968):505–510
55. Ellington AD, Szostak JW (1990) In vitro selection of RNA molecules that bind specific ligands. Nature 346(6287):818–822
56. Blank M et al (2001) Systematic evolution of a DNA aptamer binding to rat brain tumor microvessels, selective targeting of endothelial regulatory protein pigpen. J Biol Chem 276(19):16464–16468
57. Morris KN et al (1998) High affinity ligands from in vitro selection: complex targets. Proc Natl Acad Sci U S A 95(6):2902–2907
58. Beigelman L et al (1995) Chemical modification of hammerhead ribozymes. Catalytic activity and nuclease resistance. J Biol Chem 270(43):25702–25708
59. Aurup H, Williams DM, Eckstein F (1992) 2′-Fluoro- and 2′-amino-2′-deoxynucleoside 5′-triphosphates as substrates for T7 RNA polymerase. Biochemistry 31(40):9636–9641
60. Pieken WA et al (1991) Kinetic characterization of ribonuclease-resistant 2′-modified hammerhead ribozymes. Science 253(5017):314–317
61. Eulberg D, Klussmann S (2003) Spiegelmers: biostable aptamers. Chembiochem 4(10):979–983
62. Wang DL et al (2014) Selection of DNA aptamers against epidermal growth factor receptor with high affinity and specificity. Biochem Biophys Res Commun 453(4):681–685
63. Farokhzad OC et al (2004) Nanoparticle-aptamer bioconjugates: a new approach for targeting prostate cancer cells. Cancer Res 64(21):7668–7672
64. Rosenberg JE et al (2014) A phase II trial of AS1411 (a novel nucleolin-targeted DNA aptamer) in metastatic renal cell carcinoma. Invest New Drugs 32(1):178–187
65. Gref R et al (1994) Biodegradable long-circulating polymeric nanospheres. Science 263(5153):1600–1603
66. Langer R, Peppes NA (2003) Advances in biomaterials, drug delivery, and bionanotechnology. AICHE J 49(12):2990–3006
67. Gref R et al (1997) Poly(ethylene glycol)-coated nanospheres: potential carriers for intravenous drug administration. Pharm Biotechnol 10:167–198
68. Santini JT Jr, Cima MJ, Langer R (1999) A controlled-release microchip. Nature 397(6717):335–338
69. Chertok B et al (2013) Drug delivery interfaces in the 21st century: from science fiction ideas to viable technologies. Mol Pharm 10(10):3531–3543

70. Langer R, Tirrell DA (2004) Designing materials for biology and medicine. Nature 428(6982):487–492
71. Brigger I et al (2004) Negative preclinical results with stealth nanospheres-encapsulated Doxorubicin in an orthotopic murine brain tumor model. J Control Release 100(1):29–40
72. Garcia-Carbonero R, Supko JG (2002) Current perspectives on the clinical experience, pharmacology, and continued development of the camptothecins. Clin Cancer Res 8(3):641–661
73. Khandare J, Minko T (2006) Polymer-drug conjugates: progress in polymeric prodrugs. Prog Polym Sci 31(4):359–397
74. Bangham AD, Horne RW (1964) Negative staining of phospholipids and their structural modification by surface-active agents as observed in the electron microscope. J Mol Biol 12:660–668
75. Bangham AD, Standish MM, Watkins JC (1965) Diffusion of univalent ions across the lamellae of swollen phospholipids. J Mol Biol 13(1):238–252
76. Horne RW, Bangham AD, Whittaker VP (1963) Negatively stained lipoprotein membranes. Nature 200:1340
77. Forssen EA, Ross ME (1994) Daunoxome® treatment of solid tumors: preclinical and clinical investigations. J Liposome Res 4(1):481–512
78. Chang HI, Yeh MK (2012) Clinical development of liposome-based drugs: formulation, characterization, and therapeutic efficacy. Int J Nanomedicine 7:49–60
79. Gabizon AA (2001) Stealth liposomes and tumor targeting: one step further in the quest for the magic bullet. Clin Cancer Res 7(2):223–225
80. Gabizon AA (2001) Pegylated liposomal doxorubicin: metamorphosis of an old drug into a new form of chemotherapy. Cancer Invest 19(4):424–436
81. Safra T et al (2000) Pegylated liposomal doxorubicin (doxil): reduced clinical cardiotoxicity in patients reaching or exceeding cumulative doses of 500 mg/m^2. Ann Oncol 11(8):1029–1033
82. Olson F et al (1979) Preparation of liposomes of defined size distribution by extrusion through polycarbonate membranes. Biochim Biophys Acta 557(1):9–23
83. Slingerland M, Guchelaar HJ, Gelderblom H (2012) Liposomal drug formulations in cancer therapy: 15 years along the road. Drug Discov Today 17(3–4):160–166
84. Torchilin VP (2005) Recent advances with liposomes as pharmaceutical carriers. Nat Rev Drug Discov 4(2):145–160
85. Perche F, Torchilin VP (2013) Recent trends in multifunctional liposomal nanocarriers for enhanced tumor targeting. J Drug Deliv 2013:705265
86. Blume G et al (1993) Specific targeting with poly(ethylene glycol)-modified liposomes: coupling of homing devices to the ends of the polymeric chains combines effective target binding with long circulation times. Biochim Biophys Acta 1149(1):180–184
87. Allen TM, Mumbengegwi DR, Charrois GJ (2005) Anti-CD19-targeted liposomal doxorubicin improves the therapeutic efficacy in murine B-cell lymphoma and ameliorates the toxicity of liposomes with varying drug release rates. Clin Cancer Res 11(9):3567–3573
88. Park JW et al (2002) Anti-HER2 immunoliposomes: enhanced efficacy attributable to targeted delivery. Clin Cancer Res 8(4):1172–1181
89. Gao J et al (2009) Tumor-targeted PE38KDEL delivery via PEGylated anti-HER2 immunoliposomes. Int J Pharm 374(1–2):145–152
90. Gabizon A et al (2010) Improved therapeutic activity of folate-targeted liposomal doxorubicin in folate receptor-expressing tumor models. Cancer Chemother Pharmacol 66(1):43–52
91. Eavarone DA, Yu X, Bellamkonda RV (2000) Targeted drug delivery to C6 glioma by transferrin-coupled liposomes. J Biomed Mater Res 51(1):10–14
92. Maruyama K (2011) Intracellular targeting delivery of liposomal drugs to solid tumors based on EPR effects. Adv Drug Deliv Rev 63(3):161–169
93. Jones RA (2004) Tough and smart. Nat Mater 3(4):209–210
94. Sawant RM et al (2006) "SMART" drug delivery systems: double-targeted pH-responsive pharmaceutical nanocarriers. Bioconjug Chem 17(4):943–949
95. Ghanbarzadeh S et al (2014) Improvement of the antiproliferative effect of rapamycin on tumor cell lines by poly (monomethylitaconate)-based pH-sensitive, plasma stable liposomes. Colloids Surf B Biointerfaces 115:323–330

96. Ducat E et al (2011) Nuclear delivery of a therapeutic peptide by long circulating pH-sensitive liposomes: benefits over classical vesicles. Int J Pharm 420(2):319–332
97. Mizrahy S et al (2011) Hyaluronan-coated nanoparticles: the influence of the molecular weight on CD44-hyaluronan interactions and on the immune response. J Control Release 156(2):231–238
98. Peer D et al (2007) Nanocarriers as an emerging platform for cancer therapy. Nat Nanotechnol 2(12):751–760
99. Landesman-Milo D et al (2013) Hyaluronan grafted lipid-based nanoparticles as RNAi carriers for cancer cells. Cancer Lett 334(2):221–227
100. Peer D, Margalit R (2004) Tumor-targeted hyaluronan nanoliposomes increase the antitumor activity of liposomal Doxorubicin in syngeneic and human xenograft mouse tumor models. Neoplasia 6(4):343–353
101. Eliaz RE, Szoka FC Jr (2001) Liposome-encapsulated doxorubicin targeted to CD44: a strategy to kill CD44-overexpressing tumor cells. Cancer Res 61(6):2592–2601
102. Peer D et al (2008) Systemic leukocyte-directed siRNA delivery revealing cyclin D1 as an anti-inflammatory target. Science 319(5863):627–630
103. Kandra P, Kalangi HP (2015) Current understanding of synergistic interplay of chitosan nanoparticles and anticancer drugs: merits and challenges. Appl Microbiol Biotechnol 99(5):2055–2064
104. Sahu SK et al (2011) Hydrophobically modified carboxymethyl chitosan nanoparticles targeted delivery of paclitaxel. J Drug Target 19(2):104–113
105. Yang R et al (2009) Lung-specific delivery of paclitaxel by chitosan-modified PLGA nanoparticles via transient formation of microaggregates. J Pharm Sci 98(3):970–984
106. Malhotra M et al (2013) Systemic siRNA delivery via peptide-tagged polymeric nanoparticles, targeting PLK1 gene in a mouse xenograft model of colorectal cancer. Int J Biomater 2013:252531
107. Wang X et al (2014) Delivery of platinum(IV) drug to subcutaneous tumor and lung metastasis using bradykinin-potentiating peptide-decorated chitosan nanoparticles. Biomaterials 35(24):6439–6453
108. Marty JJ, Oppenheim RC, Speiser P (1978) Nanoparticles—new colloidal drug delivery system. Pharm Acta Helv 53(1):17–23
109. Alonso MJ et al (1994) Biodegradable microspheres as controlled-release tetanus toxoid delivery systems. Vaccine 12(4):299–306
110. Qaddoumi MG et al (2002) Molecular mechanisms mediating the cEdocytosis of biodegradable PLGA nanoparticles in rabbit conjunctival epithelial cell layers. Invest Ophthalmol Vis Sci 43:U875
111. Deng JS et al (2003) In vitro characterization of polyorthoester microparticles containing bupivacaine. Pharm Dev Technol 8(1):31–38
112. Molpeceres J et al (1999) A polycaprolactone nanoparticle formulation of cyclosporin-a improves the prediction of area under the curve using a limited sampling strategy. Int J Pharm 187(1):101–113
113. Sommerfeld P, Sabel BA, Schroeder U (2000) Long-term stability of PBCA nanoparticle suspensions. J Microencapsul 17(1):69–79
114. Gao JM et al (1998) Surface modification of polyanhydride microspheres. J Pharm Sci 87(2):246–248
115. Huang G et al (2004) Controlled drug release from hydrogel nanoparticle networks. J Control Release 94(2–3):303–311
116. Eastoe J, Warne B (1996) Nanoparticle and polymer synthesis in microemulsions. Curr Opin Colloid Interface Sci 1(6):800–805
117. Roy I et al (2003) Ceramic-based nanoparticles entrapping water-insoluble photosensitizing anticancer drugs: a novel drug-carrier system for photodynamic therapy. J Am Chem Soc 125(26):7860–7865
118. Morawski AM, Lanza GA, Wickline SA (2005) Targeted contrast agents for magnetic resonance imaging and ultrasound. Curr Opin Biotechnol 16(1):89–92

119. Bergen JM et al (2006) Gold nanoparticles as a versatile platform for optimizing physicochemical parameters for targeted drug delivery. Macromol Biosci 6(7):506–516
120. Kohler N et al (2005) Methotrexate-modified superparamagnetic nanoparticles and their intracellular uptake into human cancer cells. Langmuir 21(19):8858–8864
121. Loo C, Lowery A, Halas N, West J, Drezek R (2005) Immunotargeted nanoshells for integrated cancer imaging and therapy. Nano Lett 5(4):709–711
122. Hirsch LR et al (2003) Nanoshell-mediated near-infrared thermal therapy of tumors under magnetic resonance guidance. Proc Natl Acad Sci U S A 100(23):13549–13554
123. Tomalia DA et al (1985) A new class of polymers—starburst-dendritic macromolecules. Polym J 17(1):117–132
124. Longmire M, Choyke PL, Kobayashi H (2008) Clearance properties of nano-sized particles and molecules as imaging agents: considerations and caveats. Nanomedicine (Lond) 3(5):703–717
125. Bielinska A et al (1996) Regulation of in vitro gene expression using antisense oligonucleotides or antisense expression plasmids transfected using starburst PAMAM dendrimers. Nucleic Acids Res 24(11):2176–2182
126. Hong SP et al (2004) Interaction of poly(amidoamine) dendrimers with supported lipid bilayers and cells: hole formation and the relation to transport. Bioconjug Chem 15(4):774–782
127. Duncan R, Izzo L (2005) Dendrimer biocompatibility and toxicity. Adv Drug Deliv Rev 57(15):2215–2237
128. Khan MK et al (2005) In vivo biodistribution of dendrimers and dendrimer nanocomposites—implications for cancer imaging and therapy. Technol Cancer Res Treat 4(6):603–613
129. Lee CC et al (2005) Designing dendrimers for biological applications. Nat Biotechnol 23(12):1517–1526
130. Gillies ER, Frechet JMJ (2005) Dendrimers and dendritic polymers in drug delivery. Drug Discov Today 10(1):35–43
131. Malik N et al (2000) Dendrimers: relationship between structure and biocompatibility in vitro, and preliminary studies on the biodistribution of I-125-labelled polyamidoamine dendrimers in vivo. J Control Release 65(1–2):133–148
132. Maeda H (2001) SMANCS and polymer-conjugated macromolecular drugs: advantages in cancer chemotherapy. Adv Drug Deliv Rev 46(1):169–185
133. Petersen WC Jr et al (2014) Comparison of allergic reactions to intravenous and intramuscular pegaspargase in children with acute lymphoblastic leukemia. Pediatr Hematol Oncol 31(4):311–317
134. Duncan R (2014) Polymer therapeutics: top 10 selling pharmaceuticals—what next? J Control Release 190:371–380
135. Markovsky E et al (2012) Administration, distribution, metabolism and elimination of polymer therapeutics. J Control Release 161(2):446–460
136. Pisal DS, Kosloski MP, Balu-Iyer SV (2010) Delivery of therapeutic proteins. J Pharm Sci 99(6):2557–2575
137. Hays JL et al (2013) A phase II clinical trial of polyethylene glycol-conjugated L-asparaginase in patients with advanced ovarian cancer: early closure for safety. Mol Clin Oncol 1(3):565–569
138. Zhang R et al (2016) N-(2-hydroxypropyl)methacrylamide copolymer-drug conjugates for combination chemotherapy of acute myeloid leukemia. Macromol Biosci 16(1):121–128
139. Duncan R et al (1987) Anticancer agents coupled to N-(2-hydroxypropyl)methacrylamide copolymers: I. Evaluation of daunomycin and puromycin conjugates in vitro. Br J Cancer 55(2):165–174
140. Satchi-Fainaro R et al (2004) Targeting angiogenesis with a conjugate of HPMA copolymer and TNP-470. Nat Med 10(3):255–261
141. Seymour LW et al (2009) Phase II studies of polymer-doxorubicin (PK1, FCE28068) in the treatment of breast, lung and colorectal cancer. Int J Oncol 34(6):1629–1636
142. Chipman SD et al (2006) Biological and clinical characterization of paclitaxel poliglumex (PPX, CT-2103), a macromolecular polymer-drug conjugate. Int J Nanomedicine 1(4): 375–383

143. Galic VL et al (2011) Paclitaxel poliglumex for ovarian cancer. Expert Opin Investig Drugs 20(6):813–821
144. O'Brien ME et al (2008) Randomized phase III trial comparing single-agent paclitaxel Poliglumex (CT-2103, PPX) with single-agent gemcitabine or vinorelbine for the treatment of PS 2 patients with chemotherapy-naive advanced non-small cell lung cancer. J Thorac Oncol 3(7):728–734
145. Patel LN, Zaro JL, Shen WC (2007) Cell penetrating peptides: intracellular pathways and pharmaceutical perspectives. Pharm Res 24(11):1977–1992
146. Lindgren M et al (2000) Cell-penetrating peptides. Trends Pharmacol Sci 21(3):99–103
147. Li ZJ, Cho CH (2012) Peptides as targeting probes against tumor vasculature for diagnosis and drug delivery. J Transl Med 10(Suppl 1):S1
148. Zhao Y et al (2014) CD44-tropic polymeric nanocarrier for breast cancer targeted rapamycin chemotherapy. Nanomedicine (Lond) 10(6):1221–1230
149. Journo-Gershfeld G et al (2012) Hyaluronan oligomers-HPMA copolymer conjugates for targeting paclitaxel to CD44-overexpressing ovarian carcinoma. Pharm Res 29(4): 1121–1133
150. Bonnet ME et al (2013) Systemic delivery of sticky siRNAs targeting the cell cycle for lung tumor metastasis inhibition. J Control Release 170(2):183–190
151. Shen J et al (2013) Simultaneous inhibition of metastasis and growth of breast cancer by co-delivery of twist shRNA and paclitaxel using pluronic P85-PEI/TPGS complex nanoparticles. Biomaterials 34(5):1581–1590
152. Rajeev KG et al (2015) Hepatocyte-specific delivery of siRNAs conjugated to novel non-nucleosidic trivalent N-acetylgalactosamine elicits robust gene silencing in vivo. Chembiochem 16(6):903–908
153. Sehgal D, Vijay IK (1994) A method for the high efficiency of water-soluble carbodiimide-mediated amidation. Anal Biochem 218(1):87–91
154. Majoros IJ et al (2006) PAMAM dendrimer-based multifunctional conjugate for cancer therapy: synthesis, characterization, and functionality. Biomacromolecules 7(2):572–579
155. Majoros IJ et al (2005) Poly(amidoamine) dendrimer-based multifunctional engineered nanodevice for cancer therapy. J Med Chem 48(19):5892–5899
156. Klibanov AL, Torchilin VP, Zalipsky S (2003) Chemical conjugation. In: Torchilin VP, Weissig V (eds) Liposomes: practical approach. Oxford University Press, Oxford, pp 193–265
157. Gupta B et al (2005) Monoclonal antibody 2C5-mediated binding of liposomes to brain tumor cells in vitro and in subcutaneous tumor model in vivo. J Drug Target 13(6):337–343
158. Spragg DD et al (1997) Immunotargeting of liposomes to activated vascular endothelial cells: a strategy for site-selective delivery in the cardiovascular system. Proc Natl Acad Sci U S A 94(16):8795–8800
159. Prabhakar U et al (2013) Challenges and key considerations of the enhanced permeability and retention effect for nanomedicine drug delivery in oncology. Cancer Res 73(8):2412–2417
160. Yhee JY et al (2013) Tumor-targeting transferrin nanoparticles for systemic polymerized siRNA delivery in tumor-bearing mice. Bioconjug Chem 24(11):1850–1860
161. Park HK et al (2015) Smart nanoparticles based on hyaluronic acid for redox-responsive and CD44 receptor-mediated targeting of tumor. Nanoscale Res Lett 10(1):981
162. Rashidi LH et al (2015) Investigation of the strategies for targeting of the afterglow nanoparticles to tumor cells. Photodiagnosis Photodyn Ther 2015. doi:10.1016/j.pdpdt.2015.08.001

Chapter 8
The Importance of Particle Geometry in Design of Therapeutic and Imaging Nanovectors

Matthew J. Ware, Jenolyn F. Alexander, Huw D. Summers, and Biana Godin

Abstract One of the main objectives in nanomedicine is to enable specific delivery of therapeutics and imaging agents to the disease loci. While previously the main tool for targeting was incorporating entities with biological recognition on the surface of the nanovector, recently, the focus has shifted to the ability of the physical characteristics of particles to guide them crossing numerous barriers to the site of action. In this chapter we will focus on how the geometry of nanovectors affects their interaction with biological milieu, and as a result their therapeutic and diagnostic potential. The chapter is divided into sections describing interactions of various particles with extracellular components, whole cells and various cell organelles. In the intracellular interactions subsection, we focus on the initial interaction between particle and cell membrane, the uptake mechanism and intracellular trafficking of particles with various geometries. Furthermore, the importance of charge density and the zeta potential parameter is discussed. The chapter concludes with a discussion of challenges and outlook on future developments of theranostic particle delivery using particle geometry as the rational design feature.

Keywords Particle size • Particle shape • Zeta potential • Surface • Interactions with cells • In vivo • In vitro • Geometry

---

M.J. Ware
Centre for Nanohealth, College of Engineering, Swansea University, Swansea, UK

Department of Nanomedicine, Houston Methodist Research Institute, Houston, TX, USA

J.F. Alexander • B. Godin (✉)
Department of Nanomedicine, Houston Methodist Research Institute, Houston, TX, USA
e-mail: bgodin@houstonmethodist.org; bianagodinv@gmail.com

H.D. Summers
Centre for Nanohealth, College of Engineering, Swansea University, Swansea, UK

K.A. Howard et al. (eds.), *Nanomedicine*, Advances in Delivery Science and Technology, DOI 10.1007/978-1-4939-3634-2_8

8.1 Introduction

Due to the numerous physical and biological barriers to drug and contrast agent transport in the body, the minimization of side-effects with simultaneous targeted delivery of active agents to specific tissues of interest continues to pose a challenge. The use of nanoscale vectors to deliver drugs, genetic material, and imaging contrast agents has been an intensively investigated niche in medicine, as achieving desirable bioavailability of therapeutic agents been especially challenging in the fields of gene therapy [1] and oncology [2–4]. Furthermore, nanovectors may support the transport of hydrophobic drugs [5], as dissolution of such drugs in the bloodstream is difficult, and in cases where the drug or imaging agent is too fragile, insoluble, or toxic for direct in-vivo administration.

It is becoming increasingly apparent that the design of contrast agent and drug nanovectors involves the manipulation of several critical factors. These include their chemical functionality and mechanical properties, the degree of dispersion or encapsulation of the drug, on or within, the nanovector, the size and shape of the particle, and the surface charge.

The ability to easily fabricate drug or contrast agent-loaded spherical particles with a narrow size distribution by a number of bottom-up methods including emulsion, dispersion, suspension, polymerization, precipitation and spraying techniques has meant that the role of particle size and composition play in biodistribution [6], cellular binding [7], cell entry pathways [8], cell uptake [9–11], and tumor bed penetration [12, 13] have been extensively studied for spherical particles. However, the majority of naturally occurring (and blood-born) biological entities are non-spherical and biological processes typically occur under dynamic conditions in which the motion of spherical and non-spherical objects will differ. For example, Decuzzi and Ferrari's theoretical work predicts that, under conditions of linear shear flow such as that in the bloodstream, oblate particles adhere more strongly to biological substrate than spherical particles, and hence, the use of non-spherical particles for delivery is predicted to improve therapeutic and imaging efficacy [14]. Additionally, shape is crucial to the particle's mechanism of cell entry [15–17], and the release rate of the therapeutic cargo [18, 19]. Past limitations in the control of particle fabrication have been overcome as novel methods such as photolithography have been developed. This means researchers are now able to access a multitude of different shapes at the sub-micron size range on a sufficient volume scale to allow for extensive in vitro and in vivo biological studies (Fig. 8.1). This will allow a more complete understanding of how the combined effects of particle size, shape, charge, chemical composition and mechanical properties affect biodistribution, target tissue interaction and cell entry which will lead to the development of increasingly efficient drug and image contrast based nanovectors. While numerous studies further discussed in this chapter have shown that the geometry of the particle is an important parameter when the particles are administered intravenously, interactions with mucosal barriers have been shown not to depend on the shape of particulates [20]. In contrast, the size of nano- and micro-particles does play a role in permeation across the mucosal barrier. As an example, Hanes's group has shown that even when nanoparticles are larger than the mucus mesh size (e.g. 500 nm), they can

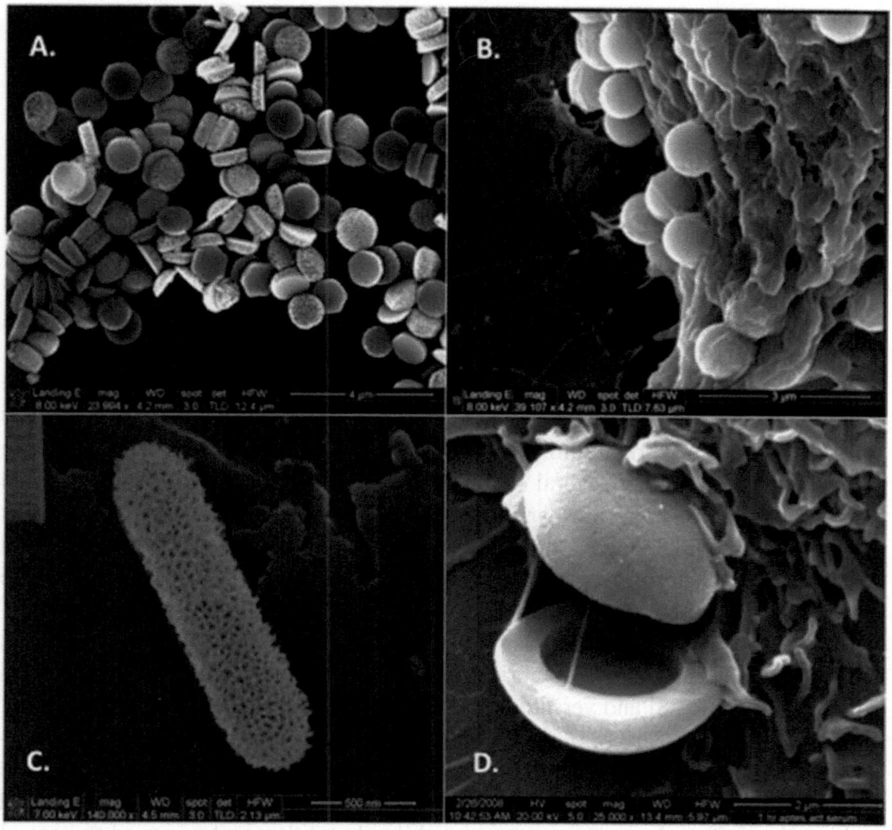

Fig. 8.1 Scanning electron micrographs of silicon particles fabricated by top-down approach using photolithography and electrochemical etching: (**a**) Discoidal porous silicon particles. (**b**) Spherical silica beads (fabricated by bottom-up synthesis) on a pancreatic cancer cell (PANC-1, ATCC). (**c**) Porous silicon rods. (**d**) Quasi-hemispherical porous silicon particles

be mobile in the mucus if coated with hydrophilic polymers [21]. Further in this chapter we will focus on systems that are intended for intravenous administration.

This review aims to summarize the various fabrication approaches, including 'Top Down', 'Bottom Up', and novel combinational fabrication techniques, which are able to create nanoscale particles with various intricate geometries tailored towards a specific biological function. We attempt to relate aspects of particle design with their systemic behavior in vivo and also their action at the target cell. We discuss the importance of the surface charge of nanoscale vectors and common assumptions regarding its measurement on non-spherical particulates, which affect understanding of the initial particle-cell membrane interaction as well as with other biological entities in vivo. We summarize the approaches to improve efficiency in particle function, including geometry and surface charge manipulation, the use of targeting moieties, ligand binding, PEG conjugation and other functional groups. The chapter concludes with an overview of future perspectives relating to the improved use of nanoscale vectors in active agent transport.

8.2 Particle Fabrication Approaches

8.2.1 'Top Down' Fabrication

8.2.1.1 Photolithography and Electrochemical Etching

Top down fabrication of therapeutic and diagnostic particles of various shapes and sizes uses the same principles developed and refined in the microelectronics industry throughout the past several decades: Photolithography and electrochemical etching. These techniques are characterized by high reproducibility and yield, and allow particles to be fabricated with homogenous and precise dimensions "by-design". Rationally designed particles have an increased ability to negotiate the numerous bio-barriers associated with the delivery of active agents in the body. For instance, in recent works multistage vectors (MSV) are presented [22, 23]. These vectors are comprised of several distinct nano-elements or 'stages'; each designed to negotiate one or more bio-barriers from the site of administration to the target lesion. The discoidal stage 1 mesoporous silicon particles (S1MP) not only hold and protect the cargo drug or image contrast agent (Fig. 8.2), but also enable increased margination and interaction with vasculature walls and cell surface adhesion properties during their transportation in the blood [24, 25]. Porous silicon was used as the material for S1MP fabrication due to its biodegradability [26, 27], biocompatibility [27, 28] and versatile fabrication methods which allow the control of shape, size and porosity [23]. S1MPs are fabricated by an integration of two top-down approaches; photolithography and electrochemical etching for particle porosification. Concentration of the etching solution, doping, electrical current and etching time affect the porous structure, while photolithography is used to control the pattern and geometry of the particles via silicon micro-fabrication, which is well known in the silicon micromachining industry [23, 29, 30]. These techniques enable precision, reproducibility and scalability.

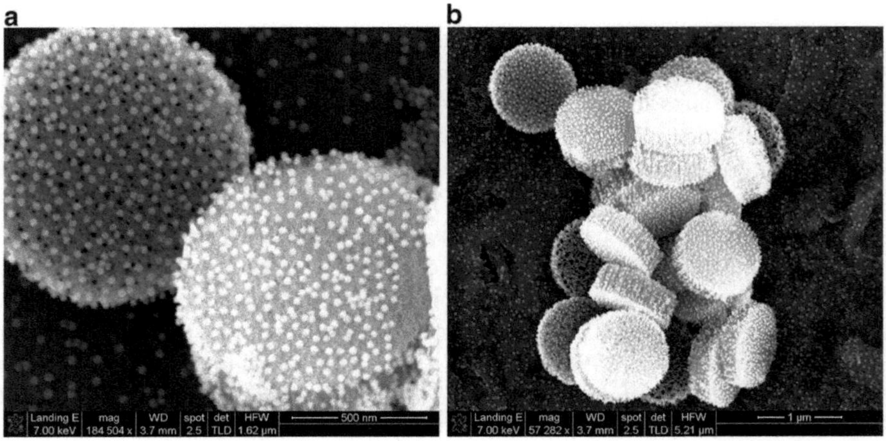

Fig. 8.2 Scanning electron micrographs of silicon mesoporous particles (S1MP) loaded with iron oxide nanoparticles for image contrast applications. Scale bar: (**a**) 500nm; (**b**) 1 micron

Chiappini et al. [29] outlined the fabrication protocol of quasi-hemispherical S1MPs yielding monodisperse particles 900 nm–3.2 μm in diameter, 30–65 % porosity and 3–55 nm pores. Briefly, starting with a silicon wafer coated with a dielectric silicon nitride (SiN) film, an array of particle patterns is first transferred to a dielectric layer by a photolithography process and Reactive Ion Etch. A silicon etch is then performed which defines the trenches that nucleate S1MPs with different profiles [29]. A successive, two-step electrochemical etch process is applied to yield the S1MP of desired porosity, pore size and thickness. A highly porous release layer is formed at the wafer-S1MP interface retaining S1MP on the substrate and allowing for differential chemical modifications of the two sides of S1MP if required (Fig. 8.3).

Discoidal vectors are fabricated by photolithographic patterning of a porous silicon film, a process that eliminates potential distortion effects introduced by non-uniform current distribution during electrochemical etch, simultaneously granting independent control of the porous structure. A novel sealing strategy to pattern pSi films using low-pressure chemical vapor deposition of LTO was implemented to overcome the problem of photoresist adsorption within the pores. After electrochemical etch of double layered porous silicon films, the porous films are sealed by LTO, and an array of circles is patterned on LTO film. A CF_4 Reactive ion etch is then performed to etch through LTO and two-layered porous silicon producing monodisperse discoidal S1MPs with diameter as small as 400 nm and pore size ranging 3–150 nm. The S1MP fabrication process is robust and well characterized. Using 1000 nm diameter disks as an example, more than 2 billion S1MP patterned on a 4 wafer can be obtained. Recent studies [22] point towards the possibility for a multi-layer fabrication method of discoidal particles, which should increase the production yield by tens times.

8.2.1.2 Step and Flash Imprint Lithography

Step-and-flash imprint lithography is a technique that replicates the topography of a rigid mold using a photocurable prepolymer solution [31]. In this technique a low viscosity, photocurable liquid or solution fills the void spaces of the mold. The solution consists of a low-molecular-weight monomer and a photo-initiator. UV light exposure polymerizes and, therefore, solidifies the precursor while in contact with the mold. Removing the mold leaves a topographically patterned inverse replica on the substrate.

Step and flash imprint lithography has recently been used to fabricate a range of non-spherical particles such as cubic, triangular and pentagonal cylindrical particles. From the numerous scientific studies involving enzymatically degradable peptide sequences for drug delivery [32–35], Glangchai et al. [36] were successful in using cathepsins to initiate the release of DNA plasmids from cubic, triangular and pentagonal cylindrical particles in vitro, by combining the superior shape control provided by lithographic methods with novel drug loading and release approaches.

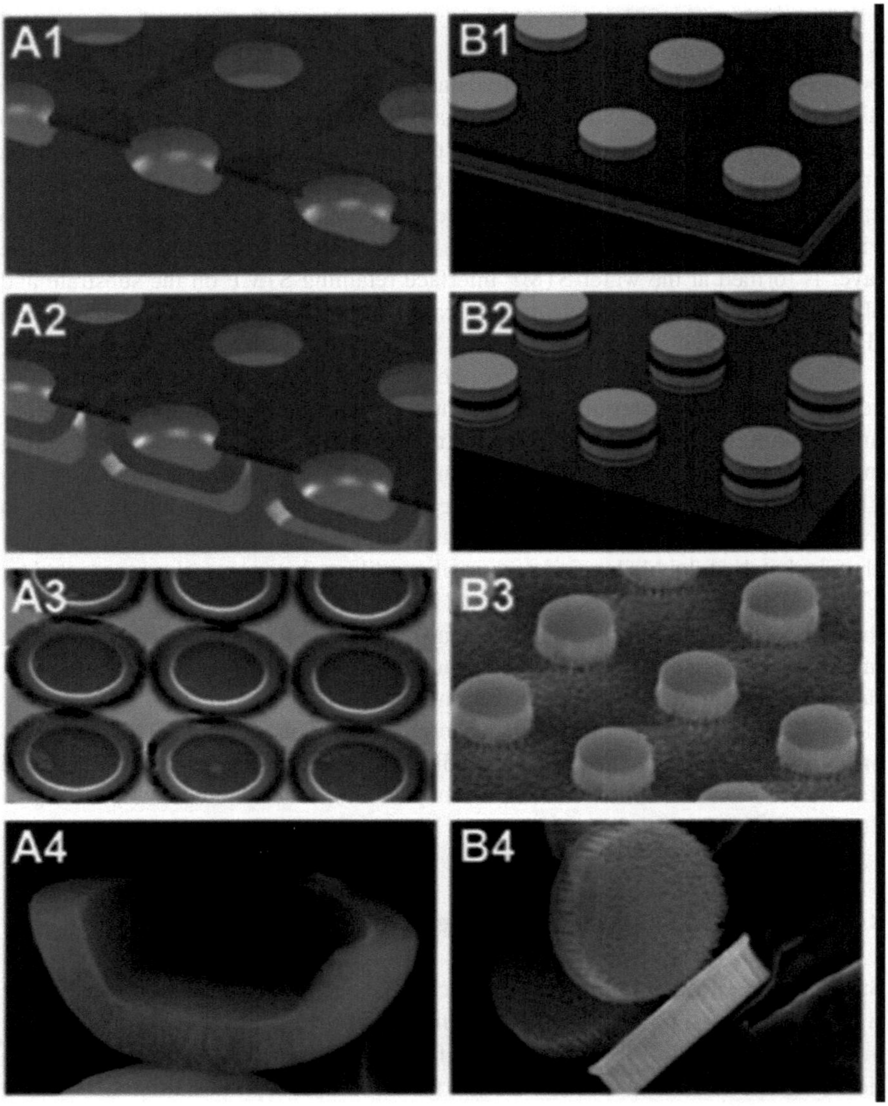

Fig. 8.3 Fabrication of silicon mesoporous particles (S1MP). (**a1**) Patterned silicon nitride (SiN) layer and trenches etched into silicon. (**a2**) Electrochemically etched S1MP with release layer. (**a3**) S1MP array on wafer after removal of SiN. (**a4**) Cross-section of hemispherical S1MP. (**b1**) Photoresist pattern on Low Temperature Oxide (LTO) capped porous silicon film with release layer. (**b2**) Particle array on wafer after RIE. (**b3**) Discoidal S1MP array on wafer after LTO removal. (**b4**) Released discoidal S1MP [22]

8.2.1.3 Particle Replication in Non-Wetting Templates (PRINT®)

The PRINT® (Liquidia Technologies) process [15, 37–45] is a modification of traditional lithographic methods and can produce polymeric micro and nano vectors. PRINT® provides precision, control and flexibility in the fabrication of rationally designed particles and can regulate particle size and shape independently of composition. Since PRINT® uses elastomeric fluoropolymers, instead of traditionally used silicones it provides a selective filling of nano-scale cavities in the mold using an organic liquid whilst avoiding the formation of interconnecting "flash layers" which are a common problem in traditional imprint lithography [46]. PRINT® particles can be produced with diverse surface chemistries, degradation characteristics and elastic moduli because organic liquids and sol-gel metal oxide precursors do not swell fluoropolymers in the same fashion as they swell silicones. Furthermore, photocurable perfluoropolyethers (PFPEs) used in the PRINT® process maintain the desirable flexibility of Polydimethylsiloxane (PDMS) whilst exhibiting an improved resistance to solvents [47, 48]. The PFPE molding materials are able to accurately pattern fragile or soft objects at the nanoscale, such as block copolymer micelles and naturally occurring entities, such as viruses [38]. The PRINT® process has constructed particles from natural matrices such as biologically relevant proteins [44] and synthetic matrix materials, such as highly cross-linked hydrogels and linear polyesters [37, 39] and has fabricated particles which have been shown to be effective delivery agents for therapeutic drugs and contrast imaging agents [42, 43], such as for the delivery of doxorubicin to cervical HeLa cancer cells [43].

Despite its uses, fluorinated molding matrices are more expensive than silicone materials. However, cost has been mitigated by the development of continuous mold manufacturing processes that use inexpensive backing materials coated with a fine film of the more expensive fluorinated molding material [49]. This strategy has enabled the PRINT® process to become economically more feasible and has lead it to become a scalable particle manufacturing process.

8.2.1.4 PDMS-Molding Processes

Lightly cross-linked PDMS, also known as silicone rubber, is one of the most widely studied molding material due to its low toxicity, low modulus, high gas permeability, which allows dead-end filling and low surface energy which enables facile release from master pattern templates as well as replication of nanoscale objects [50]. However, for some applications, the swelling of PDMS molds in organic solvents or PDMS-fragment transfer to the sample are major weaknesses. Other silicone-based materials and filled molding material formulations have been developed that limit the mechanical distortion of the stamp [51]. Additionally, several mold surface treatments [52, 53], such as oxygen plasma and fluorosilane grafting, have been investigated to ensure the mold surface is more hydrophilic or hydrophobic, respectively, which effects wetting during mold filling and the release of the particles from the mold. Lastly, cleaning procedures have been developed which limit sample contamination from small fragments originating from the PDMS-stamp [54].

Hansford et al. have used PDMS stamps to produce both thermoset and thermoplastic microparticles and medical devices with various geometries [55–57] for drug delivery applications. Furthermore, Guan and colleagues highlighted the numerous possibilities created by the precise geometric control on the micro and nanometer scale by the design of unique geometries with extensions that can self-fold to imbed in intestinal tissue [57] after oral delivery. The same group also fabricated microcapsules for intravenous delivery, which contained sucrose as the cargo and a thin poly(vinyl alcohol) (PVA) or poly(lactic-co-glycolic acid) (PLGA) based outer layer which swells in the presence of water due to osmosis. This releases the drug via diffusion through the outer membrane rather than dissolution of the thermoplastic matrix, and, thus, creating a novel drug transport mechanism with higher drug loading capacities than conventionally prepared PGLA microcapsules [58]. Additionally, Oudshoorn et al. have explored various molding techniques including rigid micro-molding, soft micro-molding, and photolithography to fabricate uniform methacrylated hyperbranched polyglycerol microparticles [52]. Square and hexagonal microparticles particles were obtained in high yields, however, PDMS-based molding techniques produced a higher number of less well defined particles in relation to shape when compared to the rigid, epoxy-based micro-mold.

In summary of this section, the ability to prepare particles in a manner than allows independent manipulation of one geometric parameter at a time has insights into shape dependent biological processes such as phagocytosis to be developed. Many techniques, such as stretching thermoplastic particles [16] or flash imprint lithography preparation of cross-linked particles as previously described [36], can be used to fabricate small amounts of shape-specific particles. Practical, large scale manufacturing techniques for fabrication of geometrically precise particles constitute a major development gained by the application of the PRINT® process to biomedical and other disciplines. Traditional soft lithography using PDMS-based molding materials offers an additional technique in the design of micron-sized particles with specific geometry. However, for applications such as intracellular drug and imaging agent delivery, where nanoscale vectors are often most efficacious, opportunities exist to improve on PDMS-based soft lithography using various other molding techniques.

8.2.2 'Bottom Up' Design of Biomimetic Assemblies

Bottom up fabrication is a more classic method to produce nanoscale and microscale particles starting from molecules, which are the building blocks in such case. Since spherical shape is, generally, the most thermodynamically favorable one, most of the bottom-up produced particles are spheres. It is difficult to develop synthetic systems that mimic biological macromolecules and objects, as their structures and geometries are created with highly specific interactions that work at extremely small dimensions. However, using various building blocks and fabrication techniques, it is possible to manipulate the geometry of bottom-up produced particles to some extent. The use of self-assembling molecules, where the structure contains at least two distinctive moieties, a biologically active portion and a fraction that directs assembly, presents one particular design strategy that does emulate nature. This approach fits into a larger framework known as 'bottom up' design, where complex functional structures are generated by assembling a group of molecularly interlocking parts [59]. Such systems mimic biological architectures by using a particular set of primary molecules that are capable of associating into secondary and tertiary structures. This process is initiated by incorporating particular peptides, nucleotides, sugars or binding synthetic molecules to the primary molecule to provide it with the appropriate biologically specific information. Secondly, a molecular region must be included to direct the assembly of the molecules into a highly controlled tertiary structure, such as micelles, vesicles, tubules, and other more complex architectures. The assembly provides potency to the biological information it carries by several possible mechanisms, which include the control of spatial relationships, and concentrating locally the bio-functional entities, as well as providing opportunities for combinatorial selectivity. However, the ability to precisely design for a specific function remains a significant challenge. Currently, an ever more-growing number of interesting new structures can be designed with function in mind; however, accurate structure function relationships in this type of assembly are warranted to further accelerate the effective design process. Additionally, many of these new biomimetic particle designs are comprised of relatively new molecules and the balance between therapeutic efficacy and toxicity is poorly understood. Several unknown parameters for these nascent molecular assemblies can result in the restriction of use, primarily toxic side-effects, but also the sterility, stability and scale-up of the complicated drug carrier systems [60].

Bottom-up strategies are frequently used by nature to combine proteins into quaternary structures to create systems with a specific function. Nature is adept at maximizing utility whilst minimizing the use of resources and economizing synthesis by assembling molecules in a variety of ways to give a multitude of binding targets by the integration of heteromeric assemblies. Dimers such as antibodies, integrins, and leucine zippers, trimers such as collagen, and larger multimeric assemblies such as viral capsids are examples of molecules that act in this fashion. Leucine zippers show that homodimers and heterodimers give quaternary structures where several molecules are coupled together in a combinatorial fashion to give greater target diversity and specificity. Another mechanism of control is provided by the coupling reaction itself, where the monomeric uncoupled assembly does not possess the desired binding pattern or function and is, therefore, in a dormant phase. Cellular networks comprised of these modular components of a structurally comparable class may selectively self-assemble to produce subtle functional correlations. This concept is found in a viral capsid, where the process of self-assembly has formed unique protein architecture to create a functional icosahedral cage and would be extremely useful to mimic for drug delivery applications. Several attempts have been made to imitate biological structure and function by using bottom-up strategies to design molecules that are capable of self-assembly. However, successful synthetic emulation of this type of assembly would require a comprehensive knowledge of the precise forces that facilitate self-assembly in a variety of nanoscale molecules that can be tailored to facilitate the delivery of drugs by emulating the characteristics of natural biological systems.

In summary, several physical parameters dictate the assembly of macro-molecular structures and nano-crystals/metallic particles, and predicting and controlling these aggregates is extremely complex. The synthesis of simple molecular building blocks can lead to intricate self-assembled architectures much more complex than the original target structure [61–63], which can be subsequently rationalized based on the physical parameters and chemical principles of self-assembly, which include, but are not limited to, intermolecular forces and thermodynamics which dictate the mechanisms of assembly [64].

8.2.2.1 Hydrophobic Interactions

Hydrophobic interactions play a major role in the rational design of self-assembling molecular aggregates. Amphiphilic molecules, which comprise both hydrophilic and hydrophobic moieties, use the hydrophobic interaction to form a variety of aggregated structures. The hydrophobic moiety provides a driving force for assembly, however, assembly can stem from several factors that make aggregation either favourable or unfavourable. Understanding these forces will help describe the general driving forces for aggregate formation [65, 66]. These forces include; the transfer of hydrophobic tails from an aqueous environment into the aggregate, the assembly and interaction of the alkyl groups with the solvent as a function of the micelle area, hydrophobic packing, which causes an entropic decrease that is associated with less conformational freedom in the aggregate and finally, forces involved with the attractive, or more commonly, the repulsive interactions of the head-groups, such as steric interactions of polymeric head-groups, or electrostatic repulsions in charged head-groups.

Self-assembling amphiphilic molecules have attracted attention for the novel design of complex structures with biological specific functionalities. Luk and Abbott [67] review a variety of head-group interactions that are biologically relevant, whereas, several other groups have explored the conjugation of carbohydrates [68], peptides [63, 69], polymerizable elements in the hydrophilic head-groups [70] and hydrophobic cores [71] for stability that affect the structure and function of amphiphiles. Nature's use of amphiphiles follows two specific design goals; the formation of organelles and other compartmentalized environments and simultaneously the precise presentation of specific signals at the interface of these enclosures. This is comparable to the applications pursued by chemists, who control assembly by the subtle manipulation of chemical structures and hydrophobic interactions which leads to the production of a variety of nanoscale assemblies, including vesicles, micelles, sheets, tubes and bundles [67].

8.2.2.2 Electrostatic Interactions

Electrostatic interactions provide a useful tool to manipulate the dynamics of self-assembly as they provide both attractive and repulsive forces, which can be controlled by adjusting salt concentrations within the system. For instance, coiled-coil heterotrimeric assemblies have recently been designed via the manipulation of both hydrophobic and electrostatic interactions to drive assembly and to precisely match primary structures, respectively [72]. The specificity of assembly can further be controlled via the manipulation of charge-to-charge interaction, and provides a design that is selective for trimeric a-helical coiled-coils as building blocks. This coiled-coils design strategy demonstrates that specificity can be included in the assembly process via electrostatic interactions and these types of biologically relevant assemblies can be used to develop therapeutics with tightly controlled geometries.

Hydrogen bonding originates between positively charged hydrogen atoms and negatively charged electronegative atoms and, thus, may be considered a subset of electrostatic interaction [65]. The positive charge in hydrogen atoms is a result of a covalent bond to electronegative atoms, and quantification of this interaction is difficult as it is dependent on the chemical components and the angle of bonding between the atoms involved [73].

To conclude this sub-section, self-assembling systems must be designed for particular applications based on the class of molecules used and the forces that dominate their assembly. A combination of forces that describe self-assembly as a complex milieu with several local energy minima, much like protein folding, should be considered when making realistic predictions of self-assembled structure. This will enable evermore-complex drug and contrast agent delivery vectors with biologically relevant dimensions and functions to be designed, which take advantage of the elegant and intricate process of self-assembly.

8.2.2.3 Metallic and Carbon Nanoparticles

In addition to molecular assemblies, other nanoscale materials of different shapes, such as metallic and carbon nanoparticles can be synthesized via bottom up methods. Metallic nanoparticles are unique materials for optical, electronic, catalytic, and sensing applications. There is a great deal of interest in using metallic nanoparticles as building blocks in the development of more complex nanostructures through the use of a 'bottom-up' approach due to the flexibility in controlling the surface chemistry of these particles through functionalization. Using self-assembly techniques, one can exploit spontaneous chemical interactions to build complex nanoconstructs. Gold nanorods have been synthesized and modified with various polymers, inorganic oxides and organic ligands to establish principles for self-assembly of these unique nanomaterials, and are of great interest due to their strong optical absorption in the visible and near infrared regions, which can be tuned through material preparation and modification of the surrounding environment. Gold nanorods can be prepared via: (1) the template method, where gold is electrochemically deposited within the pores of either a polycarbonate or alumina template membrane; (2) the electrochemical method where an electrical potential is placed across a gold anode and a platinum cathode. The two electrodes are placed in an electrolytic solution containing cationic surfactants, which form cylindrical micelles. As gold ions from the anode are deposited onto the platinum cathode, the surfactant stabilizes the rod allowing it to grow in a cylindrical manner and (3) the seed-mediated growth method, where a small amount of spherical particles, which act as seeds, are added to a growth solution containing additional metal salt and a structure-direction agent and is one of the most reliable and versatile methods to control the shapes of noble metal nanocrystals.

Several research groups have succeeded in fabricating organized CNT arrays by growing carbon nanotubes on patterned catalyst printed substrates or by post-synthesis alignment [74–76]. Controlled CNT arrays are usually fabricated either by growing CNTs on catalyst pre-patterned substrates, such as iron, nickel or cobalt substrates, in a chemical vapour deposition (CVD) reactor [75–77] or by self-assembling functionalized CNTs on the substrate [78, 79]. Aligned CNT structures by self-assembly approaches [78–80] present some solutions for the limitations of organized CNT structures prepared by existing approaches, including the ability to fabricate CNTs at room temperature and flexible substrate selections. Additionally, CNT arrays have been manipulated with functionalized CNTs by self-assembly. Shortened CNTs by chemical oxidation were self-assembled on silver coated silicon substrate [80], and Fe ion coated silicon substrates by interactions with carboxylic groups of CNTs and substrates [78]. Thiol functionalized shortened CNTs have been used for self-assembled arrays on gold-coated Si substrates [79]. For bulk production of aligned CNTs, self-assembly pyrolytic methods [81], thermolytic processes with sprayed solutions [82], and pyrolyze systems with sophisticated aerosol generators [83] have been suggested. Though self-assembly approaches demonstrate possibilities for fabricating CNT arrays, it is still challenging to precisely control nano-size dimension CNT arrays during their fabrication. Ultimately the combination of well-defined self-assembly techniques and nano-size patterning processes to obtain nano-size well organized CNT patterns for applications remains the goal.

8.2.3 Combination Fabrication Approaches

Several types of nanovectors with various geometries combining the top-down and bottom-up methods of fabrication have been recently developed with a view to combine the possibilities of both and/or overcome the limitations of either one of the independent approaches as discussed in this section [84, 85]. Examples include particles fabricated by photolithography followed by a self-assembling coverage of polymers or biological entities like phage [86–94].

Recent particles evolving in the world of nanotherapeutics are the hollow microcapsules composed of polymeric film, which have shown to possess extended circulation time [95], controllable degradation [96] and tunable drug release [97–99], capable of surface functionalization for enhanced cellular association, and function as suitable drug delivery agents to target cells [100–110]. These microcapsules can be generally produced using a self-assembly of hydrophobic-hydrophilic block copolymers resulting in polymerosomes or by combining top-down and bottom up approaches. Microcapsules formed by self-assembly are commonly poly-disperse and limited to be spherical or worm-like in shape [111]. These two limitations are overcome by fabricating microcapsules by combining top-down and bottom-up methods; Rigid core-templates which are homogeneous in size and of desired shape are manufactured by methods that involve photolithography (top-down) and then are deposited with multiple layers of a polymeric monolayer or polymeric bilayers (bottom-up), depending on the functionality, on a rigid core template, following by dissolution of the core which forms hollow elastic microcapsules [90, 91, 93, 112, 113].

Using non-porous silicon discs as the rigid, core templates, erythrocyte-mimicking elastic capsule could be fabricated [114]. The sacrificial silicon discoid cores were made by photolithography combined with a reactive ion etching of Polycrystalline Silicon (PolySi). The particles upon detachment from the wafer were subject to PEI adsorption and then hydrogen bonded, pH sensitive copolymer layers of poly(N-vinylpyrrolidone) or poly(methacrylic acid) were assembled in a bottom-up fashion and cross-linked with carbodiimide. The rigid core-templates were dissolved in an aqueous solution of hydrofluoric acid, and dialyzed for 2–3 days. These cubical and discoid (erythrocyte-mimicking) capsules were hence programmed to exhibit change in dimension and volume based on environmental pH. Several such hollow polymeric particles have been engineered in the last decade, comprising of close-to-monodisperse, rigid sacrificial core-templates chemically synthesized from nano-seed solutions ammonium bicarbonate and manganese sulfate. By controlling the time, temperature and stirring conditions the shape of the rigid cores were manipulated to be spherical or cubical. The manganese sulfate concentration determined the size of the particles [93].

Polyelectrolyte microcapsules are capable of multiple functionalities by incorporation of quantum dots for imaging, superparamagnetic nanoparticles for application using magnetic fields, enzymes for pro-drug activation and intracellular sensing, pharmaceutical agents for therapy, metallic nanoparticles for remote drug release and surface functionalization for targeting purposes. The remarkable advantage of such systems is that all these functional properties can be potentially incorporated in a single nanoparticle system [108].

A recent study showed the design and therapeutic capability of a Multistage Nanovector based on Bacteriophage Associated Silicon Particles (BASP) [86]. This system, consisted of porous silicon particles manufactured by photolithography to produce quasi-hemispherical particles, and electrochemically etched in a hydrofluoric acid-ethanol mixture, to impart porosity. The particles were oxidized, and then functionalized with 2 % 3-aminopropyltriethoxysilane to confer a positive charge. Approximately 45 nm gold nanoparticles (AuNP) were prepared from Gold(III) chloride and trisodium citrate as per the Turkevich method [115]. Bacteriophages with suitable peptides were used to form AuNP-bacteriophage networks by overnight self-assembly [116]. The final BASP were obtained by mixing AuNP-bacteriophage networks with silicon particles for 1 h. By integrating the top-down and self-assembly approaches, it was possible to develop a unique theranostic multistage nanovector. Porous silicon particles, loaded with SPIONS coated with chitosan using combinational fabrication technology promise intercellular and intracellular targeting as well as transport of drugs and imaging agents [94].

8.3 Biological Interactions of Active Particles

In general, any drug molecule or imaging payload must reach its intended site of action to exert an efficient therapeutic or diagnostic action and to prevent adverse side-effects [117]. The delivery of such particles to the target organ or diseased loci generally requires a systemic administration through the circulatory system. Consequently, several biological barriers must be overcome to achieve efficient drug or contrast agent delivery [118]. These barriers involve systemic hurdles, including interactions with extracellular matrix and non-target cells, mononuclear phagocytic capture (MPC) by the mononuclear phagocytic system (MPS), and non-specific biodistribution, as well as cellular hurdles such as the target cell membrane, avoiding complete lysosomal degradation, and in some cases, overcoming the nuclear envelope. Particles must also remain stable and soluble, whilst not aggregating in the blood, or degrading which may expose the payload to degradation enzymes within the blood or inter-tissue fluid [119].

The manipulation of particle design is important for successful delivery to the target loci, and affects particles transport in the bloodstream, their extravasation out of the blood vessels, the diffusion through tissue, and finally, the degree of cellular internalization and subsequent uptake pathway and long-term fate. The appropriate design of a therapeutic/diagnostic particle requires a comprehensive understanding of all the characteristics of the vectors, as well as the mechanisms by which they interact within the biological environment and targeted cells.

8.3.1 Systemic Interactions

Regardless of whether the drug delivery vector is being administered via inhalation, intramuscular injection, gavage, or intravascular injection, it will unavoidably come into contact with the extracellular environment. Within this environment, multiple barriers exist which can result in rapid clearance and/or degradation of the nanovector before it ever reaches the intended site [120]. Furthermore, therapeutic efficacy of intravenously administered particles may be compromised by anatomical constraints and non-specific interactions with biological membranes, such as vasculature endothelium and connective tissue. Particle aggregation and disassembly may be caused by unwanted interactions with blood components, biological tissue and immune cells which are specialized in inactivation and removal of foreign particles, can lead to non-specific organ accumulation and low targeting efficiency and adverse side-effects [121, 122]. For example, blood circulatory half-life is highest for neutral particles, whereas cationic nanoparticles (NPs) possess the lowest half-life in the bloodstream and cause several complications such as hemolysis and platelet aggregation [123].

8.3.2 Interactions with Immune Cells

Generally, opsonisation of serum proteins, such as IgM, fibronectin and C-reactive proteins, to the surface of particle systems stimulates recognition by receptors on the surface of macrophages which enhances mononuclear phagocyte system (MPS) capture [124]. This leads to rapid removal from the blood and accumulation within non-target organs which contain a high population of resident tissue macrophages such as the liver [124], spleen and lungs. Most nanomaterials, when administered into the blood, are taken up within minutes or hours by phagocytic cells and are seen to possess a general predisposition for phagocytic clearance by the MPS. This is probably due to the fact that particles smaller than 5 μm in their largest dimension are comparable in size to the pathogens combated during the evolution of the immune system [125]. Micro-scale sized particles (0.5 μm and larger) have been described as invoking "bacteria-like responses" primarily being removed from circulation via phagocytosis, while NPs up to 200 nm in diameter are more likely to invoke "virus-like responses" [125]. Exploiting this trend, NPs have been intentionally designed to invoke, suppress, or avoid the immune response [126]. The use of nanovectors to target MPS organs has proven relevant in major therapeutic areas such as oncology, for the treatment of tumors, especially those that are situated in an organ of the MPS, such as hepatocarcinoma and liver metastasis and secondly in the treatment of infectious diseases, as many of pathogens that need to be eliminated are located in the MPS macrophages. Still the therapeutic moiety should escape phagocytic capture to exert therapeutic effect in other parts of the liver. Invocation of the immune response has driven size-dependence research in particulate vaccine vectors [127–129], suppression of an immune response could also make particulate carriers useful during organ transplant therapy, while avoidance of the immune system and reduced systemic toxicity using targeted particulate carriers is most often desirable in the case of imaging and therapeutic agents.

The emerging understanding of the importance of particle shape in phagocytosis by macrophages has been described and carefully studied by Champion et al. [16, 17]. The recognition that naturally occurring immunological targets vary widely in both size and shape provided motivation for this work. By carefully varying shape at constant size, the authors concluded that it is indeed particle shape rather than size, which plays a dominant role for determining the complexity of the local actin structure, and ultimately whether phagocytic processes occur. Lipidic disks have been developed as alternatives to liposomes, having a diameter of 120 nm [130] to 250 nm [131], and a thickness of only a few nanometers, showing efficient uptake by MPS macrophages. Other lipidic assemblies such as cube-shaped so-called cubosomes [132] have been also proposed as drug nanocarriers. However, the impact of the shape of such systems on phagocytosis as compared to liposomes remains to be fully elucidated. As for polymeric systems, a recent study performed with polystyrene ellipsoids and discoidal particles has shown that, independently of opsonization, the local particle shape at the point of contact dictates whether macrophages initiate phagocytosis or simply spread on particles [16]. For example, a macrophage attached to an ellipse at the pointed end will internalize it in a few minutes, while a macrophage attached to a flat region of the same ellipse will not internalize it for over 12 h. Interestingly, the same study has also shown that particle shape, but not size, plays a dominant role in phagocytosis. While initiation of phagocytosis in at least one orientation happened for particles of all shapes, the local particle shape, measured by tangent angles, at the point of initial contact is responsible for an ability of phagocytic cells to initiate/complete the phagocytosis or to "spread" the particle. On the other hand, particle size is sculpturing the ability of phagocytes to complete the process of the particle uptake [16].

Deformability, or rigidity, can also play a significant role in a particle phagocytic predisposition. As far as interaction with the cell membrane is concerned, macrophages tend to show a strong preference for rigid particles. One study showed that soft polyacrylamide particles were unable to stimulate the assembly of actin filaments required for the proper formation phagosomes, as opposed to rigid particles which had the same total polymer mass and surface properties [133]. Macrophage's increased sensitivity to bead stiffness can be connected to the fact that bacteria and other pathogens have cell walls usually more rigid than the surrounding tissue they invade. On the other hand, particle rigidity can have an opposite effect on opsonization. Rigid liposomal membranes, composed of cholesterol and saturated phospholipids with a high melting point, are known to decrease complement activation and hence decrease phagocytosis [134]. Similarly, core-shell nanoparticles having a rigid polystyrene core were significantly less prone to uptake by MPS than NPs made of a more flexible core, based on fluid-like poly(methyl acrylate) PMA [6]. Particles with low elastic modulus are thought to provide a greater number of surface interactions with the biological environment. Though the research in this area is ongoing, there is not a well-defined relationship between particle rigidity and phagocytosis as of yet.

Clearance by phagocytes, a high blood perfusion and discontinuous sinusoidal endothelium are features that contribute to a general tendency for non-modified NPs to accumulate in the liver. Early work by Fidler et al. [135] reported clearance of liposomes to the liver, whilst Illum and Davis [136] showed a short circulatory half-life and liver accumulation in rabbits of non-modified polystyrene particles which showed the biodistribution could be changed with poloxamer 388 functionalization. This rapid clearance can be avoided by adding poly(ethylene)glycol (PEG) to the surface of NPs. Dunn et al. [136] showed removal within 1 h and sequestration of 100 nm polystyrene particles in the resident Kupffer cells of the liver in rats that could be overcome by particle PEGylation. The prevention of opsonisation with the addition of PEG drastically increases the blood half-life of all nanomaterials regardless of surface charge. Generally, the blood half-life of gold NPs is also increased by increasing the length of PEG, which causes the protective layer to thicken [137]. The synthesis of these long circulating "Stealth" NPs improves accumulation in target tissue. In addition to PEGylation of NP surfaces, blood half-life also depends on a NP's shape and size and surface chemistry. For example, rod-shaped micelles have a circulation lifetime ten times longer than that of spherical micelles [138].

8.3.3 Particle Dynamics in the Blood Stream

NPs encounter numerous physical and biological barriers created by normal and abnormal physiology en route to their target site. Regardless of the targeting strategy, it is desirable for blood-borne particles to navigate within the vasculature, identify the abnormal blood vessels and there, either adhere firmly to the walls withstanding the dislodging hydrodynamic forces (vascular targeting) or extravasate, leaving the vessels through inter-endothelial openings, and seek the diseased cells within the surrounding tissue. Interestingly, and perhaps not surprisingly, this is the strategy used by circulating cells such as leukocytes, to marginate and adhere to the inflamed endothelium as well as platelets which are also mostly found in close proximity to the vessel walls, where they can periodically sense and adhere to sites of vascular injury. This characteristic platelet behaviour is ascribed to their small size (3× smaller than RBCs) and quasi-discoidal shape [139, 140]. Conversely, healthy red blood cells (RBCs) can continuously navigate the circulatory system for several weeks while delivering oxygen to the tissues. The size, shape, and physiochemical properties of the RBC membrane enable these cells to avoid entrapment within the complex microvasculature and, in larger vessels, to repel from the endothelial walls and preferentially stay within the vessel center [141, 142]. Despite these intricate design strategies exhibited by nature, limited investigations have focused on predicting the transport behaviour of synthetic particles with different sizes and shapes within the vascular system. Such a feature can be exploited in the rational design of particle-based intravascular and pulmonary delivery systems. Ferrari and Decuzzi [143] presented a general model to predict the dynamics of nano-/micro-particles immersed in a linear laminar flow. They showed that non-spherical particles under the simultaneous action of inertial and hydrodynamic forces marginate and, therefore, increase the interaction with the wall surface. Among the shapes considered, discoidal particles with low aspect ratio exhibit the largest propensity to marginate. The effect of shape has also been shown to play a critical role in the margination of NPs [143, 144]. It has also been predicted [14] that particle shape can dramatically affect firm adhesion under flow, proposing the use of oblate spheroidal rather than classical spherical particles, as oblate-shaped particles are subjected to torques resulting in tumbling and rotation [143, 145]. These complex dynamics of non-spherical particles cause translational as well as rotational motions. These theoretical predictions have been supported by both in vitro experiments with silica particles [144] and by in vivo studies with silicon discoidal particles [146] and discoidal polystyrene particles targeted against ICAM-1 molecules, that both exhibited higher tumor accumulation/targeting specificity compared to spherical particles [147]. Furthermore, the particle density also has a strong influence on margination, as generally, lighter particles marginate further than denser counterparts [148].

8.3.4 Biodistribution

For intravenously administered NPs, diameter is an important determinant of pharmokinetics and biodistribution owing to the variable size of fenestrations lining blood vessels. Hydrophilic NPs with diameters smaller than 6 nm, can be eliminated from the body as they can be excreted by the kidneys [149]. However, the majority of injected nanomaterials are larger in size and, thus, accumulate in the organs of the MPS, namely liver, spleen and lungs. For instance, Godin et al. [146] described quantitative analysis of organ distribution of different discoidal particle sizes as compared to spherical particles of the same diameter (Fig. 8.4). They found that nanovectors with diameter of 600 nm accumulated more prominently in the liver, but to a lesser extent in the lungs when compared to their 1000 nm in diameter counterparts ($p<0.05$). For 1700 nm particles the amount of Si increased dramatically in the lungs, liver and spleen, the MPS organs, as well as the heart [150]. A total of 66.4 %, 63.1 % and 68.2 % ID were recovered in the above five major organs and tumor for 600 nm × 400 nm, 1000 nm × 400 nm, and 1700 nm × 400 nm in diameter nanovectors after 4 h. They attributed these numbers to a partial degradation of the discoidal Si particles leading to elevated levels of Si in kidneys without

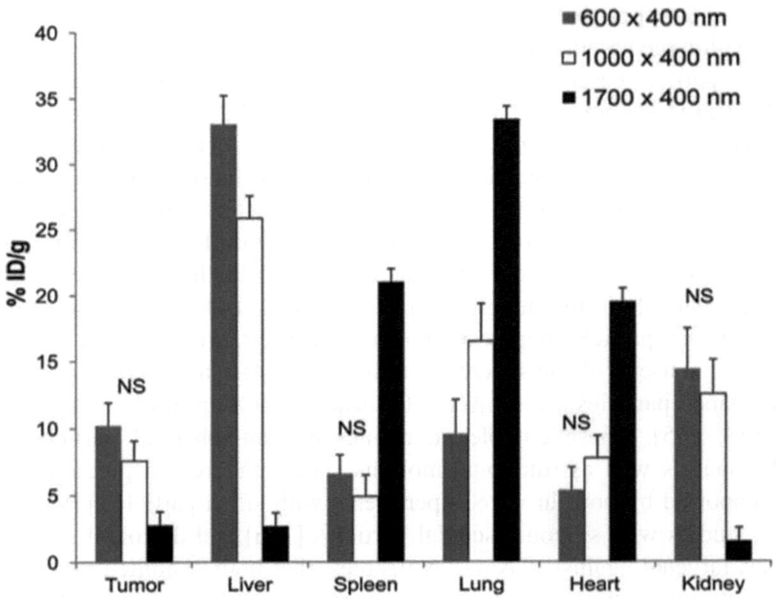

Fig. 8.4 Biodistribution of 600, 1000, and 1700 nm in diameter and 400 nm in thickness discoidal silicon particles 4 h following intravenous administration at the dose of 7.5–9 mg/kg to the tail vein of 4T1 orthotopic breast tumor bearing mice. Silicon analysis normalized to basal silicon levels in the tissues as evaluated by Inductive Coupled Plasma Atomic Emission Spectroscopy (ICP-AES). The data is presented as % of injected dose/g organ. NS, $p<0.05$; NS—not significant [146]

apparent presence of the nanovectors. Another possibility is that a distribution of the nanovectors to other organs/systems such as gastrointestinal tract and pancreas (through entero-hepatic circulation), and lymphatic system occurred [146].

However, the long term fate of non-degradable nanomaterial that are too large to be excreted by the kidneys, remain in the tissues inside the body for up to 8 months [151] and may cause toxicity. Resident macrophages in MPS organs will phagocytose NPs, degrade a small fraction of them, and eventually exocytose both the degraded and intact NPs [152]. The long-term distribution and toxicity (>1 year) of non-degradable particles have not been evaluated, raising concern about toxicity. Cadmium-Selenium (CdSe) quantum dots exhibit hepatotoxicity in cell cultures but do not appear to be toxic in Sprague-Dawley rats on the basis of histopathology as well as liver and kidney marker analysis. In addition, pristine and PEGylated single-walled carbon nanotubes, persisted within liver and spleen macrophages for 4 months without apparent toxicity in rats and rabbits despite their reported toxicity within in vitro culture models. The differential toxicity may be related to a localized concentration of NPs [153]. In vivo, NPs seem to be transported from organ to organ, whereas in vitro, the NPs are confined within a limited cell population. Therefore, the differential toxicity between in vivo and in vitro may be due to exposure concentration.

A study of the biodistribution and pharmacokinetic study of [125I]-labeled non-targeted, cylindrical particles prepared by the PRINT® process in the 200 nm size range in healthy mice showed the expected uptake primarily in the liver and spleen [40, 41]. The conventional strategies to reduce the rapid clearance from the bloodstream and uptake by the liver and spleen have been to increase hydrophilicity of the particle surface and reduce particle size. In an alternative approach, however, Geng and coworkers have compared soft spherical assemblies with flexible filaments and found that the in vivo circulation time for non-spherical filomicelles was about ten times longer than the analogous spherical counterparts [138]. They extended their study to the delivery of paclitaxel and showed significant tumor shrinkage in a xenograft mouse tumor model, and showed that an increase in the filomicelle length had the same relative therapeutic effect as a similar increase in the paclitaxel dosage. These results show that, in applications where a prolonged circulation time is desired, a long, worm-like structure can be more effective than a sphere.

Though particle size has long been considered dominant in determining in vivo behaviours such as circulation time and biodistribution profile, it stands to reason that physical filtration barriers in the body could be navigated by larger, but more deformable particles [154, 155]. Deformable particles that resembled RBCs in size and shape have been shown to deform in restricted channels [156] or capillaries [112] that were smaller than the particle diameter, though the modulus of these particles was not characterized [156] or was poorly matched to RBCs [112] and was restricted to in vitro testing in both cases. In a simulation of renal filtration of soft particles, microgels translocated through pores that were 1/10th of the particle diameter under physiologically relevant pressures [157]. Merkel et al. [158] recently reported the fabrication of a series of discoidal, monodisperse, low-modulus hydrogel PRINT particles with diameters ranging from 0.8 to

8.9 μm, spanning sizes smaller than and larger than RBCs, which were injected into healthy mice. They tracked their concentration in the blood and their distribution into major organs and found that the particles demonstrated some hold up in filtration tissues like the lungs and spleen, followed by release back into the circulation, characterized by decreases in particles in these tissues with concomitant increases in particle concentration in blood. They concluded that particles similar to RBC in size demonstrated longer circulation times, suggesting that this size and shape of deformable particle is uniquely suited to avoid clearance. Another study [138] found filamentous worm-like micelles to have increased circulation times with respect to their spherical counterparts, though this effect was attributed to the size and shape of these particles as well as the deformability. Inspired by nature's example of long-circulating microparticles, Merkel et al., sought to mimic the size and shape of RBCs while varying the modulus across a physiologically relevant range to probe the physical barriers encountered in vivo by soft microparticles. They showed that increasing the deformability of RBC-sized particles increased their circulation times beyond that of conventional microparticles and significantly altered their biodistributions. As a consequence of their low modulus, these discoid microparticles bypassed several in vivo filtration mechanisms, illustrated by animal survival and dramatic increases in elimination half-lives of particles with decreasing modulus [95]. Such deformable and long-circulating particles may find utility in the fields of drug delivery and medical imaging, where long-circulation times and varied biodistributions are often desirable characteristics [159–161] and, hence, particle deformability must be considered as a vital design parameter during the fabrication of therapeutic or diagnostic particles.

8.3.5 Tumor Accumulation

Passive accumulation and physiological effects have been exploited for disease treatment in preclinical models for inflammation [162] and cancer [163] without surface modification or active targeting by ligand attachment. This may reflect the neovasculature associated with these conditions and the suggested enhanced permeability and retention effect (EPR) in tumors in which nanoscale particles migrate and accumulate across leaky vasculature in disease tissue. NPs can accumulate in tumors and be used to deliver therapeutic compounds and contrast agents for imaging. Tumors possess large fenestrations between the endothelial cells of blood vessels produced by angiogenesis and can retain particulates found in the blood. EPR effect allows NPs to accumulate inside the tumor if they are not cleared by the liver or spleen or excreted through the kidney. Longer circulation time in the bloodstream allows repeated passes of nanomaterials through tumor blood vasculatures and favors passive tumor targeting via an EPR effect [164]. To produce long-circulating NPs that can accumulate inside tumor tissues a diameter between 30 nm and 200 nm is desired [165]. Using this approach, researchers may induce passive accumulation of nanomaterials inside a tumor. In fact, it is possible to control the overall accumulation and penetration depth into the tumor by changing the NPs diameter [137] and shape [166]. Researchers determined that the NPs capacity to navigate between the tumor interstitium after extravasation increased with decreasing size. By contrast, larger NPs (100 nm) do not extravasate far beyond the blood vessel because they remain trapped in the extracellular matrix between cells. Thus, the smallest NPs (20 nm) penetrate deep into the tumor tissue but are not retained beyond 24 h.

Surprisingly, adding a targeting moiety on the surface of the NPs does not appear to increase accumulation inside the tumor or change biodistribution. Active targeting of NPs changes the intra-tissue localization of NPs and their increased internalization into cancer cells [167, 168]. However, some studies have shown active targeting increases tumor accumulation of NPs. For instance, Godin et al. [22] discuss how their MSV particle is able to release 'Stage 2' NP (S2NP), such as cytotoxic agents, contrast enhancing imaging agents, bioactives or metal nanoparticles after the S1MP has adhered to the diseased vasculature. In the case of metal S2NP (e.g. gold, silver and iron oxide), the activation of the system can be triggered with an external energy source, such as radio-frequency [169] or near-infrared [170] energy. Another possible mode of action includes targeting specific cell elements, such as macrophages and myofibroblasts, in the disease location to achieve preferential localization. The intracellular payload can then be released to destinations influenced by S1MP and S2NP surface chemistry. However, owing to the many contradictory findings in the literature, it is not certain how active targeting affects NP accumulation in the target tissue [168], however, Frieboes et al. [171] recently presented a computational model which predicts vascular accumulation of blood-borne NPs. The fraction of injected NPs depositing within the diseased vasculature and their spatial distribution was computed as a function of tumor stage, from 0 to 24 days post-tumor inception. They explain that NP vascular affinity, interpreted as the likelihood for a NP in the blood to firmly adhere to the vessel walls, is a fundamental parameter and depends on NP size and ligand density, as well as vascular receptor expression. They go onto show that for high vascular affinities, NPs tend to accumulate mostly at the inlet tumor vessels leaving the inner and outer vasculature depleted of NPs, whereas, for low vascular affinities, NPs distribute quite uniformly intra-tumorally but exhibit low accumulation doses.

Unfortunately, tumor accumulation generally represents a relatively small fraction of total injected dose, due to a number of factors, such as high intra-tumoral pressures and the architecture of vasculature networks. For instance Frieboes et al. [171] have recently demonstrated that for insufficiently developed vascular networks, such as those seen within tumors, NPs are transported preferentially through the healthy, pre-existing vessels, thus, bypassing the tumor mass. However, a few recent studies with discoidal particles have shown enhanced accumulation in the tumor based on particle geometry [146, 166]. For instance, a high accumulation of discoidal porous silicon (pSi) nanovectors into orthotopically grown breast tumors in mice has been recently reported [146]. For 600 nm × 400 nm pSi particles, 10.2 % ID/g was found in the tumor, which is five times higher than their spherical

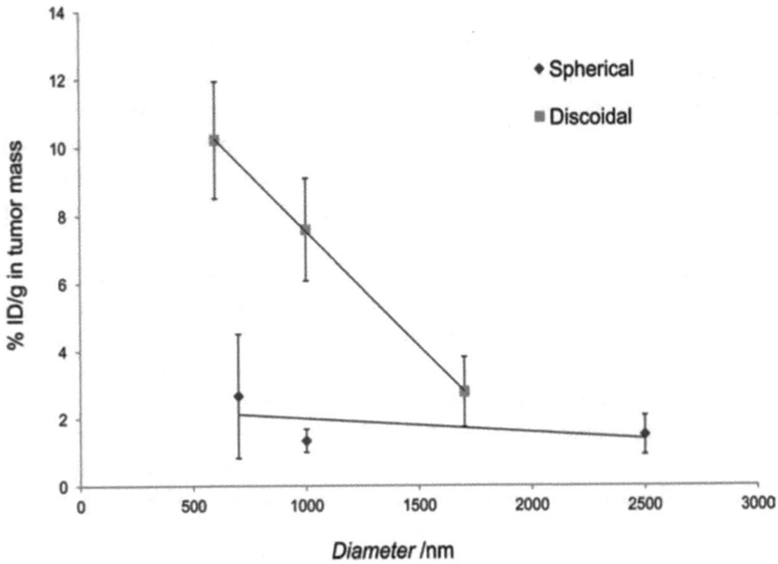

Fig. 8.5 Effect of particle diameter on tumor accumulation of discoidal particles as compared to spherical particles. For discoidal particles, the height is 400 nm [146]

counterparts (Fig. 8.5). Their results were achieved without the use of any active targeting strategy, and, therefore, emphasized the effect of geometry on in vivo behaviour of therapeutic particles.

In an additional study Van deVen et al. [166] used intravital microscopy and elemental analysis to compare the in vivo localization of particles with different geometries and demonstrated that discoidal (plateloid) particles preferentially accumulate within the tumor vasculature at unprecedented levels, independent of the EPR effect. They found that in melanoma-bearing mice, 1000 × 400 nm plateloid particles adhered to the tumor vasculature at about 5 % and 10 % of the injected dose per gram organ (ID/g) for untargeted and RGD-targeted particles respectively, and exhibited the highest tumor-to-liver accumulation ratios (0.22 and 0.35), however smaller and larger plateloid particles, as well as cylindroid particles, were more extensively sequestered by the liver, spleen, and lungs. The authors concluded that plateloid particles appeared well-suited for taking advantage of hydrodynamic forces and interfacial interactions required for efficient tumoritropic accumulation, even without using specific targeting ligands [166].

Improvements in fabrication protocols have allowed to create a potential toolbox for increased targeting efficacy. For instance, in vitro and in vivo studies have demonstrated that liposomal carriers slightly improved the delivery of siRNA to melanoma, lung cancer, breast cancer, and ovarian cancer [172, 173]. The efficacy of hemispherical 1.6 μm pSi particles loaded with neutral dioleoyl phosphatidylcholine nanoliposomes (DOPC) containing EphA2-specific siRNA was evaluated by Tanaka and colleagues in two independent orthotopic mouse models of ovarian cancer [174]. EphA2 is an oncoprotein overexpressed in most malignancies including ovarian tumors. After a single treatment with EphA2-targeted-MSV and without simultaneous chemotherapy, gene silencing and decrease in tumor burden was observed, evaluated by cell proliferation and angiogenesis. A dose twice as high and administered twice a week for 3 weeks was required to achieve a similar effect with siRNA-DOPC, alone. The mechanism of the efficient and sustained liposomal siRNA delivery was due to surface modification, tissue distribution, and slow biodegradation of the pSi particles (S1MP). S1MP not only served as storage for liposomal siRNA, but also shielded siRNA oligos extensive and prompt degradation by nucleases in the bloodstream. This novel approach enables new avenues in personalization of siRNA therapeutics through controlled delivery of synergistic payloads in a time-dependent manner.

8.4 Interactions with Cells

8.4.1 Binding to the Cell Membrane

Once the theranostic particle has reached the target cell, it faces the plasma membrane, which is the first barrier for the intracellular delivery of drug or imaging payloads. NPs were initially considered as benign carriers, but multiple studies have demonstrated that NPs' interaction with serum proteins and cell membrane receptors is determined by their design, in effect, influencing cell uptake, gene expression, and toxicity. Interactions between NP-bound ligands and cellular receptor depend on the engineered geometry, NP arrival rate, surface charge and the ligand density of a nano or micro-material (Fig. 8.6).

The particle size and shape dictates the number of ligands that interact with the receptor target. A multivalent effect occurs when multiple ligands on the NPs interact with the multiple receptors on the cells. The binding strength of complexed ligands is more than the sum of the individual affinities and is measured as the avidity for the whole complex. Similarly, on NP surfaces, the presence of ligands at a given density over a specific curvature will contribute to the overall avidity of the NP-linked ligands for available receptors. To illustrate this, the binding affinity of Herceptin to the ErbB2 receptor is 10^{-10} M in solution, 5.5×10^{-12} M on a 10 nm NP, and 1.5×10^{-13} M on a 70 nm NP [175]. This example shows how a ligand's binding affinity increases proportionally to the size of a NP owing to a higher protein density on the NP surface. However, when viewed in terms of the downstream signalling via the ErbB2 receptor, 40–50 nm gold NPs induced the strongest effect, suggesting other factors beyond binding affinity must play a role. Nevertheless, several studies have shown that NP design can cause differential cell signalling when compared with free ligand in solution. For example, the previously mentioned 40–50 nm Herceptin-coated gold NPs altered cellular apoptosis by influencing the activation of caspase enzymes [175]. Similarly, receptor-specific peptides improved their ability to induce angiogenesis

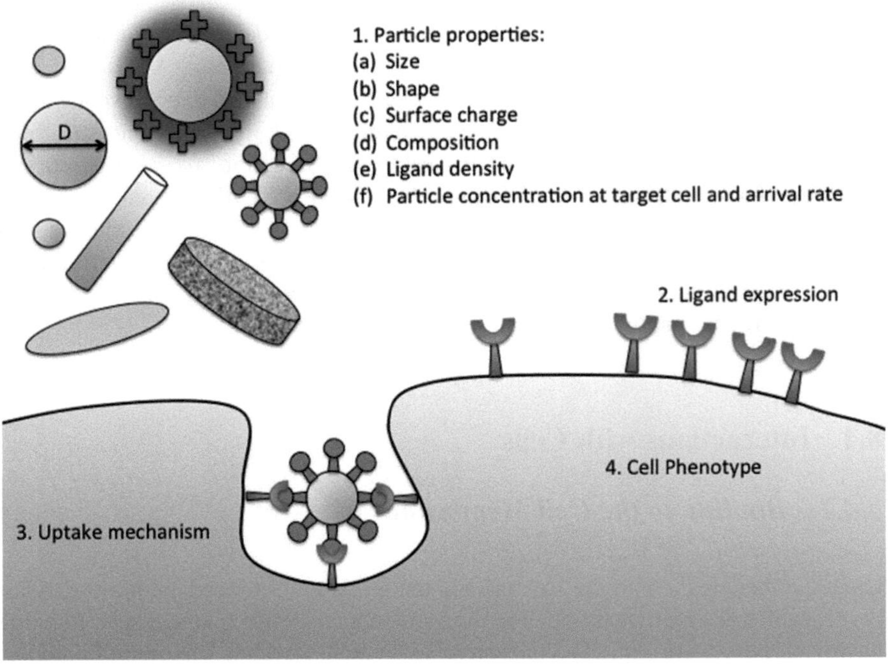

Fig. 8.6 Schematic presentation of the effect of physicochemical features of the particles on cellular uptake. Size, shape, surface charge, composition, ligand density and arrival rate influences a particles interaction with the surface of a specific cell

when conjugated to a NPs surface [176]. These findings highlight the advantage of having a ligand bound to a NP as opposed to it being free in solution. The NP surface creates a region of highly concentrated ligand, which increases avidity and, potentially, alters cell signalling.

Despite their advantages, a general concern with NP-ligand complexes is the potential denaturation of proteins when bound to the engineered NP surface. The denaturation of a protein can affect binding to its receptor, increase nonspecific interactions or provoke inflammation. For instance, when lysosome is bound to gold NPs it denatures and interacts with other lysozyme molecules and induces the formation of protein-NP aggregates [177]. Fibrinogen also unfolds when bound to the surface of polyacrylic acid-coated gold NPs, which means it is able to bind to the integrin receptor Mac-I which leads to an inflammatory response [178].

Surface charge is also important in dictating membrane interaction and subsequent cellular fate. Compared with NPs with a neutral or negative charge, positively charged NPs are taken up at a faster rate [179, 180]. It has been suggested that the cell membrane possesses a slight negative charge and cell uptake is driven by electrostatic interactions [181, 182]. A recent study demonstrated that this electrostatic attraction between the cell membrane and positively charged NPs favours adhesion onto a cell's surface, leading to uptake. Negatively charged heparan sulfate proteoglycans and integrins play a role in the cellular binding of positively charged particles. For example, Mislick and Baldeschweiler showed that the binding of cationic particles to the cell membrane was significantly reduced in proteoglycan-deficient mutant cells [122]. Even so, heparan sulfate proteoglycans may induce non-specific binding but the exact role they play in cellular uptake is not clear, and also non-adherent cells do not contain such molecules, which seems to result in lower uptake [183]. The clustering of trans-membrane proteins and syndecans at the plasma membrane during binding to cationic particles may stimulate interaction with the actin cytoskeleton and result in the formation of tension fibers. This tension provides the energy required for internalization to take place [184].

For small NPs (2 nm), a positive charge can perturb the cell membrane potential, causing Ca2+ influx into cells and the inhibition of cell proliferation [185]. For larger NPs (4–20 nm), surface charge induces the reconstruction of lipid bilayers [186]. It has been reported that NPs conjugated with polycationic molecules interact with fluid-phase domains of lipid bilayers [187] and can induce the development of plasma membrane pores leading to membrane disruption. Binding of negatively charged NPs to a lipid bilayer causes local gelation, whereas binding positively charged NPs induces fluidity. Several studies have confirmed the pivotal role surface charge plays in downstream biological responses to NPs.

Additionally, Particle-membrane interactions can be altered by the development of a protein corona which is caused by particle interactions with serum proteins such as lipoproteins, immunoglobulin, complement and coagulation factors, acute phase proteins and metal-binding and binding proteins [188] during circulation in the blood. This is particularly relevant for NPs with high surface charge density. As an example, Dawson and colleagues have demonstrated that positively charged polystyrene particles adsorbed a layer of plasma proteins changing the biological fate of the particles and treated cells. Namely, the corona enabled the protection of the cells from the damage induced by the highly positive NP surface, thus, significantly altering any particle-cell interaction [189].

Therefore, the measurement of surface charge plays a key role in the design of novel theranostic NP approaches. The Zeta-potential (ζ) is one of the most important micro or nano scale parameters in the interaction between particle and cell. Whilst ζ is relatively straightforward to measure, its relation to the structure and charge content of an agent can be complex and so care must be taken when interpreting ζ readings. Zeta potential is an energy potential that arises due to electrical charge on the particle. In particular the particle charge produces a distance dependent potential, ϕ that is described in 1-D by the Poisson equation:

$$\frac{d^2\phi}{dx^2} = -\frac{\rho}{\varepsilon} \tag{8.1}$$

where ρ is the charge density on the particle and ε the permittivity of the medium. Solution of the Poisson equation describes a potential, whose value is dependent upon the charge density on the particle and which decays exponentially with increasing distance, x from the particle. In the majority of practical situations there is further complexity due to interactions of ions within the local environment with the particle surface charge. Ions of opposite charge to that of the particle will firmly bind due to electrostatic attraction to form a layer known as the Stern layer. Beyond this there are further electrostatic interactions between ions and particle but these are mediated by the presence of the bound ions within the Stern layer. It is the electric potential at the outer edge of the Stern layer that is defined as the ζ-potential. These interactions of solvent ions with particle make the ζ value dependent upon the properties of the medium (and hence experiment specific) as well as those of the particle.

ζ is usually measured indirectly via the electrophoretic mobility, μ, which quantifies the motion of the particle due to forces induced by an applied electric field. In practice a voltage is applied and particle motion measured, usually by optical means. The value of ζ is then given by:

$$\xi = \frac{\mu n}{\varepsilon_0 \varepsilon_r} \tag{8.2}$$

where η is the viscosity of the medium.

Often ζ is used as a ready 'look up' value to assess the effects of particle charge, however this can be misleading. As we have already seen the medium is intrinsically linked to the ζ value and so realistic conditions must be used, e.g. in in-vitro studies measurements in water will not accurately reflect the value pertaining to exposure within a cell culture medium containing a multitude of charged molecules. The size of the charged particle must also be carefully considered. Equation 1 shows that the value of ζ is dependent upon charge density and so different sized particles with the same ζ value will have different total charge. This may be important if considering effects where the sum charge amount is the determining factor. For example, if cationic particles of 2 nm and 10 nm diameter and the same ζ are used to create a proton sponge effect for endosomal escape, very different results should be expected as the total charge of the 10 nm particles is 25 times greater than that of the 2 nm.

The interaction of charged particles is known to be size dependent and so again the zeta potential is not the only parameter of importance. Consider the attraction energy per unit area due to Van der Waal's forces, for two identical spheres of radius, r and charge density, ρ_S this is:

$$W = -\frac{\pi^2 C \rho_s^2}{12d} \cdot r \tag{8.3}$$

where d is the separation distance of the spheres and C is a constant of proportionality. Thus the strength of interaction is determined by r, irrespective of the charge (and hence ζ) properties of the particle. This is true for a range of geometries, for instance consider the case for a sphere interacting with a charged plate, i.e. a physical model akin to membrane-particle interaction. Here the attraction energy is again dependent upon r and given by:

$$W = -\frac{\pi^2 C \rho_s \rho_p}{6d} \cdot r \tag{8.4}$$

In summary, ζ is a useful, and hence widely used, parameter that characterizes the electrical potential of charged particles. However care must be used when using ζ measurements in studies involving multiple particle types for they relate to charge *density* on the surface, i.e. particle size is factored out through this *per unit area* measure. The particle radius will still determine the total charge and the spatial arrangement of that charge in respect to structural components of the cell. Thus different sized particles may exhibit very different effects on cells despite having similar ζ values.

A complete understanding the measurement of ζ will foster a more a comprehensive understanding of the initial interaction between the particle and the cell membrane and will enable us to reduce cytotoxic effects associated with cationic polymers and simultaneously increase therapeutic efficacy. Furthermore, it is generally assumed that the initial interaction of the particle with the cell membrane is closely linked with the uptake pathway and subsequent intra-cellular trafficking and, hence, its therapeutic efficacy within the cell. Therefore, a promising strategy for increasing the efficiency of a drug or gene delivery vector is to target specific internalization pathways that improve the intra-cellular fate of the particle vectors.

8.4.2 Cellular Uptake

The plasma membrane presents a substantial barrier to particles successfully gaining entry to the cell, as it is a dynamic and a relatively lipophilic structure that restricts the admittance of large, hydrophilic, or charged molecules As is well documented, NPs can be transported into cells through a process called endocytosis [190]. In particular, endocytosis occurs in all cell types and is mediated by at least four basic mechanisms: macropinocytosis (Fig. 8.7), clathrin-mediated endocytosis, caveolae-mediated endocytosis, and clathrin–caveolae and dynamin-independent endocytosis [191] (Fig. 8.8).

It is generally accepted that systems which are less than 200 nm in size [8, 192] and that have not been functionalized with cell surface receptors, such as transferrin or low-density lipoprotein receptors for specific binding [193–196], are internalized via clathrin-mediated endocytosis (CME) [8, 192]. CME is considered kinetically the most effective [197, 198] endocytic uptake pathway for various particles. However, a limitation of utilizing this pathway for theranostic delivery is that the internalized therapeutic drug or imagine contrast agent is typically destined for acidic and/or enzymatic degradation in the late endosome or lysosome in the final stages of the process [199, 200]. Thus, when nanovectors are taken up via CME, it is advantageous for them to be used in conjunction with a polymer material, such as PEI [201], polyamidoamine (PAMAM) dendrimers [201] or an imidazole-containing

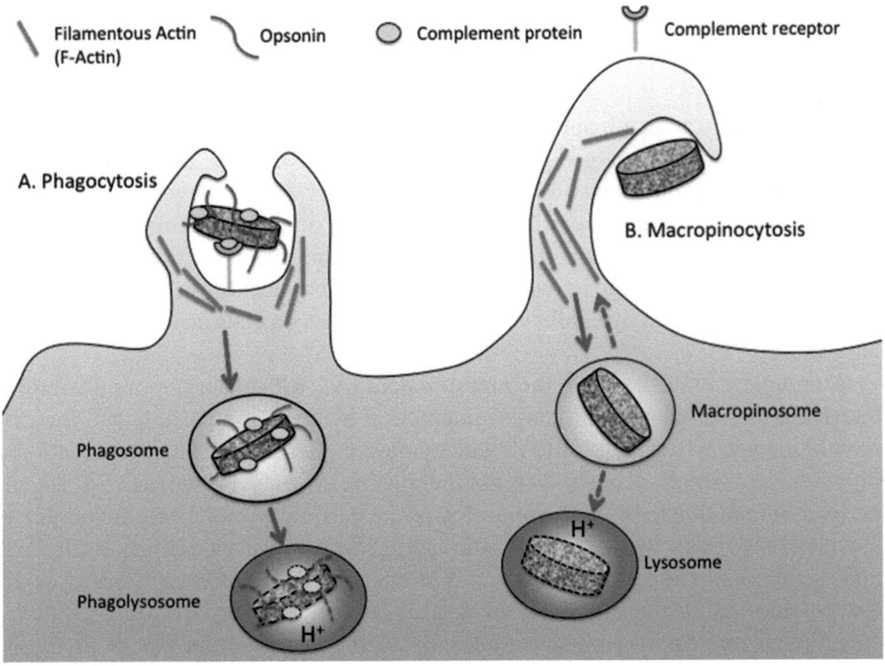

Fig. 8.7 Intracellular uptake of particles. (**a**) Phagocytosis is an actin-based mechanism occurring primarily in cells associated with Mononuclear Phagocytic System (MPS), such as macrophages, and is closely associated with opsonisation of complement proteins and other opsonins. However, fibroblasts, epithelial and endothelial cells may also display some lower phagocytic activity [235]. (**b**) Macropinocytosis is an actin-based pathway, engulfing NPs and other biological constituents with poor selectivity

polymers [202], that are capable of inducing endosomal escape of the polyplex into the cytoplasm of the cell before significant degradation occurs. Other desirable internalization pathways include macropinocytosis [201, 203], micropinocytosis [203] and caveolae-mediated endocytosis. Caveolae-mediated endocytosis maybe an advantageous internalization mechanism for gene transfection strategies as the vesicles that result from the internalization in this pathway do not develop into lysosomes and, therefore, avoid significant degradation [204].

The endocytic process occurs through either specific or nonspecific cellular uptake depending on the surface properties of NPs. In many cases, NPs enter the cell after binding to the receptor target. Once bound, the internalization of particle systems can vary greatly depending on a number of factors such as particle chemistry, cell type, cell polarization state and stage of cell cycle [205]. For instance, a NPs shape can directly influence uptake into cells, Gratton et al. [15] determined that rods were taken up the most, followed by spheres, cylinders, and cubes in synthesized NPs larger than 100 nm. In studies with sub-100 nm particles, spheres show an appreciable advantage over rods [206, 207]. In fact, at this size range, increasing the aspect ratio of nanorods seems to decrease total cell uptake. The effects of shape

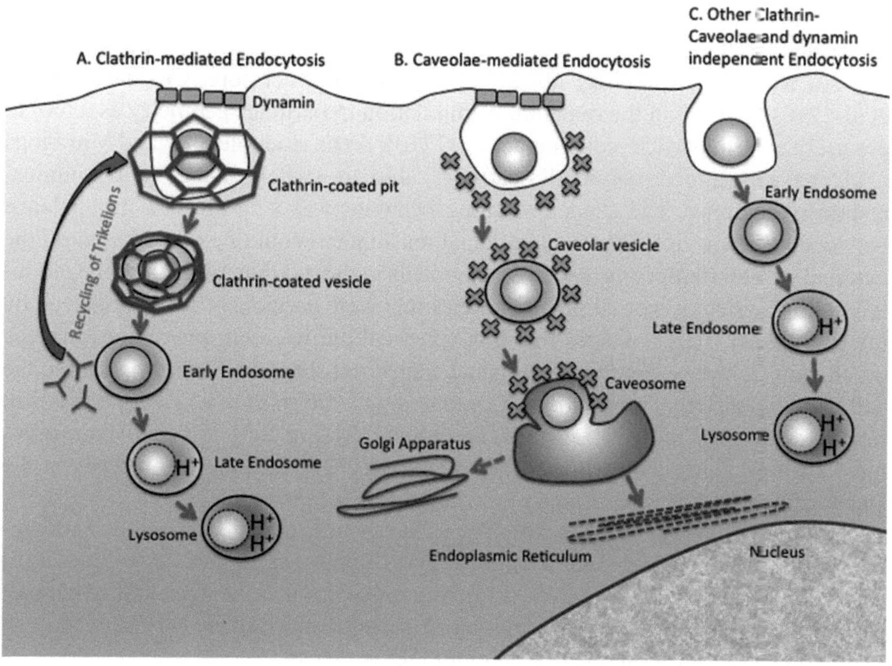

Fig. 8.8 Intracellular trafficking of nanovectors following internalization via clathrin-mediated endocytosis, caveolae-mediated endocytosis and other clathrin-caveolae and dynamin independent endocytosis. (**a**) Clathrin mediated endocytosis. The assembly of clathrin triskelions into a polygonal matrix helps deform the overlying plasma membrane to form a clathrin coated pit After assembly of the cup-like clathrin matrix, dynamin is employed at the neck of the pit to mediate membranal fission. This leads to the release of the clathrin-coated vesicle to the cytoplasm. The following un-coating of the vesicle allows clathrin triskelion recycling. Clathrin-coated vesicles then form early endosomes, which are acidified and fuse with pre-lysosomal vesicles containing enzymes to give rise to late endosomes and eventually lysosomes. This pathway leads to substantial nanovector and drug degradation mediated by the acidic environment and enzyme action. (**b**) Caveolae-mediated endocytosis of a nanocarrier gives rise to a caveolar vesicle that is delivered to a caveosome, which does not have a degradative acidic and enzyme-rich environment

and size of metallic particles on cellular internalization were also reported by Xu et al., using layered double hydroxide Mg6 Al2 nanoparticles with two distinct particle morphologies: (1) rods that were 30–60 nm in width and 100–200 nm in length, and (2) hexagonal sheets that were 50–150 nm wide (laterally) and 10–20 nm thick [208, 209]. The group found that both shapes are quickly taken up by CHO-K1, NIH 3T3, and HEK 293T cell and that rod-like particles specifically target the nucleus while the sheet-like particles are retained in the cytoplasm [208, 209]. In a study focusing on the uptake rates of length-fractionated single-walled carbon nanotubes (SWNT) 130–660 nm in diameter, it was suggested that nanoparticulate aggregates on the cell membrane form a cluster which is sufficient in size to generate a large enough enthalpic contribution for overcoming the elastic and entropic energy barriers associated with membrane vesicle formation. Endocytosis rate for nanotubes was 1000 times higher than for spherical gold nanoparticles, while similar exocytosis kinetics was measured for poly(D,L-lactide-co-glycolide), SWNT, and Au nanoparticles across distinct cells [182]. Furthermore, Caldorera-Moore et al. [210] focused on the dynamic manipulation in particles geometry as a tool to precisely control particle-cell interactions [210]. As an example, Yoo and Mitragotri [211] have designed polymeric particles able to switch shape in a stimulus-responsive manner. The shape-switching behavior was a result of a fine balance between polymer viscosity and interfacial tension and could be tuned based on the external stimuli (temperature, pH, or chemicals in the medium), enabling the modulation of phagocytosis of elliptical particles that were previously not internalized by the cells [211]. This indicates that non-spherical particle-cell interactions may be much more complex. For instance, ligand coated rod shaped NPs may present to the cell with two different orientations. Compared with the short axis, the long axis will interact with many more cell surface receptors [212] (Fig. 8.9). For spiky nanostructures such as gold nanourchins, whether the ligand is located on or between the spikes affects how it is presented to the target cell receptors [213].

Serda et al. have shown that discoidal S1MP vectors are internalized by endothelial cells through a combination of phagocytosis and macropinocytosis [214, 215] (Fig. 8.6). Both mechanisms are actin-driven processes that involve extensive membrane reorganization with the formation of pseudopodia and extension of the cell membrane to surround and actively engulf the particles.

Within a given geometric shape, a nanomaterial's dimensions are a strong determinant of total cell uptake. For gold spherical NPs, silica particles, single walled carbon nanotubes, and quantum dots, a 50 nm diameter is optimal to maximize the rate of uptake and intracellular concentration in certain mammalian cells [182, 212, 213]. In addition to size and shape, the composition of the nanomaterials also affects the uptake as single-walled carbon nanotubes and gold NPs, each 50 nm in diameter, possess endocytosis rates of 10^{-3} min^{-1} and 10^{-6} min^{-1}, respectively. This 1000-fold difference may be due to the intrinsic properties of carbon versus gold. Which ligand is used to coat the NP will also affect the downstream biological responses. For example, the uptake and cytotoxicity of NPs were significantly altered when the NPs were coated with two different proteins targeting the same receptor [181].

If a NP is bound to a receptor on the cell membrane then it will most likely enter the cell via receptor-mediated endocytosis [175, 207, 216]. The binding of the NPs-ligand conjugate to the receptor produces a localized decrease in Gibbs free energy, which induces the membrane to wrap around the NP to form a closed vesicle structure [217]. The vesicle eventually buds off the membrane and fuses with other vesicles to form endosomes, which fuse with lysosomes where degradation occurs. The size dependent uptake of NPs is likely related to the membrane wrapping processes. Small NPs coated with 50 kDa proteins may interact with only one or two cell receptors. By contrast, a 100 nm NP has many more ligand-receptor interactions per particle. Several small-ligand coated NPs must bind to receptors in close proximity to one another to produce enough free energy to drive membrane wrapping. Larger NPs can act as a cross-linking agent to cluster receptors and induce uptake. Thermodynamically, a 40–50 nm NP is capable of recruiting and binding enough

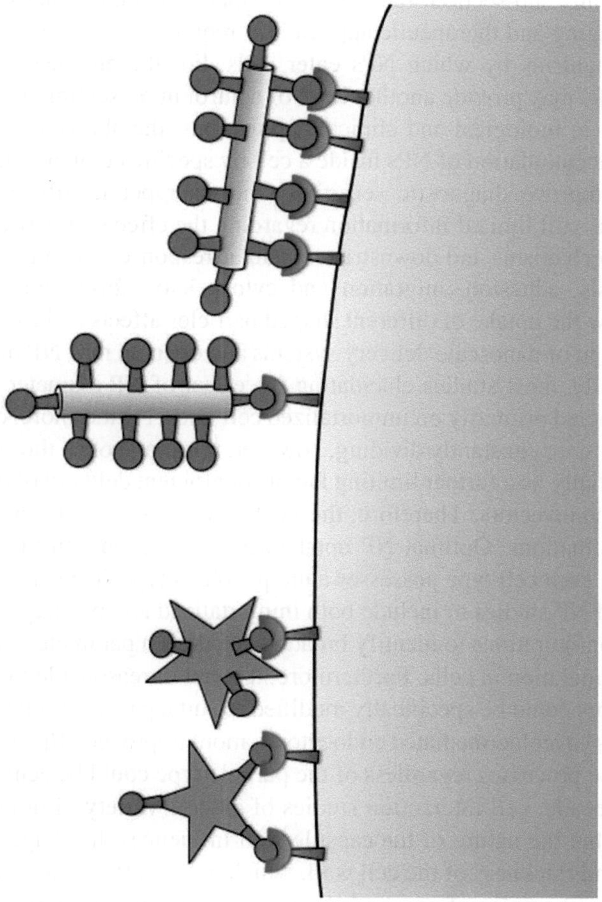

Fig. 8.9 Geometry of the nano- and micro-particles has a dramatic impact on the presentation of the NP at the cell interface, which affects cell membrane interaction, uptake and long-term fate of the particle

receptors to successfully produce membrane wrapping. Above, 50 nm, NPs bind such a large number of receptors that uptake is limited by the redistribution of receptors on the cells surface via diffusion to compensate for local depletion. NPs larger than 50 nm bind with high affinity to a great number of receptors and may limit binding of additional particles. Mathematically modeling this phenomenon has demonstrated that optimal endocytosis occurs when there is no ligand shortage on the NP, and no localized receptor shortage on the cell surface [218], which occurs in NPs 30–50 nm in diameter where ligand density is optimal.

To conclude this sub-section, the design of smart functional nanosystems for intracellular imaging and therapeutic applications requires a thorough understanding of the mechanisms by which NPs enter cells. For the engineering process, asymmetrical NPs may provide another level of control in presenting ligands to the target receptors. In biological and clinical applications, the ability to control and manipulate the accumulation of NPs inside a cell by specific cellular uptake makes it possible to improve diagnostic sensitivity and therapeutic efficiency [219]. However, there is still limited information regarding the effects of particle geometry on uptake mechanisms and downstream cellular responses, such as cell proliferation, apoptosis, adhesion, migration and cytoskeleton formation. Therefore, investigating how the uptake of different shaped particles affects cell functions will assist in the design of nanoscale delivery systems and open up new NP bio-applications. Additionally, most studies elucidating the effect of NP diameter on uptake have been conducted primarily on immortalized cell lines. Furthermore, cells under culture conditions are constantly dividing, however, this may not be the case in vivo and could potentially be a further limiting factor for efficient delivery of therapeutic and diagnostic nanovectors. Therefore, the nuclear envelope cannot be neglected within in vivo situations. Optimal NP uptake size may depend on the cell being assayed because each cell type possesses a unique phenotype. Thus, it will be necessary to expand NP studies to include both immortalized and primary cells in different culture configurations to identify broad-scope design parameters for optimal uptake and accumulation in cells. Furthermore, it is not unreasonable to generalize that if nanovectors could be specifically modified to suit a particular internalization pathway, such as caveolae-mediated endocytosis, more improved efficiencies in the endosome escape processes regardless of the particle type could be achieved.

In the past decade, cell interaction studies of diverse variety of hollow microcapsules show that the nature of the capsule wall influences the shape, elasticity, deformability and thickness of the cell wall, which in turn affects the cell recognition and uptake. Examples of some of the factors influencing microcapsule-cell interaction are depicted here.

Spherical poly(methacrylic acid) capsules were internalized by HeLa cells at a faster rate than rod-shaped capsules with a higher aspect ratio, but both capsule types were subject to lysosomal compartmentalization [220]. It must be noted that in case of sized rigid particles, rod-shaped particles with higher aspect ratio showed a higher internalization rate [15].

In terms of particle elasticity, theoretical analysis and molecular modeling of cell uptake suggested that rigid particles are enfolded by membrane more entirely and more easily than softer particles implying that elasticity of the particles play a crucial role in cell internalization [221]. Polystyrenesulfonate/polyallylamine hydrochloride multilayer capsules were shown to have no impact on normal rat kidney (NRK) cells at low concentrations of about 5 particles/cell. At higher concentrations, however, the capsules were found to settle over the cells and affect their viability. When these capsules were embedded with inorganic ions such as cadmium telluride (CdTe), toxic ions were released and impacted the cell viability drastically [222].

Humanized A33 monoclonal antibody (huA33 mAb) functionalized poly(allylamine hydrochloride)/poly(sodium 4-styrenesulfonate) capsules of ~500 nm in diameter, have shown to specifically target colorectal cancer cells and trigger internalization via receptor-mediated endocytosis [105]. In case of mixed cell populations, huA33 mAb functionalized PVPON capsules of diameter around 800 nm, showed significant uptake suggesting antibody-independent internalization such as macropinocytosis [104]. Micron sized capsules were taken up by cancer cells be means of phagocytosis, while the nature of the polymer decided the integrity and intracellular localization of the capsules—non-biodegradable, synthetic PSS/PAH microcapsules remained intact and in the cytoplasmic region after 3 h incubation with cells, while the biodegradable DXS/PRM microcapsules were found to be degraded and excluded from nuclei in the perinuclear region [106].

Hollow microcapsules fabricated with layer-by-layer assemblies of poly-(sodium 4-styrene sulfonate) [PSS] and poly-(allylamine hydrocholoride) [PAH] have shown to be biocompatible and applicable for 3D tumor tissue studies comprising of cells such as C6 glioma and 3T3 fibroblasts [223]. Polyelectrolyte microcapsules consisting of layer-by-layer coatings of dextran sulfate and poly-L-arginine layers, with the number of bilayers, contributing to the integrity of the capsules in vivo, have been subcutaneously injected into mice eliciting a moderate immune reaction, with a majority of the injected capsules internalized within 16 days and also initiate degradation [96]. Multilayer hollow capsules thus show promising potential as drug delivery agents capable of cell-specific targeting, by manipulating their polymeric configurations and geometry.

8.4.3 Intracellular Trafficking

After internalization via endocytosis, the particles are predominately localized within endosomes which either fuse with lysosomes for degradation or recycle their contents back to the cell surface. The behavior of the particles in endo-lysosomal vesicles is not fully elucidated. Whenever the particles are decorated with targeting ligands, these can be cleaved by the protease cathepsin L inside endo-lysosomal vesicles [224]. In some cases (e.g. quantum dots), there is an evidence of the slow particle core structure decomposition by the lysosomal enzymes [152]. In general, the escape from endosomes is favorable, since the therapeutic payload is intended to act in the cell compartments such as nucleus and cytoplasm and the accumulation of particles in endosomes followed by the degradation in endo-lysosomes strongly limits the efficiency of a payload transfer.

If a NP is engineered to escape the endo-lysosomal system, it can enter the cytoplasm, where it may interact with intracellular organelles potentially affecting cell behavior. Cytoplasm is not enzymatically-inert and contains nucleases and proteases that can neutralize the active compound. For instance, it was shown that plasmid DNA in is degraded in the cytoplasm of HeLa and COS cells with a half-life of 50–90 min [225]. Additionally, the cytoplasm is a viscous environment crowded with molecules, characterized by the decreased mobility of macromolecules [226–228].

There are controversial reports on the effect of the particle size on the intracellular trafficking. As an example, compartmentalization of cadmium-telluride (CdTe) quantum dots into subcellular organelles or their ability to transverse the nucleus was reported to be affected by the particle size [229]. On contrary, the uptake of silver NPs and quantum dots into macrophages was shown to induce the expression of inflammatory mediators such as TNF-α, MIP-2 and IL-1β independent on size [152, 230]. In addition to modulating NPs size for controlling their intra-cellular fate, coating a NP with cationic polymers, such as polyethylenimine (PEI), may initiate endosomal escape via so-called "proton sponge effect" due to the buffering inside endosomes. PEI causes additional pumping of protons into the endosome, along with a concurrent influx of chloride ions to neutralize the net charge, this causes an increase in ionic strength inside the endosome and leads to osmotic swelling, physical rupture of the endosome, and the escape of the vesicle material into the cytosol avoiding the degradative lysosomal trafficking pathway [231]. This ability for osmolytic endosomolysis has been shown to be dependent on the particle size, with small PEI800 or PEI25/DNA complexes formed in water having lower transfection efficacy compared to larger complexes formed at higher ionic strength. The localization of NPs in the intracellular space may also be directed to mitochondria. When in the cytosol, NP can elicit biological responses by disrupting mitochondrial function, eliciting production of reactive oxygen species and activation of the oxidative stress mediates signaling cascade [232]. The production of reactive oxygen species can have detrimental effects on the mitochondrial genome, induce oxidative DNA damage, and promote micronuclei formation [233]. Furthermore, certain types of NPs can induce nuclear DNA damage, leading to gene mutations, cell cycle arrest, cell death, or carcinogenesis. Demonstrating the later, hydrophilic titanium oxide NPs will persist unless they are sorted back into the endolysosomal system where they can be exocytosed. NPs that persist in the cytosol during mitosis will be distributed within the daughter cells [234]. However, the effect of the NPs on subsequent cell generations remains unclear. To date, there is no consensus on the toxicity and properties on NPs inside the cytoplasm. Additional work is still required before researchers fully understand the fate and toxicity of internalized NPs.

8.5 Concluding Remarks

Whilst the biologically driven interactions of particulates based on specific recognition moieties are being investigated for the past few decades, effects of physical parameters and especially geometry of the nanovectors still pose more questions than answers. Recent advances in fabrication of particles with various geometries, based on above described top-down and bottom-up approaches, allowed to isolate these parameters for careful systematic analysis. For example, lithography methods used in microelectronics industry are becoming valuable tools for imparting highly reproducible size, shape, and composition of the produced particulates. It is commonly agreed that the expected use of novel nano and micro-particles will lead to a substantial improvement in the delivery of therapeutic and diagnostic payloads to diseased loci. Though a large number physicochemical compositions have been identified and studied for this purpose, there are still questions about how the physical and chemical structure of the particle affects its interaction with biological milieu. Namely, the particle shape determines the kinetics it displays within the bloodstream and its uptake at the target cell. Studies have shown that nanovectors can be efficiently rationally designed based on theoretical models of flow in the bloodstream and interactions with cells at the target site. This rational design is based on changes in the physical factors such as flow, pressure, porosity in the tissue of interest. Moreover, a combination of physical and biological features of the particles can synergistically attenuate their accumulation in the desired organ. In this regard, the thorough advanced characterization of the particles to determine surface properties such as graft density of coatings or ligands (e.g. with X-ray photoelectron spectroscopy) is very important to determine the elemental composition as the surface properties have to be standardised between particles of different geometry. A detailed understanding of particle-cell interaction from molecular to tissue level as a result of the particle's geometry will allow improved particle-based drug and imaging payload delivery systems to be developed.

To conclude, recent developments in micro and nano-fabrication of rationally designed shape and size-specific vectors, show great promise in overcoming bio-barriers en route to the disease loci. Further incorporation of stimuli-responsive biomaterials into the rationally designed vectors will enable the development of conceptually novel drug delivery systems.

Acknowledgments The authors acknowledge the Swansea University and Houston Methodist Research Institute Joint PhD Graduate Program. BG and MW acknowledge the financial support from NIH U54CA143837 Physical Sciences and Oncology grant. BG acknowledges the financial support from NIH 1U54CA151668-01 Cancer Centre for Nanotechnology Excellence grant. The authors would also like to acknowledge James Gu, Xuewu Liu and Rita Serda for providing SEM images.

References

1. Wong S, Pelet J, Putnam D (2007) Polymer systems for gene delivery—past, present, and future. Prog Polym Sci 32:799–837
2. Zamboni W (2008) Concept and clinical evaluation of carrier-mediated anticancer agents. Oncologist 13:248–260
3. Peer D et al (2007) Nanocarriers as an emerging platform for cancer therapy. Nat Nanotechnol 2(12):751–760

4. Godin B et al (2010) Nanoparticles for cancer detection and therapy. Nanotechnology. Wiley, Weinheim
5. Nie S et al (2007) Nanotechnology: applications in cancer. Annu Rev Biomed Eng 9:257–288
6. Sun X et al (2005) An assessment of the effects of shell cross-linked nanoparticle size, core composition, and surface PEGylation on in vivo biodistribution. Biomacromolecules 6(5):2541–2554
7. Cortez C et al (2007) Influence of size, surface, cell line, and kinetic properties on the specific binding of A33 antigen-targeted multilayered particles and capsules to colorectal cancer cells. ACS Nano 1(2):93–102
8. Rejman J et al (2004) Size-dependent internalization of particles via the pathways of clathrin- and caveolae-mediated endocytosis. Biochem J 377:159–169
9. Zauner W, Farrow N, Haines A (2001) In vitro uptake of polystyrene microspheres: effect of particle size, cell line, and cell density. J Control Release 71(1):39–51
10. Win K, Feng S-S (2005) Effects of particle size and surface coating on cellular uptake of polymeric nanoparticles for oral delivery of anticancer drugs. Biomaterials 26:2713–2722
11. Hu Y et al (2007) Effect of PEG conformation and particle size on the cellular uptake efficiency of nanoparticles with the HepG2 cells. J Control Release 188:7–17
12. Goodman T, Olive P, Pun S (2007) Increased nanoparticle penetration in collagenase-treated multicellular spheroids. Int J Nanomedicine 2(2):265–274
13. Dreher M et al (2006) Tumor vascular permeability, accumulation, and penetration of macromolecular drug carriers. J Natl Cancer Inst 98(5):335–344
14. Decuzzi P, Ferrari M (2006) The adhesive strength of non-spherical particles mediated by specific interactions. Biomaterials 27(30):5307–5314
15. Gratton S et al (2008) Effect of particle design on cellular internalization pathways. Proc Natl Acad Sci U S A 105(33):11613–11618
16. Champion J, Mitragotri S (2006) Role of target geometry in phagocytosis. Proc Natl Acad Sci U S A 103(13):4930–4934
17. Champion J, Katare Y, Mitragotri S (2007) Particle shape: a new design parameter for micro- and nanoscale drug delivery carriers. J Control Release 121(1):2–9
18. Mader K et al (1997) Noninvasive in vivo monitoring of drug release and polymer erosion from biodegradable polymers by EPR spectroscopy and NMR imaging. J Pharm Sci 86:126–134
19. Goldberg M, Langer R, Jia X (2007) Nanostructured materials for applications in drug delivery and tissue engineering. J Biomater Sci Polym Ed 18(3):241–268
20. Kirch J et al (2012) Mucociliary clearance of micro- and nanoparticles is independent of size, shape and charge—an ex vivo and in silico approach. J Control Release 159(1):128–134
21. Lai SK et al (2007) Rapid transport of large polymeric nanoparticles in fresh undiluted human mucus. Proc Natl Acad Sci U S A 104(5):1482–1487
22. Godin B et al (2011) Multistage nanovectors: from concept to novel imaging contrast agents and therapeutics. Acc Chem Res 44(10):979–989
23. Tasciotti E et al (2008) Mesoporous silicon particles as a multistage delivery system for imaging and therapeutic applications. Nat Nanotechnol 3(3):151–157
24. Godin B et al (2010) An integrated approach for the rational design of nanovectors for biomedical imaging and therapy. Adv Genet 69:31–64
25. Decuzzi P, Ferrari M (2008) Design maps for nanoparticles targeting the diseased microvasculature. Biomaterials 29(3):377–384
26. Park JH et al (2009) Biodegradable luminescent porous silicon nanoparticles for in vivo applications. Nat Mater 8(4):331–336
27. Godin B et al (2010) Tailoring the degradation kinetics of mesoporous silicon structures through PEGylation. J Biomed Mater Res A 94(4):1236–1243
28. Alvarez SD et al (2009) The compatibility of hepatocytes with chemically modified porous silicon with reference to in vitro biosensors. Biomaterials 30(1):26–34
29. Chiappini C et al (2010) Tailored porous silicon microparticles: fabrication and properties. Chemphyschem 11(5):1029–1035
30. Ananta JS et al (2010) Geometrical confinement of gadolinium-based contrast agents in nanoporous particles enhances T1 contrast. Nat Nanotechnol 5(11):815–821

31. Wolfe D, Love J, Whitesides G (2004) Nanostructures replicated by polymer molding. In: Dekker encyclopedia of nanoscience and nanotechnology. Marcel Dekker, Inc., Harvard University, Cambridge, MA
32. Kopecek J, Rejmanova P, Chytry V (1981) Polymers containing enzymatically degradable bonds: 1. Chymotrypsin catalyzed hydrolysis of p-nitroanilides of phenylalanine and tyrosine attached to side-chains of copolymers of N-(2-hydroxypropyl)methacrylamide. Makromol Chem 182:799–809
33. Binaschi M et al (2006) Human and murine macrophages mediate activation of MEn 4901/t-01298: a new promising camptothecin analogue-polysaccharide complex. Anticancer Drugs 17(10):1119–1126
34. Schmid B et al (2007) Albumin-binding prodrugs of camptothecin and doxorubicin with an ala-leu-ala-leu linker that are cleaved by cathepsin B: synthesis and antitumor efficacy. Bioconjug Chem 18(3):702–716
35. Ulbrich K, Strohalm J, Kopecek J (1982) Polymers containing enzymatically degradable bonds: VI. Hydrophilic gels cleavable by chymotrypsin. Biomaterials 3(3):150–154
36. Glangchai LC et al (2008) Nanoimprint lithography based fabrication of shape-specific, enzymatically-triggered smart nanoparticles. J Control Release 125(3):263–272
37. Rolland J et al (2005) Direct fabrication and harvesting of monodisperse, shape-specific nanobiomaterials. J Am Chem Soc 127(28):10096–10100
38. Maynor B et al (2007) Supramolecular nanomimetics: replication of micelles, viruses, and other naturally occurring nanoscale objects. SMALL 3(5):845–849
39. Olson D et al (2006) Amorphous linear aliphatic polyesters for the facile preparation of tunable rapidly degrading elastomeric devices and delivery vectors. J Am Chem Soc 128(41):13625–13633
40. Gratton S et al (2007) Nanofabricated particles for engineered drug therapies: a preliminary biodistribution study of PRINT nanoparticles. J Control Release 121(1–2):10–18
41. Gratton S et al (2008) The pursuit of a scalable nano-fabrication platform for use in material and life science applications. Acc Chem Res 41(12):1685–1695
42. Herlihy K, DeSimone J (2007) Magneto-polymer composite particles fabricated utilizing patterned perfluoropolyether elastomer molds. SPIE Preprint 6517(2):1685–1695
43. Petros R, Ropp P, DeSimone J (2008) Reductively labile PRINT particles for the delivery of doxorubicin to HeLa cells. J Am Chem Soc 130(15):5008–5009
44. Kelly J, DeSimone J (2008) Shape-specific, mono-disperse nano-molding of protein particles. J Am Chem Soc 130(16):5438–5439
45. DeSimone J et al (2007) Organic delivery vehicles for probing and treating biological systems: adapting fabrication processes from the electronics industry for use in nano-medicine. Polym Mater Sci Eng 96:268
46. Dumond J, Low H (2008) Residual layer self-removal in imprint lithography. Adv Mater 20(7):1291–1297
47. Rolland J et al (2004) High-resolution soft lithography: enabling materials for nanotechnologies. Angew Chem Int Ed 43(43):5796–5799
48. Rolland J et al (2004) Solvent-resistant photocurable liquid fluoropolymers for microfluidic device fabrication. J Am Chem Soc 126(8):2322–2323
49. Dave B et al (2012) Epithelial-mesenchymal transition, cancer stem cells and treatment resistance. Breast Cancer Res 14(1):202
50. Hua F et al (2004) Polymer imprint lithography with molecular-scale resolution. Nano Lett 4(12):2467–2471
51. Schmid H, Michel B (2000) Siloxane polymers for high-resolution, high-accuracy soft lithography. Macromolecules 33(8):3042–3049
52. Oudshoorn M et al (2007) Preparation and characterization of structured hydrogel microparticles based on cross-linked hyperbranched polyglycerol. Langmuir 23(23):11819–11825
53. Moran I et al (2008) High-resolution soft lithography of thin film resists enabling nanoscopic pattern transfer. Soft Matter 4:168–176
54. Thibault C et al (2007) Poly(dimethylsiloxane) contamination in microcontract printing and its influence on patterning oligonucleotides. Langmuir 23(21):10706–10714

55. Guan J, Chakrapani A, Hansford D (2005) Polymer microparticles fabricated by soft lithography. Chem Mater 17(25):6227–6229
56. Guan J et al (2006) Fabrication of polymeric microparticles for drug delivery by soft lithography. Biomaterials 27(221):4034–4041
57. Guan J et al (2007) Fabrication of particulate reservoir-containing, capsulelike, and self-folding polymer microstructures for drug delivery. Small 3(3):412–418
58. Astete C, Sabliov C (2006) Synthesis and characterization of PLGA nanoparticles. J Biomater Sci Polym Ed 17(3):247–289
59. Shimomura M, Sawadaishi T (2001) Bottom-up strategy of materials fabrication: a new trend in nanotechnology of soft materials. Curr Opin Colloid Interface Sci 6(1):11–16
60. Lasic D (1993) Liposomes: from physics to applications. Elsevier, Amsterdam
61. Nowak A et al (2002) Rapidly recovering hydrogel scaffolds from self-assembling diblock copolypeptide amphiphiles. Nature 417(6887):424–428
62. Zhang J et al (1994) Nanoarchitectures: 5. Geometrically-controlled and site-specifically-functionalized phenylacetylene macrocycles. J Am Chem Soc 116(10):4227–4239
63. Gore T et al (2001) Self-assembly of model collagen peptide amphiphiles. Langmuir 17(17):5352–5360
64. Tu R, Tirrel M (2004) Bottom-up design of biomimetic assemblies. Adv Drug Deliv Rev 56:1537–1563
65. Stokes R, Evans D (1997) Fundamentals of interfacial engineering. In: Advances in interfacial engineering. VCH Publishers, New York
66. Evans D, Wennerstrom H (1994) The colloidal domain: where physics, chemistry, biology and technology meet. VCH Publishers, New York
67. Luk Y, Abbott N (2002) Applications of functional surfactants. Curr Opin Colloid Interface Sci 7(5–6):267–275
68. Matsumoto Y et al (2000) Highly specific inhibitory effect of three-component hybrid liposomes including sugar surfactants on the growth of glioma cells. Bioorg Med Chem Lett 10(23):2617–2619
69. Yu Y-C et al (1996) Self-assembling amphiphiles for construction of protein molecular architecture. J Am Chem Soc Rev 118(50):12515–12520
70. Hartgerink J, Beniash E, Stupp S (2001) Self-assembly and mineralization of peptide-amphiphile nanofibers. Science 294(5547):1684–1688
71. Pindzola B, Hoag B, Gin D (2001) Polymerization of a phosphonium diene amphiphile in the regular hexagonal phase with retention of mesostructure. J Am Chem Soc Rev 123(19):4617–4618
72. Schnarr N, Kennan A (2003) Specific control of peptide assembly with combined hydrophilic and hydrophobic interfaces. J Am Chem Soc 125(3):667–671
73. Creighton T (1993) Proteins: structures and molecular properties. W. H. Freeman, New York, p 507
74. De Heer W et al (1995) Aligned carbon nanotube films—production and optical and electronic-properties. Science 268:845
75. Franklin N, Dai H (2000) An enhanced CVD approach to extensive nanotube networks with directionality. J Adv Mater 12:890
76. Terrones M et al (1997) Controlled production of aligned-nanotube bundles. Nature 388:52
77. Li W et al (1996) Large-scale synthesis of aligned carbon nanotubes. Science 274:1701
78. Chattopadhyay D, Galeska I, Papadimitrakopoulos F (2001) Metal-assisted organization of shortened carbon nanotubes in monolayer and multilayer forest assemblies. J Am Chem Soc 123:9451
79. Liu Z et al (2000) Organizing single-walled carbon nanotubes on gold using a wet chemical self-assembling technique. Langmuir 16:3569
80. Wu B et al (2001) Chemical alignment of oxidatively shortened single-walled carbon nanotubes on silver surface. J Phys Chem B 105:5075
81. Huang S, Dai L, Mau A (1999) Patterned growth and contact transfer of well-aligned carbon nanotube films. J Phys Chem B 103:4223

82. Kamalakaran R et al (2000) Synthesis of thick and crystalline nanotube arrays by spray pyrolysis. Appl Phys Lett 77:3385
83. Mayne M et al (2001) Pyrolytic production of aligned carbon nanotubes from homogeneously dispersed benzene based aerosols. Chem Phys Lett 338:101
84. Euliss LE et al (2006) Imparting size, shape, and composition control of materials for nanomedicine. Chem Soc Rev 35(11):1095–1104
85. Riehemann K et al (2009) Nanomedicine—challenge and perspectives. Angew Chem Int Ed 48(5):872–897
86. Srinivasan S et al (2013) Bacteriophage associated silicon particles: design and characterization of a novel theranostic vector with improved payload carrying potential. J Mater Chem B 1(39):5218–5229
87. Kharlampieva E et al (2008) Hydrogen-bonded polymer multilayers probed by neutron reflectivity. Langmuir 24(20):11346–11349
88. Kharlampieva E et al (2010) Co-cross-linking silk matrices with silica nanostructures for robust ultrathin nanocomposites. ACS Nano 4(12):7053–7063
89. Kozlovskaya V et al (2012) Hydrogen-bonded multilayers of silk fibroin: from coatings to cell-mimicking shaped microcontainers. ACS Macro Lett 2012:384–387
90. Kozlovskaya V et al (2011) Shape switching of hollow layer-by-layer hydrogel microcontainers. Chem Commun 47(29):8352–8354
91. Kozlovskaya V et al (2010) Responsive microcapsule reactors based on hydrogen-bonded tannic acid layer-by-layer assemblies. Soft Matter 6(15):3596–3608
92. Kozlovskaya V et al (2010) pH-controlled assembly and properties of LbL membranes from branched conjugated poly(alkoxythiophene sulfonate) and various polycations. Langmuir 26(10):7138–7147
93. Kozlovskaya V et al (2012) pH-triggered shape response of cubical ultrathin hydrogel capsules. Soft Matter 8(38):9828–9839
94. Serda RE et al (2010) Logic-embedded vectors for intracellular partitioning, endosomal escape, and exocytosis of nanoparticles. Small 6(23):2691–2700
95. Merkel TJ et al (2011) Using mechanobiological mimicry of red blood cells to extend circulation times of hydrogel microparticles. Proc Natl Acad Sci U S A 108(2):586–591
96. De Koker S et al (2007) In vivo cellular uptake, degradation, and biocompatibility of polyelectrolyte microcapsules. Adv Funct Mater 17(18):3754–3763
97. Liu X et al (2005) Multilayer microcapsules as anti-cancer drug delivery vehicle: deposition, sustained release, and in vitro bioactivity. Macromol Biosci 5(12):1209–1219
98. Ng SL et al (2011) Controlled release of DNA from poly(vinylpyrrolidone) capsules using cleavable linkers. Biomaterials 32(26):6277–6284
99. Kozlovskaya VA et al (2009) Single-component layer-by-layer weak polyelectrolyte films and capsules: loading and release of functional molecules. Polym Sci Ser A 51(6):719–729
100. De Cock LJ et al (2010) Polymeric multilayer capsules in drug delivery. Angew Chem Int Ed 49(39):6954–6973
101. Tao L et al (2011) Shape-specific polymeric nanomedicine: emerging opportunities and challenges. Exp Biol Med 236(1):20–29
102. Pridgen EM, Langer R, Farokhzad OC (2007) Biodegradable, polymeric nanoparticle delivery systems for cancer therapy. Nanomedicine 2(5):669–680
103. Vergaro V et al (2011) Drug-loaded polyelectrolyte microcapsules for sustained targeting of cancer cells. Adv Drug Deliv Rev 63(9):847–864
104. Johnston APR et al (2012) Targeting cancer cells: controlling the binding and internalization of antibody-functionalized capsules. ACS Nano 6(8):6667–6674
105. Cortez C et al (2006) Targeting and uptake of multilayered particles to colorectal cancer cells. Adv Mater 18(15):1998–2003
106. Palamà IE et al (2010) Multilayered polyelectrolyte capsules and coated colloids: cytotoxicity and uptake by cancer cells. Sci Adv Mater 2(2):138–150
107. Patel T et al (2012) Polymeric nanoparticles for drug delivery to the central nervous system. Adv Drug Deliv Rev 64(7):701–705

108. Sukhorukov GB et al (2007) Multifunctionalized polymer microcapsules: novel tools for biological and pharmacological applications. Small 3(6):944–955
109. Yan Y et al (2010) Uptake and intracellular fate of disulfide-bonded polymer hydrogel capsules for doxorubicin delivery to colorectal cancer cells. ACS Nano 4(5):2928–2936
110. Wang K et al (2007) Encapsulated photosensitive drugs by biodegradable microcapsules to incapacitate cancer cells. J Mater Chem 17(38):4018–4021
111. Cui J et al (2014) Emerging methods for the fabrication of polymer capsules. Adv Colloid Interface Sci 207:14–31
112. Doshi N et al (2009) Red blood cell-mimicking synthetic biomaterial particles. Proc Natl Acad Sci U S A 106(51):21495–21499
113. Kharlampieva E, Kozlovskaya V, Sukhishvili SA (2009) Layer-by-layer hydrogen-bonded polymer films: from fundamentals to applications. Adv Mater 21(30):3053–3065
114. Kozlovskaya V et al (2014) Internalization of red blood cell-mimicking hydrogel capsules with pH-triggered shape responses. ACS Nano 8(6):5725–5737
115. Handley DA (1989) Colloidal gold: principles, methods and applications. In: Hayat MA (ed) Colloidal gold: principles, methods and applications. Academic Press, San Diego, pp 23–27
116. Souza GR et al (2006) Networks of gold nanoparticles and bacteriophage as biological sensors and cell-targeting agents. Proc Natl Acad Sci U S A 103(5):1215–1220
117. Khalil IA et al (2006) Uptake pathways and subsequent intracellular trafficking in nonviral gene delivery. Pharmacol Rev 58(1):32–45
118. Bally MB et al (1999) Biological barriers to cellular delivery of lipid-based DNA carriers. Adv Drug Deliv Rev 38(3):291–315
119. Kircheis R, Wagner E (2000) Polycation/DNA complexes for in vivo gene delivery. Gene Ther Regul 1(1):95–114
120. Kircheis R, Wightman L, Wagner E (2001) Design and gene delivery activity of modified polyethylenimines. Adv Drug Deliv Rev 53(3):341–358
121. Hunter AC, Moghimi SM (2010) Cationic carriers of genetic material and cell death: a mitochondrial tale. Biochim Biophys Acta 1797(6–7):1203–1209
122. Mislick KA, Baldeschwieler JD (1996) Evidence for the role of proteoglycans in cation-mediated gene transfer. Proc Natl Acad Sci U S A 93(22):12349–12354
123. Albanese A, Tang P, Chan C (2012) The effect of nanoparticle size, shape, and surface chemistry on biological systems. Annu Rev Biomed Eng 14:1–16
124. Owens Iii DE, Peppas NA (2006) Opsonization, biodistribution, and pharmacokinetics of polymeric nanoparticles. Int J Pharm 307(1):93–102
125. Xiang S et al (2006) Pathogen recognition and development of particulate vaccines: does size matter? Methods Enzymol 40(1):1–9
126. Dobrovolskaia M, McNeil S (2007) Immunological properties of engineered nanomaterials. Nat Nanotechnol 2:469–478
127. Raychaudhuri S, Rock K (1998) Fully mobilizing host defense: building better vaccines. Nat Biotechnol 16(11):1025–1031
128. Fifis T et al (2004) Size-dependent immunogenicity: therapeutic and protective properties of nano-vaccines against tumors. J Immunol 173(5):3148–3154
129. O'Hagan D, Singh M, Ulmer J (2006) Microparticle-based technologies for vaccines. Methods 40(1):10–19
130. Guo LS (2001) Amphotericin B colloidal dispersion: an improved antifungal therapy. Adv Drug Deliv Rev 47(2–3):149–163
131. Larabi M et al (2003) Toxicity and antileishmanial activity of a new stable lipid suspension of amphotericin B. Antimicrob Agents Chemother 47(12):3774–3779
132. Drummond C, Fong C (1999) Surfactant self-assembly objects as novel drug delivery vehicles. Curr Opin Colloid Interface Sci 4:449–456
133. Beningo KA, Wang YL (2002) Fc-receptor-mediated phagocytosis is regulated by mechanical properties of the target. J Cell Sci 115(Pt 4):849–856
134. Allen TM et al (1991) Uptake of liposomes by cultured mouse bone marrow macrophages: influence of liposome composition and size. Biochim Biophys Acta 1061(1):56–64

135. Fidler J et al (1980) Pulmonary localization of intravenously injected liposomes. Recent Results Cancer Res 75:246–251
136. Dunn S et al (1994) Polystyrene-poly (ethylene glycol) (PS-PEG2000) particles as model systems for site specific drug delivery: 2. The effect of PEG surface density on the in vitro cell interaction and in vivo biodistribution. Pharm Res 11:1016–1022
137. Perrault S et al (2009) Mediating tumor targeting efficiency of nanoparticles through design. Nano Lett 9(5):1909–1915
138. Geng Y et al (2007) Shape effects of filaments versus spherical particles in flow and drug delivery. Nat Nanotechnol 2:249–255
139. Yeh C, Eckstein EC (1994) Transient lateral transport of platelet-sized particles in flowing blood suspensions. Biophys J 66(5):1706–1716
140. Saadatmand M et al (2011) Fluid particle diffusion through high-hematocrit blood flow within a capillary tube. J Biomech 44(1):170–175
141. Kim S et al (2009) The cell-free layer in microvascular blood flow. Biorheology 46(3):181–189
142. Lipowsky HH (2005) Microvascular rheology and hemodynamics. Microcirculation 12(1):5–15
143. Lee S-Y, Ferrari M, Decuzzi P (2009) Shaping nano-/micro-particles for enhanced vascular interaction in laminar flows. Nanotechnology 20(49):495101
144. Gentile F et al (2008) The effect of shape on the margination dynamics of non-neutrally buoyant particles in two-dimensional shear flows. J Biomech 41:2312–2318
145. Gavze E, Shapiro M (1998) Motion of inertial spheroidal particles in a shear flow near a solid wall with special application to aerosol transport in microgravity. J Fluid Mech 371:59–79
146. Godin B et al (2012) Discoidal porous silicon particles: fabrication and biodistribution in breast cancer bearing mice. Adv Funct Mater 22(20):4225–4235
147. Muro S et al (2008) Control of endothelial targeting and intracellular delivery of therapeutic enzymes by modulating the size and shape of ICAM-1-targeted carriers. Mol Ther 16(8):1450–1458
148. Randall T et al (2011) The effects of particle size, density and shape on margination of nanoparticles in microcirculation. Nanotechnology 22(11):115101
149. Choi H et al (2007) Renal clearance of quantum dots. Nat Biotechnol 25(10):1165–1170
150. Decuzzi P et al (2010) Size and shape effects in the biodistribution of intravascularly injected particles. J Control Release 141(3):320–327
151. Ballou B et al (2004) Noninvasive imaging of quantum dots in mice. Bioconjug Chem 15(1):79–86
152. Fischer H et al (2010) Exploring primary liver macrophages for studying quantum dot interactions with biological systems. Adv Mater 22(23):2520–2524
153. Schipper M et al (2008) A pilot toxicology study of single-walled carbon nanotubes in a small sample of mice. Nat Nanotechnol 3:216–221
154. Fox M, Szoka F, Frechet J (2009) Soluble polymer carriers for the treatment of cancer: the importance of molecular architecture. Acc Chem Res 42:1141–1151
155. Mitragotri S, Lahann J (2009) Physical approaches to biomaterial design. Nat Mater 8:15–23
156. Haghgooie R, Toner M, Doyle P (2009) Squishy non-spherical hydrogel microparticles. Macromol Rapid Commun 31:128–134
157. Hendrickson G, Lyon L (2010) Microgel translocation through pores under confinement. Angew Chem Int Ed 49:2193–2197
158. Merkel TJ et al (2012) The effect of particle size on the biodistribution of low-modulus hydrogel PRINT particles. J Control Release 162(1):37–44
159. Kohane D (2007) Microparticles and nanoparticles for drug delivery. Biotechnol Bioeng 96:203–209
160. Alexis F et al (2008) Factors affecting the clearance and biodistribution of polymeric nanoparticles. Mol Pharmacol 5:505–515
161. Canelas D, Herlihy K, DeSimone J (2009) Top-down particle fabrication: control of size and shape for diagnostic imaging and drug delivery. Wiley Interdiscip Rev Nanomed Nanobiotechnol 1(4):391–404

162. Ishimoto T et al (2008) Downregulation of monocyte chemoattractant protein-1 involving short interfering RNA attenuates hapten-induced contact hypersensitivity. Mol Ther 16:387–395
163. Urban-Klein B et al (2005) RNAi-mediated gene-targeting through systemic application of polyethylenimine (PEI) complexed siRNA in vivo. Gene Ther 12:461–466
164. Li W et al (2013) Shape design of high drug payload nanoparticles for more effective cancer therapy. Chem Commun 49:10989
165. Jain R, Stylianopoulos T (2010) Delivering nanomedicine to solid tumors. Nat Rev Clin Oncol 7(11):653–664
166. van de Ven A et al (2013) Rapid tumoritropic accumulation of systemically injected plateloid particles and their biodistribution. J Control Release 158(1):148–155
167. Choi C et al (2009) Mechanism of active targeting in solid tumors with transferrin-containing gold nanoparticles. Nat Rev Clin Oncol 107(3):1235–1240
168. Lee H et al (2010) The effects of particle size and molecular targeting on the intratumoral and subcellular distribution of polymeric nanoparticles. Mol Pharm 7(4):1195–1208
169. Glazer ES et al (2010) Noninvasive radiofrequency field destruction of pancreatic adenocarcinoma xenografts treated with targeted gold nanoparticles. Clin Cancer Res 16(23):5712–5721
170. Loo C et al (2004) Nanoshell-enabled photonics-based imaging and therapy of cancer. Technol Cancer Res Treat 3(1):33–40
171. Frieboes H et al (2013) A computational model for predicting nanoparticle accumulation in tumor vasculature. PLoS One 8(2):e56876
172. Whitehead KA, Langer R, Anderson DG (2009) Knocking down barriers: advances in siRNA delivery. Nat Rev Drug Discov 8(2):129–138
173. Mangala LS et al (2009) Liposomal siRNA for ovarian cancer. Methods Mol Biol 555:29–42
174. Tanaka T et al (2010) Sustained small interfering RNA delivery by mesoporous silicon particles. Cancer Res 70(9):3687–3696
175. Jiang W et al (2008) Nanoparticle-mediated cellular response is size-dependent. Nat Nanotechnol 3(3):145–150
176. Kanaras A et al (2011) Receptor mediated interactions between colloidal gold nanoparticles and human umbilical vein endothelial cells. Small 7:388–394
177. Zhang D et al (2009) Gold nanoparticles can induce the formation of protein-based aggregates at physiological pH. Nano Lett 9(2):666–671
178. Deng Z et al (2011) Nanoparticle-induced unfolding of fibrinogen promotes Mac-1 receptor activation and inflammation. Nat Nanotechnol 6(1):39–44
179. Thorek DL, Tsourkas A (2008) Size, charge and concentration dependent uptake of iron oxide particles by non-phagocytic cells. Biomaterials 29(26):3583–3590
180. Slowing I, Trewyn BG, Lin VS (2006) Effect of surface functionalization of MCM-41-type mesoporous silica nanoparticles on the endocytosis by human cancer cells. J Am Chem Soc 128(46):14792–14793
181. Wang J et al (2010) The complex role of multivalency in nanoparticles targeting the transferrin receptor for cancer therapies. J Am Chem Soc 132(32):11306–11313
182. Jin H et al (2009) Size-dependent cellular uptake and expulsion of single-walled carbon nanotubes: single particle tracking and a generic uptake model for nanoparticles. ACS Nano 3(1):149–158
183. Mounkes L et al (1998) Proteoglycans mediated cationic liposome-DNA complex-based gene delivery in vitro and in vivo. J Biol Chem 273:26164–26170
184. Woods A, Couchman J (1994) Syndecan 4 heparan sulfate proteoglycan is a selectively enriched and widespread focal adhesion component. Mol Biol Cell 5:183–192
185. Arvizo R et al (2010) Effect of nanoparticle surface charge at the plasma membrane and beyond. Nano Lett 10(7):2543–2548
186. Wang B et al (2008) Nanoparticle-induced surface reconstruction of phospholipid membranes. Proc Natl Acad Sci U S A 105(47):18171–18175
187. Mecke A et al (2005) Synthetic and natural polycationic polymer nanoparticles interact selectively with fluid-phase domains of DMPC lipid bilayers. Langmuir 21(19):8588–8590

188. Cedervall T et al (2007) Understanding the nanoparticle-protein corona using methods to quantify exchange rates and affinities of proteins for nanoparticles. Proc Natl Acad Sci U S A 104(7):2050–2055
189. Wang F et al (2013) The biomolecular corona is retained during nanoparticle uptake and protects the cells from the damage induced by cationic nanoparticles until degraded in the lysosomes. Nanomedicine 9(8):1159–1168
190. Conner S, Schmid S (2003) Regulated portals of entry into the cell. Nature (Lond) 422:37–44
191. Hild W, Breunig M, Goepferich A (2008) Quantum dots—nano-sized probes for the exploration of cellular and intracellular targeting. Eur J Pharm Biopharm 68(2):153–168
192. Grosse S et al (2005) Potocytosis and cellular exit of complexes as cellular pathways for gene delivery by polycations. J Gene Med 7(10):1275–1286
193. Dautryvarsat A, Ciechanover A, Lodish H (1983) Ph and the recycling of transferrin during receptor-mediated endocytosis. Proc Natl Acad Sci U S A 80(8):2258–2262
194. Brodsky F et al (2001) Biological basket weaving: formation and function of clathrin-coated vesicles. Annu Rev Cell Dev Biol 17:517–568
195. Schmid S (1997) Clathrin-coated vesicle formation and protein sorting: an integrated process. Annu Rev Biochem 66:511–548
196. Takei K, Haucke V (2001) Clathrin-mediated endocytosis: membrane factors pull the trigger. Trends Cell Biol 11(9):385–391
197. Goncalves C et al (2004) Macropinocytosis of polyplexes and recycling of plasmid from clathrin-dependent pathway impair the transfection efficiency into human hepatocarcinoma cells. Mol Ther 9:S317
198. Zhou X, Huang L (1994) DNA transfection mediated by cationic liposomes containing lipopolylysine characterization and mechanism of action. Biochim Biophys Acta 1189(2):195–203
199. Goldstein J et al (1985) Receptor-mediated endocytosis—concepts emerging from the Ldl receptor system. Annu Rev Cell Biol 1:1–39
200. Maxfield F, McGraw T (2004) Endocytic recycling. Nat Rev Mol Cell Biol 5(2):121–132
201. Sonawane N, Szoka F, Verkman A (2003) Chloride accumulation and swelling in endosomes enhances DNA transfer by polyamine-DNA polyplexes. J Biol Chem 278(45):44826–44831
202. Pack D, Putnam D, Langer R (2000) Design of imidazole-containing endosomolytic biopolymers for gene delivery. Biotechnol Bioeng 67(2):217–223
203. Bishop N (1997) An update on non-clathrin-coated endocytosis. Rev Med Virol 7(4):199–209
204. Gabrielson N, Pack D (2009) Efficient polyethylenimine-mediated gene delivery proceeds via a caveolar pathway in HeLa cells. J Control Release 136(1):54–61
205. Midoux P et al (2008) Polymer-based gene delivery: a current review on the uptake and intracellular trafficking of polyplexes. Curr Gene Ther 8(5):335–352
206. Qiu Y et al (2010) Surface chemistry and aspect ratio mediated cellular uptake of Au nanorods. Biomaterials 31(30):7606–7619
207. Chithrani B, Ghazani A, Chan W (2006) Determining the size and shape dependence of gold nanoparticle uptake into mammalian cells. Nano Lett 6(4):662–668
208. Ladewig K, Xu ZP, Lu GQ (2009) Layered double hydroxide nanoparticles in gene and drug delivery. Expert Opin Drug Deliv 6(9):907–922
209. Xu ZP et al (2008) Subcellular compartment targeting of layered double hydroxide nanoparticles. J Control Release 130(1):86–94
210. Caldorera-Moore M et al (2010) Designer nanoparticles: incorporating size, shape and triggered release into nanoscale drug carriers. Expert Opin Drug Deliv 7:479–495
211. Yoo J, Mitragotri S (2010) Polymer particles that switch shape in response to a stimulus. Proc Natl Acad Sci U S A 107:11205–11210
212. Chithrani B, Chan W (2007) Elucidating the mechanism of cellular uptake and removal of protein-coated gold nanoparticles of different sizes and shapes. Nano Lett 7(6):1542–1550
213. Hutter E et al (2010) Microglial response to gold nanoparticles. ACS Nano 4(5):2595–2606

214. Serda R et al (2009) The association of silicon microparticles with endothelial cells in drug delivery to the vasculature. Biomaterials 30(13):2440–2448
215. Muro S, Koval M, Muzykantov V (2004) Endothelial endocytic pathways: gates for vascular drug delivery. Curr Vasc Pharmacol 2(3):281–299
216. Gao H, Shi W, Freund L (2005) Mechanics of receptor-mediated endocytosis. Proc Natl Acad Sci U S A 102(27):9469–9474
217. Jiang J et al (2008) Does nanoparticle activity depend upon size and crystal phase? Nanotoxicology 1(2):33–42
218. Yuan H et al (2010) Variable nanoparticle-cell adhesion strength regulates cellular uptake. Phys Rev Lett 105(13):138101
219. Xu Z et al (2009) The performance of docetaxelloaded solid lipid nanoparticles targeted to hepatocellular carcinoma. Biomaterials 30(2):226–232
220. Shimoni O et al (2012) Shape-dependent cellular processing of polyelectrolyte capsules. ACS Nano 7(1):522–530
221. Yi X, Shi X, Gao H (2011) Cellular uptake of elastic nanoparticles. Phys Rev Lett 107(9):098101
222. Kirchner C et al (2005) Cytotoxicity of nanoparticle-loaded polymer capsules. Talanta 67(3):486–491
223. An Z et al (2009) Polyelectrolyte microcapsule interactions with cells in two- and three-dimensional culture. Colloids Surf B Biointerfaces 70(1):114–123
224. See V et al (2009) Cathepsin L digestion of nanobioconjugates upon endocytosis. ACS Nano 3(9):2461–2468
225. Lechardeur D et al (1999) Metabolic instability of plasmid DNA in the cytosol: a potential barrier to gene transfer. Gene Ther 6(4):482–497
226. Kao H, Abney J, Verkman A (1993) Determinants of the translational mobility of a small solute in cell cytoplasm. J Cell Biol 120(1):175–184
227. Lukacs G et al (2000) Size-dependent DNA mobility in cytoplasm and nucleus. J Biol Chem 275(3):1625–1629
228. Dauty E, Verkman A (2005) Actin cytoskeleton as the principal determinant of size-dependent DNA mobility in cytoplasm: a new barrier for non-viral gene delivery. J Biol Chem 280(9):7823–7828
229. Williams Y et al (2009) Probing cell-type-specific intracellular nanoscale barriers using size-tuned quantum dots. Small 5(22):2581–2588
230. Carlson C et al (2008) Unique cellular interaction of silver nanoparticles: size-dependent generation of reactive oxygen species. J Phys Chem B 112(43):13608–13619
231. Behr J (1997) The proton sponge: a trick to enter cells the viruses did not exploit. Chimia 51:34–36
232. AshaRani PV et al (2009) Cytotoxicity and genotoxicity of silver nanoparticles in human cells. ACS Nano 3(2):279–290
233. Berneburg M et al (2006) 'To repair or not to repair—no longer a question': repair of mitochondrial DNA shielding against age and cancer. Exp Dermatol 15(12):1005–1015
234. Rees P et al (2011) A transfer function approach to measuring cell inheritance. BMC Syst Biol 5(1):31
235. Rabinovitch M (1995) Professional and non-professional phagocytes: an introduction. Trends Cell Biol 5(3):85–87

Chapter 9
Delivery of Peptides and Proteins to the Brain Using Nano-Drug Delivery Systems and Other Formulations

David Stepensky

Abstract Peptides and proteins can potently modulate processes in the central nervous system (CNS) and can be used for management of CNS disorders and other diseases. However, peptides and proteins permeate to a limited extent into the CNS, due to the blood-brain and blood-cerebrospinal fluid barriers, and undergo rapid clearance by endogenous proteases and via other pathways. These pharmacokinetic drawbacks prevent accumulation of peptide and protein drugs in the CNS and render them pharmacologically inefficient. Different approaches for enhanced delivery of peptide/protein drugs to the brain have been developed in order to overcome these drawbacks and to increase their therapeutic efficiency. These approaches include: (a) focal administration into the brain tissue or fluids (via intracerebral, intrathecal, or intracerebroventricular injection), (b) incorporation of drugs into specialized nano-drug delivery systems (DDSs) and other formulations, (c) disruption of CNS barriers, (d) use of immune cells as 'Trojan horses' for peptide/protein brain deliver, and others. In this chapter, the major approaches used to deliver peptide and protein drugs to the brain are described and factors that affect the systemic and local drug disposition (distribution and elimination) are summarized. Current problems and limitations in the field of brain delivery of the peptide and protein drugs are presented, and recommendations for further brain delivery enhancement of peptide/protein drugs are given.

Keywords Peptide and protein drugs • Central nervous system disorders • Blood-brain barrier • Drug targeting to the brain • Drug delivery systems • Nano-formulations

---

D. Stepensky (✉)
Department of Clinical Biochemistry and Pharmacology, Faculty of Health Sciences
Ben-Gurion University of the Negev, POB 653, Beer-Sheva 84105, Israel
e-mail: davidst@bgu.ac.il

K.A. Howard et al. (eds.), *Nanomedicine*, Advances in Delivery Science and Technology, DOI 10.1007/978-1-4939-3634-2_9

Abbreviations

BBB	Blood-brain barrier
B-CSF-B	Blood-cerebrospinal fluid barrier
BDNF	Brain-derived neurotrophic factor
bFGF	Basic fibroblast growth factor
CSF	Cerebrospinal fluid
DDS	Drug delivery system
EGF	Epidermal growth factor
MPS	Mononuclear phagocyte system
MRI	Magnetic resonance imaging
NGF	Nerve growth factor
PLGA	Poly(lactic-co-glycolic) acid
PEG	Polyethylene glycol
STL	Solanum tuberosum lectin
TNF	Tumor necrosis factor
VIP	Vasoactive intestinal peptide
WGA	Wheat germ agglutinin

9.1 Peptides and Proteins for Treatment of Brain Diseases and Barriers for Their Use as Drugs

Central nervous system (CNS) disorders comprise a group of diseases (neurodegenerative diseases, epilepsy, neoplastic and inflammatory CNS diseases, acute and chronic pain, and others) with complex etiology and pathophysiology that have a major impact on human life duration and quality. Despite many efforts to develop drugs for management of CNS disorders, the incidence of these diseases is constantly increasing during the last decades, in part due to prolongation of human life span, and new treatments are required for their better management. Due to advances in the understanding of the regulation of the CNS functions in physiological and pathophysiological conditions, peptides and proteins have emerged as promising therapeutic modalities for management of CNS disorders. Moreover, via their pharmacological activities in the CNS, peptides and proteins can modify activity of other organs and tissues and can be used for management of non-CNS diseases, such as gastrointestinal, metabolic, and hormonal disorders. Selected examples of peptides and proteins that can be used for management of CNS and non-CNS diseases are listed in Table 9.1.

In spite of their clear therapeutic potential, many peptides and proteins are clinically ineffective and cannot be applied for management of patients with specific diseases. This lack of efficiency stems from the pharmacokinetic limitations of peptides and proteins; low permeability to the CNS tissue and fluids, and high clearance (by endogenous proteases and via other pathways, see below) that prevent their accumulation in the CNS and render them pharmacologically inefficient. Permeability of drugs from the site of administration to the CNS is limited by the blood-brain barrier (BBB) that is formed by several types of cells and is strengthened by the tight junctions (see Fig. 9.1b).

Table 9.1 Examples of peptides and proteins that can be used for management of central nervous system disorders and of other diseases

Group	Examples
Neuropeptides	Neurotensin, melanocortins, brain natriuretic peptide
Opioid peptides	Enkephalin, kyotorphin, dynorphin
Ingestive peptides	Leptin, ghrelin, orexin
Cytokines	Interleukins, tumor necrosis factor α, granulocyte/monocyte colony stimulating factor, macrophage inflammatory protein-1
Neurotropic factors	Brain derived neurotrophic factor, nerve growth factor, epidermal growth factor, insulin-like growth factor I, neurotrophins
Proteins	Therapeutic antibodies, insulin, transferrin
Hormones	Oxytocin, luteinizing hormone-releasing hormone, corticotropin releasing hormone, somatostatin, vasopressin

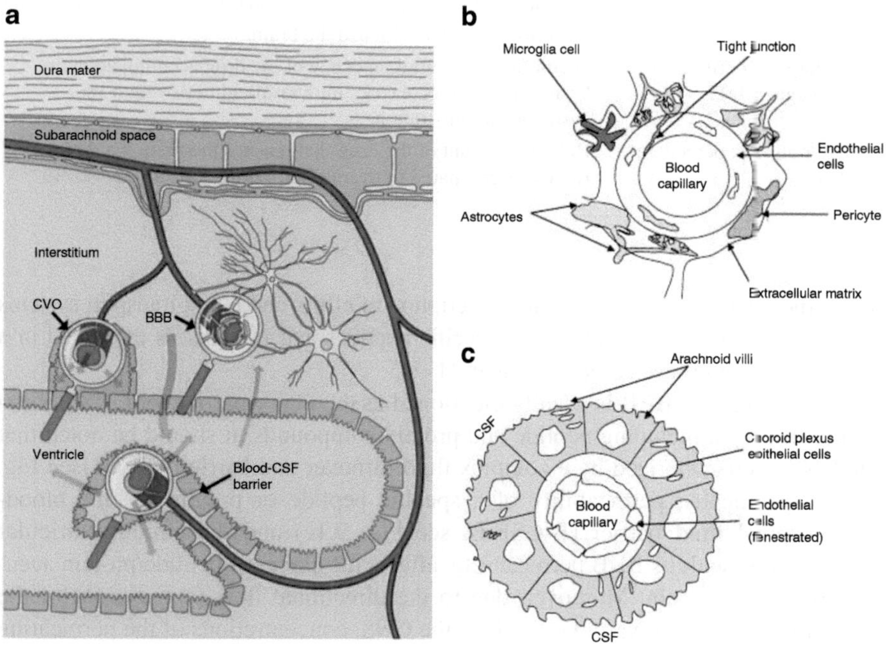

Fig. 9.1 Barriers for delivery of peptide and protein drugs to the brain. A. Brain fluid turnover and barrier system. Cerebral interstitial fluid is secreted at the cerebral capillaries (at the blood brain barrier) and flows from brain to CSF via perivascular spaces. The brain-cerebrospinal fluid barrier is formed by choroid epithelium where CSF is formed. In certain regions of the brain (circumventricular organs, CVO) capillaries are leaky and neuronal projections have direct access to solutes of the plasma. Reprinted from [70] with permission from John Wiley and Sons. B. Morphology of the blood–brain barrier (BBB) that is formed by the brain capillary endothelium, pericytes and astrocytes and strengthened by the tight junctions between the endothelial cells. Reprinted from [2] with permission from Elsevier. C. Morphology of the blood-cerebrospinal fluid barrier (B-CSF-B) situated at the choroid plexus and meninges. The barrier is formed by the choroid plexus epithelial cells and is strengthened by the tight junctions between these cells. Reprinted from [2] with permission from Elsevier

Table 9.2 Quantitative parameters for analysis of brain delivery of peptide and protein drugs [46, 71–74]

Parameter	Description
Brain uptake index	Relative uptake of a drug compared to a reference substance (compound freely diffusible across the BBB, such as ^{14}C-butanol)
Relative concentration	Concentration of the drug in the brain vs. other organs or body fluids; e.g., brain/plasma brain-plasma partition coefficient, K_p
Influx rate constant, K_{in}	The unidirectional influx constant from blood to brain, is most commonly determined after intravenous injection or after in situ perfusion of the compound
Permeability-surface area product, PS product	The unidirectional clearance from blood to brain, is most commonly determined after intravenous injection or after in situ perfusion of the compound
Concentration above threshold value	Concentration of the drug in the brain in comparison to a specific threshold value (e.g., the lower boundary of therapeutic window)
Fraction of dose	% of drug dose that reached the brain
Drug targeting efficiency percentage, %DTE	Quantifies the overall tendency of the drug accumulate in the brain following local (e.g., intranasal) administration vs. systemic administration
Direct transport percentage, %DTP	Relative amount of the drug that reaches the brain via direct routes, as compared to overall brain drug delivery

The structure of the BBB, including description of efflux and influx transport systems that affect the CNS permeability of specific peptides and proteins, is explained in a detailed fashion in several recent reviews [1–3].

Despite that the BBB is usually mentioned as the major barrier for CNS permeability of drugs, including peptide and protein compounds, it should be noted that the CNS is characterized by a complex fluid turnover and barrier system (see Fig. 9.1a). Specifically, permeability of a specific peptide or protein via the blood-cerebrospinal fluid barrier (B-CSF-B, see Fig. 9.1c) and the circumventricular organs, along with its BBB permeability, affects the extent of peptide/protein accumulation in the brain. Moreover, due to the directional flow (convection) of the interstitial and cerebrovascular fluids in the CNS, concentrations of the permeating compounds e.g., peptide/protein drugs are unequal in different parts of the brain. These BBB permeability and CNS convection pathways are altered in many pathological conditions (brain trauma or stroke, inflammation, tumor, epilepsy, neurodegenerative diseases, etc.) [4–6], affecting the patterns of peptides/proteins accumulation and distribution in the CNS [7–9].

Several parameters can be used to quantify the brain delivery efficiency of peptides/proteins, and of other types of drugs (see Table 9.2). There is no general consensus on the best quantitative parameter for analysis of brain drug permeability, and some of these parameters are better suited for clinical needs, while others are used predominantly in preclinical settings (for analysis of data obtained in animal experiments). Nevertheless, it is commonly accepted that peptides and proteins are largely unable to reach the CNS following systemic administration [2, 10] due to inability to permeate the CNS barriers and rapid clearance, leading to low values of all the parameters listed in Table 9.2. Subsequently, numerous approaches have been proposed to increase the CNS accumulation of peptides and proteins for enhancement of their therapeutic efficiency.

9.2 Drug Delivery Systems for Enhanced Delivery of Peptide and Protein Drugs to the Brain

Different approaches for enhanced delivery of peptide/protein drugs to the brain have been developed over the last several decades (see Table 9.3). The drug can be focally injected into the brain tissue or fluids (as intracerebral, intrathecal, or intracerebroventricular injection) to bypass the BBB and B-CSF-B barriers. This form of drug administration is invasive and, due to safety concerns, is usually restricted to severe clinical conditions that require single drug dosing or rare drug administrations.

Alternatively, for enhanced delivery of peptide/protein drugs to the brain, the drugs' physicochemical properties can be masked by its encapsulation, adsorption or chemical conjugation to different types of nano-drug delivery systems and other formulations (e.g., nanoparticles, liposomes, polymer conjugates etc., see Table 9.3). As a result, the clearance of peptide/protein drug can be decreased by its protection from proteolytic cleavage. Indeed, the half-life of the drug/DDS in the central circulation is

Table 9.3 Approaches that can be applied for enhancement of brain delivery of peptides and proteins

Formulation types	Administration routes	Barriers disruption	Other approaches
Free drug (solution) Nanoparticles Liposomes Micelles Bolavesicles Polymer conjugates Dendrimers Pasty polymers Hydrogels ...	Intravenous Intraarterial Intranasal Intracerebral Intrathecal Intracerebroventricular	Hyperosmolar agents Proinflammatory agents Permeation enhancers Ultrasound Local heating	Decoration with residues that can enhance BBB crossing via: • adsorptive transcytosis • transporters and pumps • receptor-mediated transcytosis Inhibition of efflux pumps and transporters Use of immune cells (monocytes, macrophages, etc.) as 'Trojan horses' Magnetic field for magnetite-containing formulations

usually much longer than that of the free drug, especially if the DDS is decorated with polyethylene glycol (PEG) residues or other 'inert' protective groups [11, 12].

As a result of incorporation of peptide/protein drugs into the DDSs, the dimensions of the drug-containing particles are larger than those of the original free drug. Consequently, the permeability of drug encapsulated into the DDS via the CNS barriers is usually decreased. Several approaches have been proposed to overcome this problem and to enhance the brain permeability and accumulation of systemically administered DDSs (see Table 9.3). These approaches can be based on disruption of the CNS barriers by chemical compounds (e.g., mannitol of inflammatory cytokines) or external signals (e.g., ultrasound or local heating), inhibition of CNS drug efflux mechanisms, use of specific targeting residues or cell-mediated brain delivery pathways.

Nowadays, the most intensively investigated approaches for increased drug/DDS permeability to the CNS are based on their decoration with specific targeting residues. For instance, decoration with transferrin or with anti-transferrin receptor antibodies (such as OX26 murine monoclonal antibody) was proposed for enhanced transport of the nerve growth factor across the BBB into the CNS [13, 14]. Less frequently investigated approach involves 'feeding' drug/DDS into monocytes, macrophages or other immune cells that can subsequently permeate via the CNS barriers and deliver the drug into the brain (the 'Trojan horses' approach) [15–18].

It should be noted that some of the above-mentioned approaches can be used in combination e.g., DDSs decorated specific surface residues can be administered together with barrier disrupting agent or treatment. However, use of a specific approach, or combination of approaches, does not guarantee more efficient drug/DDS brain accumulation (see Sect. 9.3). Apparently, only in some cases the approaches listed in Table 9.3 are efficient enough to increase the values of the quantitative brain delivery parameters (see Table 9.2) for peptide/protein drugs. Only in rare cases, brain targeting of peptide/protein drugs is attained (i.e., increased drug accumulation in the brain, as compared to other organs and tissues). In part, these outcomes stem from the complex pathways of drug/DDS disposition (i.e., distribution and elimination) in the body (following systemic or local administration) and diverse effects of the specific brain targeting approaches on these pathways. For instance, application of focused ultrasound to disrupt the CNS barriers increases local permeability of the drug/DDS into the brain, but also enhances the drug transport in the opposite direction, from the brain into the systemic circulation, and can induce local release of the drug from the DDS (i.e., release of the drug with its subsequent degradation, not at the target site). As a result of these processes, focused ultrasound will not necessarily enhance the brain drug accumulation efficiency. Therefore, detailed analysis of the drug disposition pathways and of the resulting targeting efficiency of the individual targeting approaches and of their combinations is required for rational development and successful application of peptide/protein drugs intended for brain delivery. In the next sections of this chapter, detailed analysis of the major factors that affect the disposition and efficiency of systemically- and focally-administered peptide and protein drugs is presented and current limitations in the brain delivery of these drugs are summarized.

9.3 Disposition Pathways of Peptide and Protein Drugs Intended for Brain Delivery

9.3.1 Systemic Administration

Systemic parenteral administration (intravenous, subcutaneous, intraperitoneal, etc.) is the most commonly applied way to administer peptide/protein drugs intended for brain delivery. Enteral administration routes, oral, buccal, rectal, etc. are usually not applied for these drugs due to their rapid degradation in the gastrointestinal tract that leads to low bioavailability. Approaches for brain drug delivery via systemic parenteral administration routes are extensively described in several recent reviews [3, 19–21]. In addition, several reviews focused specifically on brain delivery of peptide/protein drugs following systemic administration [2, 10, 22].

Many strategies have been proposed for the purpose of brain delivery of peptide/protein drugs. Most commonly, these strategies are based on liposomes, nanoparticles, polymer conjugates, dendrimers, and other DDSs that encapsulate peptide/protein drugs, usually applied in combination with targeting approaches (based on CNS barrier disruption or transport modulation, see Table 9.3). For the purpose of brain-targeted drug delivery, these DDSs should encapsulate substantial amounts of peptide/protein drug, should be stable in the systemic circulation for sufficient amount of time (usually several hours or longer), should efficiently reach the target tissue (the brain), and eventually release the encapsulated drug in the brain.

The choice of the DDS type and the targeting approaches may significantly affect the systemic and local (brain) drug/DDS pharmacokinetics and pharmacodynamics. Following systemic administration and permeation into the bloodstream, peptide/protein drugs and their formulations are exposed to a plethora of processes which can interfere with their targeting efficiency (see Table 9.4). Specifically, drug/DDS can undergo elimination by proteases or via hepatic and renal elimination pathways, endocytosis and metabolism by the cells that are circulating in the blood, or the cells that are lining the blood vessels. DDSs that encapsulate the peptide/protein drugs can aggregate in the bloodstream (with subsequent deposition in capillaries or uptake by the phagocytic cells in the bloodstream), and can release the encapsulated drugs (decapsulate) in the systemic circulation, prior to reaching the target tissue. The remaining fraction of the free and encapsulated peptide/protein drugs that managed to avoid the above-mentioned processes can extravasate to reach the brain or other organs/tissues. For detailed analysis of the factors that affect the disposition pathways of free and encapsulated drugs, the readers are referred to several excellent recent reviews [23, 24].

It can be seen that the chances of the individual drug/DDS to reach the target tissue are low and drug targeting to the site of action in the brain requires blocking or avoiding the above-mentioned interfering pathways. Efficiency of the individual disposition pathways listed in Table 9.4 can be substantially affected by the dimensions (size and shape) and the surface properties (charge, surface residues) of the drug/DDS. For instance, the cells of the mononuclear phagocyte system (MPS)

Table 9.4 Major pathways of peptide and protein drugs disposition (distribution and elimination) following systemic (intravenous) administration (in a free form or encapsulated into drug delivery systems)

Pathway
Degradation in the systemic circulation
Elimination by liver and kidneys
Aggregation in the bloodstream
Interaction with the non-immune endogenous compounds (formation of 'corona')
Interaction with the soluble components of the immune system (antibodies, complement)
Uptake by mononuclear phagocyte system (MPS) cells
Uptake by the endothelial cells
Uptake by the red blood cells
Permeation to the target organ (the brain)
Permeation to other organs and tissues
Release of the encapsulated drug from the DDS

cells have their 'preferences' for the size, shape, and charge of the DDSs and will preferably endocytose charged (with positive or negative surface charge) drugs/DDSs with diameter bigger than 200 nm [23, 25]. Therefore, for the purpose of preferential drug/DDS delivery to the brain, systemically-administered formulations with the diameter of less than 200 nm should preferably be used.

It should be noted that during their presence in the systemic circulation, drugs/DDSs can interact with and adsorb endogenous compounds such as albumin, transferrin, lipoproteins, and other plasma components [23]. As a result of this interaction, coating or 'corona' is formed on the surface of drug/DDS which affects its surface properties and efficiency of its disposition pathways (see Table 9.4), and can further reduce the chances for efficient drug targeting to the brain. A popular strategy to block these pathways and to avoid 'corona' formation is drug/DDS decoration (via non-covalent adsorption or covalent conjugation) with polyethylene glycol (PEG) residues. PEGylated drugs/DDSs are often referred to as 'stealth' formulations, and PEGylation technology that forms 'inert' steric barrier on the drug/DDSs' surface can indeed be used to prolong their residence in the bloodstream [11, 12, 26]. However, repeated administration of PEGylated drug/DDS can provoke immune reactions of the host against the PEG residues, via formation of anti-PEG IgM antibodies and complement activation, that can increase the drug/DDS clearance and diminish its brain accumulation [27, 28]. These immunogenic aspects of PEGylation need more detailed investigation, and combination of PEGylation with 'active' targeting approaches is recommended for enhanced drug/DDS accumulation in the brain.

Despite that brain drug delivery using systemic administration routes was analyzed in many studies, only in few of them quantitative data on efficiency of drug targeting to the brain was reported [29]. Examples of the studies that analyzed brain delivery of peptide/protein drugs following systemic administration and reported quantitative brain targeting data are presented below.

Radiolabeled brain-derived neurotrophic factor (BDNF) was attached to the murine OX26 monoclonal antibody (that recognizes the rat transferrin receptor) using avidin-biotin technology [30]. Brain accumulation of this conjugate in experimental rats was threefold higher than that of the free peptide (0.49 brain/plasma ratio for the conjugate, as compared to the 0.16 ratio for the BDNF peptide) [29]. Subsequently, brain accumulation of PEGylated form of this conjugate was analyzed and was found to be approximately 17-fold higher than that of the free BDNF peptide [31].

Effect of decoration with the same OX26 antibody on brain accumulation of amyloid beta 1-40 peptide [32] and of PEGylated epidermal growth factor (EGF) [33] was investigated by the same research group. Brain accumulation of the amyloid beta 1-40 peptide decorated with the OX26 antibody was approximately 16-fold higher than that of the free peptide (based on the calculated values of % of the dose that reached the brain) [29]. Brain accumulation of the EGF-PEG3400-OX26 conjugate was 1.4-fold higher than that of the free peptide (0.31 brain/plasma ratio for the conjugate, as compared to the 0.22 ratio for the BDNF peptide) [29] In a different study, decoration with the same OX26 antibody increased by eightfold the brain accumulation of chimeric opioid peptide DALDA (Tyr-DArg-Phe-Lys; 0.057 brain/plasma ratio for the conjugate, as compared to the 0.0063 ratio for the free DALDA peptide) [34].

Increased brain accumulation of a nerve growth factor (NGF) using two different approaches has been reported. In one study, NGF was encapsulated into PEGylated liposomes decorated with RMP-7, a ligand to the B2 receptor, leading to 12.5-fold its higher accumulation in the rat brain (0.30 brain/blood ratio for the decorated liposomes, as compared to the 0.024 ratio for the free NGF peptide) [35]. In another study, NGF encapsulation into poly(butyl cyanoacrylate) nanoparticles coated with polysorbate 80 induced a 2.8-fold increase in its brain accumulation, as compared to the free peptide [29], and significantly reduced the basic symptoms of Parkinsonism in a rat disease model [36].

The above-described examples summarize the findings in selected publications only, and for description of other studies and the major research trends in the field of brain delivery of peptide/protein drugs following systemic administration the readers are referred to several recent reviews [2, 10, 22]. Results of recent quantitative analysis indicate that the efficiency of brain drug delivery is dependent on the drug formulation: nanoparticle-based DDSs appear to be more efficient than the liposome- or conjugate-based DDSs [29]. Incorporation of specialized reagents, such as PEG residues and surface-active agents, decoration with specific targeting residues, and use of external targeting signals apparently can increase the efficiency of drug/DDS delivery to the brain and can lead to brain-targeted drug delivery. However, more detailed quantitative analysis of the drug/DDS disposition and its mechanisms is required to identify the limiting factors for drug brain delivery and the drug formulations that are the most suitable for this purpose [29].

9.3.2 Intranasal Administration

Nasal administration of drugs has been increasingly applied in recent years in an attempt to deliver drugs to the brain. The reasons for this are: (a) relative inefficiency of drug delivery to the brain via systemic routes [29], (b) non-invasive nature of nasal drug administration, as compared to the intravenous and other systemic parenteral administration routes, (c) possibility of direct drug delivery to the brain following intranasal administration, bypassing the BBB and avoiding the gastrointestinal and hepatic first-pass metabolism, (d) intranasal delivery may fasten the onset of action of CNS drugs, which can be of special clinical importance for the anti-pain and anti-migraine medications [37]. Detailed description of the brain drug delivery approaches via nasal route is provided in several recent reviews [38–42]. Additional publications describe brain delivery of peptide and protein drugs using nasal route [43–45].

The mechanisms of drug delivery to the brain following intranasal administration are understood only partially. It appears that following intranasal administration drugs/DDSs can reach the brain via several direct and indirect pathways (see Fig. 9.2). Specifically, some drugs/DDSs can apparently reach the CNS via the olfactory pathway, that consists of the olfactory epithelium, olfactory tract, anterior olfactory nucleus, piriform cortex, amygdala, and hypothalamus [40, 44]. An additional direct pathway that can mediate drug/DDS delivery to the brain from the

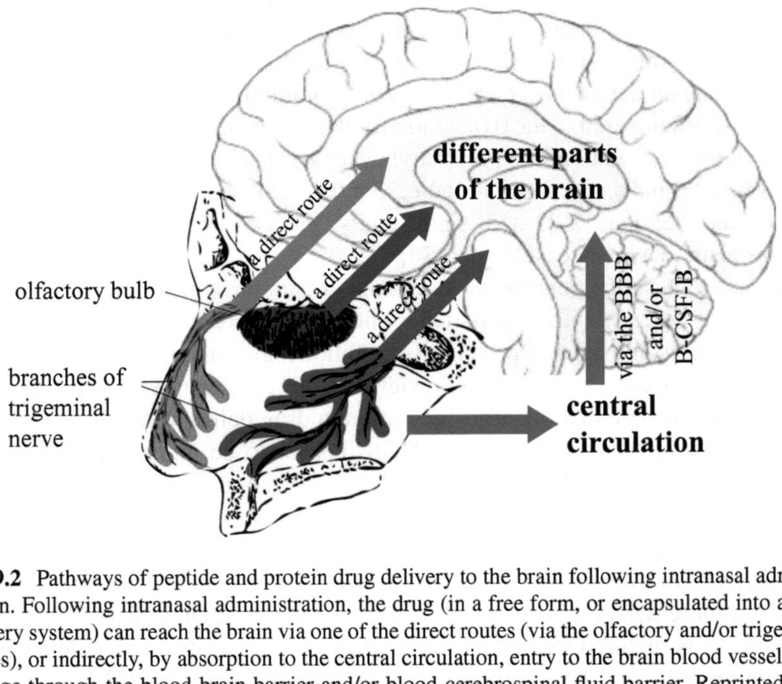

Fig. 9.2 Pathways of peptide and protein drug delivery to the brain following intranasal administration. Following intranasal administration, the drug (in a free form, or encapsulated into a drug delivery system) can reach the brain via one of the direct routes (via the olfactory and/or trigeminal nerves), or indirectly, by absorption to the central circulation, entry to the brain blood vessels, and passage through the blood-brain barrier and/or blood-cerebrospinal fluid barrier. Reprinted from [46] with permission from Elsevier

nasal cavity is via the three branches of the trigeminal nerve (ophthalmic division, maxillary division, and mandibular division) which merge at the trigeminal ganglion, enter the CNS in the pons, and terminate in the spinal trigeminal nuclei in the brainstem [40].

In addition to these direct pathways, drug/DDS can enter the brain indirectly, via the central circulation (and the BBB and B-CSF-B barriers, refer to Fig. 9.2), following its absorption from the nasal cavity to the systemic circulation. Indeed, drugs and DDSs can undergo efficient absorption into the local blood and lymph vessels of the highly vascularized nasal mucosa. This absorption can be increased in some pathological conditions e.g., flu, allergic diseases and other diseases accompanied by local inflammation and vessel permeability in the nasal cavity, or if drug formulation contains permeation enhancers such as penetrating peptides, surface-active agents, pro-inflammatory compounds, etc. In addition, efficiency of both direct and indirect nose-to-brain pathways can increase if the drug or its formulation induces local toxic effects that compromise the function of physiological barriers. This factor is overlooked by many researchers and insufficient amount of efforts is directed towards safety characterization of intranasally-administered drugs and DDSs.

Direct and indirect nose-to-brain routes apparently comprise minor routes of drug/DDS disposition following intranasal administration [46], especially for the peptide and protein drugs. The absolute bioavailability of the nasal desmopressin (nine amino acid peptide), buserelin (nine amino acid peptide) and calcitonin (32 amino acid peptide) in human subjects was reported to be lower than 10 % [47–49], reflecting low efficiency of drug absorption into the central blood circulation (directly and via the brain). Majority of the peptide/protein drug/DDS that is applied intranasally is not reaching the brain, but undergoes other disposition processes (see Table 9.5). Specifically, part of the administered drug/DDS is inhaled or/and ingested. Other part is undergoing enzymatic degradation and clearance by the mucociliary system (the self-clearing mechanism of the nasal cavity and of the airways). DDSs that are deposited in the nasal cavity may release the encapsulated drug which becomes available for enzymatic degradation in the nasal cavity or/and acquire coating or 'corona' which reduces the efficiency of the direct and indirect brain delivery pathways.

Table 9.5 Major pathways of peptide and protein drugs disposition (distribution and elimination) following intranasal administration (in a free form or encapsulated into drug delivery systems)

Pathway
Degradation in the nasal cavity
Discharge with mucus (e.g., due to ciliary movement)
Interaction with the endogenous compounds (formation of 'corona')
Release of the encapsulated drug from the DDS
Passage into the airways and lungs (inhalation)
Passage into the gastrointestinal system (ingestion)
Absorption into the central circulation
Brain permeation via the branches of the trigeminal nerve (ophthalmic division, maxillary division, and mandibular division) in the nasal cavity
Brain permeation via the olfactory bulb and olfactory nerve

Overall, it appears that only a small fraction of the intranasal-administered drug reaches the brain and exerts pharmacological effects in the CNS. Efficiency of drug delivery to the brain following intranasal administration is usually reported using the relative targeting indexes (such as brain/plasma concentrations ratio, drug targeting efficiency percentage and direct transport percentage, see Table 9.2) since it is difficult to measure directly the fraction of the dose that reached the brain in preclinical and clinical studies. In our recent publication, we analyzed the available quantitative data on drug delivery to the brain via nasal route and the relative brain targeting efficiency of different types of intranasal-administered DDSs [46]. Based on quantitative experimental data from 73 publications (that reported data on 82 compounds, only part of them from the peptide/protein group), we analyzed the efficiency of brain drug delivery following intranasal drug/DDS administration. Results of studies that investigated brain delivery of peptide/protein drugs following intranasal administration and reported quantitative brain targeting data are described below.

Brain-targeted delivery of peptide/protein drugs that were administered into the nasal cavity in the form of drug solution was investigated quantitatively in several studies. Following intranasal administration of hypocretin-1 neuropeptide (orexin-A, a 33 amino acid peptide) to rats, drug targeting to the brain and spinal cord increased five- to eightfold, as compared to the intravenous drug administration [50]. It was estimated that 77 % of hypocretin-1 reached the brain following intranasal administration via direct (olfactory and trigeminal) transport routes. The brain and cerebrospinal fluid levels achieved following intranasal administration of hexarelin (a 9 amino acid growth hormone releasing neuropeptide) solution to rabbits were approximately 1.6 times greater than those attained after intravenous administration, despite that the plasma drug levels were significantly lower after intranasal drug administration, as compared to intravenous drug injection [51]. Intranasal administration resulted in significantly different spatial distribution patterns of hexarelin in various regions of brain consistent with substantial drug traffic via olfactory route, whereas, intravenous administration yielded nearly homogeneous drug distribution in the brain. Intranasal administration of a TNF-alpha inhibitory single-chain antibody fragment (scFv, molecular weight of 26.3 kDa) solution to mice resulted in 0.3 % systemic bioavailability [52]. Addition of a penetration enhancing peptide to the formulation enhanced delivery of this drug to the olfactory bulb and to the cerebrum (estimated 98 % of brain drug delivery via direct transport routes [46]), without increasing systemic exposure.

In two studies, brain delivery of peptides/proteins encapsulated in nanoparticle formulations was investigated in quantitative fashion. Brain accumulation of vasoactive intestinal peptide (VIP, a 28 amino acid peptide hormone) encapsulated into poly(ethylene glycol)-poly(lactic acid) nanoparticles modified with wheat germ agglutinin (WGA) was investigated in comparison to VIP-loaded unmodified nanoparticles and drug solution [53]. It was found that higher amounts of VIP have reached the brain following administration of the drug encapsulated into the WGA nanoparticles (estimated 84 % of brain drug delivery via direct transport routes [46]), as compared to the unmodified nanoparticles and drug solution. The same improvements in spatial memory in ethylcholine aziridium-treated rats were observed following intranasal administration of VIP encapsulated into unmodified nanoparticles and twofold lower dose of VIP encapsulated into wheat germ agglutinin-modified nanoparticles (25 μg/kg and 12.5 μg/kg, respectively). In another study, nasal delivery of basic fibroblast growth factor (bFGF) encapsulated in polyethylene glycol-polylactide-polyglycolide (PEG-PLGA) nanoparticles conjugated with Solanum tuberosum lectin (STL, potato lectin), which selectively binds to N-acetylglucosamine on the nasal epithelial membrane, was investigated [54]. The overall brain exposure to the bFGF in the olfactory bulb, cerebrum, and cerebellum of rats following nasal application of STL modified nanoparticles was 1.79- to 5.17-fold of that of rats with intravenous administration, and 0 61- to 2.21-fold and 0.19- to 1.07-fold higher compared with intranasal solution and unmodified nanoparticles, respectively. These findings indicate substantial contribution of direct nose-to-brain routes to overall brain delivery of bFGF following nasal administration (estimated 72 % for the STL-decorated nanoparticle formulation [46]). The neuroprotective effects of different bFGF formulations were determined using a rat model of Alzheimer's disease, and it was found that the STL-decorated nanoparticle bFGF formulation significantly improved spatial learning and memory of diseased rats following intranasal administration, while effects of other bFGF formulations (undecorated nanoparticles and solution) were less prominent.

Overall, the experimental data indicate that nasal administration of peptide/protein drugs can be used for management of brain diseases, but the efficiency of brain drug delivery following intranasal drug/DDS administration appears to be low. We were not able to identify correlations between the drug physico-chemical characteristics, formulation properties, or their combination, and the values of the analyzed brain targeting indexes [46]. Outcomes of data analysis indicate that particle- and gel-based DDSs offer limited advantage for brain drug delivery in comparison to the intranasal administration of drug solution. On the other hand, incorporation of specialized reagents (e.g., absorption enhancers, mucoadhesive compounds, targeting residues, etc.) can increase the efficiency of drug delivery to the brain via nasal route [46]. Overall, more elaborate and detailed investigations of drug/DDSs disposition following intranasal administration (see Table 9.5) are needed to reveal the mechanisms of drug targeting to the brain and to identify the most suitable drugs and DDSs for the management of brain diseases via nasal route.

9.3.3 Local Administration

Drugs can be administered directly into the CNS via the intracerebral, intrathecal or intracerebroventricular routes, as drug solution or encapsulated into DDSs or specialized implants. Due to the invasive nature of these administration routes, and safety concerns, clinical application of these administration techniques is limited to delivery of selected anticancer drugs. Indeed, specialized approaches have been developed for brain injection of anticancer drug solutions via different routes [55, 56] and implantation of drug-eluting polymers during the brain tumor resection

Table 9.6 Disposition (distribution and elimination) pathways following intracerebral administration of peptide and protein drugs (in a free form or encapsulated into drug delivery systems)

Pathway
Diffusion
Convection
Degradation
Aggregation
Interaction with the endogenous compounds (formation of 'corona')
Release of the encapsulated drug from the DDS
Uptake by the target cells (e.g., the neurons)
Uptake by non-target cells

procedure [57, 58]. Drug-eluting polymer formulations can be designed to release the encapsulated drug over the desired period of time, usually several hours—up to several weeks or months, with the desired kinetics. For example, release of carmustine from the Gliadel® wafers in the brain exhibits two phases: the induction period ('burst' release by diffusion of ~60 % of the drug over the first ~10 h) followed by the erosion period (~5 days) [59]. In addition, convection-enhanced delivery technique can be used to modulate the local drug concentrations in the brain tissue (e.g., within the brain tumor and its vicinity) [60, 61].

Local administration of peptide and protein drugs into the brain tissue and fluids has been investigated much less extensively as compared to that of the anticancer drugs, and no such treatments have been approved for clinical use. The reader is referred to the selected experimental and review manuscripts that analyzed drug peptide/protein delivery to the brain via local administration routes [62, 63]. Due to the analytical limitations, local disposition of peptide and protein drugs in the brain following their local administration is seldom analyzed, and pharmacological responses are usually reported in the experimental research studies. Within the brain tissue, locally injected or released drug is exposed to several disposition pathways, which are summarized in Table 9.6. The drug/DDS can traffic in the brain via simple diffusion, and with flow of the interstitial fluid and of the CSF, direction and extent of which varies in the different parts of the brain [7, 64]. In addition, the drug/DDS can undergo degradation by local proteases, interaction with the endogenous compounds, aggregation, and uptake by the target and non-target cells within the brain. As a result of these disposition pathways, the drug/DDS is not staying at the site of local injection or implantation, and a gradient of drug/DDS concentrations is formed in the brain tissue.

From the studies with anti-cancer drugs administered locally into the brain, it is known that the major limitation of the local drug administration approaches is that therapeutic drug concentrations are attained only in a close vicinity to the drug/DDS administration site [65, 66]. The thickness of this layer of therapeutic drug concentrations in the brain is up to several millimeters [58], as revealed by the non-invasive imaging-based or destructive analytical techniques (e.g., fluorescence, MRI, radiolabeling, etc.) [65, 66], reflecting the balance between the release and disposition kinetics of the specific drug/DDS combination. This rapid drug disposition from the administration site and inefficient exposure of the brain to the locally-administered drug poses a major limitation for the clinical effectiveness of peptide and protein drugs. More detailed analysis of the individual disposition pathways of peptide and protein drugs and their dependence on the formulation properties is required for devising efficient local brain administration strategies for these drugs. On the other hand, locally-administered drugs/DDSs are usually characterized by high safety (high therapeutic index) due to the rapid drug disposition from the administration site and dilution of the drug in the brain and other organs, tissues and fluids [58, 66].

9.4 Current Limitations in the Field of Brain Delivery of the Peptide and Protein Drugs

9.4.1 *Limited Targeting Efficiency*

Based on the available data, it appears that peptide and protein drugs do not reach efficiently the intended site of action following administration via systemic and intranasal routes, and are rapidly disposed from the injection or release site following local administration into the brain. Development of specialized drug formulations that incorporate specific reagents (or their combinations), such as PEG residues and surface-active agents, decoration with specific targeting residues and use of external targeting signals, apparently can enhance the drug/DDS delivery to the brain, elevate and prolong the exposure of different parts of the brain to therapeutic drug concentrations, and increase brain drug targeting efficiency. However, rational development of such drug formulations should be based on detailed understanding of drug disposition pathways inside and outside the brain (in the systemic circulation, in the nasal cavity, etc.). For this purpose, sensitive and reliable analytical methods should be used to quantify the peptide and protein drugs and DDSs in organs, tissues, and body fluids.

Unfortunately, many publications that investigate brain delivery of peptide and protein drugs report qualitative data and/or pharmacological effects of the investigated drugs. In the majority of studies with quantitative data, only a short description of the analytical methods is provided and it is not clear whether the specific method had been validated, and whether it was able to quantify reliably and separately the drug at it different forms in the analyzed samples (free, aggregated, protein-bound, encapsulated in the DDS, etc.). This problem is especially acute for the imaging-based analytical methods that are becoming the common way for drug quantification in preclinical and clinical studies [67, 68]. It is possible that insufficient characterization of the analytical methods could have introduced bias in the data that were reported in the individual manuscripts, and lead to erroneous conclusions regarding the drug/DDS disposition [29, 46]. It is recommended to perform detailed assay validation, including quantitative analysis of the different forms of the analyzed drug, in the future studies that will assess the drug/DDSs brain delivery and targeting via different routes.

9.4.2 Heterogeneous Drug Distribution in the Brain

The drug that permeates the brain following systemic or intranasal administration, or that is locally administered into the brain, is exposed to complex disposition processes (see Table 9.6) that lead to heterogeneous drug distribution in different parts of the brain. For instance, uptake to the brain via olfactory route leads to increased drug concentrations in the amygdala and hypothalamus; uptake via trigeminal nerve leads to increased drug concentrations in the trigeminal nucleus and thalamus, etc. From these sites, the drug can subsequently permeate into other parts of the brain by simple diffusion and according to the directions of convection of the interstitial fluid and of the CSF flow [7, 64]. Indeed, in several publications, drug concentrations in different parts of the brain (cerebrum, hippocampus, etc.), and/or in the CSF were reported, and substantial differences in drug concentrations in the analyzed samples were found (see the raw data from the individual studies that are summarized in the Supplementary Tables of [29, 46]).

For selective and efficient treatment of specific CNS diseases, intra-brain drug/DDS targeting to the specific parts and structures of the brain may be required (e.g., substantia nigra in Parkinson's disease, lesions sites in Alzheimer's disease, etc.). Thus, brain drug delivery strategies should preferably take into account the three-dimensional structure of the brain, the directions of CSF flow and of the convection-mediated drug/DDS disposition pathways in the brain [7, 64]. Future drug/DDS brain delivery and targeting approaches will efficiently rely on the heterogeneous nature of drug distribution in the brain, but more detailed knowledge of brain drug/DDS disposition pathways will be required for this purpose.

9.4.3 Toxicity Risks

Drugs and DDS components can exert toxic effects on the physiological barriers (the nasal epithelium, the brain endothelium, and others). Specifically, the BBB and B-CSF-B barriers function can be compromised by specific formulation components and treatments (permeation enhancers, hyperosmolar agents, efflux pumps inhibitors, ultrasound, etc., see Table 9.3). These alterations in brain barrier functions can lead to permeation of undesired compounds into the brain and subsequent alterations of brain function or/and toxicity. Extent, duration and consequences of these alterations in brain barrier permeability should be determined for the individual drugs/DDSs to pave the way for their safe clinical application. To this end, specific markers (e.g., methylene blue) or imaging approaches [69] can be applied in preclinical and clinical settings.

Depending on the administration route and the type of DDS, additional patterns and mechanisms of toxicity, in addition to the toxicity at the BBB and B-CSF-B barriers, should be investigated. For instance, local drug/DDS injection or implantation into the brain can exert direct toxic effects on neurons and other types of cells. On the other hand, nasal-administered drugs/DDSs can be toxic to the nasal epithelium, and/or to the neurons of the olfactory and trigeminal nerves. These types of toxicity have not been studied in sufficient fashion, and their dependence on the drug/DDS composition (e.g., presence and content of permeation enhancers), dosing frequency and application volume is largely unknown. For the purpose of safe and efficient application of brain delivery of peptide and protein drugs, all these potential toxicity pathways should be investigated and ruled out.

9.5 Summary

Peptide and protein compounds can be potentially used for management of CNS disorders and of some non-CNS diseases. However, pharmacological efficiency of drugs from this class is limited by their inefficient permeation into the CNS and rapid clearance by endogenous proteases and via other pathways. Different types of nano-drug delivery systems and other formulations, formulation components, targeting approaches and administration techniques have been developed to overcome these pharmacokinetic limitations of peptide and protein drugs. During the recent decades, substantial advances in development of peptide/protein drugs and drug delivery systems were made. However, the currently available delivery approaches apparently offer limited brain exposure to the administered drugs and are characterized by low pharmacological effectiveness. More detailed analysis of the individual drug/DDS disposition pathways in the brain and in other organs/tissues/fluids is required to devise more efficient brain drug delivery approaches. In addition, the mechanisms and magnitude of toxic effects induced by the peptide and protein drugs and drug delivery systems, in the brain and in other organs and tissues, should be investigated and ruled out to pave the way for efficient clinical application of these drugs.

References

1. Pardridge WM (2003) Blood-brain barrier drug targeting: the future of brain drug development. Mol Interv 3:90–105, 51
2. Brasnjevic I, Steinbusch HW, Schmitz C, Martinez-Martinez P (2009) European NanoBioPharmaceutics Research Initiative. Delivery of peptide and protein drugs over the blood-brain barrier. Prog Neurobiol 87:212–251
3. Alam MI, Beg S, Samad A, Baboota S, Kohli K, Ali J et al (2010) Strategy for effective brain drug delivery. Eur J Pharm Sci 40:385–403
4. Oby E, Janigro D (2006) The blood-brain barrier and epilepsy. Epilepsia 47:1761–1774
5. Sandoval KE, Witt KA (2008) Blood-brain barrier tight junction permeability and ischemic stroke. Neurobiol Dis 32:200–219
6. Storkebaum E, Quaegebeur A, Vikkula M, Carmeliet P (2011) Cerebrovascular disorders: molecular insights and therapeutic opportunities. Nat Neurosci 14:1390–1397
7. Arifin DY, Lee KY, Wang CH (2009) Chemotherapeutic drug transport to brain tumor. J Control Release 137:203–210

8. Arifin DY, Lee KY, Wang CH, Smith KA (2009) Role of convective flow in carmustine delivery to a brain tumor. Pharm Res 26:2289–2302
9. Wolak DJ, Thorne RG (2013) Diffusion of macromolecules in the brain: implications for drug delivery. Mol Pharm 10:1492–1504
10. Bickel U, Yoshikawa T, Pardridge WM (2001) Delivery of peptides and proteins through the blood-brain barrier. Adv Drug Deliv Rev 46:247–279
11. Parveen S, Sahoo SK (2006) Nanomedicine: clinical applications of polyethylene glycol conjugated proteins and drugs. Clin Pharmacokinet 45:965–988
12. Perry JL, Reuter KG, Kai MP, Herlihy KP, Jones SW, Luft JC et al (2012) PEGylated PRINT nanoparticles: the impact of PEG density on protein binding, macrophage association, biodistribution, and pharmacokinetics. Nano Lett 12:5304–5310
13. Zhang Y, Pardridge WM (2001) Conjugation of brain-derived neurotrophic factor to a blood-brain barrier drug targeting system enables neuroprotection in regional brain ischemia following intravenous injection of the neurotrophin. Brain Res 889:49–56
14. Liao GS, Li XB, Zhang CY, Shu YY, Tang SX (2001) Pharmacological actions of nerve growth factor-transferrin conjugate on the central nervous system. J Nat Toxins 10:291–297
15. Afergan E, Epstein H, Dahan R, Koroukhov N, Rohekar K, Danenberg HD et al (2008) Delivery of serotonin to the brain by monocytes following phagocytosis of liposomes. J Control Release 132:84–90
16. Park K (2008) Trojan monocytes for improved drug delivery to the brain. J Control Release 132:75
17. Choi M-R, Bardhan R, Stanton-Maxey KJ, Badve S, Nakshatri H, Stantz KM et al (2012) Delivery of nanoparticles to brain metastases of breast cancer using a cellular Trojan horse. Cancer Nanotechnol 3:47–54
18. Klyachko NL, Haney MJ, Zhao Y, Manickam DS, Mahajan V, Suresh P et al (2014) Macrophages offer a paradigm switch for CNS delivery of therapeutic proteins. Nanomedicine (Lond) 9:1403–1422
19. Gabathuler R (2010) Approaches to transport therapeutic drugs across the blood-brain barrier to treat brain diseases. Neurobiol Dis 37:48–57
20. Chen Y, Liu L (2012) Modern methods for delivery of drugs across the blood-brain barrier. Adv Drug Deliv Rev 64:640–665
21. Gao H, Pang Z, Jiang X (2013) Targeted delivery of nano-therapeutics for major disorders of the central nervous system. Pharm Res 30:2485–2498
22. Pardridge WM, Boado RJ (2012) Reengineering biopharmaceuticals for targeted delivery across the blood-brain barrier. Methods Enzymol 503:269–292
23. Moghimi SM, Hunter AC, Andresen TL (2012) Factors controlling nanoparticle pharmacokinetics: an integrated analysis and perspective. Annu Rev Pharmacol Toxicol 52:481–503
24. Ruenraroengsak P, Cook JM, Florence AT (2010) Nanosystem drug targeting: facing up to complex realities. J Control Release 141:265–276
25. Decuzzi P, Godin B, Tanaka T, Lee SY, Chiappini C, Liu X et al (2010) Size and shape effects in the biodistribution of intravascularly injected particles. J Control Release 141:320–327
26. Barenholz Y (2012) Doxil—the first FDA-approved nano-drug: lessons learned. J Control Release 160:117–134
27. Ishida T, Kiwada H (2008) Accelerated blood clearance (ABC) phenomenon upon repeated injection of PEGylated liposomes. Int J Pharm 354:56–62
28. Schellekens H, Hennink WE, Brinks V (2013) The immunogenicity of polyethylene glycol: facts and fiction. Pharm Res 30:1729–1734
29. Kozlovskaya L, Stepensky D (2013) Quantitative analysis of the brain-targeted delivery of drugs and model compounds using nano-delivery systems. J Control Release 171:17–23
30. Pardridge WM, Kang YS, Buciak JL (1994) Transport of human recombinant brain-derived neurotrophic factor (BDNF) through the rat blood-brain barrier in vivo using vector-mediated peptide drug delivery. Pharm Res 11:738–746
31. Pardridge WM, Wu D, Sakane T (1998) Combined use of carboxyl-directed protein pegylation and vector-mediated blood-brain barrier drug delivery system optimizes brain uptake of brain-derived neurotrophic factor following intravenous administration. Pharm Res 15:576–582

32. Saito Y, Buciak J, Yang J, Pardridge WM (1995) Vector-mediated delivery of 125I-labeled beta-amyloid peptide A beta 1-40 through the blood-brain barrier and binding to Alzheimer disease amyloid of the A beta 1-40/vector complex. Proc Natl Acad Sci U S A 92:10227–10231
33. Kurihara A, Deguchi Y, Pardridge WM (1999) Epidermal growth factor radiopharmaceuticals: 111In chelation, conjugation to a blood-brain barrier delivery vector via a biotin-polyethylene linker, pharmacokinetics, and in vivo imaging of experimental brain tumors. Bioconjug Chem 10:502–511
34. Kang YS, Voigt K, Bickel U (2000) Stability of the disulfide bond in an avidin-biotin linked chimeric peptide during in vivo transcytosis through brain endothelial cells. J Drug Target 8:425–434
35. Xie Y, Ye L, Zhang X, Cui W, Lou J, Nagai T et al (2005) Transport of nerve growth factor encapsulated into liposomes across the blood-brain barrier: in vitro and in vivo studies. J Control Release 105:106–119
36. Kurakhmaeva KB, Djindjikhashvili IA, Petrov VE, Balabanyan VU, Voronina TA, Trofimov SS et al (2009) Brain targeting of nerve growth factor using poly(butyl cyanoacrylate) nanoparticles. J Drug Target 17:564–574
37. Veldhorst-Janssen NM, Fiddelers AA, van der Kuy PH, Neef C, Marcus MA (2009) A review of the clinical pharmacokinetics of opioids, benzodiazepines, and antimigraine drugs delivered intranasally. Clin Ther 31:2954–2987
38. Mistry A, Stolnik S, Illum L (2009) Nanoparticles for direct nose-to-brain delivery of drugs. Int J Pharm 379:146–157
39. Wong YC, Zuo Z (2010) Intranasal delivery—modification of drug metabolism and brain disposition. Pharm Res 27:1208–1223
40. Dhuria SV, Hanson LR, Frey WH 2nd (2010) Intranasal delivery to the central nervous system: mechanisms and experimental considerations. J Pharm Sci 99:1654–1673
41. Illum L (2012) Nasal drug delivery—recent developments and future prospects. J Control Release 161:254–263
42. Gizurarson S (2012) Anatomical and histological factors affecting intranasal drug and vaccine delivery. Curr Drug Deliv 9:566–582
43. Malerba F, Paoletti F, Capsoni S, Cattaneo A (2011) Intranasal delivery of therapeutic proteins for neurological diseases. Expert Opin Drug Deliv 8:1277–1296
44. Lochhead JJ, Thorne RG (2012) Intranasal delivery of biologics to the central nervous system. Adv Drug Deliv Rev 64:614–628
45. Lalatsa A, Schatzlein AG, Uchegbu IF (2014) Strategies to deliver peptide drugs to the brain. Mol Pharm 11:1081–1093
46. Kozlovskaya L, Abou-Kaoud M, Stepensky D (2014) Quantitative analysis of drug delivery to the brain via nasal route. J Control Release 189:133–140
47. Harris AS, Ohlin M, Lethagen S, Nilsson IM (1988) Effects of concentration and volume on nasal bioavailability and biological response to desmopressin. J Pharm Sci 77:337–339
48. Suprefact Nasal Spray, professional prescribing information. Aventis Pharma Ltd., updated on 19 Feb 2015
49. Calcitonin-Salmon Nasal Spray. http://www.drugs.com/pro/calcitonin-salmon-nasal-spray.html. Accessed 24 Sept 2015
50. Dhuria SV, Hanson LR, Frey WH 2nd (2009) Intranasal drug targeting of hypocretin-1 (orexin-A) to the central nervous system. J Pharm Sci 98:2501–2515
51. Yu H, Kim K (2009) Direct nose-to-brain transfer of a growth hormone releasing neuropeptide, hexarelin after intranasal administration to rabbits. Int J Pharm 378:73–79
52. Furrer E, Hulmann V, Urech DM (2009) Intranasal delivery of ESBA105, a TNF-alpha-inhibitory scFv antibody fragment to the brain. J Neuroimmunol 215:65–72
53. Gao X, Wu B, Zhang Q, Chen J, Zhu J, Zhang W et al (2007) Brain delivery of vasoactive intestinal peptide enhanced with the nanoparticles conjugated with wheat germ agglutinin following intranasal administration. J Control Release 121:156–167

54. Zhang C, Chen J, Feng C, Shao X, Liu Q, Zhang Q et al (2014) Intranasal nanoparticles of basic fibroblast growth factor for brain delivery to treat Alzheimer's disease. Int J Pharm 461:192–202
55. Ningaraj NS (2006) Drug delivery to brain tumours: challenges and progress. Expert Opin Drug Deliv 3:499–509
56. Laquintana V, Trapani A, Denora N, Wang F, Gallo JM, Trapani G (2009) New strategies to deliver anticancer drugs to brain tumors. Expert Opin Drug Deliv 6:1017–1032
57. Wadee A, Pillay V, Choonara YE, du Toit LC, Penny C, Ndesendo VM et al (2011) Recent advances in the design of drug-loaded polymeric implants for the treatment of solid tumors. Expert Opin Drug Deliv 8:1323–1340
58. Wolinsky JB, Colson YL, Grinstaff MW (2012) Local drug delivery strategies for cancer treatment: gels, nanoparticles, polymeric films, rods, and wafers. J Control Release 159:14–26
59. Fleming AB, Saltzman WM (2002) Pharmacokinetics of the carmustine implant. Clin Pharmacokinet 41:403–419
60. Debinski W, Tatter SB (2009) Convection-enhanced delivery for the treatment of brain tumors. Expert Rev Neurother 9:1519–1527
61. Allard E, Passirani C, Benoit JP (2009) Convection-enhanced delivery of nanocarriers for the treatment of brain tumors. Biomaterials 30:2302–2318
62. Soderquist RG, Mahoney MJ (2010) Central nervous system delivery of large molecules: challenges and new frontiers for intrathecally administered therapeutics. Expert Opin Drug Deliv 7:285–293
63. Calias P, Banks WA, Begley D, Scarpa M, Dickson P (2014) Intrathecal delivery of protein therapeutics to the brain: a critical reassessment. Pharmacol Ther 144:114–122
64. Liu Y, Paliwal S, Bankiewicz KS, Bringas JR, Heart G, Mitragotri S et al (2010) Ultrasound-enhanced drug transport and distribution in the brain. AAPS PharmSciTech 11:1005–1017
65. Fung LK, Shin M, Tyler B, Brem H, Saltzman WM (1996) Chemotherapeutic drugs released from polymers: distribution of 1,3-bis(2-chloroethyl)-1-nitrosourea in the rat brain. Pharm Res 13:671–682
66. Weinberg BD, Blanco E, Gao J (2008) Polymer implants for intratumoral drug delivery and cancer therapy. J Pharm Sci 97:1681–1702
67. North AJ (2006) Seeing is believing? A beginners' guide to practical pitfalls in image acquisition. J Cell Biol 172:9–18
68. Liu Y, Tseng Y-C, Huang L (2012) Biodistribution studies of nanoparticles using fluorescence imaging: a qualitative or quantitative method? Pharm Res 29:3273–3277
69. Chassidim Y, Veksler R, Lublinsky S, Pell GS, Friedman A, Shelef I (2013) Quantitative imaging assessment of blood-brain barrier permeability in humans. Fluids Barriers CNS 10:9
70. Lichota J, Skjorringe T, Thomsen LB, Moos T (2010) Macromolecular drug transport into the brain using targeted therapy. J Neurochem 113:1–13
71. Bickel U (2005) How to measure drug transport across the blood-brain barrier. NeuroRx 2:15–26
72. van Rooy I, Cakir-Tascioglu S, Hennink WE, Storm G, Schiffelers RM, Mastrobattista E (2011) In vivo methods to study uptake of nanoparticles into the brain. Pharm Res 28:456–471
73. Chow HS, Chen Z, Matsuura GT (1999) Direct transport of cocaine from the nasal cavity to the brain following intranasal cocaine administration in rats. J Pharm Sci 88:754–758
74. Zhang Q, Jiang X, Jiang W, Lu W, Su L, Shi Z (2004) Preparation of nimodipine-loaded microemulsion for intranasal delivery and evaluation on the targeting efficiency to the brain. Int J Pharm 275:85–96

Chapter 10
Polymer-Based DNA Delivery Systems for Cancer Immunotherapy

Ayelet David and Adi Golani-Armon

Abstract The use of gene delivery systems for the expression of antigenic proteins is an established means for activating a patient's own immune system against the cancer they carry. Since tumor cells are poor antigen-presenting cells, cross-presentation of tumor antigens by dendritic cells (DCs) is essential for the generation of tumor-specific cytotoxic T-lymphocyte responses. A number of polymer-based nanomedicines have been developed to deliver genes into DCs, primarily by incorporating tumor-specific, antigen-encoding plasmid DNA with polycationic molecules to facilitate DNA loading and intracellular trafficking. Direct in vivo targeting of plasmid DNA to DC surface receptors can induce high transfection efficiency and long-term gene expression, essential for antigen loading onto major histocompatibility complex molecules and stimulation of T-cell responses. This chapter summarizes the physicochemical properties and biological information on polymer-based non-viral vectors used for targeting DCs, and discusses the main challenges for successful in vivo gene transfer into DCs.

Keywords Antigen presenting cells • Cancer immunotherapy • Cationic polymers • Chitosan • Dendrimers • Dendritic cells • DNA vaccine • Immunization • Gene delivery • Polyamidoamine • Polyethylenimine • Poly(lactic acid) • Poly(lactide-co-glycolide) • Poly(glycolic acid) • Poly(lysine)

Abbreviations

Ag	Antigen
APC	Antigen-presenting cell
BMDCs	Bone marrow-derived dentritic cells
CRD	Carbohydrate recognition domain
CLR	C-type lectin receptor

A. David (✉) • A. Golani-Armon
Department of Clinical Biochemistry and Pharmacology, Faculty of Health Sciences
Ben-Gurion University of the Negev, Beer-Sheva 84105, Israel
e-mail: ayeletda@bgu.ac.il

K.A. Howard et al. (eds.), *Nanomedicine*, Advances in Delivery Science and Technology, DOI 10.1007/978-1-4939-3634-2_10

CTL	Cytotoxic T-lymphocytes
DCs	Dendritic cells
ER	Endoplasmic reticulum
GM-CSF	Granulocyte-macrophage colony-stimulating factor
HLA	Human leukocyte antigen
iDCs	Immature DCs
MΦs	Macrophages
MHC	Major histocompatibility complex
MR	Mannose receptor
MRD	Minimal residual disease
NK	Natural killer
NLS	Nuclear localization signal
OVA	Ovalbumin
PAMAM	Polyamidoamine
PEG	Poly(ethylene glycol)
PEI	Poly(ethylenimine)
PIC	Polyion complexes
PLGA	Poly(lactide-co-glycolide)
PLL	Poly(L-lysine)
PS	Polystyrene
SiRNA	Small interfering RNA
TAA	Tumor-associated antigen
TAP	Transporter associated with antigen presentation
WTC	Whole tumor cells

10.1 Introduction

Cancer has long represented a major burden on health and longevity [1]. The high prevalence of the disease has made it the basis of a major research focus, with the disease being investigated in many contexts. Intensive research has yielded a large body of information that has served to uncover biological, biochemical and pathophysiological aspects of the disease, as well as their underlying mechanisms. Based on the information gathered by the efforts of numerous researchers, novel nanosized medicines (nanomedicines), that can be effectively localized to tumors and actively taken up by cancer cells have been developed, and are expected to replace traditional chemotherapy and possibly irradiation due to increased efficiencies and attenuated adverse side-effects [2].

Unfortunately, recent years have seen only minor improvements in cancer morbidity and mortality rates, with the most commonly used cancer therapies still being conventional chemotherapy, despite their serious side-effects. This has encouraged researchers to seek answers in the field of immunotherapy [3]. Whereas exogenous anti-cancer agents have failed to generate strong and effective responses against tumors and induce regression, the endogenous activity of the immune system was hoped to elicit specific anti-tumor immunity that would result in tumor regression and elimination. In particularly, dendritic cell (DC) vaccinations were expected to induce immunological memory that would enable recognition and destruction of residual tumor cells that evaded earlier treatment, thereby avoiding recurrence of the disease [4]. Previous studies employing mouse models indicated that the establishment of anti-tumor immunity required the presentation of tumor-associated antigens (TAA) by DCs [5, 6]. Two unique properties make DCs the most potent antigen-presenting cells (APCs), namely their ability to present exogenous antigens on major histocompatibility complex (MHC)-I molecules to prime $CD8^+$ T-cells (i.e., cross-presentation), and their capability to initiate, activate and modulate the various arms of the immune system in a coordinated manner. Indeed, in vivo targeting of tumor antigens into DCs was shown to elicit strong TAA-specific $DC4^+$ and $CD8^+$ immune responses [7]. Therefore, DCs loaded with TAA or tumor antigen-encoding plasmid DNA may facilitate the development of new immunotherapies for cancer treatment. This chapter highlights the repertoire of non-viral, nanosized polymeric DNA delivery systems (polyplexes) available to achieve efficient gene transfer into DCs for immunotherapeutic applications in cancer therapy. The physicochemical characteristics and surface properties of polyplexes required for efficient gene transfer and gene expression in DCs are discussed. The influence of the ligand valency on the targeting of DCs via dendritic cell surface receptors is also described.

10.1.1 Cancer Immunotherapy

The field of immunotherapy aims at activating a patient's own immune system against a host disease. Immunotherapy includes the investigation of different mechanisms leading to activation of the immune response, and the development of methods to control such responses in a desired manner so as to combat disease, either as a single strategy or in combination with other treatments. In the specific case of cancer immunotherapy, it is speculated that the in vivo activity of endogenous immune cells would be far more sensitive and specific than would any exogenous treatment [8]. Anti-cancer vaccines have, thus been designed to induce both tumor-specific effector T cells and tumor-specific memory T-cells [4]. Such anti-cancer vaccination, it is hoped, will induce regression of established tumors, as well as prevent the onset of secondary tumors and metastases, possibly by addressing the issue of minimal residual disease (MRD). In this case, sparse cancerous cells that evaded irradiation or chemotherapy will be detected and destroyed by the immune system cells that, unlike the temporary exogenous treatment, are present in the body at all times. A major requirement for an efficient anti-cancer vaccination is the generation of a cytotoxic T-lymphocyte (CTL) response, but manipulation of the other arms of the immune response, mainly $CD4^+$ T helper cells, is critical for a robust and long lasting CTL response [9–11]. CTL response generation depends on antigen (Ag) presentation on MHC to naïve $DC8^+$ T cells. The poor Ag presentation activity demonstrated by tumor cells [12] highlights the need for professional APCs to generate the desired CTL response. Among the several types of APCs in the body, DCs are the most suitable for this purpose.

10.2 Dendritic Cells

10.2.1 Dendritic Cell Biology

In 2011, Ralph Steinman was awarded the Nobel Prize for his 1973 discovery of DCs. DCs are a family of professional APCs. As the most powerful APCs, they are equipped with specialized machinery that enables them to regulate the initiation of a primary immune response [13–15]. Originally derived from bone marrow, DCs are seeded in all tissues, where they sample their environment in an attempt to detect tissue damage, pathogen entry, inflammation and malignantly transformed cells by taking up particles and molecules, processing them into short peptides and presenting the resulting peptides on MHC molecules. DCs use different routes to capture Ags, including phagocytosis, micro- and macropinocytosis, and receptor-mediated endocytosis [9, 15, 16]. Importantly, receptor-mediated endocytosis can be exploited for the targeting of TAAs into DCs.

Immature DCs in peripheral tissues are characterized by high Ag capture activity, low surface MHC levels and co-stimulatory molecules, as well as a tendency to induce tolerance to self-antigens. Upon recognition of "danger" signals, DCs undergo a maturation process characterized by a reduction of Ag capture activity and an increase of MHC and co-stimulatory molecule expression on the cell surface [17]. Mature DCs migrate to adjacent lymph nodes, where they present antigenic peptides to naïve T-cells [4]. Ag presentation, co-stimulatory molecules expression, and the secretion of an appropriate set of cytokines and chemokines by DCs facilitate the differentiation of naïve $CD4^+$ and $CD8^+$ T-cells into effector T-helper or T-cytotoxic cells, respectively [18, 19]. DCs can, thus, simultaneously activate both a CTL response to directly kill pathogens or affected cells and a T-helper response to recruit and enhance the activity of other immune cells in the right context. This unique ability of DCs to activate the various arms of the immune system to achieve a powerful and efficient immune response makes them the most promising candidates for immunotherapeutic manipulations.

10.2.2 Antigen Presentation Pathways in Dendritic Cells

Three distinct pathways are used by different cell types for the presentation of antigens, namely MHC class I and class II antigen presentation, and cross-presentation.

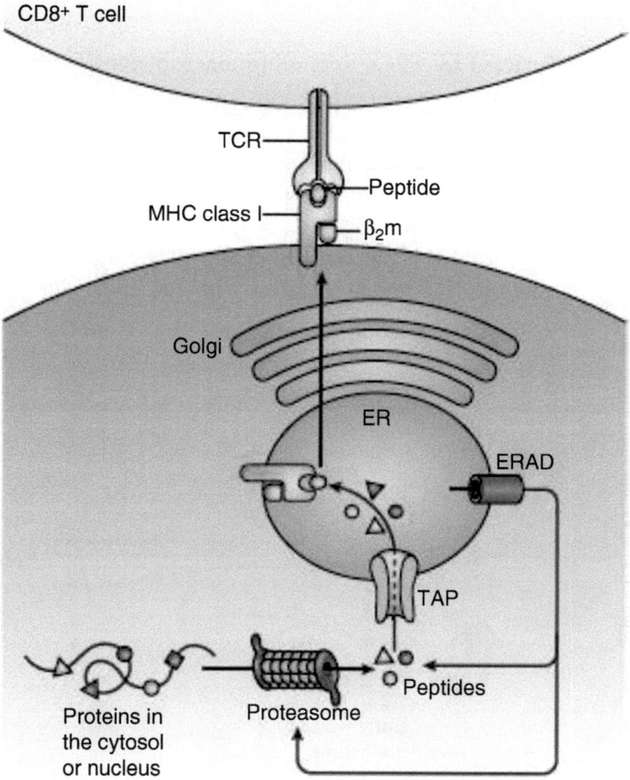

Fig. 10.1 MHC class I processing and presentation pathway. Taken with permission from Ref. 18

10.2.2.1 MHC Class I Antigen Presentation

This pathway is present in all nucleated cells, and is used to present endogenous antigens derived from the nucleus or the cytoplasm. MHC class I antigen presentation can, therefore, report intracellular bacterial or viral infections, as well as malignant transformation in the presenting cell. The resulting intracellular products are degraded in the cytoplasm by the proteasome, translocated into the endoplasmic reticulum (ER) via the transporter associated with antigen presentation (TAP), and loaded on newly formed MHC class I molecules. The MHC class I/peptide complexes are then exported along the constitutive secretory pathway to the cell membrane, where they are presented to naïve $CD8^+$ T-cells. The interaction between MHC class I/peptide complexes on DCs and T-cell receptors (TCR) on $CD8^+$ T-cells induce the differentiation of naïve $CD8^+$ T-cells into cytotoxic T-cells that are capable of killing the infected or transformed cells (Fig. 10.1) [14, 18, 20].

10.2.2.2 MHC Class II Antigen Presentation

This pathway is restricted to APCs, including macrophages (MΦs), DCs and B lymphocytes, and is used to present exogenous antigens. The exogenous antigens are taken up into these cells by endocytosis, where they are found in early endosomes, a mildly acidic cell compartment containing a small amount of proteases. Early endosomes develop into late endosomes that are more acidic and present higher proteolytic activity, which process the captured Ag into short peptides suitable for MHC presentation [21]. The α and β chains of the MHC class II molecule are assembled in the ER and are associated with the Invariant chain (Ii) that

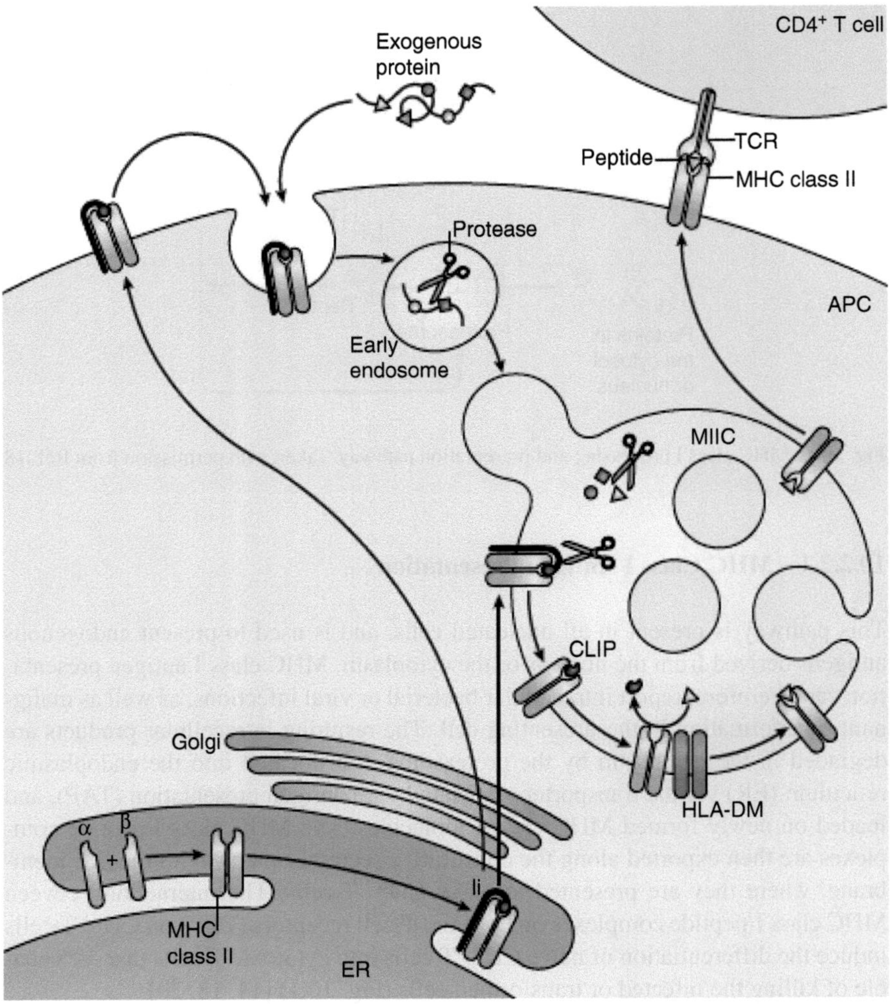

Fig. 10.2 MHC class II processing and presentation pathway. Taken with permission from Ref. 18

occupies the peptide-binding groove of the complex. The MHC class II molecule is then transported to the late endosome, also referred to as the MHC class II compartment, where, following partial digestion of Ii by cathepsins, it finally binds the exogenous peptide. The MHC class II/peptide complexes are transported to the cell membrane for peptide presentation to CD4⁺ T-cells, and induction of T-helper response to support and enhance the activity of other immune cells (Fig. 10.2) [18].

10.2.2.3 Cross-Presentation

The generation of a CTL response against virally infected or transformed cells is possible through the MHC class I pathway described above. However other kinds of threats can be addressed by the immune system APCs through the MHC class II pathway, but this pathway exclusively generates a T-helper response. As it is obvious that not every viral infection or malignant transformation involves APCs, and that many pathogens can impair antigen presentation by their host cells, an alternative mechanism must exist that would enable APCs to generate a cytotoxic T-cell response against the various exogenous threats they encounter [22]. Such a pathway indeed exists and is a unique property of DCs. In addition to the MHC class I and MHC class II antigen presentation pathways, DCs can present exogenous antigens on MHC class I molecules to CD8⁺ T-cells. This pathway is termed cross-presentation, implying that it involves mechanisms from both the MHC class I and the MHC class II pathways [23]. Although first described more than three decades ago, the details of this pathway remain only poorly understood, with two main theories attempting to explain the mechanisms leading to MHC class I loading of endocytosed exogenous antigens. The cytosolic track suggests that the antigen is translocated from the endosome into the cytosol, where it enters the cytosolic processing pathway. The endocytic track, on the other hand, suggest that MHC I molecules are recycled from the cell membrane to the endosome, where they are loaded with antigenic peptides processed by endosomal proteases [21, 24].

10.2.3 *The Role of DCs in Anti-Tumor CTL-Response Generation*

As mentioned earlier, anti-cancer immunity requires the generation of a specific anti-tumor CTL response that, in turn, requires proper antigen presentation. As part of the many techniques tumor cells use to avoid or weaken the immune system, they can substantially reduce the expression of MHC class I/peptide complexes on their cell surface. Since tumor cells are, thus, poor Ag-presenting cells, cross-presentation of tumor antigens by DCs is essential for the generation of tumor-specific CTL responses [4]. In addition, given the unique ability of DCs to initiate and orchestrate the various arms of the immune system, including recruitment and activation of macrophages, natural killer (NK) cells, T-helper cells and B-cells, a strong and comprehensive anti-cancer immune response can hopefully be achieved by inducing such cells to present an appropriate tumor antigen [10].

10.3 Dendritic Cell-Based Cancer Vaccination

10.3.1 DC Immunization

DC immunization requires that an appropriate antigen be presented to DCs in a proper manner and in the right immune context so as to allow Ag uptake, processing and presentation on MHC molecules, parallel to DC maturation and T-cell priming. The first method developed for DC immunization involved the use of whole tumor cells (WTC) or tumor cell lysates, either alone, mixed with an adjuvant, or genetically modified to express an adjuvant. The major advantage of this method is that no Ag identification is required, and multiple Ags are being delivered simultaneously. Nevertheless, all WTC vaccine methods showed limited efficiency in clinical trials, probably due to insufficient interaction between the tumor antigens and the DCs [25]. Pulsing DCs with antigenic proteins or peptides corresponds to another approach. Protein-pulsed DCs are capable of presenting Ags on both MHC class I and MHC class II molecules, with a long half-life of MHC class I presentation and no HLA restriction (i.e., no need for patient selection), but they require an appropriate processing of the protein by DCs, which may be difficult to achieve. Peptide-pulsed DCs, on the other hand, require no processing. Still, they show limited MHC class II presentation, a short half-life of MHC class I presentation, and a HLA restriction. Specific or total tumor mRNA can be processed and presented on both MHC class I and class II molecules. This, however, requires mRNA extraction from a tumor sample and is, thus, patient-specific [15, 26]. Lately, the delivery of tumor Ag-encoding plasmid DNA has emerged as a promising method for DC immunization [19, 27].

10.3.2 DNA Vaccines

DNA vaccines are DCs that were genetically modified to express TAAs. DNA vaccines are advantageous for several reasons. They enable the presentation of multiple epitopes of full-length TAAs on MHC class I and class II molecules, and since processing occurs within the cell, they are not HLA-restricted. In addition, efficient gene transfer allows for a continuous supply of peptides in the modified DC [10]. Genetic modification of DCs can be achieved either by ex vivo gene delivery methods, including use of a gene gun, electroporation, ultrasound and microinjection, or by in vivo approaches, including naked DNA delivery, and viral or synthetic vectors [8, 27]. Ex vivo delivery offers many advantages. The extracted cells are cultured and undergo maturation under controlled conditions and in the absence of inhibitory signals provided by the tumor cells, such that maturation status can be determined before re-administration. Since transfection occurs in culture, high specificity is achieved as only the selected cells are transfected. Nevertheless, ex vivo gene delivery carries substantial limitations. The process is laborious, costly and time-consuming [12]. Patients must undergo cytopheresis, followed by culturing and maturation of the acquired cells, steps that must be performed for each patient separately. Reproducibility is very low and different quality control methods are used. Thus, only a limited number of patients would be expected to benefit from this approach. Hence, ex vivo gene delivery is not likely to become widely marketable. Moreover, transfected mature DCs can show poor distribution from the injection site and ex vivo maturation can impair DC trafficking to lymph nodes, a critical requirement for cross-priming [28, 29]. In contrast, in vivo gene delivery may be expected to become "off the shelf" therapy. As a single product suitable for all patients, in vivo gene delivery can be produced on a large scale with lower costs. Simple and uniform manufacturing and product control procedures will enable reproducibility and control over product quality. In addition, DCs can be targeted at different sites and in their natural environment, thereby not impairing their natural course of maturation and activation [25, 29]. For these reasons, in vivo gene delivery is considered by many to be the best strategy for Ag delivery into DCs [8, 10, 19].

10.3.3 In Vivo Gene Delivery

In vivo gene carriers must meet important requirements. First, they should be able to incorporate their plasmid DNA cargo into the core of the nanoparticle and be stable enough to carry such cargo in the circulation and protect it from degradation. Second, they should be able to selectively target the desired cell type and be properly internalized. Third, they must facilitate escape from the endosome cytoplasm trafficking, nuclear transport and DNA unpacking [30]. To date, the most efficient gene delivery systems are viral vectors. Using small amounts of DNA, viral vectors can induce high transfection efficiency and stable, long-term gene expression. Unfortunately, viral vectors possess some serious safety issues, including toxicity, immunogenicity and oncogenicity, with numerous clinical trials having been terminated because of this [15, 31, 32]. Restricted gene size is another major limitation of viral vectors. Synthetic vectors correspond to cationic lipids or cationic polymers, respectively termed lipoplexes or polyplexes, which electrostatically bind the negatively charged DNA. The main drawback of such vectors is their low transfection efficiency, as compared to viral vectors. Yet, synthetic vectors are simple, safe, easy to manufacture on a large scale and can carry plasmids of unrestricted size. Moreover, they can be easily modified to possess desirable properties, including targetability, serum stability, reactivity to external signals and an ability to stimulate an immune response. In fact, cationic polymers themselves may provide immune stimulation and mimic nuclear localization signals (NLS), facilitating nuclear transport of their cargo [9, 30].

10.4 Polymer-Based Systems for DC-Targeted Gene Delivery

Various polymeric carriers for DNA and RNA delivery have been developed in the last decade as alternatives to viral vectors. Cationic polymers with free primary, secondary and/or tetiary amines were the earliest DNA carriers investigated as transfection reagents in mammalian cells. Such polymers can condense large genes into smaller structures, protect DNA from enzymatic degradation, and can mask the negative charges of DNA, a prerequisite for successful transfection of most types of cells. Some cationic polymers, such as poly-(L-lysine) (PLL), L-polyethylenimine (L-PEI) and chitosan, are linear, while others, like B-PEI and polyamidoamine (PAMAM) dendrimers, are highly branched chains [33]. Branched cationic polymers, such as PEI and PAMAM, exhibit high transfection efficiency in mammalian tissues. Micro- and nanoparticulate systems which adsorb or encapsulate oligonucleotides or genes, based on, for example, poly(lactide-co-glycolide) (PLGA), polycyanoacrylate, polyorthoesters, gelatin, alginate or chitosan, are also under investigation as sustained release matrices for genetic drugs. Some of the most important polymer-based systems utilized for DNA delivery into APCs are discussed below.

10.4.1 Poly(L-Lysine)

Poly(L-lysine) (PLL) was the first cationic polymer developed for gene delivery. PLL is a linear polypeptide presenting L-lysine residues in repeat units (Fig. 10.3c). This first generation cationic polymer bears ε-amino groups that electrostatically bind the negatively charged nucleic acids, and interact with the negatively charged cell membrane. The large number of active functional amine groups allows for easy modification with targeting ligands [34]. However, the transfection efficiency of PLL is very low because it cannot mediate escape from the endosomal compartment and release into the cytosol. To overcome this barrier, PLL is usually used in combination with chloroquine, although use of this agent is limited due to its cytotoxicity [35]. Another disadvantage of PLL polyplexes is that transfection efficiency is significantly influenced by serum, probably due to the rapid binding to negatively-charged serum components [36]. Furthermore, in vivo applications of PLL polyplexes are complicated by a high level of cytotoxicity [37] and lack of in vivo stability [38, 39]. Because of low transfection efficiency, imidazole groups (pKa ~ 6.5) were introduced into PLL to improve buffer capacity, thereby, enhancing transfection efficiency [40]. PLL with high imidazole content mediated high

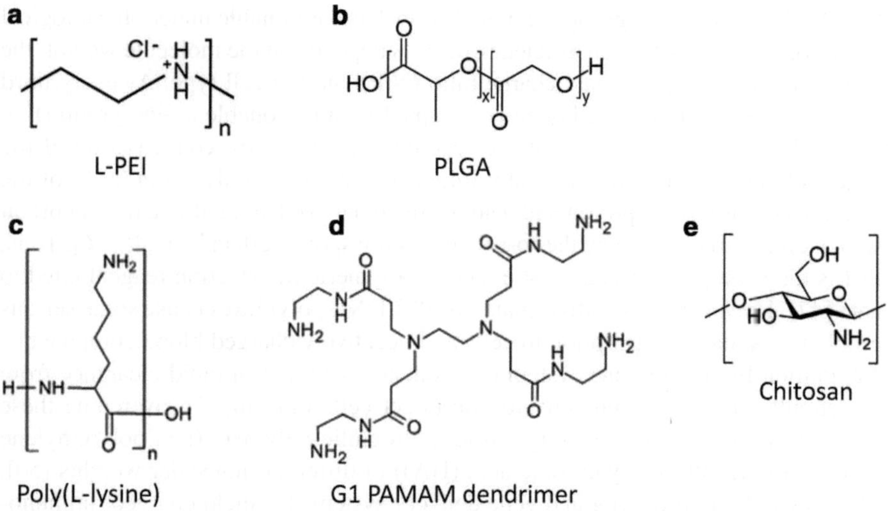

Fig. 10.3 Structures of (**a**) linear PEI, (**b**) PLGA, (**c**) poly(L-lysine), (**d**) first generation (G1) PAMAM dendrimer and (**e**) chitosan

gene transfection efficiency, with gene expression levels close to those of PEI (discussed below), and exhibited low cytotoxicity. The PEGylation of PLL can also improve its cytotoxicity profile [36].

Finally, whereas it has been found that PLL/DNA complexes can be taken up by DCs, PLL/DNA did not alter DC phenotype through surface marker expression [41]. Complexation of plasmid DNA encoding for chicken egg ovalbumin (OVA) with PLL-coated polystyrene (PS) particles induced high levels of CD8+ T-cells as well as OVA-specific antibodies in C57BL/6 mice, and further inhibited tumor growth after challenge with OVA expressing tumor cells [42]. PLL-based microspheres displaying mannan or mannoside-modified surfaces (for targeting the mannose receptor, discussed below) were readily phagocytosed by both DCs and macrophages, however neither surface-assembled mannan- nor mannoside-modified microspheres could stimulate DC maturation [43]. Thus, despite the promising results shown in early studies, PLL is not likely to find clinical applications [30].

10.4.2 Poly(ethylenimine)

Poly(ethylenimine) (PEI) is a second generation and one of the most useful polycations for gene delivery [44]. Linear PEI has only secondary amino groups that are almost all protonated under physiological conditions. Branched PEI presents not only primary and secondary amines but also tertiary amines. As such, only about two-thirds of the amino groups in branched PEI are protonable under physiological conditions. The transfection efficiency of PEIs depends on the molecule weight, the PEI nitrogen/DNA phosphate charge ratio (N/P) and the cell type. As every third position on the polymer backbone is occupied by a protonable amino group (Fig. 10.3a), its cationic charge density is very high, allowing for condensation of the negatively charged DNA. Under physiological conditions, only about 20 % of the PEI nitrogen atoms are protonated, leaving the other 80 % to facilitate the important step of endosomal escape by the so-called "proton sponge effect" [30, 45, 46]. Thus, PEI is the most popular and most effective polymeric transfection reagent cited to date [47]. However, the positive charge of PEI/DNA polyplexes cause some serious problems, including adsorption to cells and negatively charged blood components, recognition by the immune system components, resulting in rapid clearance from the circulation, and cytotoxicity to non-target cells [44, 48]. To overcome these limitations, PEI has been conjugated to hydrophilic polymers (i.e., poly(ethylene glycol) (PEG) [49], or hyaluronic acid (HA)) of different molecular weights [50]. PEG is wildly used in drug and gene delivery systems to shield charged, immunogenic or toxic segments, resulting in less toxic "stealth" particles that can evade the immune system and, thus, avoid rapid clearance from the circulation [51, 52]. The conjugation of PEG to PEI was previously shown to increase polyplex solubility and serum stability and reduce cytotoxicity by shielding the high positive charge of PEI [49]. PEGylated PEI, however, showed lower transfection efficiency, as compared to PEI/DNA complexes, and also lacked cell-specificity [53]. Ligation of PEI/DNA complexes to molecules targeting DC uptake receptors can significantly increase transfection efficiency.

Approaches for targeting DC receptors have generally involved either natural receptor ligands or the use of antibodies raised against specific receptors. Targeting DC cell surface receptors may also provide cell activation signals [4, 54]. Different DC receptors have been used to facilitate targeted delivery of gene and drug carriers into DCs, including β-integrins (CD11b and CD11c) [55], CD40 [56] and the Fc receptor [57], but the most studied proteins for this purpose are members of the C-type lectin receptor (CLR) family [58]. CLR is a family of receptors sharing structural homology in their carbohydrate recognition domain (CRD), where specific sugar residues are bound in a calcium-dependent manner. One of the CLRs, the mannose receptor (MR, CD206), is the most widely used receptor for targeting DCs, using both mannose and mannan [59]. While several studies demonstrated that mannosylated-PEI conjugates are effective in gene delivery via MR [60–62], their transfection potential for primary human and mouse DC was found to be rather low [60]. This might be due to the low affinity of carbohydrate ligands to their receptors, which is usually in the low millimolar range [63], thus, limiting in vivo transfection efficiency. The eight adjacent CRDs in MR may help to increase the binding affinity and specificity of polyplexes containing mannosylated glycans in a multivalent display [59, 64]. We have recently described the design of multivalent mannosylated PEI/DNA complexes bearing mono- and trivalent mannose as a ligand for targeting MR-positive DCs [65]. Complexes bearing mono- and trivalent mannose (Man-PEG-b-PEI/DNA and Man_3-PEG-b-PEI/DNA, respectively) were safe and demonstrated significantly higher in vitro transfection efficiency in DCs. The mannosylated

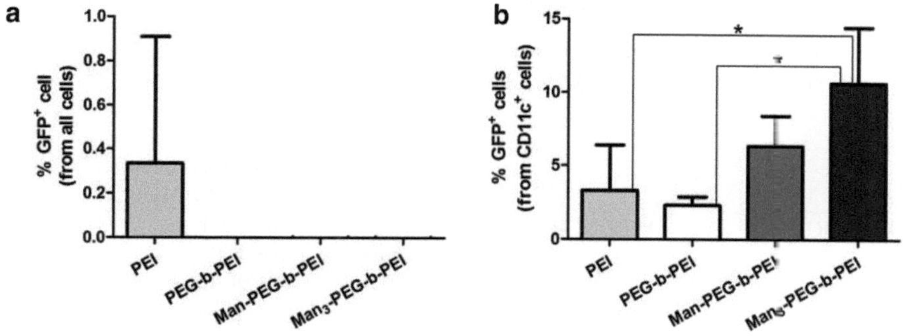

Fig. 10.4 In vivo uptake of mannosylated-PEI/DNA complexes by cells in the draining lymph nodes of C57/BL6 mice 24 h after subcutaneous injection. Results represent the percentage of GFP+ cells in the entire cell population (**a**) and in the CD11c+ cell population (**b**). Taken from Ref. 65

complexes were injected into the tail base of C57/BL6 mice, and 24 h post-injection, the percentages of GFP positive cells in the entire cell population and in the CD11c + cells taken from the inguinal lymph nodes were measured. When examining the entire cells population extracted from the lymph node, only PEI/DNA complexes showed detectable GFP activity, but no significant change was observed between the treatments (Fig. 10.4a). However, Man_3-PEG-*b*-PEI/DNA was significantly more efficient in transfecting CD11c+ DCs collected from inguinal lymph nodes, as compared to polyplexes prepared with PEG-b-PEI/DNA or PEI/DNA (Fig. 10.4b).

One challenge of targeting the MR, as holds true for all of the other above-mentioned receptors, is that they are not exclusively expressed on DCs, and are also on several other cell types [57]. MR, for example, is also expressed on monocytes, subsets of endothelial cells and tumor-associated macrophages [9]. Furthermore, the synthesis of carbohydrate ligands and analogs, especially multivalent or complex carbohydrates, often requires many time-consuming, low yielding steps [66]. The stereochemistry of such carbohydrate ligands is difficult to control, and the products are difficult to purify. A frequently used alternative for carbohydrate ligands are antibodies raised against DC receptors (i.e., integrin CD11c/CD18 [55], Fc receptors [67], DEC-205 [7], DC-SIGN [68] and MR [69]), but these, even when humanized, may still elicit adverse immune responses that decrease the efficiency of treatment and induce auto-immune side-effects [70].

The use of peptide ligands remains, nonetheless, a promising method to target DC receptors. Such ligands are easily synthesized, do not possess immunogenicity, and show high selectivity to DCs (Table 10.1). Accordingly, peptide ligands that could serve as DC-targeting moieties have been sought.

A DC3-nona-arginine fusion peptide (DC3-9dR) that binds nucleic acids by electrostatic interactions was previously exploited for the specific delivery of siRNA to DCs in vitro and in vivo [78] and was also used to silence immunosuppressive molecules in DCs so as to induce strong human T-cell immune responses [79]. The DC3 targeting peptide (Table 10.1) was recently studied by our group to mediate the specific delivery of PEGylated-PEI/DNA polyplexes into DCs [80]. Polyplexes

Table 10.1 Examples for peptide ligands used for targeting DCs

Peptide name	Amino acid sequence	Characteristics	Reference
C-GRWSGWPADL-C	C-GRWSGWPADL-C	A circular peptide that binds to human CD11c/CD18, which shares homology with the D4 domain of intracellular adhesion molecule (ICAM)-1	[71]
APEDNGRSFS	APEDNGRSFS	Derived from ICAM-1, and shares homology with C-GRWSGWPADL-C	[72]
DC3	FYPSYHSTPQRP	Identified using a phage display peptide library. Binds specifically to CD11C+ cells but not monocytes, T and B lymphocytes, or NK, endothelial or fibroblast cells. The cognate receptor for DC3 on the DC surface is unknown	[73]
P-D2	VTLTYEFAAGPRD	Derived from the Ig-like domain 2 of intercellular adhesion molecule 4 (ICAM-4)	[74]
TP	TPAFRYS	Identified by phage display method. The counterpart receptor for TP on BMDC surface is unknown	[75]
NW	NWYLPWLGTNDW	Identified by phage libraries on monocyte-derived immature DCs (iDCs)	[76]
Pan HLA DR-binding peptide	PADRE	MHC class II-binding pan-DR peptide, acts directly on iDCs to induce differentiation into mature DCs, with potent T cell-stimulating capacity	[77]

show significant transfection efficiency in DCs when decorated with DC3 peptide but not in endothelial cells. The transfection efficiency observed was higher than that of PEI/DNA, signifying the potential use of DC3-bearing polyplexes for immunotherapy via DCs.

10.4.3 Polyamidoamine (PAMAM) Dendrimers

Polyamidoamine (PAMAM) dendrimers are hyper-branched, symmetrical, flexible and monodisperse polymeric molecules (Fig. 10.3d). Szoka et al. firstly investigated PAMAM cascade polymers as non-viral gene delivery vectors [81]. These polymers have an ammonia initiator core and amido-amine repeat units of different generations (G2–G10). Their stepwise synthesis results in products that are uniform in terms of structure and size [82] and bear a well-defined number of primary and tertiary amines. The surface primary amines are protonated, resulting in an extremely positively charged surface that enables DNA binding in stable complexes, and interaction with the cell membrane. The tertiary amines in the interior of the polyplexes are protonable, endowing the dendrimers with pH buffering capacity that enables endosomal escape of the polyplexes following cellular uptake [83]. All these properties, in addition to their non-immunogenic nature, make PAMAM dendrimers an alternative to the highly efficient and highly immunogenic viral vectors. However, a major drawback of PAMAM dendrimers is that low generation dendrimers (G5 or lower) show poor transfection efficiencies, while dendrimers of high generation (G6 and higher) are efficient transfection agents, yet possess serious cytotoxicity [82]. In addition, synthesis of high generation dendrimers is a high cost, labor-consuming process that last several days, making them less likely to be designed for large scale production [83].

Mannosylated G4-PAMAM dendrimers conjugated to OVA specifically targeted DCs and induced cross-presentation in vivo [84]. Moreover, pre-immunization with mannosylated PAMAM-OVA leads to delayed onset of B16-OVA melanoma development, slower kinetics of tumor growth and increased survival of OVA-immunized mice. G4-PAMAM dendrimers bearing DC-SIGN ligand in multivalent presentation also achieved efficient DC targeting properties, however, did not affect DCs maturation [20]. It has been postulated that mannosylated dendrimers, as opposed to DC-SIGN-modified dendrimers, may trigger not only DC-SIGN, but also other mannose-specific CLRs that contributes to DC maturation and activation. G5-PAMAM dendrimers conjugated to MHC class II-targeting peptide (PADRE, Table 10.1) and surface-loaded DNA have been shown to effectively transfect murine and human APCs in vitro [85]. When applied subcutaneously this conjugate preferentially transfected DCs in draining lymph nodes, promoted generation of high affinity T-cells, and elicited rejection of established B16 tumors.

10.4.4 Chitosan

Chitosan is a linear polysaccharide (Fig. 10.3e) obtained by deacetylation of its parent polymer chitin, a compound that is widely distributed in nature. Chitosan is composed of randomly distributed ß-(1-4)-linked D-glucosamines (deacetylated units) and N-acetyl-D-glucosamines (acetylated units) and has an apparent pKa value of 6.5 [86]. Chitosan is a biodegradable and biocompatible polymer, with low or no immunogenicity and antibacterial activity [86, 87]. It was previously shown to be non-toxic in both test animals [88] and humans [89]. Due to these properties, chitosan is attractive for drug and gene delivery. The high density of positive charges along the polymeric chain contributes to the condensation of the negatively charged DNA, and more frequently, siRNA molecules, into compact structures, thus protecting them from degradation by blood nucleases promoting their cellular uptake [90]. Chitosan also possess adhesive properties by interacting with glycoproteins in the mucus [91], and is thus used in muco-adhesive drug delivery systems as an adsorption enhancer [92]. DNA plasmid incorporated into chitosan nanoparticles was able to induce DCs maturation and increase IFN-γ secretion from T cells after pulmonary mucosal immunization [93]. However, unless modified, chitosan has low solubility at physiological pH, which significantly limits its use in many applications [87, 92]. Chemical modifications of chitosan have been performed to improve its solubility at physiological pH, including PEGylation or quaternization of the amine groups of chitosan. However, the transfection efficiency of chitosan-based derivatives reported so far is generally not superior to that of PEI [94].

Various DC-targeted chitosan-based DNA delivery formulations has been designed to overcome the major obstacles facing the clinical development of chitosan, namely, the lack of cell-specificity and low transfection efficiency. Mannosylated chitosan/DNA complexes were more efficient in transfecting DCs, as compared to water-soluble chitosan/DNA, and induced better INF-γ production from DCs [95]. Mannosylated-chitosan-entrapping PEI/HBV-DNA complexes induced significantly enhanced serum antibody production and CTL levels after intramuscular immunization [96]. Biotinylated chitosan nanoparticles were modified with bifunctional fusion protein (bfFp) vectors for achieving DC-selective targeting. bfFp is a recombinant fusion protein consisting of truncated core-streptavidin fused to an anti-DEC-205 single chain antibody (scFv). Intranasal administration of plasmid DNA-loaded bfFp/chitosan nanoparticles, along with anti-CD40 DC maturation stimuli, enhanced the amount of mucosal IgAs, as well as systemic IgGs, against nucleocapsid (N) protein of severe acute respiratory syndrome coronavirus (SARS-CoV) [97]. Finally, to improve low transfection efficiency, chitosan-linked-PEI/DNA complexes have been designed, and showed high transfection efficiency and low cytotoxicity towards DCs [98]. Vaccination with DCs transfected with chitosan-linked-PEI/DNA encoding gp100 (melanoma-associated antigen) slightly improved resistance to the B16BL6 melanoma challenge.

10.4.5 Micro- and Nano-Particulate DNA Delivery Vectors

Polymeric particulates have been shown to be efficient in delivering plasmid DNA into APCs (reviewed in [99–101]). A key advantage of particulate vectors relative to other non-viral gene delivery systems is their superior in vivo stability. The principal types of polymers studied in this context include those made of poly(lactide) (PLA; reviewed in [99]), poly(lactide-co-glycolide) (PLGA; reviewed in [99, 100, 102]), polyorthoesters [103], polystyrene (PS) [104, 105] and poly(ε-caprolactone) [106]. Of these, PLGA has been studied most extensively in terms of its capacity to stimulate APCs.

10.4.5.1 Poly(lactide-co-glycolide)

Poly(lactide-co-glycolide) (Fig. 10.3b) is an FDA- and European Medicine Agency-approved polymer for drug delivery systems for parenteral administration [107]. This biocompatible and biodegradable polymer slowly degrades in vivo by hydrolysis, and its byproducts (lactic and glycolic acid) are easily metabolized and excreted. DCs and macrophages appear to have high affinity for PLGA particles, as a high level of internalization of antigen-loaded particles has been demonstrated in both in vitro and in vivo settings [108, 109]. PLGA microspheres carrying protein antigens or antigen-encoding plasmid DNA are capable of eliciting potent antigen-specific immune responses [100, 110, 111]. PLGA also affected expression maturation markers and cytokine production in DCs [112]. This capacity appears to be driven by unique physical features of PLGA particles (i.e., surface charge, shape and the rate of polymer degradation [100]). The mechanisms leading to maturation induction in DCs, however, remain unclear. With respect to their stability in PLGA particles, antigen-encoding plasmid DNA offers advantages over protein-based immuno-modulators, which can lose their biological activity in response to small changes in their tertiary and quaternary structures during formulation.

DNA delivery into DCs can be improved by using ligand-decorated PLGA particles. Mannose-grafted PLGA nanoparticles lead to a significant enhancement of OVA accumulation in the inflamed colon compared to the healthy one, underlining the benefit of active targeting of macrophages and dendritic cells in diseased tissues [113]. Mannan-decorated OVA-loaded PLGA nanoparticles simultaneously enhanced antigen-specific CD4+ and CD8+ T-cell responses in vaccinated mice [114]. Modification of PLGA micro-particles with P-D2 peptide (Table 10.1) significantly improved DC antigen presentation in vitro, and increased the rate and extent of microsphere translocation by DCs and macrophages in vivo [110].

PLGA has been also utilized in combination with other cationic polymers for gene expression in DCs, with [115] or without targeting ligands [116, 117]. PLGA scaffolds encapsulating PEI/DNA and granulocyte-macrophage colony-stimulating factor (GM-CSF) led to a significant increase in gene expression, and high levels of expression that persisted for a period of time [116]. Yet, since PLGA nanoparticles are hydrophobic in nature, they tend to form aggregates that reduce the efficiency of the system [108]. In addition, PLGA particles are opsonized by the immune system components, and degraded before reaching their destination. Another drawback of PLGA for delivery of genetic vaccines stems from their sustained release property [102]. PLGA particles release their cargo very slowly, over days or even weeks, but in vivo they may be exocytosed from the cell or degraded in the lysosome over much shorter periods of time, before sufficient amounts of cargo are released [35]. Moreover, most DCs die within 7 days after activation and migration to draining lymph nodes [118], hence even fast degrading PLGA systems, which fully release the encapsulated DNA within few weeks, cannot meet with the rapid release kinetic criteria and thus fail to induce high levels of target gene expression [108]. Finally, hydrolysis of PLGA leads to low pH within the particle and thus to DNA degradation [102]. For these reasons, despite the promising results obtained with animal models [114] and clinical trials [119, 120], PLGA has been of limited use in this sense [121].

10.5 Conclusion

The delivery of antigens into DCs carries tremendous potential for immune modulation, and specifically, for cancer immunotherapy manipulations. Various polymeric nanomedicines for DNA delivery, ranging from linear homopolymers to block copolymers, branched polymers, as well as combinations of different types of polymers, selected to reduce cytotoxicity and facilitate endocytosis of particles into DCs, are now been routinely examined for their ability to modulate DCs and macrophages. Ultimately, DNA delivery systems that can also induce DC maturation and activation would be advantageous. The main challenges for successful in vivo gene transfer into DCs are the cytotoxicity associated with many cationic polymer gene carriers, the lack of cell specificity and relatively low transfection efficiency, when compared to viral vectors. Together with the fact that DCs compose only 1–3 % of cells in peripheral tissues, such gene systems will have limited success unless targeted. Given the number of targeting molecules, immune-modulatory agents, chemokines, growth factors and antigens that can be considered for DC-specific delivery, large numbers of potentially useful formulations for DC manipulation are available. Targeting DCs in vivo with tumor antigen-encoding plasmid DNA can elicit effective and long-lasting tumor antigen-specific immunity, with minimal inconvenience to the patient. Furthermore, DNA delivery formulations that are stable for extended periods of time and which can enhance antigen presentation on both MHC-I and MHC-II molecules are preferable. With respect to clinical translation, efficacious non-viral gene delivery into DCs will depend on the combination of intelligent material design, the appropriate tumor specific antigen-encoding DNA and immuno-stimulatory molecules to promote DC maturation and activation.

Acknowledgments This work was supported by a grant from the Israeli National Nanotechnology Initiative (INNI), Focal Technology Area (FTA) program, Nanomedicine for Personalized Theranostics.

References

1. Siegel RL, Miller KD, Jemal A (2015) Cancer statistics, 2015. CA Cancer J Clin 65(1):5–29
2. Davis ME, Chen ZG, Shin DM (2008) Nanoparticle therapeutics: an emerging treatment modality for cancer. Nat Rev Drug Discov 7(9):771–782
3. Cross D, Burmester JK (2006) Gene therapy for cancer treatment: past, present and future. Clin Med Res 4(3):218–227
4. Palucka K, Banchereau J (2012) Cancer immunotherapy via dendritic cells. Nat Rev Cancer 12(4):265–277
5. Diamond MS, Kinder M, Matsushita H, Mashayekhi M, Dunn GP, Archambault JM, Lee H, Arthur CD, White JM, Kalinke U, Murphy KM, Schreiber RD (2011) Type I interferon is selectively required by dendritic cells for immune rejection of tumors. J Exp Med 208(10):1989–2003
6. Fuertes MB, Kacha AK, Kline J, Woo SR, Kranz DM, Murphy KM, Gajewski TF (2011) Host type I IFN signals are required for antitumor CD8+ T cell responses through CD8{alpha}+ dendritic cells. J Exp Med 208(10):2005–2016

7. Bonifaz LC, Bonnyay DP, Charalambous A, Darguste DI, Fujii S, Soares H, Brimnes MK, Moltedo B, Moran TM, Steinman RM (2004) In vivo targeting of antigens to maturing dendritic cells via the DEC-205 receptor improves T cell vaccination. J Exp Med 199(6): 815–824
8. Chen YZ, Yao XL, Tabata Y, Nakagawa S, Gao JQ (2010) Gene carriers and transfection systems used in the recombination of dendritic cells for effective cancer immunotherapy. Clin Dev Immunol 2010:565643
9. Joshi MD, Unger WJ, Storm G, van Kooyk Y, Mastrobattista E (2012) Targeting tumor antigens to dendritic cells using particulate carriers. J Control Release 161(1):25–37
10. Boudreau JE, Bonehill A, Thielemans K, Wan Y (2011) Engineering dendritic cells to enhance cancer immunotherapy. Mol Ther 19(5):841–853
11. Vanneman M, Dranoff G (2012) Combining immunotherapy and targeted therapies in cancer treatment. Nat Rev Cancer 12(4):237–251
12. Williams BJ, Bhatia S, Adams LK, Boling S, Carroll JL, Li XL, Rogers DL, Korokhov N, Kovesdi I, Pereboev AV, Curiel DT, Mathis JM (2012) Dendritic cell based PSMA immunotherapy for prostate cancer using a CD40-targeted adenovirus vector. PLoS One 7(10), e46981
13. Banchereau J, Briere F, Caux C, Davoust J, Lebecque S, Liu YJ, Pulendran B, Palucka K (2000) Immunobiology of dendritic cells. Annu Rev Immunol 18:767–811
14. Steinman RM, Banchereau J (2007) Taking dendritic cells into medicine. Nature 449(7161): 419–426
15. Robson NC, Hoves S, Maraskovsky E, Schnurr M (2010) Presentation of tumour antigens by dendritic cells and challenges faced. Curr Opin Immunol 22(1):137–144
16. Wagner CS, Grotzke JE, Cresswell P (2012) Intracellular events regulating cross-presentation. Front Immunol 3:138
17. Pardoll DM (2002) Spinning molecular immunology into successful immunotherapy. Nat Rev Immunol 2(4):227–238
18. Neefjes J, Jongsma ML, Paul P, Bakke O (2011) Towards a systems understanding of MHC class I and MHC class II antigen presentation. Nat Rev Immunol 11(12):823–836
19. Smits EL, Anguille S, Cools N, Berneman ZN, Van Tendeloo VF (2009) Dendritic cell-based cancer gene therapy. Hum Gene Ther 20(10):1106–1118
20. Garcia-Vallejo JJ, Ambrosini M, Overbeek A, van Riel WE, Bloem K, Unger WW, Chiodo F, Bolscher JG, Nazmi K, Kalay H, van Kooyk Y (2012) Multivalent glycopeptide dendrimers for the targeted delivery of antigens to dendritic cells. Mol Immunol 53(4):387–397
21. Neefjes J, Sadaka C (2012) Into the intracellular logistics of cross-presentation. Front Immunol 3:31
22. Kurts C, Robinson BW, Knolle PA (2010) Cross-priming in health and disease. Nat Rev Immunol 10(6):403–414
23. Segura E, Villadangos JA (2011) A modular and combinatorial view of the antigen cross-presentation pathway in dendritic cells. Traffic 12(12):1677–1685
24. Zehner M, Burgdorf S (2013) Regulation of antigen transport into the cytosol for cross-presentation by ubiquitination of the mannose receptor. Mol Immunol 55(2):146–148
25. Le DT, Pardoll DM, Jaffee EM (2010) Cellular vaccine approaches. Cancer J 16(4): 304–310
26. Chiang CL, Benencia F, Coukos G (2010) Whole tumor antigen vaccines. Semin Immunol 22(3):132–143
27. Bolhassani A, Safaiyan S, Rafati S (2011) Improvement of different vaccine delivery systems for cancer therapy. Mol Cancer 10:3
28. Frankenberger B, Schendel DJ (2012) Third generation dendritic cell vaccines for tumor immunotherapy. Eur J Cell Biol 91(1):53–58
29. Tacken PJ, de Vries IJ, Torensma R, Figdor CG (2007) Dendritic-cell immunotherapy: from ex vivo loading to in vivo targeting. Nat Rev Immunol 7(10):790–802
30. Pack DW, Hoffman AS, Pun S, Stayton PS (2005) Design and development of polymers for gene delivery. Nat Rev Drug Discov 4(7):581–593

31. Ulmer JB, Mason PW, Geall A, Mandl CW (2012) RNA-based vaccines. Vaccine 30(30):4414–4418
32. Hu YL, Fu YH, Tabata Y, Gao JQ (2010) Mesenchymal stem cells: a promising targeted-delivery vehicle in cancer gene therapy. J Control Release 147(2):154–162
33. De Smedt SC, Demeester J, Hennink WE (2000) Cationic polymer based gene delivery systems. Pharm Res 17(2):113–126
34. Du J, Sun Y, Shi QS, Liu PF, Zhu MJ, Wang CH, Du LF, Duan YR (2012) Biodegradable nanoparticles of mPEG-PLGA-PLL triblock copolymers as novel non-viral vectors for improving siRNA delivery and gene silencing. Int J Mol Sci 13(1):516–533
35. Little SR, Kohane DS (2008) Polymers for intracellular delivery of nucleic acids. J Mater Chem 18:832–841
36. Ahn CH, Chae SY, Bae YH, Kim SW (2004) Synthesis of biodegradable multi-block copolymers of poly(L-lysine) and poly(ethylene glycol) as a non-viral gene carrier. J Control Release 97(3):567–574
37. Toncheva V, Wolfert MA, Dash PR, Oupicky D, Ulbrich K, Seymour LW, Schacht EH (1998) Novel vectors for gene delivery formed by self-assembly of DNA with poly(L-lysine) grafted with hydrophilic polymers. Biochim Biophys Acta 1380(3):354–368
38. Kwoh DY, Coffin CC, Lollo CP, Jovenal J, Banaszczyk MG, Mullen P, Phillips A, Amini A, Fabrycki J, Bartholomew RM, Brostoff SW, Carlo DJ (1999) Stabilization of poly-L-lysine/DNA polyplexes for in vivo gene delivery to the liver. Biochim Biophys Acta 1444(2): 171–190
39. Oupicky D, Howard KA, Konak C, Dash PR, Ulbrich K, Seymour LW (2000) Steric stabilization of poly-L-lysine/DNA complexes by the covalent attachment of semitelechelic poly[N-(2-hydroxypropyl)methacrylamide]. Bioconjug Chem 11(4):492–501
40. Putnam D, Gentry CA, Pack DW, Langer R (2001) Polymer-based gene delivery with low cytotoxicity by a unique balance of side-chain termini. Proc Natl Acad Sci U S A 98(3): 1200–1205
41. Dhanoya A, Chain BM, Keshavarz-Moore E (2012) Role of DNA topology in uptake of polyplex molecules by dendritic cells. Vaccine 30(9):1675–1681
42. Minigo G, Scholzen A, Tang CK, Hanley JC, Kalkanidis M, Pietersz GA, Apostolopoulos V, Plebanski M (2007) Poly-L-lysine-coated nanoparticles: a potent delivery system to enhance DNA vaccine efficacy. Vaccine 25(7):1316–1327
43. Wattendorf U, Coullerez G, Voros J, Textor M, Merkle HP (2008) Mannose-based molecular patterns on stealth microspheres for receptor-specific targeting of human antigen-presenting cells. Langmuir 24(20):11790–11802
44. Yue Y, Jin F, Deng R, Cai J, Dai Z, Lin MC, Kung HF, Mattebjerg MA, Andresen TL, Wu C (2011) Revisit complexation between DNA and polyethylenimine—effect of length of free polycationic chains on gene transfection. J Control Release 152(1):143–151
45. Behr JP (1997) The proton sponge: a trick to enter cells the viruses did not exploit. Chimia 51:34–36
46. Cho YW, Kim JD, Park K (2003) Polycation gene delivery systems: escape from endosomes to cytosol. J Pharm Pharmacol 55(6):721–734
47. Wightman L, Kircheis R, Rossler V, Carotta S, Ruzicka R, Kursa M, Wagner E (2001) Different behavior of branched and linear polyethylenimine for gene delivery in vitro and in vivo. J Gene Med 3(4):362–372
48. Di Gioia S, Conese M (2009) Polyethylenimine-mediated gene delivery to the lung and therapeutic applications. Drug Des Devel Ther 2:163–188
49. Ahn CH, Chae SY, Bae YH, Kim SW (2002) Biodegradable poly(ethylenimine) for plasmid DNA delivery. J Control Release 80(1–3):273–282
50. Nawwab Al-Deen FM, Selomulya C, Kong YY, Xiang SD, Ma C, Coppel RL, Plebanski M (2014) Design of magnetic polyplexes taken up efficiently by dendritic cell for enhanced DNA vaccine delivery. Gene Ther 21(2):212–218
51. Petersen H, Fechner PM, Martin AL, Kunath K, Stolnik S, Roberts CJ, Fischer D, Davies MC, Kissel T (2002) Polyethylenimine-graft-poly(ethylene glycol) copolymers: influence of

copolymer block structure on DNA complexation and biological activities as gene delivery system. Bioconjug Chem 13(4):845–854
52. Wang W, Xiong W, Wan J, Sun X, Xu H, Yang X (2009) The decrease of PAMAM dendrimer-induced cytotoxicity by PEGylation via attenuation of oxidative stress. Nanotechnology 20(10):105103
53. Mishra S, Webster P, Davis ME (2004) PEGylation significantly affects cellular uptake and intracellular trafficking of non-viral gene delivery particles. Eur J Cell Biol 83(3):97–111
54. Palucka K, Ueno H, Banchereau J (2011) Recent developments in cancer vaccines. J Immunol 186(3):1325–1331
55. Berry JD, Licea A, Popkov M, Cortez X, Fuller R, Elia M, Kerwin L, Kubitz D, Barbas CF 3rd (2003) Rapid monoclonal antibody generation via dendritic cell targeting in vivo. Hybrid Hybridomics 22(1):23–31
56. Schjetne KW, Fredriksen AB, Bogen B (2007) Delivery of antigen to CD40 induces protective immune responses against tumors. J Immunol 178(7):4169–4176
57. Tacken PJ, Torensma R, Figdor CG (2006) Targeting antigens to dendritic cells in vivo. Immunobiology 211(6–8):599–608
58. Burgdorf S, Kautz A, Bohnert V, Knolle PA, Kurts C (2007) Distinct pathways of antigen uptake and intracellular routing in CD4 and CD8 T cell activation. Science 316(5824): 612–616
59. Stahl PD (1992) The mannose receptor and other macrophage lectins. Curr Opin Immunol 4(1):49–52
60. Diebold SS, Kursa M, Wagner E, Cotten M, Zenke M (1999) Mannose polyethylenimine conjugates for targeted DNA delivery into dendritic cells. J Biol Chem 274(27): 19087–19094
61. Sun X, Chen S, Han J, Zhang Z (2012) Mannosylated biodegradable polyethyleneimine for targeted DNA delivery to dendritic cells. Int J Nanomedicine 7:2929–2942
62. Li M, Jiang Y, Xu C, Zhang Z, Sun X (2013) Enhanced immune response against HIV-1 induced by a heterologous DNA prime-adenovirus boost vaccination using mannosylated polyethyleneimine as DNA vaccine adjuvant. Int J Nanomedicine 8:1843–1854
63. Varki A (1993) Biological roles of oligosaccharides: all of the theories are correct. Glycobiology 3(2):97–130
64. Taylor PR, Martinez-Pomares L, Stacey M, Lin HH, Brown GD, Gordon S (2005) Macrophage receptors and immune recognition. Annu Rev Immunol 23:901–944
65. Raviv L, Jaron-Mendelson M, David A (2015) Mannosylated polyion complexes for in vivo gene delivery into CD11c(+) dendritic cells. Mol Pharm 12(2):453–462
66. Oldenburg KR, Loganathan D, Goldstein IJ, Schultz PG, Gallop MA (1992) Peptide ligands for a sugar-binding protein isolated from a random peptide library. Proc Natl Acad Sci U S A 89(12):5393–5397
67. Schuurhuis DH, Ioan-Facsinay A, Nagelkerken B, van Schip JJ, Sedlik C, Melief CJ, Verbeek JS, Ossendorp F (2002) Antigen-antibody immune complexes empower dendritic cells to efficiently prime specific CD8+ CTL responses in vivo. J Immunol 168(5):2240–2246
68. Tacken PJ, de Vries IJ, Gijzen K, Joosten B, Wu D, Rother RP, Faas SJ, Punt CJ, Torensma R, Adema GJ, Figdor CG (2005) Effective induction of naive and recall T-cell responses by targeting antigen to human dendritic cells via a humanized anti-DC-SIGN antibody. Blood 106(4):1278–1285
69. He LZ, Crocker A, Lee J, Mendoza-Ramirez J, Wang XT, Vitale LA, O'Neill T, Petromilli C, Zhang HF, Lopez J, Rohrer D, Keler T, Clynes R (2007) Antigenic targeting of the human mannose receptor induces tumor immunity. J Immunol 178(10):6259–6267
70. Wang J, Zou ZH, Xia HL, He JX, Zhong NS, Tao AL (2012) Strengths and weaknesses of immunotherapy for advanced non-small-cell lung cancer: a meta-analysis of 12 randomized controlled trials. PLoS One 7(3), e32695
71. Frick C, Odermatt A, Zen K, Mandell KJ, Edens H, Portmann R, Mazzucchelli L, Jaye DL, Parkos CA (2005) Interaction of ICAM-1 with beta 2-integrin CD11c/CD18: characterization of a peptide ligand that mimics a putative binding site on domain D4 of ICAM-1. Eur J Immunol 35(12):3610–3621

72. Akazawa T, Ohashi T, Nakajima H, Nishizawa Y, Kodama K, Sugiura K, Inaba T, Inoue N (2014) Development of a dendritic cell-targeting lipopeptide as an immunoadjuvant that inhibits tumor growth without inducing local inflammation. Int J Cancer 135(12):2847–2856
73. Curiel TJ, Morris C, Brumlik M, Landry SJ, Finstad K, Nelson A, Joshi V, Hawkins C, Alarez X, Lackner A, Mohamadzadeh M (2004) Peptides identified through phage display direct immunogenic antigen to dendritic cells. J Immunol 172(12):7425–7431
74. Ihanus E, Uotila LM, Toivanen A, Varis M, Gahmberg CG (2007) Red-cell ICAM-4 is a ligand for the monocyte/macrophage integrin CD11c/CD18: characterization of the binding sites on ICAM-4. Blood 109(2):802–810
75. Jung SN, Kang SK, Yeo GH, Li HY, Jiang T, Nah JW, Bok JD, Cho CS, Choi YJ (2015) Targeted delivery of vaccine to dendritic cells by chitosan nanoparticles conjugated with a targeting peptide ligand selected by phage display technique. Macromol Biosci 15(3): 395–404
76. Sioud M, Skorstad G, Mobergslien A, Saeboe-Larssen S (2013) A novel peptide carrier for efficient targeting of antigens and nucleic acids to dendritic cells. FASEB J 27(8): 3272–3283
77. Agadjanyan MG, Ghochikyan A, Petrushina I, Vasilevko V, Movsesyan N, Mkrtichyan M, Saing T, Cribbs DH (2005) Prototype Alzheimer's disease vaccine using the immunodominant B cell epitope from beta-amyloid and promiscuous T cell epitope pan HLA DR-binding peptide. J Immunol 174(3):1580–1586
78. Subramanya S, Armant M, Salkowitz JR, Nyakeriga AM, Haridas V, Hasan M, Bansal A, Goepfert PA, Wynn KK, Ladell K, Price DA, N M, Kan-Mitchell J, Shankar P (2010) Enhanced induction of HIV-specific cytotoxic T lymphocytes by dendritic cell-targeted delivery of SOCS-1 siRNA. Mol Ther 18(11):2028–2037
79. Subramanya S, Kim SS, Abraham S, Yao J, Kumar M, Kumar P, Haridas V, Lee SK, Shultz LD, Greiner D, N M, Shankar P (2010) Targeted delivery of small interfering RNA to human dendritic cells to suppress dengue virus infection and associated proinflammatory cytokine production. J Virol 84(5):2490–2501
80. Golani-Armon A, Golan M, Shamay Y, Raviv L, David A (2015) DC3-decorated polyplexes for targeted gene delivery into dendritic cells. Bioconjug Chem 26(2):213–224
81. Haensler J, Szoka FC Jr (1993) Polyamidoamine cascade polymers mediate efficient transfection of cells in culture. Bioconjug Chem 4(5):372–379
82. Ziraksaz Z, Nomani A, Ruponen M, Soleimani M, Tabbakhian M, Haririan I (2012) Cell-surface glycosaminoglycans inhibit intranuclear uptake but promote post-nuclear processes of polyamidoamine dendrimer-pDNA transfection. Eur J Pharm Sci 48(1–2):55–63
83. Liu H, Wang H, Yang W, Cheng Y (2012) Disulfide cross-linked low generation dendrimers with high gene transfection efficacy, low cytotoxicity, and low cost. J Am Chem Soc 134(42):17680–17687
84. Sheng KC, Kalkanidis M, Pouniotis DS, Esparon S, Tang CK, Apostolopoulos V, Pietersz GA (2008) Delivery of antigen using a novel mannosylated dendrimer potentiates immunogenicity in vitro and in vivo. Eur J Immunol 38(2):424–436
85. Daftarian P, Kaifer AE, Li W, Blomberg BB, Frasca D, Roth F, Chowdhury R, Berg EA, Fishman JB, Al Sayegh HA, Blackwelder P, Inverardi L, Perez VL, Lemmon V, Serafini P (2011) Peptide-conjugated PAMAM dendrimer as a universal DNA vaccine platform to target antigen-presenting cells. Cancer Res 71(24):7452–7462
86. Dasha FCM, Ottenbriteb RM, Chiellini E (2011) Chitosan—a versatile semi-synthetic polymer in biomedical applications. Prog Polym Sci 36(8):981–1014
87. Liu Y, Kong M, Feng C, Yang KK, Li Y, Su J, Cheng XJ, Park HJ, Chen XJ (2013) Biocompatibility, cellular uptake and biodistribution of the polymeric amphiphilic nanoparticles as oral drug carriers. Colloid Surface B 103:345–353
88. de Moura MR, Aouada FA, Mattoso LH (2008) Preparation of chitosan nanoparticles using methacrylic acid. J Colloid Interf Sci 321(2):477–483
89. Aspden TJ, Mason JD, Jones NS, Lowe J, Skaugrud O, Illum L (1997) Chitosan as a nasal delivery system: the effect of chitosan solutions on in vitro and in vivo mucociliary transport rates in human turbinates and volunteers. J Pharm Sci 86(4):509–513

90. Al-Qadi S, Alatorre-Meda M, Zaghloul EM, Taboada P, Remunán-López C (2013) Chitosan–hyaluronic acid nanoparticles for gene silencing: the role of hyaluronic acid on the nanoparticles' formation and activity. Colloid Surface B 103:615–623
91. Zhao K, Chen G, Shi XM, Gao TT, Li W, Zhao Y, Zhang FQ, Wu J, Cui X, Wang YF (2012) Preparation and efficacy of a live newcastle disease virus vaccine encapsulated in chitosan nanoparticles. PLoS One 7(12), e53314
92. Gaware VS, Hakerud M, Leosson K, Jonsdottir S, Hogset A, Berg K, Masson M (2013) Tetraphenylporphyrin tethered chitosan based carriers for photochemical transfection. J Med Chem 56(3):807–19
93. Bivas-Benita M, van Meijgaarden KE, Franken KL, Junginger HE, Borchard G, Ottenhoff TH, Geluk A (2004) Pulmonary delivery of chitosan-DNA nanoparticles enhances the immunogenicity of a DNA vaccine encoding HLA-A*0201-restricted T-cell epitopes of Mycobacterium tuberculosis. Vaccine 22(13–14):1609–1615
94. Illum L (1998) Chitosan and its use as a pharmaceutical excipient. Pharm Res 15(9): 1326–1331
95. Kim TH, Nah JW, Cho MH, Park TG, Cho CS (2006) Receptor-mediated gene delivery into antigen presenting cells using mannosylated chitosan/DNA nanoparticles. J Nanosci Nanotechnol 6(9–10):2796–2803
96. Zhou X, Liu B, Yu X, Zha X, Zhang X, Chen Y, Wang X, Jin Y, Wu Y, Shan Y, Liu J, Kong W, Shen J (2007) Controlled release of PEI/DNA complexes from mannose-bearing chitosan microspheres as a potent delivery system to enhance immune response to HBV DNA vaccine. J Control Release 121(3):200–207
97. Raghuwanshi D, Mishra V, Das D, Kaur K, Suresh MR (2012) Dendritic cell targeted chitosan nanoparticles for nasal DNA immunization against SARS CoV nucleocapsid protein. Mol Pharm 9(4):946–956
98. Chen YZ, Yao XL, Ruan GX, Zhao QQ, Tang GP, Tabata Y, Gao JQ (2012) Gene-carried chitosan-linked polyethylenimine induced high gene transfection efficiency on dendritic cells. Biotechnol Appl Biochem 59(5):346–352
99. O'Hagan DT, Singh M, Ulmer JB (2004) Microparticles for the delivery of DNA vaccines. Immunol Rev 199:191–200
100. Meng WS, Butterfield LH (2005) Activation of antigen-presenting cells by DNA delivery vectors. Expert Opin Biol Ther 5(8):1019–1028
101. Jilek S, Merkle HP, Walter E (2005) DNA-loaded biodegradable microparticles as vaccine delivery systems and their interaction with dendritic cells. Adv Drug Deliv Rev 57(3): 377–390
102. Walter E, Moelling K, Pavlovic J, Merkle HP (1999) Microencapsulation of DNA using poly(DL-lactide-co-glycolide): stability issues and release characteristics. J Control Release 61(3):361–374
103. Nguyen DN, Raghavan SS, Tashima LM, Lin EC, Fredette SJ, Langer RS, Wang C (2008) Enhancement of poly(orthoester) microspheres for DNA vaccine delivery by blending with poly(ethylenimine). Biomaterials 29(18):2783–2793
104. Chamarthy SP, Jia L, Kovacs JR, Anderson KR, Shen H, Firestine SM, Meng WS (2004) Gene delivery to dendritic cells facilitated by a tumor necrosis factor alpha-competing peptide. Mol Immunol 41(8):741–749
105. Kovacs JR, Zheng Y, Shen H, Meng WS (2005) Polymeric microspheres as stabilizing anchors for oligonucleotide delivery to dendritic cells. Biomaterials 26(33):6754–6761
106. Silva JM, Vandermeulen G, Oliveira VG, Pinto SN, Rodrigues C, Salgado A, Afonso CA, Viana AS, Jerome C, Silva LC, Graca L, Preat V, Florindo HF (2014) Development of functionalized nanoparticles for vaccine delivery to dendritic cells: a mechanistic approach. Nanomedicine (Lond) 9(17):2639–2656
107. Danhier F, Ansorena E, Silva JM, Coco R, Le Breton A, Preat V (2012) PLGA-based nanoparticles: an overview of biomedical applications. J Control Release 161(2):505–522
108. Walter E, Dreher D, Kok M, Thiele L, Kiama SG, Gehr P, Merkle HP (2001) Hydrophilic poly(DL-lactide-co-glycolide) microspheres for the delivery of DNA to human-derived macrophages and dendritic cells. J Control Release 76(1–2):149–168

109. Newman KD, Elamanchili P, Kwon GS, Samuel J (2002) Uptake of poly(D,L-lactic-co-glycolic acid) microspheres by antigen-presenting cells in vivo. J Biomed Mater Res 60(3):480–486
110. Lewis JS, Zaveri TD, Crooks CP 2nd, Keselowsky BG (2012) Microparticle surface modifications targeting dendritic cells for non-activating applications. Biomaterials 33(29):7221–7232
111. Silva AL, Rosalia RA, Varypataki E, Sibuea S, Ossendorp F, Jiskoot W (2015) Poly-(lactic-co-glycolic-acid)-based particulate vaccines: particle uptake by dendritic cells is a key parameter for immune activation. Vaccine 33(7):847–854
112. Jilek S, Ulrich M, Merkle HP, Walter E (2004) Composition and surface charge of DNA-loaded microparticles determine maturation and cytokine secretion in human dendritic cells. Pharm Res 21(7):1240–1247
113. Coco R, Plapied L, Pourcelle V, Jerome C, Brayden DJ, Schneider YJ, Preat V (2013) Drug delivery to inflamed colon by nanoparticles: comparison of different strategies. Int J Pharm 440(1):3–12
114. Hamdy S, Haddadi A, Shayeganpour A, Samuel J, Lavasanifar A (2011) Activation of antigen-specific T cell-responses by mannan-decorated PLGA nanoparticles. Pharm Res 28(9):2288–2301
115. Moffatt S, Cristiano RJ (2006) Uptake characteristics of NGR-coupled stealth PEI/pDNA nanoparticles loaded with PLGA-PEG-PLGA tri-block copolymer for targeted delivery to human monocyte-derived dendritic cells. Int J Pharm 321(1–2):143–154
116. Ali OA, Mooney DJ (2008) Sustained GM-CSF and PEI condensed pDNA presentation increases the level and duration of gene expression in dendritic cells. J Control Release 132(3):273–278
117. Kanazawa T, Takashima Y, Murakoshi M, Nakai Y, Okada H (2009) Enhancement of gene transfection into human dendritic cells using cationic PLGA nanospheres with a synthesized nuclear localization signal. Int J Pharm 379(1):187–195
118. Garg S, Oran A, Wajchman J, Sasaki S, Maris CH, Kapp JA, Jacob J (2003) Genetic tagging shows increased frequency and longevity of antigen-presenting, skin-derived dendritic cells in vivo. Nat Immunol 4(9):907–912
119. Fournier C, Hecquet B, Bouffard P, Vert M, Caty A, Vilain MO, Vanseymortier L, Merle S, Krikorian A, Lefebvre JL et al (1991) Experimental studies and preliminary clinical trial of vinorelbine-loaded polymeric bioresorbable implants for the local treatment of solid tumors. Cancer Res 51(19):5384–5391
120. Sartor O, Dineen MK, Perez-Marreno R, Chu FM, Carron GJ, Tyler RC (2003) An eight-month clinical study of LA-2575 30.0 mg: a new 4-month, subcutaneous delivery system for leuprolide acetate in the treatment of prostate cancer. Urology 62(2):319–323
121. Putnam D (2006) Polymers for gene delivery across length scales. Nat Mater 5(6):439–451

Chapter 11
The Use of Silk in Nanomedicine Applications

Raymond Chiasson, Moaraj Hasan, Qusai Al Nazer, Omid C. Farokhzad, and Nazila Kamaly

Abstract Biopolymers made up of silk proteins have been used in numerous drug delivery applications and represent an excellent source of natural biomaterials. In particular silk fibroin has proved valuable as a building block for nanomedicines and drug delivery implants, owing to its favorable biocompatibility, degradation, stabilization and controllability. In this chapter we will discuss the various sources of silk biomaterial and how this naturally occurring biopolymer has been utilized in the development of nanomedicines and implantable drug delivery systems, demonstrating how silk is a unique biological template which has opened up many possibilities for the generation of functional biomaterials and drug delivery systems in a green and cost-effective manner.

Keywords Silk • Polymers • Silk fibroin • Nanomedicine • Drug delivery • Silk biomaterials • Nanoparticle

11.1 Introduction

Since the lucrative trade of silk began along the Silk Road during the Chinese Han Dynasty (206 BC–220 AD), it has been regarded as a valuable commodity. Spiders and silkworms produce silk protein fibers that have unmatched mechanical strengths in comparison to synthetic biomaterials, and in addition to wide-ranging applications in textiles, silk proteins and fibers have also been utilized in biomedical research for the development of nanomedicines and drug delivery scaffolds. Silk is a non-toxic biomaterial that can degrade at favorable rates for drug delivery

R. Chiasson • M. Hasan • Q. Al Nazer • O.C. Farokhzad (✉)
Laboratory of Nanomedicine and Biomaterials, Brigham and Women's Hospital, Harvard Medical School, Boston, MA 02115, USA
e-mail: ofarokhzad@bwh.harvard.edu; nazk@nanotech.dtu.dk

N. Kamaly
Laboratory of Nanomedicine and Biomaterials, Brigham and Women's Hospital, Harvard Medical School, Boston, MA 02115, USA

Department of Micro- and Nanotechnology, Technical University of Denmark, DTU Nanotech, 2800 Kgs., Lyngby, Denmark

K.A. Howard et al. (eds.), *Nanomedicine*, Advances in Delivery Science and Technology, DOI 10.1007/978-1-4939-3634-2_11

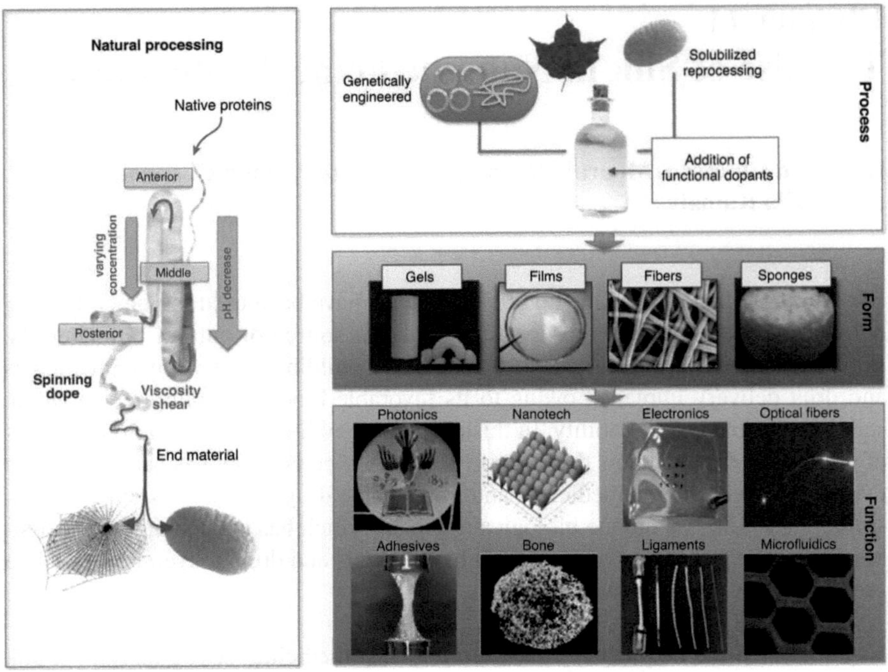

Fig. 11.1 Silk generation and materials development. *Left*: process of natural silk production. *Right*: silk produced from reconstituted native silk proteins or genetically engineered silk, and the variety of silk materials and technology platforms generated from silk. Figure taken from Omenetto et al. [12]

applications, and has ideal mechanical and chemical properties [1], which have enabled its use in a range of wound dressings and sutures [2–11]. Silk proteins have led to the development of new materials, technology platforms and functional devices (Fig. 11.1) [12]. In addition to the development of macroscale products, silk polymers have also been used as building blocks for the formulation of a range of nanomedicines with biodegradability and controlled release properties. In this chapter we will firstly discuss the origins and types of silk biopolymers, and then explore the uses of silk and silk-like products for the development of nanomedicines and healthcare products such as drug depots and implants.

11.1.1 Spider Silk

Orb weaving spiders, such as *N. calvipes* and *E. australis* spin complex webs to catch flying prey [13]. These spiders have seven different spinning glands with each producing a silk with different physical properties and functions; such as pray wrapping, web construction, and coating of egg cocoons [13]. The major ampullate gland produces dragline silk which is important in web construction, has the highest mechanical strength and has been the most extensively studied [14].

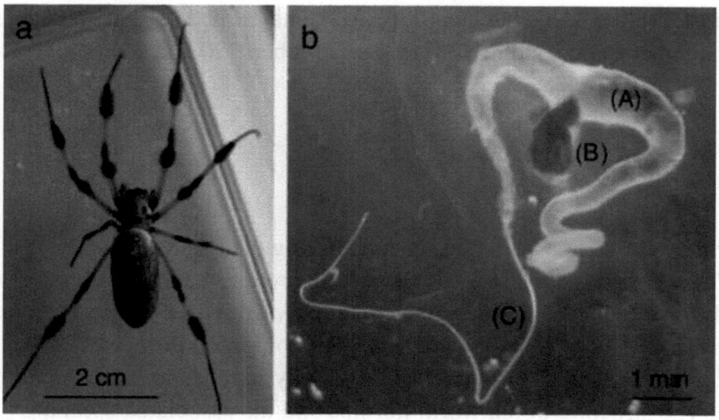

Fig. 11.2 Orb weaving spider and its ampullate. (**a**) Adult female *N. clavipes* (golden-orb) spider, (**b**) dissected major ampullate gland (A) tail (B) sac (C) duct. Figure taken from Tokareva et al. [18]

The spider silk gland contains a tail, sac and duct region (Fig. 11.2). The two major proteins produced in this gland are known as major ampullate silk protein 1 and 2 (MaSp1 and MaSp2) [14]. These are relatively large proteins with molecular masses between 260 and 350 kDa [15], with both proteins possessing a polar N- and C-termini which flank long repeating hydrophobic sequences with small interspersed hydrophilic sequences [16]. The central region of MaSps contain many repeats of certain short amino acid sequences such as; alanine and glycine repeats (poly(A), poly(GA) and GGX), proline containing regions (GPGGX/PGGQQ), and short hydrophilic spacers [17–19].

The central region of MaSps contain many sequences of short alanine and glycine repeats (poly(A), poly(GA) and GGX), proline containing regions (GPGGX/PGGQQ), and short hydrophilic spacers [17–19]. The alanine and glycine repeats have the potential to form anti-parallel beta-sheets, while the proline-containing motifs form beta-spirals [17, 20]. The N- and C-termini form dimers which link individual fibroins [20–24]. The secondary structure is determined not only by the amino acid sequences, but also by their location and the pH and ionic environment of the major ampullate gland [25]. Silk proteins are synthesized in the tail and stored in the sac. During spinning, silk is pulled through the three limbs of the duct and exits at the spigot [26]. Silk is stored in the sac as a concentrated aqueous solution (30–50 % w/v) [25], here, it lacks ordered secondary structure and appears to be randomly coiled and oriented [27]. The pH of the aqueous silk solution drops from about 6.9 in the ampulla to 6.3 by the third limb of the duct, and the sodium and chloride ion concentrations decrease while potassium, phosphorus and sulfur ion concentrations increase [25, 28, 29]. The combination of changes in pH and ionic strength and the shear forces in the narrow ducts, cause the conversion of disordered turns and helices of the liquid silk into the ordered beta-sheet conformation of spun silk fiber [25]. The dimerization of C-termini and the pH-dependant dimerization of N-termini is also an essential step in the formation of fibers [21–24]. The secondary structure of spun major ampullate silk is approximately 40 % beta-sheets which are oriented parallel to the fiber axis [30, 31]. This structure is responsible for the exceptional mechanical strength of spider silk fibers [20, 26, 32].

11.1.2 Silkworm Silk

Silkworm larvae form a protective cocoon during their transformation into an adult moth. These cocoons contain a mixture of silk proteins and have been harvested for hundreds of years. Silk is produced commercially from the cocoons of *Bombyx mori* (*B. mori*) [25]. Silk fibroins from *B. mori* are composed primarily of two peptides made of a heavy chain (300–400 KDa) covalently attached to a light chain (25 KDa) [33–35]. The structure of silkworm fibroins and the associated spinning process is similar to that of spiders [36]. The heavy chain fibroin contains 12 hydrophobic blocks of around 400 amino acids made up of small glycine and alanine-rich repeats (GAGAGS, GX, GAAS) [11, 37]. Between these blocks are hydrophilic spacers of about 30 amino acids and hydrophilic C- and N-termini flank the sequence (Fig. 11.3) [11, 37]. The heavy chain-light-chain dimer also associates non-covalently with a protein called P25 (30 kDa), forming a complex in a 6:1 ratio (dimer: P25) [33, 35, 38]. These hydrophobic proteins form insoluble fibers which make up 70–80 % of the protein in the cocoon, while the rest of the cocoon is made up of a group of proteins known as sericins [11, 39]. These are hydrophilic proteins which act as glue holding the fibroin fibers together, and have molecular masses up to 400 kDa [16, 44]. Similar to spider silk, the secondary structure of silkworm-silk is dependent on both the amino acid sequence and the pH and ionic environment of the silk gland [41, 43].

	N-terminus	repetitive part	C-terminus	light chain
charge ratio (-/+)	(20/12)	(34/5)	(1/9)	(22/15)
pI	4.6	3.80	10.5	5.1
size (aa)	151	5062	50	262
hydropathicity	-0.35	0.24	-0.48	0.05

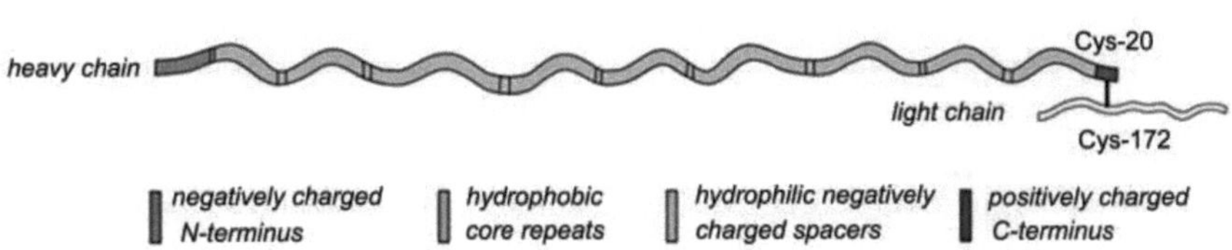

Fig. 11.3 Illustration of the size, charge and hydropathicity of silk fibroin from *B. mori*. The heavy and light chains are linked by one disulfide bond between Cys-20 and Cys-172. Figure modified from Lammel et al. [40]

Silk fibroin is synthesized in the posterior part of the silk gland and transported to the middle of the gland where sericin is added [41, 43], and as silk travels through the anterior part of the silk gland, silk fibers are formed [36, 38]. Much like spider silk, the fibroins are stored in a concentrated, aqueous state with randomly coiled secondary structure in the posterior and middle part of the gland [41, 45]. As they are released through the anterior part of the gland and spinning duct, the fibroins are transformed into an insoluble fiber dominated by beta-sheets [36, 41]. The shear stress caused by the decreasing diameter of the duct, the changes in ion composition and the drop in pH are thought to be responsible for the formation of silk fibers [42, 43]. These factors contribute to the removal of water and contraction of the structure which is necessary for the alignment of anti-parallel beta-sheets that lead to the formation of silk fibers [41–43].

11.1.3 Isolation of Silk Proteins

The majority of silk fibroin comes from silkworms and is typically isolated from the cocoons of *B. mori* [44, 45]. The cocoons are boiled in a dilute aqueous solution of $NaCO_3$ (around 0.02–0.05 M) for at least 15 min [44, 46–51] This dissolves the hydrophilic sericins and leaves undissolved silk fibroins which are rinsed with deionized water and dried, followed by solubilisation in an aqueous solution, often while being heated, using a concentrated lithium salt (9–10 M) or a concentrated mixture of $CaCl_2$, ethanol (or methanol) and water (1:2:8 molar ratio). The salts stabilize the silk in aqueous solution which is then dialyzed into deionized water to remove excess salts [44, 46–51]. The resulting aqueous silk is made up of disordered and partially digested silk fibroin proteins with molecular weights ranging from 8 to 70 kDa [50, 52].

Other methods of isolating silkworm-silk fibroins involve dissection of the silk gland prior to cocoon formation [48, 53]. The posterior silk glands are removed and rinsed with deionized water to remove traces of sericin. Tubes of the gland are squeezed to expel protein which is then dissolved in sodium dodecyl sulfate and dialyzed into deionized water [48, 53]. Generally, the farming and isolation of spider silk is challenging due to the cannibalistic nature of spiders and, therefore, the large-scale production of spider silk by natural means is not feasible [54]. Other attempts to produce spider silk involve the expression of recombinant silk genes in *E. coli*, yeast, mammalian cells, transgenic plants, and even transgenic silkworms [55–59].

11.2 Silk in Nanomedicine Applications

Nanomedicine involves the application of nanotechnology to medicine, whereby nanoscale structures are developed for medical applications such as drug delivery, controlled drug release, and imaging. The following section discusses recent work regarding the formation and potential use of silk-based nanoparticles for biomedical applications. To-date, silkworm silk and recombinant spider silks have been used as the core of nano- and micro- particles, as coatings for nanoparticles, or blended with nanoparticles or other proteins and polymers due to their enhanced biophysical and mechanical strengths [46, 60–63]. Nanoparticles made from silk fibroin have the potential to increase the efficacy of a range of therapeutics and to minimize their adverse effects.

11.2.1 Preparation of Silk Nanoparticles

Many different techniques have been successfully used to form silk nanoparticles and microparticles with the simplest being nanoprecipitation (also known as solvent displacement) and desolvation [35, 46, 60]. These techniques change the solvent conditions around solvated silk proteins to induce the precipitation of silk as nanoparticles. Like in the formation of silk fibers, precipitation of aqueous silk into nanoparticles corresponds with an increase in the beta-sheet content and decease in the random coiled content [52, 53]. Nanoprecipitation involves the drop-wise addition of aqueous silk solution into a larger volume of miscible organic solvent, such as acetone or ethanol [60]. Desolvation is the opposite process in which a miscible organic solvent or a high ionic strength solution, such as 2 M PO_4, is added to an aqueous silk solution, up to a certain concentration using dialysis or mixing [40, 46, 64]. Bulk nanoprecipitation can yield particles with diameters between 40 and 170 nm and desolvation with organic solvents has achieved diameters around 150 nm [46, 50, 53, 65, 66]. Desolvation with concentrated aqueous salts avoids the use of organic solvents, but leads to larger particles with the smallest so far reported being around 330 nm [40, 67].

Solution-enhanced dispersion by supercritical CO_2 is another notable method of preparing silk nanoparticles [68, 69]. Using this technique, nanoparticles with diameters as small as 50 nm have been made. In this method, isolated *B. mori* silk is dissolved in 1,1,1,3,3,3-hexafluoro-2-propanol and pumped under high pressure in combination with supercritical CO_2 into a high-pressure vessel forming nanoscale droplets leading to nanoparticles. Microfluidics is another promising new method of nanoparticle formation in a more controlled and reproducible manner [70, 71]. Mintopolus et al. were able to make silk fibroin spheres as small as 210 nm. Here, a stream of aqueous silk combines with a stream of poly(vinyl alcohol) to form tiny monodispersed drops, allowing for the silk to precipitate as nanoparticles [72]. Other methods of making silk nanoparticles and microparticles include spray drying, laminar jet break-up using sound waves and electrospinning dilute solutions under low voltage [35]. Loading of drugs into silk nanoparticles is typically accomplished by mixing the cargo with the silk protein prior to, or during the formation of nanoparticles [65, 66]. Particles can also be loaded by charge-based association [64]. Silk particles bear a negative charge at physiological pH, therefore, drug loading post formulation is limited to cationic species [35].

11.2.2 *Silkworm Silk-based Nanoparticles*

The majority of nanoscale silk research has involved nanoparticles made from the silk of silkworm cocoons. This section highlights particles loaded with drugs and small molecules, proteins and peptides, as well as briefly discussing the functionalization of silk particles with targeting ligands.

11.2.2.1 Small Molecule Drug Delivery with Silk Nanoparticles

Silk nanoparticles, similar to other nanomedicines, are capable of encapsulating and facilitating the controlled release of small molecules and drugs. For example, Chen et al. loaded paclitaxel into silk fibroin nanoparticles using desolvation [73]. Among different formulations, the smallest particles were 270 nm in diameter with zeta potentials between −20 and −27 mV. Drug loading was between 1 % and 7 % and the encapsulation efficiency was between 67 % and 100 %. Release of paclitaxel in vitro was relatively rapid, with the entire drug content being released in 2 h, however, the larger particles (diameters up to 520 nm), had a slower release lasting up to 300 h. Wu et al. prepared smaller paclitaxel-loaded nanoparticles, also by desolvation, with diameters between 160 and 210 nm [66]. In vitro release showed a rapid 40 % release in 8 h followed by a delayed release, totalling 47 % after 100 h. The particles were cytotoxic to two human gastric carcinoma cell lines (BGC-823 and SGC-7901) in vitro, while silk fibroin alone was not. In vivo testing on a subcutaneous human gastric cancer (BGC-823) nude mouse xenograft model showed decreased tumor volume and weight when treated with local injection of nanoparticles versus the equivalent dose of drug. Gupta et al. loaded curcumin, a potential anti-cancer drug, into silk particles prepared using the capillary microdot technique [74]. Silk and the drug mixture was dispensed on a slide, frozen, lyophilized and resuspended in methanol. Particle sizes ranged from 40 to 70 nm and in vitro release showed a burst phase followed by a slower sustained release over 8 days. The particles were shown to be taken up by and to inhibit growth in two human breast cancer cell lines (MCF-7 and MDA-MB-453) in vitro. Zheng et al. made 50 nm diameter particles using solution-enhanced dispersion by $scCO_2$ which were loaded with the anti-inflammatory drug indomethacin [69]. Drug loading was 6 % by mass with encapsulation efficiency of 31 %. In vitro release was relatively well sustained and stable: 61 % after 6 h and 87 % after 24 h.

Silk can also be blended with other polymers or macromolecules to modify properties such as drug loading [46, 75]. Subia et al. formed cross-linked albumin-silk fibroin nanoparticles loaded with the anti-cancer drug methotrexate [46]. Particles were formed by desolvation and cross-linking was done using gluteraldehyde. The particle diameters were between 100 and 200 nm with a zeta potential around −25 mV. The drug loading was 15–24 % by mass and the encapsulation which was 83-87 % was higher than that of particles made from only silk fibroin or albumin. Up to 90 % of the methotrexate was released from the various formulations after 10 days.

Both silk fibroin-albumin nanoparticles and pure silk fibroin nanoparticles inhibited the growth of the human breast cancer MDA-MB-231 cells, and were more effective than the free drug. The particles were also found to be non-toxic to human erythrocytes. They were taken up into the cytoplasm and perinuclear space in feline fibroblast cells (AH-927) as observed by confocal laser scanning microscopy after conjugation of fluorescin isothiocyanate to the nanoparticles.

Attempts have also been made to combine nanoparticles with other biocompatible structures. Numata et al. incorporated silk nanoparticles into a silk hydrogel to combine the fast release of molecules from the hydrogel with slower release from nanoparticles [76]. The particles were 175 nm in diameter with a zeta potential of about −12.5 mV. Dyes (Rhodamine B, Texas Red and fluorescein isotyiocyanate) were incorporated into both the particles and the gel, and release was measured in the presence of protease. Within 1 h, 90 % of dye was released from the hydrogel while the nanoparticles had a relatively constant release rate over 5 days which was not influenced by enzymatic degradation. No cytotoxicity was observed against human mesenchymal stem cells.

Glue-like sericin proteins are normally removed from cocoons during the degumming process [1, 49–52], and appear to be non-toxic [3, 77]. Hanjin et al. created drug-loaded microparticles by electrohydrodynamic spraying using these proteins [75]. The resulting particles, however, were large; 150–300 μm. The anti-inflammatory drug diclofenac was loaded into the particles and 60–70 % was released in 7 h. Mandal and Kundu were able to make sericin nanoparticles from *A. mylitta* silk by mixing this with poloxamers (triblock copolymers with hydrophilic-poly(propyleneoxide)-hydrophilic sequences) [63]. Their diameters ranged from 60–130 nm and both hydrophobic paclitaxel and hydrophilic insulin were successfully loaded. The growth of MCF-7 breast cancer cells was reduced when they were treated with paclitaxel-loaded nanoparticles. The authors suggested that the positively charged sericin could aid cellular uptake by charge-based association with the plasma membrane, however, in this experiment the nanoparticles were not more effective than the drug alone. Sericins might have some utility for drug delivery, however this has yet to be fully demonstrated in vivo. Silk fibroin nanoparticles on the other hand, have greater potential for the loading and controlled release (via diffusion) of drugs and small molecules.

11.2.2.2 Peptide or Protein Drug Delivery with Silk Nanoparticles

Proteins and peptides represent a further important group of pharmaceuticals and constitute therapeutic hormones, growth factors, clotting factors, anticoagulants, drug-activating enzymes, and antibodies [78]. However, these types of biologic drugs suffer from degradation by serum proteases and clearance by the immune system [78, 79]. Incorporation into nanoparticles could potentially mitigate these effects. Hai-Bo et al. formulated particles of silk fibroin cross-linked to insulin by nanoprecipitation [50]. Silk-insulin particles were cross-linked using gluteraldehyde post-formation and their diameters ranged from 40 to 120 nm. Compared to the native peptide, encapsulated peptide had improved resistance to degradation in human serum and to degradation by trypsin in vitro. Kundu et al. observed the endocytosis of particles made from *B. mori* and *A. mylitta* silk using nanoprecipitation [53]. The particles were between 150 and 180 nm in diameter with a zeta potential around −25 mV (Fig. 11.4).

The particles were endocytosed into murine squamous carcinoma cells (SCC7) and were detected in the cytoplasm and nuclear membrane. Cytotoxicity was measured using murine fibroblasts (L929) and the particles were found to be non-cytotoxic up to 100 μg/mL. The nanoparticles made from *A. mylitta* silk were loaded with vascular endothelial growth factor by charge-based association and in vitro release was approximately linear for 5 days followed by a slower, sustained release. Cao et al. prepared nanoparticles loaded with the enzyme beta-glucosidase by nanoprecipitation [65]. The particles were between 50 and 150 nm and the beta-glucosidase was incorporated by charge-based association. Wang et al. formulated nanoparticles of silk blended with poly(vinyl alcohol) using a different technique which involved drying of the silk-poly(vinyl alcohol) mixture to leave a thin film, followed by resuspension in water [51]. Rhodamine and tetramethylrhodamine conjugated to either bovine serum albumin or to dextran, was loaded into particles by mixing with silk prior to nanoparticle formation. Various silk-polymer ratios gave a large distribution of diameters with the smallest being 300–400 nm. Dextran was

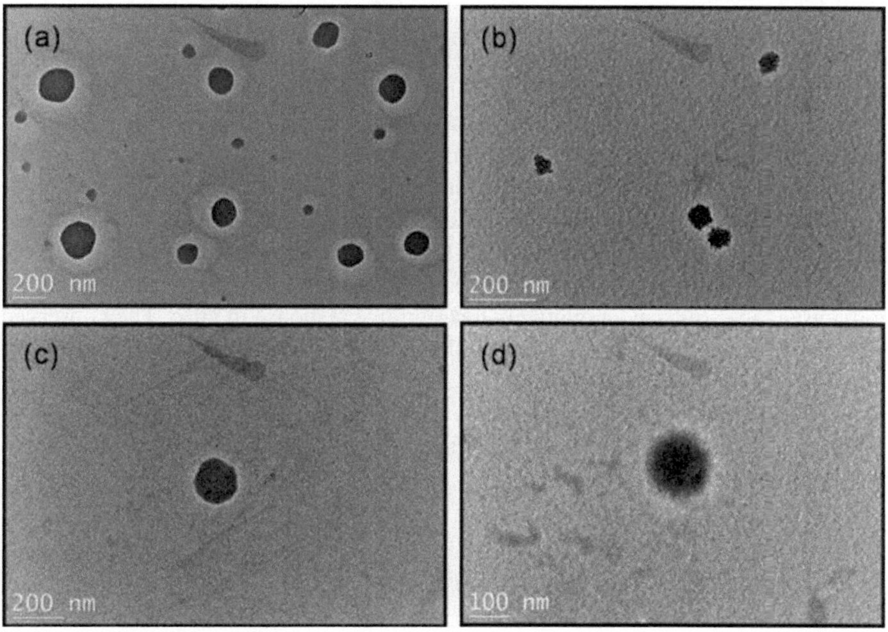

Fig. 11.4 TEM images of silk fibroin nanoparticles prepared from *A. mylitta* silk (**a**) and (**c**) and *B. mori* silk (**b**) and a single silk fibroin nanoparticles prepared from *B. mori* (**d**). Taken from Kundu et al. [53]

released from the nanoparticles at a reasonable rate: 60 % in 14 days, however, less than 5 % of the protein payload was released in the same amount of time.

Both proteins and peptides can be effectively incorporated into silk nanoparticles. Encapsulation can shield and modulate the release of proteins and peptides into their surrounding environment which would improve stability by reducing the rate of degradation by proteases [50]. More work is needed to demonstrate this in vivo and with other peptide and protein drugs.

11.2.2.3 Targeted Nanoparticles

The location and accumulation of nanoparticles in vivo is subject to blood hemodynamic forces and diffusive mechanisms [80]. Rather than relying on these passive means of accumulation, attachment of ligands, such as peptides, antibodies, or small molecules, to the surface of nanoparticles allows them to preferentially bind target cells to improve the efficiency of drug delivery [80]. Wang et al. functionalized the surface of silk microparticles (2–3 μm) with avidin using carboxydiimide coupling followed by the attachment of biotinylated antibodies to the particle surface [61]. Fluorescein isothiocyanate was bound to avidin to image the particles by confocal laser scanning microscopy. The targeted particles bound to CD3 positive human T-lymphocytic cells (ATCC®TIB-152™) with improved specificity. Another targeted silk nanoparticle was made by Subia et al. [48]. Folic acid was conjugated to nanoparticles, made from *A. mylitta* silk by nanoprecipitation, using carboxydiimide coupling. Folic acid is taken up by endocytosis and the folic acid receptor is overexpressed on many cancer cells, making it a useful tumor targeting ligand [81, 82]. The particles were around 200 nm in diameter and the anti-cancer drug doxorubicin was loaded by charge-based association. Uptake was tested in human breast cancer cells (MDA-MB-231) and was observed using confocal microscopy by conjugating rhodamine isothiocyanate to the particles. Immune response was tested by measuring cytokine expression in mouse bone marrow macrophages (RAW-264.7). Particles decorated with folic acid were taken up more effectively than non-targeted nanoparticles. Targeted nanoparticles had improved cytotoxicity to breast cancer cells and a relatively low immune response in vitro. They observed near-linear drug release over 7 days followed by slow release until 21 days [48]. Although research has been done in this area, these experiments show that it is possible to functionalize silk nanoparticles using standard carboxydiimide coupling. These experiments demonstrate that attachment of targeting ligands can improve cellular binding and drug delivery.

11.2.3 Spider Silk Nanoparticles

Spider silk proteins possessing similar chemical and physical properties as silkworm silk proteins have also been used to create a range of nanoparticles [64, 67]. Despite the many new methods of spider silk production, nanoparticles have so far been developed mostly from silk produced in *E. coli* by expression of an artificial gene construct. Hofer et al. expressed a synthetic gene, containing elements from MaSp of the spider *A. diadematus* in *E. coli* which produced a 48 kDa protein [64]. The particles were made by desolvation in 2 M KPO_4 and lysozyme was incorporated into the particles by charge-based association, with 40 % loading efficiency. Release was observed to be pH sensitive, above pH 6, lysozyme remained associated with the particles and was not released and only between pH 4 and 2 lysozyme release was observed (70–90 % after 24 h). Lammel et al. prepared recombinant spider silk and nanoparticles in the same way [67]. The average diameter was 320 nm with a large distribution and a zeta potential of −22 mV. The cationic dye methyl violet was incorporated by charged-based association and the authors showed that the dye permeated the core of the particles rather than simply associating with the surface. A release study showed a modest burst release which was accelerated in the presence of the proteases; elastase and trypsin. Xia et al. created a different gene construct containing the beta-sheet forming repeat of the *B. mori* heavy chain interspaced with the elastic motif of mammalian elastin [83]. This hybrid silk-elastin protein attempts to combine the mechanical strength of silk with the flexibility of elastin [84]. Varying ratios of silk and elastin sequences produced proteins with molecular masses between 48 and 53 kDa. The resulting micelle-like particles formed in water, were 100–200 nm in diameter. Even smaller nanoparticles with diameters 23–38 nm, have been made from silk-elastin-like proteins using an electrospraying technique [85]. Numata et al. designed a unique spider silk construct for the purpose of gene therapy [86]. The recombinant protein made by Numata et al. contained hydrophobic repeats from *N. clavipes* MaSp, 30 lysine residues, and a tumor binding peptide (Fig. 11.5) [86]. The sequence was expressed and isolated from *E. coli*. The positively charged lysines were incorporated to hold the negatively charged DNA within the particle

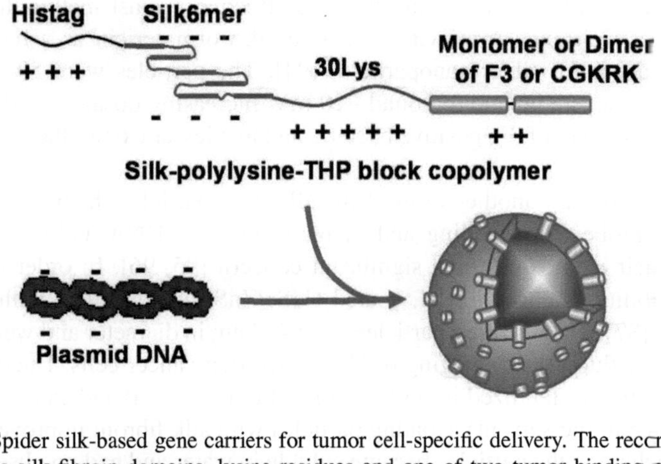

Fig. 11.5 Spider silk-based gene carriers for tumor cell-specific delivery. The recombinant peptide contains silk fibroin domains, lysine residues and one of two tumor binding peptides. The His-tag was incorporated to aid in isolation of the construct. Figure taken from Numata et al. [86]

which, in this case, was a plasmid containing the gene for luciferase. The smallest particles were 170–260 nm in diameter.

In vitro testing showed gene expression only in human melanoma cells (MDA-MB-435) and human breast cancer cells (MDA-MB-231) and not in non-cancerous mammary epithelial cells (MCF10A), indicating that the targeting peptides were effective. In vivo testing showed no toxicity in mice and significantly more luciferase expression in tumor-infected mice. Spider silk nanoparticles have yet to be studied as extensively as those produced from silkworm silk, however as evident from these preliminary investigations spider silk nanoparticles also have potential as drug delivery vehicles. Genetic engineering can further be used to incorporate other useful elements in lieu of chemical coupling [86]

11.2.4 Silk-Coated Nanoparticles

Rather than acting as the core of nanoparticles, silk can be used to coat other nanoparticles and improve their biocompatibility [62, 87]. Metallic nanoparticles have potential clinical applications in imaging and can be further stabilized by silk coatings. Liu et al. used silkworm-silk fibroin as a coat and scaffold for $ZnFe_2O_4$ nanoparticles [62]. Dissolved silk fibroin was present during the formation of the metal particles and the negative silk proteins are thought to act as a scaffold to hold Zn and Fe ions. The coated particles showed improved compatibility with P12 human brain astroglia cells versus non-coated particles. Silver and silver nanoparticles are also of interest in biomedical applications. Silver nanoparticles have been shown to be effective anti-bacterial agents against common bacteria [88, 89]. However, silver can cause significant toxic effects due to oxidative stress [90–93]. Silk has been used as a scaffold and coating for silver nanoparticles to help improve biocompatibility. Fei et al. prepared silver nanoparticles in the presence of silkworm silk fibroin [49]. The composite was effective against methicillin-resistant *Staphylococcus aureus.* Aramwit et al. used silkworm sericin as a reducing and stabilizing agent for silver nanoparticles [94]. The particles were 50–120 nm in diameter with a zeta potential around −20 mV. Increasing dosages of the particles inhibited the growth of 3 g positive bacteria and to a lesser extent, the growth of 3 g negative bacteria.

Quantum dots are another form of metallic nanoparticle which can be used as fluorescent probes for labelling and imaging proteins, DNA and cell structures, however, their cytotoxicity is a significant concern [95, 96]. In order to improve biocompatibility, Nathwani et al. coated CdSe/ZnS quantum dots with silkworm silk fibroin [87]. The resulting particles were 4–7 nm in diameter and were successfully used for fluorescent imaging of HeyA8 ovarian cancer cells. Uncoated particles could only be stabilized in organic solvent (chloroform) and caused cell death (possibly due to the solvent). Coating particles with silk fibroin improved biocompatibility, whereby the particles became soluble in water and had no observed cytotoxic effects. Instead of metallic nanoparticles, another study coated solid lipid nanoparticles, made from steric acid and ceramide, with silkworm-silk fibroin [97]. Solid lipid nanoparticles have applications in delivery of hydrophobic drugs, for example, the dermal delivery of hydrocortisone [98]. With the silk coating, the average particle diameter was around 500 nm with a large distribution [97]. Coating of the lipid particles increased the zeta potential (although still negative) which increased the skin permeation of ceramide particles in the dorsal skins of female nude mice. These examples demonstrate that silk polymers have potential to act as a coating agent to improve the biocompatibility of metallic nanoparticles [99]. The coating of solid lipid nanoparticles demonstrates that silk fibroin is a biocompatible coating agent which can enhance in vivo distribution.

11.2.5 Nanofibers and Electrospinning

Nanofibers are nanoscale structures with potential applications in controlled drug delivery, tissue engineering and wound dressings [103, 104]. This section will introduce electrospinning, the method used to generate these fibers, and research using silk fibroin nanofibers. Electrospinning mimics the spinning process that occurs in spiders and silkworms by drawing dissolved silk, with disordered and coiled structure, through a very fine stream to form insoluble fibers with aligned beta-sheet structures [100–102]. Mechanical extrusion of silk through a very fine nozzle can make fibers with diameters between 10 and 500 μm [101]. In order to create nanoscale fibers, electrical forces are used to draw and spin silk [100–102]. Silk solution is added to a container which ends in a capillary. A high voltage electrode is inserted into the solution or connected to the capillary and then to a collection plate which is several centimeters from the edge of the capillary. As charge builds up in the solution, and the attraction of the solution to the collection plate by electrical forces eventually overcomes the surface tension holding it in the capillary. The aqueous silk leaves the capillary in a very fine stream causing the solvent to evaporate. The spun fibers are collected on a plate, spool or into a solution. The applied voltage, flow rate, collector distance, and polymer and solution properties all affect the size of spun fibers [100–102]. Thangaraju et al. loaded the hydrophobic molecule curcumin into silkworm silk fibroin nanofibers [103]. The spun fibers had diameters ranging from 50 to 200 nm, and release of curcumin was monitored in vitro which totaled 80—84 % over 10 days. Jingwen et al. prepared aspirin-loaded nanofibers from a mixture of poly(lactic acid) and silkworm silk fibroin [104]. Fiber diameter was between 80 and 210 nm. The ratio of silk to polymer affected the drug release by changing the swellability and corrosion of the fibers. Drug release was 9–16 % over 10 h and only 10–20 % over 72 h. Sheikh et al. made silk nanofibers which incorporated both silver and hydroxyapetite nanoparticles [105]. The electrospun mats had significant antibacterial activity on *E. coli* and *S. aureus* cultures. Unfortunately, toxicity and attachment inhibition was observed in fibroblast cells. Based on these experiments, it is evident that electrospun silk nanofibers have potential for the encapsulation and controlled release of drugs, however, more testing is required to demonstrate their utility in vivo.

11.2.6 Potential Biomedical Applications of Silk Nanostructures

The development of improved medicines is the ultimate goal of research involving drug-loaded nanostructures. Table 11.1 summarizes promising biomedical applications of silk nanoparticles and nanofibers. These applications focus on the delivery and controlled release of anti-cancer drugs and are also used as biocompatible coatings. Although a wide range of silk nanostructures have been investigated in animal models to date, the potential utility of nanomedicines made using silk components remains to be evaluated in clinical trials [80, 106, 107].

11.3 Silk in other Drug Delivery Applications

Silk has extensively been used in biomedical applications and particularly in surgical applications. Silk sutures are mechanically superior materials with regards to yield point, toughness and elasticity moduli when compared with other types of biodegradable sutures (such as collagen, hyaluronins, alginates, absorbable synthetic polymers such as polydioxanone, polyglycolic acids and non-degrading like Teflon-coated polyester and nylon) [109]. Silk biomaterials exhibit slow biodegradation [110, 111], high biocompatibility, low incidence of acute inflammatory activity [4], and avoidance of organic solvents, surfactants and crosslinking agents in their design. Mild aqueous processing at ambient temperatures allows for the incorporation of sensitive biological cargo [10, 40], and silk encapsulation exerts a remarkable stabilizing effect on small molecules [112], protein [113], and enzyme payloads [114, 115]. Furthermore, silk produces non-toxic degradation products circumventing the need for post-surgical removal of local drug delivery implants [110]. Silk biomaterials can be processed to incorporate zero-order [116, 117], bi-phasic [118] or physiologically responsive release kinetics [119, 120], enabling desirable plasma drug levels with minimal interventions and effective long-term preventive care [121]. Drug release from silk based nanomedicines can be tuned by varying a range of parameters including; the degree of crystallinity, polymer composition and molecular weight. Silk biomaterials are mechanically robust, biocompatible high performance materials that provide an optimal platform for precise spatio-temporal control over the administration of bioactive drugs and biological signals for a variety of biomedical applications (Fig. 11.6) [35, 109, 122].

Table 11.1 Summary of biomedical applications of silk nanostructures

Nanostructure	Potential clinical application	Example	Experimental results	Particle/fiber diameter (nm)	Ref.
Nanoparticles loaded with small molecules	Delivery and controlled release of anti-tumor drugs	Anti-tumor drug paclitaxel-loaded nanoparticles	Reduced tumor size in mouse xenograft model more effectively than free drug when injected locally.	160–210	[66]
Nanoparticles loaded with proteins and peptides	Protection of peptide drugs from degradation	Insulin incorporated into silk particles	Improved resistance to protease degradation	40–120	[50]
Silk-coated nanoparticles to improve biocompatibility	Improve biocompatibility of silver nanoparticles	Sericin-coated and stabilized silver nanoparticles	Inhibited growth of six common bacteria	50–120	[94]
	Improve biocompatibility of metallic nanoparticles	Silk coated-$ZnFe_2O_4$ nanoparticles for MRI contrast agents and brain imaging	Good biocompatibility and low cytotoxicity with human brain astroglia cells in vitro		[62]
Functionalized spider or silkworm nanoparticles	Target cancer cells for delivery of anti-tumor drugs	Nanoparticle with folate as targeting ligand on surface and loaded with doxorubicin	Improved inhibition of human breast cancer cell growth in vitro	~200	[48]
	Target cancer cells for delivery of anti-tumor genes	Recombinant spider silk construct containing tumor-homing sequence loaded with reporter gene	Increased reporter gene expression in mouse breast cancer model	170–260	[86]
Nanofibers loaded with small molecules	Localized, controlled release of anti-tumor drugs	Silk nanofibers loaded with curcumin	Delayed, 10 day in vitro release	50–200	[103]
	Anti-microbial mats	Silver and hydroxyapatite-incorporated nanofibers mats	Anti-bacterial activity against *E. coli* and *S. aureus*	100–300	[108]

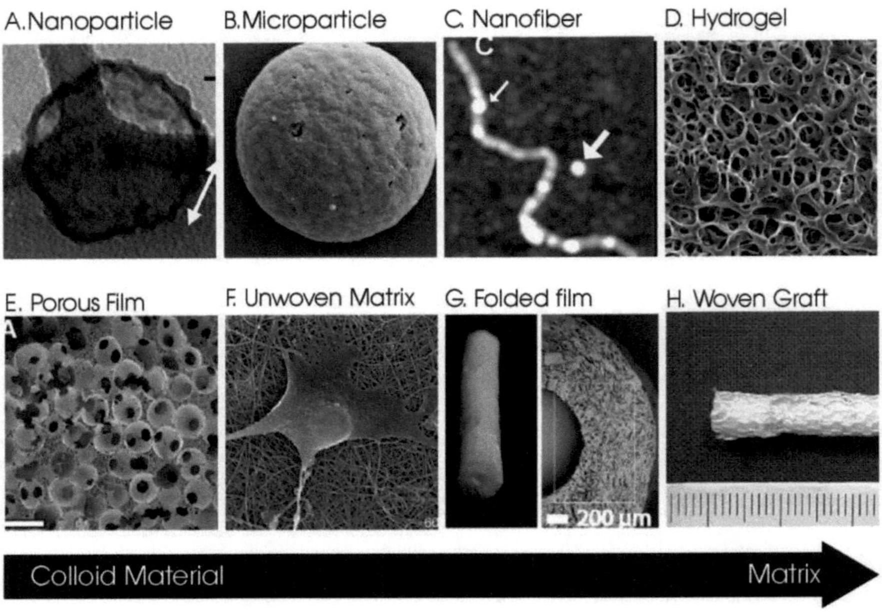

Fig. 11.6 Diverse morphologies of processed silk. (**a**) Silk elastin nanoparticle [85]. (**b**) PVA silk microparticle [51]. (**c**) AFM micrograph of dragline spider nano fiber [123]. (**d**) SEM micrograph of a pH induced *B. Mori* hydrogel [124]. (**e**) Porogen mediated porous scaffold [125]. (**f**) Osteoblast on electrospun silk matrix [126]. (**g**) Nerve conduit made from folded NGF loaded film. (**h**) Knitted miro-porous ACL scaffold [127].

11.3.1 Silk Processing

Natural silk is potentially immunogenic and can lead to IgE upregulation [128], and must firstly be degummed of hydrophilic sericin glycoproteins by boiling in alkaline sodium carbonate solution [4, 11]. Lindsay et al. demonstrated how boil times have a significant effect on silk thermal stability, decreased constituent fragment size distribution, pore-size, stiffness, porosity and in vivo degradation rate of the final fibroin, despite no appreciable disruption in ß-crystallinity [129]. The extracted core silk protein can be used as adhesion matrices in tissue engineering applications [130, 131], or dissolved in concentrated salt solutions at 50–70 °C for processing into several material formats. Dialysis yields a pure silk solution that can be lyophilized for storage or alternatively reconstituted in the organic solvent HFIP for drug integration in depot and controlled release applications [44, 110]. Fabrication protocols for microparticles, films, and hydrogels take advantage of fibroin self-assembly in techniques such as freeze-drying, gelation, Electrospinning, laminar-jet break up, and phase-separations (Fig. 11.7) [11, 109, 132]. The mechanisms of microparticle and fiber assembly are physically identical but vary in nucleation and elongation kinetics and can be committed to either morphology at critical buffer

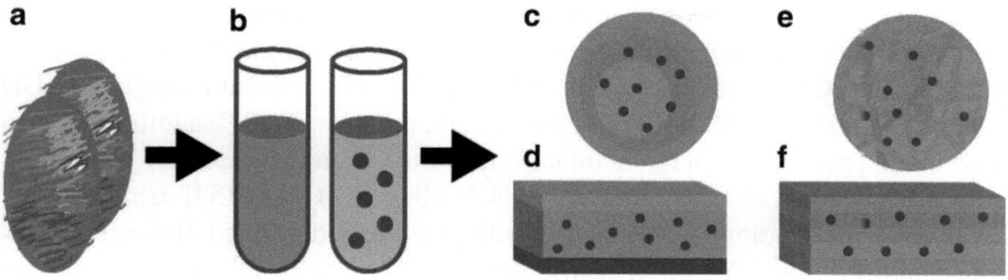

Fig. 11.7 Fabrication of silk materials. (**a**) Silk can be harvested from silk worm cocoon, recombinant cultures or from widely available textile manufacturers. It must be degummed before use. (**b**) Solubilized silk solutions (*brown*) are mixed with drug (*purple*) to be processed into; (**c**) microparticles, (**d**) films\membrane\coatings, (**e**) hydrogels, (**f**) 3D scaffolds

concentrations [133]. Template-assisted film-coating layer-by-layer deposition and film-casting techniques, at the cost of fabrication complexity, allow nanometer level control of material properties and facilitate integration of multiple drug delivery vehicles [10, 11, 35, 44].

A final step in silk processing of ß-sheet induction is biomimetically adapted to control crystallinity-dependent material properties through changes in pH [124, 134], salt concentration [40, 135, 136], alcohol mediated dehydration [137], or the application of shear forces [138], and water annealing [139], for alcohol sensitive cargo [140]. Silk can be sterilized through autoclaving, γ-radiation [141], or 70 % ethanol without compromising beta-sheet crystallinity or induced gel properties preclusive to administration, however, sterilization treatment may cause leaching or loss of activity for biological cargo.

11.3.2 Drug Integration Techniques

Silk materials can be loaded with small molecules, peptides, proteins, and DNA constructs at high encapsulation efficiencies through non-specific surface adhesion, physical entrapment and covalent modifications at ambient temperatures. Silk is inherently negative at physiological pH levels as it has a theoretical pI of 4.2. Mixing and co-incubation of fibroin and drug solution results in positively charged molecules decorating hydrophilic spacer regions of the heavy chain silk proteins while non-polar molecules [4, 142], are captured in crystalline regions through hydrophobic interactions. The efficacy of this simple encapsulation technique has been successfully validated in loading electrospun matrices [143], films [118, 144, 145], particle [51, 140, 146], and tablets with optimal retention of bioactivity. Silk and Silk-Elastin like polymer(SELP) hydrogel solutions can be loaded with drug and physically stimulated to gel, immobilizing the cargo immediately prior to administration. Highly bioactive forms of the drugs are disseminated by avoiding possible exposures to harsh hydrogel processing conditions [147–149]. Non-specific covalent coupling of amide linkages with N-hydroxysuccinimide (NHS) activated carbodiimide coupling have been used to immobilize BMP-2, linear gradients of HRP [150], recombinant PTH [151], RGD adhesion peptides [152], and biotin [61], to silk proteins. A 10 % addition of tyrosine in silk permits higher functionalization than diimide chemistry as demonstrated by Wenk et al. who achieved 99 % loading of FGF-2 [153] using sulfonic acid surface decoration in conjunction with previously described diazonium-tyrosine chemistry [154]. Yasusi et al. also successfully sulfated silk membranes using chlorosulfonic acid to exert heparin like anticoagulant activity [155]. Genetic modification of primary recombinant silk can be used to tailor silk mechanical properties and co-polymer chemistry; enriching amino acids that participate in the aforementioned coupling techniques, as well as the introduction of cysteine residues which allow disulfide coupling and sulfhydryl interfacing with colloid gold particles [156] and polylysine domains to sequester pDNA in silk-protein based gene delivery systems [157, 158].

11.3.3 Modulating Release Kinetics

Efflux of encapsulated molecules from silk fibroin occurs in a combination of extradition through diffusion—a larger factor in small molecules, and degradation (via hydrolysis or enzymatic action)—a more relevant factor in high molecular weight drugs [10]. Primary sequence and crystallinity of the fibroin can be manipulated to dictate microenvironment factors such as porosity and thickness of the silk vehicle, in order to act as diffusional barriers, as well as to exert an effect on diffusion rates of encapsulated drugs based on electrostatic and Van der Wals forces [159]. Columbic interactions influence liberation of charged compounds from negatively charged fibroin as evident by effective entrapment and slow release for cationic alkaline model drugs such as rhodamine, methylene dye [160], and burst kinetics of negatively charged model drug such as azo-casien [160], loaded in comparable amounts [40, 161]. Entropically driven coacervation of silk with mAbs and a model drug protamine, facilitates exposure and subsequent attractive interactions between hydrophobic domains. Reduction in ionic shielding between the interacting non-polar proteins can cause gradual repulsion mediated release profiles demonstrated with a hydrophobic molecule such as propranolol [113, 159], and hydrophobic antibody domains [113, 162]. Higher concentrations of silk fibroin, crosslinked with EDC/NHS and film coating [144, 146] of drug reservoir reduces material diffusivity resulting in more ideal prolonged recovery of salicylic acid [161], propranolol, theophylline [117] and buprenorphine [152] amongst other drugs [35, 130, 163, 164]. Composite materials such as microspheres loaded into films have yet to be systematically studied but potentially enable two levels of control over release kinetics for additional precision in moderating complex cellular milieus by biological signaling and drug delivery [165].

11.3.4 Direct Drug Integrated Silk Microparticles

Silk microparticles are a versatile platform for sustained release and depot applications due to readily tunable drug release profiles, ease of fabrication, potential for active targeting and high bio-compatibility (Fig. 11.8). Common fabrication methods of monolithic single layer microparticle methods include phase separations, laminar jet breaks, oil/water emulsion and spray drying, all of which exploit silk self-assembly to form microparticles from mixtures of fibroin and drug solutions. Spray drying techniques subject biological materials to high temperatures and are not as therapeutically relevant [140, 166].

11.3.4.1 Phase Separations

Silk microparticles can be produced using salting-out methods. Lemmel et al. were able to generate microparticles by salting-out silk fibroin using 500 mM KPO_4 [167] in which by exchanging chaotropic sodium and chloride ions for kosmotrophic ions, phosphate and potassium stimulates fibroin dehydration resembling salt exchange in distal spinning ducts of spiders [168]. Larger particles were generated with increased fibroin concentration and smaller particles were generated by changing the micro mixing intensity and potassium chloride concentrations in the range 250 nm–3 μm [67]. Potassium chloride and silk concentrations can modulate sizes between 0.5–4 μm, whereas lower pH levels and methanol treatments enhanced beta-sheets and encapsulation of cationic model drugs such as crystal violet alcian blue and rhodamine [40]. A mechanistic study of salt induced phase-separation in eADF4(C16) engineered silk microparticles assembly stipulates that the phase transition between nanofibril and microparticles occurs at a critical salt concentration, as a result of charge shielding that facilitates higher rates of spherical aggregation as compared to rates of elongation [133]. In a systematic investigation of charge-based release, release profiles were measured using eight cationic, three neutral and one negative model drug and loaded into salted-out particles sized between 0.17–0.7 μm. Silk

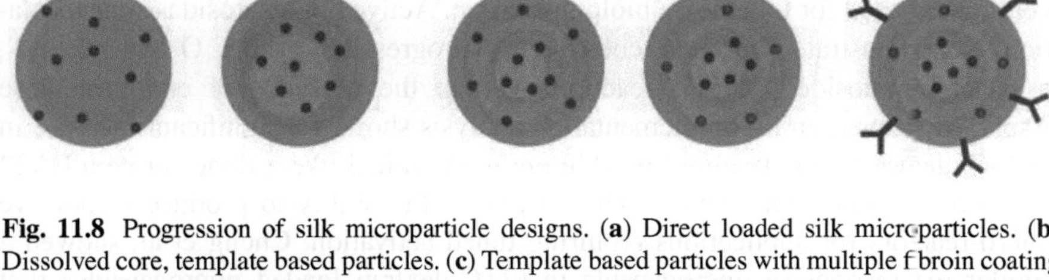

Fig. 11.8 Progression of silk microparticle designs. (**a**) Direct loaded silk microparticles. (**b**) Dissolved core, template based particles. (**c**) Template based particles with multiple fibroin coating containing drugs within and on the interfaces of the coating. (**d**) The encapsulation of sequential therapies with varying diffusional barriers. (**e**) Active targeting ligands for localization and up-take of microparticles

proteins were effective carriers of hydrophobic and basic molecules which were loaded at high efficiencies between 20.7–98.2 % and slowly released for up to 8 days, compared to 0.2–17 % encapsulation and burst release kinetics for acidic small molecules [67]. Xiaoqin et al. report an all aqueous microparticle technique based on the spontaneous phase separation of polyvinyl alcohol and drug loaded silk fibroin solutions [51]. The blended polymer solution is film cast in a petri dish and once dried, can be dissolved in water to produce silk microparticles in the range of 0.3–20 μm depending on concentration ratios of silk and PVA. The resultant microparticles doped with TMR-BSA, TMR-dextran and rhodamine-B demonstrated sustained release up to 4 weeks. This technique was adapted to co-flow microfluidics devices by Mitropoulos et al. [72], who were able to consistently generate microspheres with tunable diameters, by varying flow rates of the PVA over the silk solution, and it was demonstrated that smaller spheres exhibited larger cumulative burst release.

11.3.4.2 Laminar Jet break up

Wenk et al. outline a protocol to fabricate microparticles with sizes of 138–440 μm using a laminar jet break-up technique [161]. Consistently sized droplets force solutions of IGF-1/SF as a continuous flow through a vertically vibrating 200 μm nozzle in resonance with the Plateau-Rayleigh instability [161]. The particles were then dripped into liquid nitrogen and lyophilized. Post-treatment with methanol (or water vapor) produced diverse morphologies despite similar beta-sheet content [161]. This technique has produced impressive encapsulation efficiencies (~100 %) of the model drugs salicyclic acid, propranolol HCl and IGF-I loaded with a prolonged release of up to 7 weeks in vitro [140].

11.3.4.3 Oil/Water Emulsion

Nitchtl and Scheibel demonstrated that eADF4(C16) spider dragline silk will self-assemble at a water-toluene barrier as a permeable microcapsule that is both chemically stable to 2 % SDS, 8 M urea, and mechanically stable with a Young's modulus of 0.7–3.6 GPa [120]. Medical grade silica oil and subsequent ethanol treatments were used in the next rendition of this microcapsule to enhance beta-sheets (circumventing the need for toluene), a biological toxin. Active ß-Galactosidase encapsulation is demonstrated through colorimetric progression of the O-Nitrophenyl-, D-galactopyranoside (ONPG) reaction even in the presence of endoproteinase AspN. Moreover, an α-complementation analysis showed a significant increase in ONPG cleavage activity after the addition of a terminal like α-donor protein EA22 to a terminal deficient enzyme. This displayed the ability to produce responsive micro-reactors for applications requiring tuned activation. Cheng et al. showed a similar pH responsive microcapsule in FITC-dextran loaded microcapsules that were produced using an oil/water emulsion technique. The pH-responsive capsules transition between open and lock states as lower solution pH causes swelling in the shell walls, narrowing of the pores, and decreasing diffusion out of the capsule [169]. These microcapsule technologies form a semi-permeable enclosed reaction chamber which initiates reactions and diffuses reactants, while protecting cargo and allows for later activation, showing promise in drug delivery applications [115].

11.3.4.4 Template Based Techniques

A polyelectrolyte deposition process on a colloid template with subsequent core decomposition as described by Donath et al., has been effectively adapted to silk materials for developing precisely fabricated delivery systems [170]. This type of colloid synthesis provides structural control for encapsulation, diffusional control (addition of masking layers) of drug release and allows for the creation of sequential release therapies [35]. Wang et al. detail the use of horse radish peroxidase (HRP) loaded DOPC liposomes to generate homogenously sized microparticles around 200 μm in diameter that could deliver detectable amounts of HRP activity in vitro for over a month, depending on the processing [140]. Once drug infused liposomes are formed in solution, they are emulsified with silk solution and subjected to freeze-thaw cycles, with homogenous liposome sizes which approximately doubled the HRP loading efficiency [171]. The removal of the lipid core and induction of crystallinity is achieved through incubation in methanol or 1.7–3.6 μM NaCl solution. Moreover PLGA, and Alginate microspheres coated in silk, delayed PLGA degradation, and maintained a diffusion barrier for loaded drugs; HRP, tetramethylrhodamine conjugated BSA encapsulated in these microspheres with and without silk coating. Drug release was significantly retarded by silk coating, further hindered through methanol treatment with the mechanically stable shell now posing as a diffusion barrier to the drugs [146]. Silk integrated microspheres therefore show an effective improvement on alginate and PLGA based particles alone by reducing potential harmful levels of drugs at the local or systematic levels, by controlling burst release, while also protecting labile cargo that may degrade once released into the plasma [172].

11.3.5 Hydrogels

Hydrogels are heavily hydrated solids composed of over 95 % water content and three-dimensional polymer networks that can physically swell as they absorb water [173]. Silk hydrogels are attractive drug delivery systems because gelation kinetics are temperature dependent and the structural composition can be tuned through polymer concentration, the presence of salts and pH [174]. Significant developments have been made in tissue production using hydrogel scaffolds by promoting morphogenesis and closely reproducing signaling micro-environments, however these are exhaustively reviewed elsewhere [131, 164]. Moreover, hydrogels exhibit solid-like mechanical stability with liquid-like elasticity through a combination of osmotic and entropic structural forces [136]. Hydrogels can be formed through the addition of energy via ultra-sonication [173] and vortexing [46] the silk protein solution, which drives chain mobility that creates permanent, physical and intermolecular crosslinks. They can also be formed through decreased pH [124, 136] and surfactants [175], which facilitates the extravasation of water. Material properties such as stiffness and porosity are controlled through further dehydrating aqueous impurities from ß-sheet block regions by post-treatment with methanol [136].

11.3.5.1 Physically Induced Gels

Xiaoqin et al. demonstrated the induction of liquid-gel transition in silk hydrogels in physiological potassium salt via ultrasonication [147]. The sonication-induced gel exceeded mechanical properties and displayed enzymatic degradation resistance congruous gelatin [176], PEG-fibrinogen [177] and alginate [178], gel systems without the use of hazardous crosslinking agents or solution conditions that may deteriorate labile bioactive cargo [147]. Moreover, the ultrasonication method allows for processing hydrogels at physiological pH and ionic strengths amicable to stabilizing sensitive cargo closely juxtaposed within beta-sheets [179]. Dialysis of supersaturated SF slowly leaches silk I stabilizing salts yielding metastable α-helical silk that anneals to hydrophobic ß-sheet silk II on timescales dependent on the rate of salt diffusion [180]. Residual post-dialysis calcium concentrations in fibroin preparations reduce surface porosity of the silk material [136]. Furthermore, freezing at −20 °C as compared to freezing in liquid nitrogen allows slow nucleation of ice crystals that tend to grow larger and produce larger pores once sublimed in a lyophilization process which influences the permeation and diffusivity of drugs through the matrix [135]. Chemical cross linking with ammonium peroxodisulfate eADF4(C16) spider fibroin produce ß-sheet rich gels with higher shear and elastic moduli which develop into sponge like networks rather than pleated sheet-like structures and significantly reduce pore size [174]. Jai-You et al. induced gelation in reconstituted fibroin-buprenorphine with citric acid mediated pH reductions and showed a nearly zero order release of buprenorphine, and were able to control rates by varying the proportion of low and high molecular weight silk polymers [124].

11.3.5.2 High Performance Gels

Ultrasonicated and vortexed [148] silk allow for milk aqueous processing but exhibit relatively low resilience—deforming at strains greater than 10 % due to the brittle nature of beta sheets in high displacement [181]. Silk electrogels withstand up to a 1000 % greater compression but rely on strong voltages applied to fibroin in high pH. This reduces helical propensities through a local proton enrichment [182]. The ease of administration via hand-held applicators and the reversible adhesive nature of e-gels warrant future investigation for topical and surgical applications. Benjamin et al. proposed HRP mediated oxidative tyrosine crosslinking to form robust flexible silk hydrogel networks akin to di-tyrpsine bridges in hyaluronins and aligantes [149]. Enzymatically crosslinked silk is capable of recovering from cyclic applications of extremely compressive strain (up to 70 %) with a sheer moduli of 10 Pa–10 kPa and a tangent moduli of 15–400 kPa similar to e-gels. The mild cross-linking process incorporate anti-bodies [162], growth-factors [183, 184] and antibiotics [179] for delivery as implants dressing of dynamic tissues.

11.3.5.3 SELPs

Silk-like and Elastin-like Polymers (SELP) are recombinant block copolymers in which physiochemical properties of the material are tailored by carrying composition, length and ratios of silk and elastin amino acid sequence blocks. X-ray diffraction and differential scanning calorimetry (DSC) are consistent with crystallization of silk-like ß-sheet block thermal and chemical stability interrupted by flexible, more soluble elastin regions [84]. SELPs are liquid polymers that spontaneously undergo an irreversible liquid-gel transition in clinically useful timescales allowing the inclusion of dextrans, dansyl amino acids and other protein drugs shortly before administration as gels. SELPs can be processed in isotonic, physiologically compatible solvents allowing for the delivery of nucleic acids [185], microparticles, liposomes, live cells, and other bioactive compounds which may be sensitive to organic solvents. SELF-47 solution mixed with DNA solidified in an hour at 37 °C, released gene payloads for up to 28 days in a manner highly sensitive to ion strength of the medium that shield the charge of the phosphates in the DNA. Genetic engineering enables precise control over SELP charge densities facilitating internal control over charge-based release kinetics [186]. Controlled DNA cumulative release of only 5 % of total pFG-ERV plasmid load was seen after 28 % with preserved structure as shown by electrophoresis. Subcutaneous injection of luciferase plasmids loaded SELP gels proximal to tumours in MDA-MB-435 breast tumors in mice showed protection of DNA from nuclease activity and 3.5 times transfection of tumors as compared to naked DNA [186]. More recently silk-elastin gels loaded with adenovirus carrying ß-galactosidase and luciferase reporter genes showed 4–8 times higher gene expression levels compared to the virus alone and minimal dissemination to the liver in the head-and-neck cancer nude mice models [185]. Moreover, xenograft head-and-neck cancer JHU-012 cells targeted by gels embedded with therapeutic adenovirus encoding thymidine kinase (Ad-Tk) and ganciclovir show significant increases of Ad-Tk mRNA eliciting reductions in tumor volumes compared to PBS and ß-galactosidase controls at 7 and 14 days [185].

11.3.6 Surface Coatings for Medical Devices and Regenerative Implants

Silk coatings are an effective system as drug-eluting coatings, however silks micro-mechanical and biological properties has high utility in medical grade implants [187] and regenerative grafts as well [130]. For example, for cardiovascular surgery applications, Wang et al. demonstrated silk surface coatings of luminal stents can alleviate the potential for restenosis through prolonged release applications of heparin, paclitaxel and clopidogrel in a swine model. Coated stents reduce excessive proliferation of vascular smooth muscle cells (SMCs), thrombosis, and chronic inflammation often aggravated by foreign metallic device implantation to facilitate post-angioplasty vascular repair [188]. Surface modification of silicone breast implants with micrometer-thin layer of recombinant spider silk proteins have also shown to reduce unspecific protein adhesion and post-operative inflammation through surface hydrophobicity reduction [189] and proving a great candidate to resolve periprosthetic capsular fibrosis in silicone breast implants with reasonable effort [190].

Silk based growth factor therapies have been purposed for regenerative therapies in a plethora of tissue types [164], but show poor clinical transition due to rapid degradation and clearance in vivo, in addition to lack of understanding of how materials interface with the complex cellular milieu [191, 192]. Epidermal growth factor and sulfadiazine loaded electrospun silk nanofiber mats exceed commercially available topical wound dressing Tegadern Hydrocolloid(3 M) in terms of re-epithelialization, dermis proliferation, collagen synthesis and scar formation [193]. Schneider et al. show how EGF incorporated into silk mats with slow release (25 % in 170 h), optimizes the wound healing response in facilitating the alleviation of infection, inflammation and promotion of growth factor release from surrounding tissues, observed in previous in vitro studies using electrospun mats on wounded human skin models [194]. Silk is a prime candidate for the development of growth factor delivery scaffolds as it is resilient under strains associated with structural tissue such as bone [152], tendons and ligaments [127]. Wenk et al. showed how IGF-1, an important molecule in chondrogenesis can be incorporated into microparticles embedded within scaffolds for better chondrogenic outcome [195]. Kaplan et al. further demonstrated the dual control of silk particles loaded with osteoconductive morphogenic factors in electrospun matrices. These hybrid systems allow for adjustable loading by the protection of the encapsulated drug through the matrix's processing and modulated the release of the drug through the compositions of the loaded microparticles and their diffusion of the fibers [183].

11.4 Conclusion

The versatility of silk and silk-like products in drug delivery and tissue engineering applications has been demonstrated as evident by the large body of research publications over the last two decades. Insect and spider silk produce natural biopolymers with remarkable strength and flexibility and along with their biodegradability, biocompatibility and ease of production, make silk an exceptional biomaterial. In this review we have seen how the relationships between the chemical and molecular composition, secondary structures and mechanical properties of silk biopolymers can be used to create controlled release drug delivery nano and microparticles for the release of a range of payloads with a variety of structures and sizes.

Our understanding of the genetic coding of silk biopolymers is continuously increasing, and with improved cloning and expression of silk, we are now in a better position to understand the complex self-assembly of silk polymers. This in turn serves as an inspiration for the design, and development of future materials based on the properties of silk biopolymers for more improved nanomedicines and technology tools for healthcare.

Acknowledgements This work was supported by a Program of Excellence in Nanotechnology (PEN) Award, Contract #HHSN268201000045C, from the National Heart, Lung, and Blood Institute, National Institutes of Health (NIH). This work was also supported by NIH grants CA151884 and the David Koch-Prostate Cancer Foundation Award in Nanotherapeutics.

Conflict of interest statement Dr. Farokhzad declares financial interests in BIND Therapeutics, Selecta Biosciences, Tarveda Therapeutics and Playcon Therapeutics, four biotechnology companies developing nanoparticle technologies for medical applications. All other authors declare no conflict of interest.

References

1. Cao Y, Wang B (2009) Biodegradation of silk biomaterials. Int J Mol Sci 10:1514–1524. doi:10.3390/ijms10041514
2. Acharya C, Ghosh S, Kundu S (2008) Silk fibroin protein from mulberry and non-mulberry silkworms: cytotoxicity, biocompatibility and kinetics of L929 murine fibroblast adhesion. J Mater Sci Mater Med 19:2827–2836. doi:10.1007/s10856-008-3408-3
3. Panilaitis B et al (2003) Macrophage responses to silk. Biomaterials 24:3079–3085. doi:10.1016/S0142-9612(03)00158-3
4. Meinel L et al (2005) The inflammatory responses to silk films in vitro and in vivo. Biomaterials 26:147–155. doi:10.1016/j.biomaterials.2004.02.047
5. Kuhbier J et al (2010) Interactions between spider silk and cells—NIH/3T3 fibroblasts seeded on miniature weaving frames. PLoS One 5:e12032. doi:10.1371/journal.pone.0012032
6. Allmeling C, Jokuszies A, Reimers K, Kall S, Vogt P (2006) Use of spider silk fibres as an innovative material in a biocompatible artificial nerve conduit. J Cell Mol Med 10:770–777. doi:10.1111/j.1582-4934.2006.tb00436.x
7. Hakimi O et al (2010) Modulation of cell growth on exposure to silkworm and spider silk fibers. J Biomed Mater Res A 92:1366–1372. doi:10.1002/jbm.a.32462
8. Vollrath F, Barth P, Basedow A, Engstrom W, List H (2002) Local tolerance to spider silks and protein polymers in vivo. In Vivo 16:229–234
9. Gellynck K et al (2008) Biocompatibility and biodegradability of spider egg sac silk. J Mater Sci Mater Med 19:2963–2970. doi:10.1007/s10856-007-3330-0
10. Numata K, Kaplan D (2010) Silk-based delivery systems of bioactive molecules. Adv Drug Deliv Rev 62:1497–1508. doi:10.1016/j.addr.2010.03.009
11. Vepari C, Kaplan D (2007) Silk as a biomaterial. Prog Polym Sci 32:991–1007. doi:10.1016/j.progpolymsci.2007.05.013
12. Omenetto FG, Kaplan DL (2010) New opportunities for an ancient material. Science 329:528–531. doi:10.1126/science.1188936
13. Peakall DB (1969) Synthesis of silk, mechanism and location. Integr Comp Biol 9:71. doi:10.1093/icb/9.1.71
14. Andersson M, Holm L, Ridderstråle Y, Johansson J, Rising A (2013) Morphology and composition of the spider major ampullate gland and dragline silk. Biomacromolecules 14:2945–2952. doi:10.1021/bm400898t

15. Sponner A et al (2005) Characterization of the protein components of Nephila clavipes dragline silk. Biochemistry 44:4727–4736. doi:10.1021/bi047671k
16. Jin H-J, Kaplan D (2003) Mechanism of silk processing in insects and spiders. Nature 424:1057–1061. doi:10.1038/nature01809
17. Hayashi C, Shipley N, Lewis R (1999) Hypotheses that correlate the sequence, structure, and mechanical properties of spider silk proteins. Int J Biol Macromol 24:271–275. doi:10.1016/S0141-8130(98)00089-0
18. Tokareva O, Jacobsen M, Buehler M, Wong J, Kaplan D (2014) Structure-function-property-design interplay in biopolymers: spider silk. Acta Biomater 10:1612–1626. doi:10.1016/j.actbio.2013.08.020
19. Ayoub N, Garb J, Tinghitella R, Collin M, Hayashi C (2007) Blueprint for a high-performance biomaterial: full-length spider dragline silk genes. PLoS One 2:e514. doi:10.1371/journal.pone.0000514
20. Tokareva O et al (2014) Effect of sequence features on assembly of spider silk block copolymers. J Struct Biol 186:412. doi:10.1016/j.jsb.2014.03.004
21. Askarieh G et al (2010) Self-assembly of spider silk proteins is controlled by a pH-sensitive relay. Nature 465:236–238. doi:10.1038/nature08962
22. Gaines W, Sehorn M, Marcotte W (2010) Spidroin N-terminal domain promotes a pH-dependent association of silk proteins during self-assembly. J Biol Chem 285:40745–40753. doi:10.1074/jbc.M110.163121
23. Kronqvist N et al (2014) Sequential pH-driven dimerization and stabilization of the N-terminal domain enables rapid spider silk formation. Nat Commun 5:3254. doi:10.1038/ncomms4254
24. Schwarze S, Zwettler F, Johnson C, Neuweiler H (2013) The N-terminal domains of spider silk proteins assemble ultrafast and protected from charge screening. Nat Commun 4:2815. doi:10.1038/ncomms3815
25. Rising A, Widhe M, Johansson J, Hedhammar M (2011) Spider silk proteins: recent advances in recombinant production, structure-function relationships and biomedical applications. Cell Mol Life Sci 68:169–184. doi:10.1007/s00018-010-0462-z
26. Lefèvre T, Boudreault S, Cloutier C, Pézolet M (2008) Conformational and orientational transformation of silk proteins in the major ampullate gland of Nephila clavipes spiders. Biomacromolecules 9:2399–2407. doi:10.1021/bm800390j
27. Hronska M, van Beek J, Williamson P, Vollrath F, Meier B (2004) NMR characterization of native liquid spider dragline silk from Nephila edulis. Biomacromolecules 5:834–839. doi:10.1021/bm0343904
28. Knight D, Vollrath F (2001) Changes in element composition along the spinning duct in a Nephila spider. Naturwissenschaften 88:179–182. doi:10.1007/s001140100220
29. Vollrath F, Knight DP, Hu XW (1998) Silk production in a spider involves acid bath treatment. Proc R Soc B Biol Sci 265:817. doi:10.1098/rspb.1998.0365
30. Rousseau M-E, Lefèvre T, Pézolet M (2009) Conformation and orientation of proteins in various types of silk fibers produced by Nephila clavipes spiders. Biomacromolecules 10:2945–2953. doi:10.1021/bm9007919
31. Lefèvre T, Boudreault S, Cloutier C, Pézolet M (2011) Diversity of molecular transformations involved in the formation of spider silks. J Mol Biol 405:238–253. doi:10.1016/j.jmb.2010.10.052
32. Hu X et al (2006) Molecular mechanisms of spider silk. Cell Mol Life Sci 63:1986–1999. doi:10.1007/s00018-006-6090-y
33. Zhou C et al (2000) Fine organization of Bombyx mori fibroin heavy chain gene. Nucleic Acids Res 28:2413–2419. doi:10.1093/nar/28.12.2413
34. Zhou CZ et al (2001) Silk fibroin: structural implications of a remarkable amino acid sequence. Proteins 44:119–122
35. Wenk E, Merkle H, Meinel L (2011) Silk fibroin as a vehicle for drug delivery applications. J Control Release 150:128–141. doi:10.1016/j.jconrel.2010.11.007

36. Asakura T et al (2003) Synthesis and characterization of chimeric silkworm silk. Biomacromolecules 4:815–820. doi:10.1021/bm034020f
37. Bini E, Knight D, Kaplan D (2004) Mapping domain structures in silks from insects and spiders related to protein assembly. J Mol Biol 335:27–40. doi:10.1016/j.jmb.2003.10.043
38. Inoue S et al (2000) Silk fibroin of Bombyx mori is secreted, assembling a high molecular mass elementary unit consisting of H-chain, L-chain, and P25, with a 6:6:1 molar ratio. J Biol Chem 275:40517–40528. doi:10.1074/jbc.M006897200
39. Takasu Y et al (2007) Identification and characterization of a novel sericin gene expressed in the anterior middle silk gland of the silkworm Bombyx mori. Insect Biochem Mol Biol 37:1234–1240. doi:10.1016/j.ibmb.2007.07.009
40. Lammel A, Hu X, Park S-H, Kaplan D, Scheibel T (2010) Controlling silk fibroin particle features for drug delivery. Biomaterials 31:4583–4591. doi:10.1016/j.biomaterials.2010.02.024
41. Jin Y et al (2013) In vitro studies on the structure and properties of silk fibroin aqueous solutions in silkworm. Int J Biol Macromol 62:162–166. doi:10.1016/j.ijbiomac.2013.08.027
42. Chen X, Shao Z, Knight D, Vollrath F (2007) Conformation transition kinetics of Bombyx mori silk protein. Proteins 68:223–231. doi:10.1002/prot.21414
43. Foo CWP et al (2005) Role of pH and charge on silk protein assembly in insects and spiders. Appl Phys A 82:223. doi:10.1007/s00339-005-3426-7
44. Rockwood D et al (2011) Materials fabrication from Bombyx mori silk fibroin. Nat Protoc 6:1612–1631. doi:10.1038/nprot.2011.379
45. Banani K et al (2014) Silk proteins for biomedical applications: bioengineering perspectives. Prog Polym Sci 39:251. doi:10.1016/j.progpolymsci.2013.09.002
46. Subia B, Kundu S (2013) Drug loading and release on tumor cells using silk fibroin-albumin nanoparticles as carriers. Nanotechnology 24:35103. doi:10.1088/0957-4484/24/3/035103
47. Wadbua P, Promdonkoy B, Maensiri S, Siri S (2010) Different properties of electrospun fibrous scaffolds of separated heavy-chain and light-chain fibroins of Bombyx mori. Int J Biol Macromol 46:493–501. doi:10.1016/j.ijbiomac.2010.03.007
48. Subia B, Chandra S, Talukdar S, Kundu S (2014) Folate conjugated silk fibroin nanocarriers for targeted drug delivery. Integr Biol 6:203–214. doi:10.1039/c3ib40184g
49. Fei X et al (2013) Green synthesis of silk fibroin-silver nanoparticle composites with effective antibacterial and biofilm-disrupting properties. Biomacromolecules 14:4483–4488. doi:10.1021/bm4014149
50. Hai-Bo Y, Yu-Qing Z, Yong-Lei M, Li-Xia Z (2008) Biosynthesis of insulin-silk fibroin nanoparticles conjugates and in vitro evaluation of a drug delivery system. J Nanopart Res 11:1937. doi:10.1007/s11051-008-9549-y
51. Wang X, Yucel T, Lu Q, Hu X, Kaplan D (2010) Silk nanospheres and microspheres from silk/pva blend films for drug delivery. Biomaterials 31:1025–1035. doi:10.1016/j.biomaterials.2009.11.002
52. Yu-Qing Z, Yuan-Jing W, Hai-Yan W, Lin Z, Zhen-Zhen Z (2011) Highly efficient processing of silk fibroin nanoparticle-l-asparaginase bioconjugates and their characterization as a drug delivery system. Soft Matter 7:9728. doi:10.1039/c0sm01332c
53. Kundu J, Chung Y-I, Kim Y, Tae G, Kundu S (2010) Silk fibroin nanoparticles for cellular uptake and control release. Int J Pharm 388:242–250. doi:10.1016/j.ijpharm.2009.12.052
54. Chung H, Kim T, Lee S (2012) Recent advances in production of recombinant spider silk proteins. Curr Opin Biotechnol 23:957–964. doi:10.1016/j.copbio.2012.03.013
55. Lewis R, Hinman M, Kothakota S, Fournier M (1996) Expression and purification of a spider silk protein: a new strategy for producing repetitive proteins. Protein Expr Purif 7:400–406. doi:10.1006/prep.1996.0060
56. Teulé F et al (2009) A protocol for the production of recombinant spider silk-like proteins for artificial fiber spinning. Nat Protoc 4:341–355. doi:10.1038/nprot.2008.250
57. Hauptmann V et al (2013) Native-sized spider silk proteins synthesized in planta via intein-based multimerization. Transgenic Res 22:369–377. doi:10.1007/s11248-012-9655-6
58. Teulé F et al (2012) Silkworms transformed with chimeric silkworm/spider silk genes spin composite silk fibers with improved mechanical properties. Proc Natl Acad Sci U S A 109:923–928. doi:10.1073/pnas.1109420109

59. Wen H et al (2010) Transgenic silkworms (Bombyx mori) produce recombinant spider dragline silk in cocoons. Mol Biol Rep 37:1815–1821. doi:10.1007/s11033-009-9615-2
60. Yu-Qing Z et al (2006) Formation of silk fibroin nanoparticles in water-miscible organic solvent and their characterization. J Nanopart Res 9:885. doi:10.1007/s11051-006-9162-x
61. Wang X, Kaplan D (2011) Functionalization of silk fibroin with NeutrAvidin and biotin. Macromol Biosci 11:100–110. doi:10.1002/mabi.201000173
62. Liu J et al (2013) Preparation of $ZnFe_2O_4$ nanoparticles in the template of silk-fibroin peptide and their neuro-cytocompatibility in PC12 cells. Colloids Surf B Biointerfaces 107:19–26. doi:10.1016/j.colsurfb.2013.01.072
63. Mandal B, Kundu S (2009) Self-assembled silk sericin/poloxamer nanoparticles as nanocarriers of hydrophobic and hydrophilic drugs for targeted delivery. Nanotechnology 20:355101. doi:10.1088/0957-4484/20/35/355101
64. Hofer M, Winter G, Myschik J (2012) Recombinant spider silk particles for controlled delivery of protein drugs. Biomaterials 33:1554–1562. doi:10.1016/j.biomaterials.2011.10.053
65. Cao T-T, Zhou Z-Z, Zhang Y-Q (2014) Processing of β-glucosidase-silk fibroin nanoparticle bioconjugates and their characteristics. Appl Biochem Biotechnol 173:544. doi:10.1007/s12010-014-0861-y
66. Wu P et al (2013) Facile preparation of paclitaxel loaded silk fibroin nanoparticles for enhanced antitumor efficacy by locoregional drug delivery. ACS Appl Mater Interfaces 5:12638–12645. doi:10.1021/am403992b
67. Lammel A, Schwab M, Hofer M, Winter G, Scheibel T (2011) Recombinant spider silk particles as drug delivery vehicles. Biomaterials 32:2233–2240. doi:10.1016/j.biomaterials.2010.11.060
68. Zheng Z et al (2013) Generation of silk fibroin nanoparticles via solution-enhanced dispersion by supercritical CO_2. Ind Eng Chem Res 52:3752. doi:10.1021/ie301907f
69. Zheng Z et al (2012) Fabrication of silk fibroin nanoparticles for controlled drug delivery. J Nanopart Res 14:736. doi:10.1007/s11051-012-0736-5
70. Björnmalm M, Yan Y, Caruso F (2014) Engineering and evaluating drug delivery particles in microfluidic devices. J Control Release 190:139–149. doi:10.1016/j.jconrel.2014.04.030
71. Karnik R et al (2008) Microfluidic platform for controlled synthesis of polymeric nanoparticles. Nano Lett 8:2906–2912. doi:10.1021/nl801736q
72. Mitropoulos A et al (2014) Synthesis of silk fibroin micro- and submicron spheres using a co-flow capillary device. Adv Mater 26:1105–1110. doi:10.1002/adma.201304244
73. Chen M, Shao Z, Chen X (2012) Paclitaxel-loaded silk fibroin nanospheres. J Biomed Mater Res A 100:203–210. doi:10.1002/jbm.a.33265
74. Gupta V, Aseh A, Ríos C, Aggarwal B, Mathur A (2009) Fabrication and characterization of silk fibroin-derived curcumin nanoparticles for cancer therapy. Int J Nanomedicine 4:115–122
75. Hanjin O, Moo Kon K, Ki Hoon L (2011) Preparation of sericin microparticles by electrohydrodynamic spraying and their application in drug delivery. Macromol Res 19:266. doi:10.1007/s13233-011-0301-6
76. Numata K, Yamazaki S, Naga N (2012) Biocompatible and biodegradable dual-drug release system based on silk hydrogel containing silk nanoparticles. Biomacromolecules 13:1383–1389. doi:10.1021/bm300089a
77. Aramwit P, Kanokpanont S, Nakpheng T, Srichana T (2010) The effect of sericin from various extraction methods on cell viability and collagen production. Int J Mol Sci 11:2200–2211. doi:10.3390/ijms11052200
78. Lu Y, Yang J, Sega E (2005) Issues related to targeted delivery of proteins and peptides. AAPS J 8:78. doi:10.1208/aapsj080355
79. McGregor DP (2008) Discovering and improving novel peptide therapeutics. Curr Opin Pharmacol 8:616–619. doi:10.1016/j.coph.2008.06.002
80. Kamaly N, Xiao Z, Valencia PM, Radovic-Moreno AF, Farokhzad OC (2012) Targeted polymeric therapeutic nanoparticles: design, development and clinical translation. Chem Soc Rev 41:2971–3010. doi:10.1039/c2cs15344k

81. Wang S, Low P (1998) Folate-mediated targeting of antineoplastic drugs, imaging agents, and nucleic acids to cancer cells. J Control Release 53:39–48. doi:10.1016/S0168-3659(97)00236-8
82. Gruner B, Weitman S (1998) The folate receptor as a potential therapeutic anticancer target. Invest New Drugs 16:205–219. doi:10.1023/A:1006147932159
83. Xia X-X, Xu Q, Hu X, Qin G, Kaplan D (2011) Tunable self-assembly of genetically engineered silk—elastin-like protein polymers. Biomacromolecules 12:3844–3850. doi:10.1021/bm201165h
84. Anderson J, Cappello J, Martin D (1994) Morphology and primary crystal structure of a silk-like protein polymer synthesized by genetically engineered Escherichia coli bacteria. Biopolymers 34:1049–1058. doi:10.1002/bip.360340808
85. Anumolu R et al (2011) Fabrication of highly uniform nanoparticles from recombinant silk-elastin-like protein polymers for therapeutic agent delivery. ACS Nano 5:5374–5382. doi:10.1021/nn103585f
86. Numata K, Reagan M, Goldstein R, Rosenblatt M, Kaplan D (2011) Spider silk-based gene carriers for tumor cell-specific delivery. Bioconjug Chem 22:1605–1610. doi:10.1021/bc200170u
87. Nathwani B, Jaffari M, Juriani A, Mathur A, Meissner K (2009) Fabrication and characterization of silk-fibroin-coated quantum dots. IEEE Trans Nanobiosci 8:72–77. doi:10.1109/TNB.2009.2017295
88. Martínez-Gutierrez F et al (2012) Antibacterial activity, inflammatory response, coagulation and cytotoxicity effects of silver nanoparticles. Nanomedicine 8:328–336. doi:10.1016/j.nano.2011.06.014
89. Ratan D (2011) Preparation and antibacterial activity of silver nanoparticles. J Biomater Nanobiotechnol 2:472. doi:10.4236/jbnb.2011.24057
90. Suliman YA et al (2015) Evaluation of cytotoxic, oxidative stress, proinflammatory and genotoxic effect of silver nanoparticles in human lung epithelial cells. Environ Toxicol 30(2):149–160. doi:10.1002/tox.21880
91. Prasad R et al (2013) Investigating oxidative stress and inflammatory responses elicited by silver nanoparticles using high-throughput reporter genes in HepG2 cells: effect of size, surface coating, and intracellular uptake. Toxicol In Vitro 27:2013–2021. doi:10.1016/j.tiv.2013.07.005
92. Avalos A, Haza A, Mateo D, Morales P (2014) Cytotoxicity and ROS production of manufactured silver nanoparticles of different sizes in hepatoma and leukemia cells. J Appl Toxicol 34:413–423. doi:10.1002/jat.2957
93. Ramírez-Lee M et al (2014) Silver nanoparticles induce anti-proliferative effects on airway smooth muscle cells. Role of nitric oxide and muscarinic receptor signaling pathway. Toxicol Lett 224:246–256. doi:10.1016/j.toxlet.2013.10.027
94. Aramwit P, Bang N, Ratanavaraporn J, Ekgasit S (2014) Green synthesis of silk sericin-capped silver nanoparticles and their potent anti-bacterial activity. Nanoscale Res Lett 9:79. doi:10.1186/1556-276X-9-79
95. Bruchez M Jr (1998) Semiconductor nanocrystals as fluorescent biological labels. Science 281:2013. doi:10.1126/science.281.5385.2013
96. Jamieson T et al (2007) Biological applications of quantum dots. Biomaterials 28:4717–4732. doi:10.1016/j.biomaterials.2007.07.014
97. Teak Kwan K, Kun Bin L, Jin-Chul K (2011) Solid lipid nanoparticles coated with silk fibroin. J Ind Eng Chem 17:10. doi:10.1016/j.jiec.2010.10.001
98. Barichello J et al (2006) Inducing effect of liposomalization on the transdermal delivery of hydrocortisone: creation of a drug supersaturated state. J Control Release 115:94–102. doi:10.1016/j.jconrel.2006.07.008
99. Acharya A (2013) Luminescent magnetic quantum dots for in vitro/in vivo imaging and applications in therapeutics. J Nanosci Nanotechnol 13:3753–3768. doi:10.1166/jnn.2013.7460
100. Larrondo L, Manley RSJ (1981) Electrostatic fiber spinning from polymer melts: I. Experimental observations on fiber formation and properties. J Polym Sci Polym Phys Ed 19:909. doi:10.1002/pol.1981.180190601

101. Shin YM, Hohman MM, Brenner MP, Rutledge GC (2001) Experimental characterization of electrospinning: the electrically forced jet and instabilities. Polymer 42:9955. doi:10.1016/S0032-3861(01)00540-7
102. Sill T, von Recum H (2008) Electrospinning: applications in drug delivery and tissue engineering. Biomaterials 29:1989–2006. doi:10.1016/j.biomaterials.2008.01.011
103. Thangaraju E, Govindaswamy M, Sheeja R, Thirupathur Srinivasan N (2014) Curcumin loaded electrospun Bombyx mori silk nanofibers for drug delivery. Polym Int 63:100. doi:10.1002/pi.4499
104. Jingwen Q et al (2013) Evaluation of drug release property and blood compatibility of aspirin-loaded electrospun PLA/RSF composite nanofibers. Iran Polym J 22:729. doi:10.1007/s13726-013-0171-1
105. Sheikh F et al (2013) Facile and highly efficient approach for the fabrication of multifunctional silk nanofibers containing hydroxyapatite and silver nanoparticles. J Biomed Mater Res A 102:3459. doi:10.1002/jbm.a.35024
106. David MW, Padma S, Mark EB (2013) Injectable nanomaterials for drug delivery: carriers, targeting moieties, and therapeutics. Eur J Pharm Biopharm 84:1. doi:10.1016/j.ejpb.2012.12.009
107. Couvreur P (2013) Nanoparticles in drug delivery: past, present and future. Adv Drug Deliv Rev 65:21–23. doi:10.1016/j.addr.2012.04.010
108. Sheikh F et al (2013) A novel approach to fabricate silk nanofibers containing hydroxyapatite nanoparticles using a three-way stopcock connector. Nanoscale Res Lett 8:303. doi:10.1186/1556-276X-8-303
109. Gregory HA et al (2003) Silk-based biomaterials. Biomaterials 24:401
110. Yongzhong W et al (2008) In vivo degradation of three-dimensional silk fibroin scaffolds. Biomaterials 29:3415. doi:10.1016/j.biomaterials.2008.05.002
111. Keiji N, Peggy C, David LK (2010) Mechanism of enzymatic degradation of beta-sheet crystals. Biomaterials 31:2926. doi:10.1016/j.biomaterials.2009.12.026
112. Zhang J et al (2012) Stabilization of vaccines and antibiotics in silk and eliminating the cold chain. Proc Natl Acad Sci U S A 109:11981–11986. doi:10.1073/pnas.1206210109
113. Nicholas AG, Andrew JM, Bernardo JP-R, David LK (2013) Mechanisms of monoclonal antibody stabilization and release from silk biomaterials. Biomaterials 34(31):7766–7775
114. Shenzhou L et al (2009) Stabilization of enzymes in silk films. Biomacromolecules 10:1032. doi:10.1021/bm800956n
115. Claudia B, Alfons N, Thomas S (2014) Spider silk capsules as protective reaction containers for enzymes. Adv Funct Mater 24:763. doi:10.1002/adfm.201302100
116. Pritchard EM, Szybala C, Boison D, Kaplan DL (2010) Silk fibroin encapsulated powder reservoirs for sustained release of adenosine. J Control Release 144:159–167. doi:10.1016/j.jconrel.2010.01.035
117. Bayraktar O, Malay O, Ozgarip Y, Batigün A (2005) Silk fibroin as a novel coating material for controlled release of theophylline. Eur J Pharm Biopharm 60(3):373–381. doi:10.1016/j.ejpb.2005.02.002
118. Hofmann S et al (2006) Silk fibroin as an organic polymer for controlled drug delivery. J Control Release 111:219. doi:10.1016/j.jconrel.2005.12.009
119. Olga S, Irina D, Maneesh KG, Jeffrey L, Vladimir VT (2011) Silk-on-silk layer-by-layer microcapsules. Adv Mater 23(40):4655–4660. doi:10.1002/adma.201102234
120. Hermanson KD, Harasim MB, Scheibel T, Bausch AR (2007) Permeability of silk microcapsules made by the interfacial adsorption of protein. Phys Chem Chem Phys 9:6442–6446. doi:10.1039/b709808a
121. Langer R (1980) Invited review polymeric delivery systems for controlled drug release. Chem Eng Commun 6:148. doi:10.1080/00986448008912519
122. Pritchard EM, Hu X, Finley V, Kuo CK, Kaplan DL (2013) Effect of silk protein processing on drug delivery from silk films. Macromol Biosci 13:311–320. doi:10.1002/mabi.201200323
123. Oroudjev E et al (2002) Segmented nanofibers of spider dragline silk: atomic force microscopy and single-molecule force spectroscopy. Proc Natl Acad Sci U S A 99:6460–6465. doi:10.1073/pnas.082526499

124. Jia-You F, Jyh-Ping C, Yann-Lii L, Hsin-Yuan W (2006) Characterization and evaluation of silk protein hydrogels for drug delivery. Chem Pharm Bull 54:156
125. Uebersax L, Merkle HP, Meinel L (2008) Insulin-like growth factor I releasing silk fibroin scaffolds induce chondrogenic differentiation of human mesenchymal stem cells. J Control Release 127:12–21. doi:10.1016/j.jconrel.2007.11.006
126. Wei K, Kim I-S (2013) Fabrication of nanofibrous scaffolds by electrospinning. In: Russell M (ed) Nanotechnology and nanomaterials, Advances in nanofibers, book edited by Russell Maguire. InTech, Rijeka, Croatia. doi:10.5772/57093. ISBN 978-953-51-1209-9
127. Hongbin F, Haifeng L, Siew LT, James CHG (2009) Anterior cruciate ligament regeneration using mesenchymal stem cells and silk scaffold in large animal model. Biomaterials 30:4967
128. CeledÓN JC et al (2001) Asthma, rhinitis, and skin test reactivity to aeroallergens in families of asthmatic subjects in Anqing. China Am J Respir Crit Care Med 163:1108–1112. doi:10.1164/ajrccm.163.5.2005086
129. Wray LS et al (2011) Effect of processing on silk-based biomaterials: reproducibility and biocompatibility. J Biomed Mater Res B Appl Biomater 99:89–101. doi:10.1002/jbm.b.31875
130. Banani K, Rangam R, Subhas CK, Xungai W (2013) Silk fibroin biomaterials for tissue regenerations. Adv Drug Deliv Rev 65:457
131. Yongzhong W, Hyeon-Joo K, Gordana V-N, David LK (2006) Stem cell-based tissue engineering with silk biomaterials. Biomaterials 27:6064. doi:10.1016/j.biomaterials.2006.07.008
132. Pritchard EM, Kaplan DL (2011) Silk fibroin biomaterials for controlled release drug delivery. Expert Opin Drug Deliv 8:797–811. doi:10.1517/17425247.2011.568936
133. Ute KS, Sebastian R, Stanislav G, Thomas S (2008) An engineered spider silk protein forms microspheres. Angew Chem Int Ed Engl 47(24):4592–4594. doi:10.1002/anie 200800683
134. Yucel T, Kojic N, Leisk GG, Lo TJ, Kaplan DL (2010) Non-equilibrium silk fibroin adhesives. J Struct Biol 170:406–412. doi:10.1016/j.jsb.2009.12.012
135. Ribeiro M, de Moraes MA, Beppu MM, Monteiro FJ, Ferraz MP (2014) The role of dialysis and freezing on structural conformation, thermal properties and morphology of silk fibroin hydrogels. Biomatter 4:e28536
136. Kim UJ et al (2004) Structure and properties of silk hydrogels. Biomacromolecules 5:786–792. doi:10.1021/bm0345460
137. Minoura N, Tsukada M, Nagura M (1990) Physico-chemical properties of silk fibroin membrane as a biomaterial. Biomaterials 11:430–434
138. Jin HJ et al (2005) Water-stable silk films with reduced β-sheet content. Adv Funct Mater 15:1241. doi:10.1002/adfm.200400405
139. Hu X et al (2011) Regulation of silk material structure by temperature-controlled water vapor annealing. Biomacromolecules 12:1686–1696. doi:10.1021/bm200062a
140. Wang X et al (2007) Silk microspheres for encapsulation and controlled release. J Control Release 117:360–370. doi:10.1016/j.jconrel.2006.11.021
141. Kojthung A et al (2008) Effects of gamma radiation on biodegradation of Bombyx mori silk fibroin. Int Biodeterior Biodegradation 62:487–490. doi:10.1016/j.ibiod.2007.12.012
142. Li C, Vepari C, Jin H-JJ, Kim HJ, Kaplan DL (2006) Electrospun silk-BMP-2 scaffolds for bone tissue engineering. Biomaterials 27:3115–3124. doi:10.1016/j.biomaterials.2006.01.022
143. Kim UJ, Park J, Kim HJ, Wada M, Kaplan DL (2005) Three-dimensional aqueous-derived biomaterial scaffolds from silk fibroin. Biomaterials 26:2775–2785. doi:10.1016/j.biomaterials.2004.07.044
144. Xianyan W et al (2007) Nanolayer biomaterial coatings of silk fibroin for controlled release. J Control Release 121:190. doi:10.1016/j.jconrel.2007.06.006
145. Wang X et al (2008) Controlled release from multilayer silk biomaterial coatings to modulate vascular cell responses. Biomaterials 29:894–903. doi:10.1016/j.biomaterials.2007.10.055
146. Xiaoqin W et al (2007) Silk coatings on PLGA and alginate microspheres for protein delivery. Biomaterials 28:4161. doi:10.1016/j.biomaterials.2007.05.036
147. Wang X, Kluge JA, Leisk GG, Kaplan DL (2008) Sonication-induced gelation of silk fibroin for cell encapsulation. Biomaterials 29:1054–1064. doi:10.1016/j.biomaterials.2007.11.003

148. Tuna Y, Peggy C, David LK (2009) Vortex-induced injectable silk fibroin hydrogels. Biophys J 97:2044. doi:10.1016/j.bpj.2009.07.028
149. Benjamin PP et al (2014) Highly tunable elastomeric silk biomaterials. Adv Funct Mater 24:4615. doi:10.1002/adfm.201400526
150. Charu PV, David LK (2006) Covalently immobilized enzyme gradients within three-dimensional porous scaffolds. Biotechnol Bioeng 93(6):1130–1137. doi:10.1002/bit.20833
151. Susan S, Mary Beth M, Gloria G, David LK (2001) Functionalized silk-based biomaterials for bone formation. J Biomed Mater Res 54:139. doi:10.1002/1097-4636(200101)54:1<139::AID-JBM17>3.0.CO;2-7
152. Meinel L et al (2006) Silk based biomaterials to heal critical sized femur defects. Bone 39:922. doi:10.1016/j.bone.2006.04.019
153. Esther W et al (2010 Feb) The use of sulfonated silk fibroin derivatives to control binding, delivery and potency of FGF-2 in tissue regeneration. Biomaterials 31(6):1403–1413. doi:10.1016/j.biomaterials.2009.11.006
154. Amanda RM, Peter St J, David LK (2008 Jul) Modification of silk fibroin using diazonium coupling chemistry and the effects on hMSC proliferation and differentiation. Biomaterials 29(19):2829–2838. doi:10.1016/j.biomaterials.2008.03.039
155. Yasushi T (2004) Sulfation of silk fibroin by chlorosulfonic acid and the anticoagulant activity. Biomaterials 25:377. doi:10.1016/S0142-9612(03)00533-7
156. Mona W et al (2010) Recombinant spider silk as matrices for cell culture. Biomaterials 31:9575. doi:10.1016/j.biomaterials.2010.08.061
157. Keiji N, Juliana H, Balajikarthick S, David LK (2010) Gene delivery mediated by recombinant silk proteins containing cationic and cell binding motifs. J Control Release 146:136. doi:10.1016/j.jconrel.2010.05.006
158. Keiji N, Balajikarthick S, Heather AC, David LK (2009) Bioengineered silk protein-based gene delivery systems. Biomaterials 30:5775. doi:10.1016/j.biomaterials.2009.06.028
159. Germershaus O, Werner V, Kutscher M, Meinel L (2014) Deciphering the mechanism of protein interaction with silk fibroin for drug delivery systems. Biomaterials 35:3427–3434. doi:10.1016/j.biomaterials.2013.12.083
160. Manunya O, Ratthapol R, Sorada K, Siriporn D (2010) Preparation of Thai silk fibroin/gelatin electrospun fiber mats for controlled release applications. Int J Biol Macromol 46:544. doi:10.1016/j.ijbiomac.2010.02.008
161. Wenk E, Wandrey AJ, Merkle HP, Meinel L (2008) Silk fibroin spheres as a platform for controlled drug delivery. J Control Release 132:26–34. doi:10.1016/j.jconrel.2008.08.005
162. Guziewicz N, Best A, Perez-Ramirez B, Kaplan DL (2011) Lyophilized silk fibroin hydrogels for the sustained local delivery of therapeutic monoclonal antibodies. Biomaterials 32:2642–2650. doi:10.1016/j.biomaterials.2010.12.023
163. Lauzon MA, Bergeron E, Marcos B, Faucheux N (2012) Bone repair: new developments in growth factor delivery systems and their mathematical modeling. J Control Release 162:502–520. doi:10.1016/j.jconrel.2012.07.041
164. Naresh K, Utpal B (2012) Silk fibroin in tissue engineering advanced healthcare. Materials 1:393. doi:10.1002/adhm.201200097
165. Hardy JG, Scheibel TR (2010) Composite materials based on silk proteins. Prog Polym Sci 35:1093–1115. doi:10.1016/j.progpolymsci.2010.04.005
166. Tomoaki H, Masao T, Saburo S (2003) Change in secondary structure of silk fibroin during preparation of its microspheres by spray-drying and exposure to humid atmosphere. J Colloid Interface Sci 266:68. doi:10.1016/S0021-9797(03)00584-8
167. Lammel A, Schwab M, Slotta U, Winter G, Scheibel T (2007) Processing conditions for the formation of spider silk microspheres. ChemSusChem 1:413–416. doi:10.1002/cssc.200800030
168. Rammensee S, Slotta U, Scheibel T, Bausch AR (2008) Assembly mechanism of recombinant spider silk proteins. Proc Natl Acad Sci U S A 105:6590–6595. doi:10.1073/pnas.0709246105

169. Cheng C, Teasdale I, Brüggemann O (2014) Stimuli-responsive capsules prepared from regenerated silk fibroin microspheres. Macromol Biosci 14:807. doi:10.1002/mabi.201300497
170. Donath E, Sukhorukov G, Caruso F, Davis S, Möhwald H (1998) Novel hollow polymer shells by colloid-templated assembly of polyelectrolytes. Angew Chem Int Ed 37:2201. doi:10.1002/(SICI)1521-3773(19980904)37:16<2201::AID-ANIE2201>3.0.CO;2-E
171. Yi-You H, Ching-Hua W (2006) Pulmonary delivery of insulin by liposomal carriers. J Control Release 113:9. doi:10.1016/j.jconrel.2006.03.014
172. Huang X, Brazel CS (2001) On the importance and mechanisms of burst release in matrix-controlled drug delivery systems. J Control Release 73:121
173. Xiao H et al (2010) Biomaterials from ultrasonication-induced silk fibroin-hyaluronic acid hydrogels. Biomacromolecules 11:3178. doi:10.1021/bm1010504
174. Kristin S, Thomas S (2011) Controlled hydrogel formation of a recombinant spider silk protein. Biomacromolecules 12:2488. doi:10.1021/bm200154k
175. Gyung-Don K et al (2000) Effects of poloxamer on the gelation of silk fibroin. Macromol Rapid Commun 21:788. doi:10.1002/1521-3927(20000701)21:11<788::AID-MARC788>3.0.CO;2-X
176. Emmett PB et al (2005) Enzymatic stabilization of gelatin-based scaffolds. J Biomed Mater Res 72:37. doi:10.1002/jbm.b.30119
177. Liora A, Dror S (2005) Biosynthetic hydrogel scaffolds made from fibrinogen and polyethylene glycol for 3D cell cultures. Biomaterials 26:2467
178. Rowley JA, Madlambayan G, Mooney DJ (1999) Alginate hydrogels as synthetic extracellular matrix materials. Biomaterials 20:45
179. Pritchard EM, Valentin T, Panilaitis B, Omenetto F, Kaplan DL (2013) Antibiotic-releasing silk biomaterials for infection prevention and treatment. Adv Funct Mater 23:854–861. doi:10.1002/adfm.201201636
180. Nogueira GM, de Moraes MA, Rodas ACD, Higa OZ, Beppu MM (2011) Hydrogels from silk fibroin metastable solution: formation and characterization from a biomaterial perspective. Mater Sci Eng C 31:997–1001. doi:10.1016/j.msec.2011.02.019
181. Matsumoto A et al (2006) Mechanisms of silk fibroin sol-gel transitions. J Phys Chem B 110:21630–21638. doi:10.1021/jp056350v
182. Gary GL, Tim JL, Tuna Y, Qiang L, David LK (2010) Electrogelation for protein adhesives. Adv Mater 22:711. doi:10.1002/adma.200902643
183. Wang X et al (2009) Growth factor gradients via microsphere delivery in biopolymer scaffolds for osteochondral tissue engineering. J Control Release 134:81–90. doi:10.1016/j.jconrel.2008.10.021
184. Uebersax L et al (2007) Silk fibroin matrices for the controlled release of nerve growth factor (NGF). Biomaterials 28:4449–4460. doi:10.1016/j.biomaterials.2007.06.034
185. Greish K et al (2009) Silk-elastinlike protein polymer hydrogels for localized adenoviral gene therapy of head and neck tumors. Biomacromolecules 10:2183–2188. doi:10.1021/bm900356j
186. Megeed Z et al (2004) In vitro and in vivo evaluation of recombinant silk-elastinlike hydrogels for cancer gene therapy. J Control Release 94:433–445. doi:10.1016/j.jconrel.2003.10.027
187. Fischell TA (1996) Polymer coatings for stents. Can we judge a stent by its cover? Circulation 94:1494–1495
188. Kang Y et al (2008) Preparation of PLLA/PLGA microparticles using solution enhanced dispersion by supercritical fluids (SEDS). J Colloid Interface Sci 322:87–94. doi:10.1016/j.jcis.2008.02.031
189. Peter GC (2008) Breast reconstruction after surgery for breast cancer. N Engl J Med 359:1590. doi:10.1056/NEJMct0802899
190. Zeplin PH et al (2014) Spider silk coatings as a bioshield to reduce periprosthetic fibrous capsule formation. Adv Funct Mater 24:2658. doi:10.1002/adfm.201302813
191. Jeffrey JR et al (2013) Engineering the regenerative microenvironment with biomaterials. Adv Healthc Mater 2:57. doi:10.1002/adhm.201200197

192. Langer R, Tirrell DA (2004) Designing materials for biology and medicine. Nature 428:487–492. doi:10.1038/nature02388
193. Eun Seok G, Bruce P, Evangelia B, David LK (2013) Functionalized silk biomaterials for wound healing. Adv Healthc Mater 2:206. doi:10.1002/adhm.201200192
194. Schneider A, Wang XY, Kaplan DL, Garlick JA, Egles C (2009) Biofunctionalized electrospun silk mats as a topical bioactive dressing for accelerated wound healing. Acta Biomater 5(7):2570–2578. doi:10.1016/j.actbio.2008.12.013
195. Wenk E, Meinel AJ, Wildy S, Merkle HP, Meinel L (2009) Microporous silk fibroin scaffolds embedding PLGA microparticles for controlled growth factor delivery in tissue engineering. Biomaterials 30:2571–2581. doi:10.1016/j.biomaterials.2008.12.073

Chapter 12
Nanotoxicology and Regulatory Affairs

Christiane Beer

Abstract Like other parts of the nanotechnology revolution nanomedicines hold great promise and in the case of nanomedicines the potential for more efficient therapies. Engineered nanomaterials that are used as nanomedicines for therapeutic and diagnostic purposes are often designed to specifically interact with cells of tissues and organs of the human body. However, the unique physicochemical properties of particles at the nanoscale may contribute to adverse effects requiring nanomaterial-specific safety considerations. Therefore, before nanomedicines can be approved by organisations such as the U.S. Food and Drug Administration (FDA) or the European Medicines Agency (EMA) and reach the market, safety, efficiency and efficacy have to be shown. Beginning with some short critical remarks, this chapter addresses the toxicology of nanomaterials referred to as nanotoxicology with special attention to nanomedical applications. The second part of this book chapter will briefly describe the general drug approval process, introduce risk assessment procedures and give an overview of safety and regulatory challenges for nanomedicines.

Keywords Nanotoxicology • Pharmacokinetics • Toxicokinetics • Drug safety • Risk assessment • Nanomedicine regulation

12.1 Introduction

The application of nanotechnology offers many advantages, and products based on engineered nanomaterials (ENMs) are used in nearly all parts of our daily live [1]. Nearly everyone in the developed world have been unintentionally or intentionally in contact with nanotechnology in one way or another, either as consumers, or as workers. Nanomedicines based on nanotechnology and ENMs offer great potential in the treatment of diseases. These new developments promise, for example, improved bioimaging properties and more efficient and targeted drug delivery with fewer side-effects [2]. Due to their small size nanomedicines can cross endothelia

C. Beer (✉)
Department of Public Health, Aarhus University, Bartholins Allé 2, 8000 Aarhus C, Denmark
e-mail: cbee@ph.au.dk

K.A. Howard et al. (eds.), *Nanomedicine*, Advances in Delivery Science and Technology, DOI 10.1007/978-1-4939-3634-2_12

and epithelia biological barriers and enter organs, tissues and cells [2]. The advantageous properties of nanomedicines include a higher drug dissolution rate and enhanced drug adsorption, and increased bioavailability [2]. As much as ENMs offer the possibility to create new innovative products and new groundbreaking technologies there are, however, concerns that ENMs may pose a threat to human health and also to the environment. These concerns are based on the fact that ENMs may have unique physical, chemical and toxicological properties that differ from the parent material in a way that cannot be predicted by studying the larger-sized material. ENMs are known to be more reactive as their larger-sized counter parts, and it is, therefore, reasonable that ENMs may react with biological systems in new unpredicted ways that could lead to toxicity. Studies on natural occurring (e.g., forest fire) and unintentional (e.g., diesel exhaust) arising particles have shown that particle exposure leads to respiratory and cardiovascular diseases, cancer and possibly allergy in humans [3, 4]. Especially the particle fraction with a diameter less than 100 nm seem to play an important role for cardiovascular mortality and morbidity [3]. Due to the fast growth of nanotechnology and an increasing use of ENMs a whole new scientific branch, nanotoxicology, has emerged.

As nanomedicines consist of ENMs, the concerns regarding their toxicity due to their nano-particulate form lead to safety and regulatory challenges, especially as the field of nanotoxicology is still quite young and not all information that is needed for a comprehensive safety evaluation is available at this point.

12.2 Some Critical Remarks

Although the toxicity of particles in air pollution and their health effects have been investigated for some decades, the field of nanotoxicology with systematic toxicological investigations of ENMs is not more than 10 years old [5]. The term nanotoxicology was first mentioned in the scientific literature in a news article in Science by Robert F. Service in 2003 and was proposed as a new subcategory of toxicology by Donaldson and colleagues in 2004 [5]. Since then, the number of published nanotoxicology related articles has been increased from ~36 in 2004 to ~1919 in 2013. During the last 10 years, it has become clear that the unique physicochemical characteristics of ENMs not only make their toxicity towards biological systems difficult to predict, but that these characteristics also have a profound influence on their behavior in experimental settings. Even small changes of, e.g., their size or surface properties can have large effects on the observed toxicological outcome [6]. Over the years, it has become clear that the toxicological investigations of ENMs require high demands on the characterization of the particles and a well described experimental setting. The results from the first studies within nanotoxicology are hard to interpret due to incomplete or missing particle characterization. In addition, differences in the experimental setting due to studies in different cell line from different tissues and organs, exposure media, concentration and purity of the ENMs, dispersion protocols and media [7, 8] makes it difficult to compare the published results and to draw definite conclusions on the toxicological potential of ENMs. For example, it became clear that physicochemical properties of ENMs can change considerably in different exposure media and that the results of genotoxicity studies are influenced by the preparation of ENMs, their concentration, used dispersion agents and impurities, cell type used, bioavailability and uptake of the ENMs [7]. These parameters most likely affect the toxicity of ENMs in general.

Today the demands on the researchers and the quality of their investigations are much greater. It is difficult to publish in peer-reviewed scientific journals without a thorough characterization of the used ENMs. But not surprisingly, there is still an apparent lack of consistent results for the toxic effects of ENMs.

Another problem that has been recognized during the past years is the interference of some ENMs with toxicity tests like the Comet assay (metal oxide based ENMs) and the MTT assay (carbon nanotubes) [9–12]. Therefore, appropriate controls are of upmost importance to exclude false positive and negative results. Furthermore, when reading and interpreting the scientific literature, a critical and objective review of the available information should include an assessment if a thorough characterization of the ENMs are made, if appropriate controls are included, and if the studied concentrations are meaningful and cover realistic exposure scenarios. In many studies, animals have been exposed to an unrealistic high dose of ENMs resulting in overload scenarios inducing health effects that are not observed at concentrations that cover realistic worst case scenarios [7, 8].

However, it has to be kept in mind that the vast majority of nanotoxicological studies have been performed in vitro and that there might be a difference to the in vivo situation. One point is that in vitro cell cultures mainly are cancer or transformed cell lines that might react more or less sensitive to ENMs. Another point is that ENMs in in vitro systems interact with and possibly bind to proteins that originate in the uppermost cases from bovine serum, whereas in the in vivo situation, the ENMs are in contact with human blood including all blood cell types as well as the proteins of the human serum. The interaction of ENMs with the different serum proteins and blood cells might influence the results of toxicological studies as well.

12.3 Toxicology of Nanomedicines—Pharmacokinetics and Toxicodynamics

If not directly used as imaging agent e.g., SPIONs or Quantum dots, ENMs can be used as carriers for therapeutic drugs. In either case ENMs are a main component of nanomedicines, which by themselves could pose a potential health threat [6, 13, 14]. Based on studies on the ultrafine nanoscaled particle fraction from air pollution it is known that exposure to these particles increases the risk to develop airway and cardiovascular diseases [13, 15]. Due to increased use and exposure of consumers and workers to ENMs has resulted in concerns about potential adverse health effects of ENMs and the development of a new toxicological field, nanotoxicology. Based on the definition of toxicology by The Society of Toxicology (SOT) [16] nanotoxicology has been described by Oberdorster et al. "as the study of the adverse effects of engineered nanomaterials (ENMs) on living organisms and the ecosystems, including the prevention and amelioration of such adverse effects" [17]. In comparison to other ENMs, ENMs specifically used in nanomedicines have not been in as much focus of toxicological investigations yet, as the likelihood for an exposure of the general population by nanomedicines is considered to be low. The focus has been mainly on the development of efficient nanomedicines. However, as ENMs are a main component of nanomedicines the general nanotoxicological concepts for ENMs apply and general findings from other ENMs might be transferred to nanomedicines.

The toxicity of ENMs is crucially dependent on their physicochemical characteristics which play, therefore, a key role for the pharmacokinetics and toxicodynamics of nanomedicines. Pharmacokinetics, or toxicokinetics in the case of non-pharmaceutical substances, describes what the organism does with the nanomedicine, whereas, toxicodynamics describes what the nanomedicine does to the organism. Both, pharmacokinetics and toxicodynamics, therefore, reflect the inseparable interconnection of efficacy and toxicity of nanomedicines. This makes the task to develop nanomedicines that have the highest efficacy and, at the same time, the lowest possible toxicity not as simple as it might look at first glance. It is in fact a most difficult challenge as changes of the physicochemical properties can influence the toxicity, absorption, distribution, metabolism, and excretion of nanomedicines at the cellular but also organism level [18, 19]. Therefore, an early implementation of toxicological investigations is fundamental when developing new nanomedicines.

12.3.1 Physicochemical Properties of ENMs that Affect Toxicodynamics

Efficacy and toxicity of nanomedicines are both dependent on the physicochemical properties of the used ENMs. Through changes of their surface charge, shape, size and surface coating, one is able to control and target the drug load and delivery, influence the biodistribution and clearance from the bloodstream. However, a number of toxicological investigations of ENMs have shown that these physicochemical properties also affect the toxic potential of ENMs as they influence their interaction with the organism. The most important physicochemical properties in this connection are size including surface area, agglomeration and aggregation state and porosity; surface chemistry including surface charge and coating; shape; and chemical composition [8, 20–23].

12.3.1.1 Size, Surface Area and Shape

The physical behavior of particles changes dramatically when they reach sizes below 100 nm. Below this size and the smaller the particles are the rules of quantum physics apply more and more resulting in new chemical, mechanical, electrical, optical and/or superparamagnetic characteristics of the particles. Due to these changes of the characteristics of the particles, it is nearly impossible to extrapolate the biological reactivity and toxicity of ENMs from their larger-sized counterparts [21].

The size is undoubtedly the most important property of ENMs from a toxicological point of view as it influences a number of particle characteristics. These characteristics are high surface to volume ratio, high surface reactivity, absorption of compounds, ability to cross cellular membranes and strong interparticle forces.

How much the reduction of the size to below 100 nm size can change the physicochemical properties of the parent material becomes clear when looking at the example of gold. Gold is known to be one of the least reactive chemical elements. Larger-sized gold particles do not react with oxygen of the air or water. However, when occurring as nanoparticles with less than 10 nm in diameter, gold will burn once it is in contact with oxygen. This example shows that nanoparticles indeed may behave completely different compared to their parent material and this might as well be true for their toxic potential. First of all, the reason for that nanoparticles behave different compared to their parent material lies in the dramatic increase of the surface to volume ratio the smaller the particles become. For example, if a 1 cm cube is divided into 1 million 1 nm cubes the volume is still the same (6 cm^3) but the surface area has dramatically increased from 6 cm^2 to 60,000,000 cm^2. In addition, the percentage of molecules that are at the surface of the particles increases exponentially when the particle size is below 100 nm [13]. This and the enormous increase in the surface area enhances the possibility of ENMs to react with biological systems, including binding to cells, proteins and other biological active molecules. This characteristic of ENMs might, on one hand, be very desirable and useful for their use as nanomedicines but gives, on the other hand, reasons for concern as these interactions might be detrimental as well as uncontrollable.

The size of particles influences where the particles accumulate, how and how fast the body is able to clear particles and presumably directly influence the mechanism and level of toxicity [24]. For titanium dioxide it has been shown that the reduction of their size from 250 nm to 20 nm increases the inflammatory response in lungs of rats and mice; at least when looking at the same mass dose of the particles. At the same surface area, which ultimately means a lesser mass dose for the small particles, the large and small particles induced the same toxicity [13]. Furthermore, the surface area as a direct function of particle size has been shown to be related to inflammation and genotoxicity for nanoscaled carbon black, carbonaceous nanoparticles and titanium dioxide nanoparticles, in vitro and in vivo [25, 26]. For porous or very rough ENMs the specific surface area (Brunauer Emett Teller (BET) surface) should also be considered as this considers all surface molecules not only at the outer surface but also those in pores [21].

If the volume of the particles is kept as a constant, the surface area is dependent on the shape. Spheres have the highest volume to surface ratio, with the ratio decreasing for cubes and fibers. Therefore, the toxicity of ENMs is considered to be dependent on the shape of ENMs [27, 28]. Besides the influence on the surface area, shapes with sharp edges might be prone to faster degradation and in the case of metal-based ENMs this might be leading to the release of toxic metal ions. However, how important the shape of ENMs is for toxicity is still not clear.

When considering shape and toxicity, nano-fibers are of special concern. The global exposure of workers but also of the general population to asbestos fibers resulted in disorders of the lung and pleura like lung cancer and mesothelioma, pulmonary fibrosis and plaques at a pandemic scale. In 1997 the World Health Organization (WHO) published a definition for the so-called fiber paradigm for high aspect ratio materials [29]. Fibers with a diameter smaller than 3 μm, length greater than 5 μm and an aspect ratio greater than 3:1 are considered as harmful. This is also true for ENMs that have a high aspect ratio. Due to the similarity in their appearance to asbestos and the severity of the asbestos pandemic, high aspect ratio nanomaterials (HARNs) are generally classified as harmful. The mechanism behind the fiber pathogenicity is an incomplete engulfment of the fibers by macrophages also called frustrated phagocytosis. Fibers that are longer than 20 μg in length cannot be fully engulfed and this incomplete uptake process results in pro-inflammatory responses by the macrophages [23]. Chronic exposure to the fibers and subsequent persistent inflammation has been associated with the deposition of scar tissue in the lung (fibrosis), tissue damages and carcinogenicity of high aspect ratio fibers. However, tangled fibers that appear more like a sphere than a fiber will be completely taken up by macrophages and degraded if the material is biodegradable.

One key question, not only for toxicological investigations but also for regulatory purposes is the investigation of the aggregation and agglomeration state of ENMs. Agglomeration is a reversible process as ENM agglomerates are formed by weak bonds whereas aggregates are formed by strong covalent bonds. Aggregation of particles can change profoundly the size and size associated properties like transport, deposition and material release. Therefore, if ENMs are present as aggregates that are within the micrometer range or several hundred nanometers in all directions the material will no longer be considered as a true ENM. Agglomeration and aggregation are of special concern for nanomedicines when applied intravenously directly into the body as this can lead to thrombosis, a potentially lethal obstruction of the blood flow. Agglomeration but eventually also aggregation can occur, e.g., through binding of plasma proteins to ENMs making the surface characteristics and coating of ENMs important factors for the toxicity and intracellular fate of ENMs.

12.3.1.2 Particle Coating and Protein Corona

At the moment ENMs enter the human body they can interact theoretically with any protein of the plasma proteome that consists of ~3700 different proteins. Therefore, it is safe to say that ENMs will never exist as uncoated particles in the body. The interaction between particles and proteins results in a protein corona covering the surface of the particles. Which proteins in particular bind to the ENMs and become a part of the protein corona is, however, dependent on the chemical composition and surface charge of the particles. Several of the proteins in the protein corona of carbon black, silica, titanium dioxide and acrylamide nanoparticles particles have been identified and a number of these proteins are ligands to receptors at the cell surface [32–34]. Although binding of these proteins to the ENMs might lead to their unwanted uptake, this can also be exploited for targeted drug delivery when the ENMs are deliberately coated with ligands for receptors that are present at the cell surface of the target cells. However, unless the proteins are covalently bound to the surface of ENMs the protein corona is not a static structure and other proteins that are present in the surrounding body fluids can replace the original coating. The coating and the protein corona of ENMs have not only an influence on the clearance and targeting but also on the surface reactivity. If the surface atoms of the ENMs are covered by proteins, their surface reactivity is affected and the biological responses to the ENMs might be reduced [35]. This is exploited when adding a protein or polymer coat to the particles. For example, surface coating or surface functionalization has been shown to improve the therapeutic efficacy and minimizing adverse effects of mesoporous silica nanoparticles based nanocarriers [30]. Furthermore, coating of polystyrene nanoparticles with bovine serume albumin (BSA) has been shown to prolong the circulation time in the blood and significantly reduced the particle clearance compared to uncoated particles of the same size, although ~90 % of the particles were no longer present in the blood after 60 min [31]. However, as it takes approximately 1 min for a blood cell to circulate the body, even small increases of the circulation time will increase the likelihood that nanomedicines can reach their target organ before they are cleared by macrophages. The reason for the quite fast and efficient uptake of uncoated ENMs by macrophages lies in the binding of proteins like immunoglobulin G (IgG) and fibrinogen, so-called opsonins, to the particles. This opsonization of the particles marks the particles for destruction. Opsonins are recognized and bound by macrophages, thereby, enhancing the phagocytosis through these cells.

Taken together, coating of ENMs has several purposes: to avoid agglomeration/aggregation of the particles, to change the surface charge of the particles, target ENMs for uptake by specific cell types, change the bioavailability and degradation of the particles. All this leads to an improvement of the performance of nanomedicines as it diminishes the host defence mechanisms.

12.3.1.3 Surface Charge

Another physicochemical property that has been identified to play an important role for the toxicity of ENMs is their surface charge that can either be positive, negative or neutral. Studies showed that positively charged ENMs induce higher levels of toxicity compared to negatively charged particles of the same chemical composition. The induced toxic responses included cytotoxicity, disruption of cellular membrane integrity, apoptosis, necrosis, loss of mitochondrial membrane potential [36]. The surface charge of ENMs is dependent on their surface coating and surface functionalization. For example can the surface of carbon nanotubes (CNTs) be functionalized using acids. This treatment will result in carboxyl, carbonyl and hydroxyl groups at the surface of the nanotubes thereby leading to a negative charge of the nanotubes. A negative surface charge has been linked to an increased cytotoxicity [37–42]. On the other hand was shown that a surface functionalization that increases the water solubility of the ENMs decreases the cytotoxicity of CNTs [43].

Interestingly, just by altering the size of ENMs their hydrophobic or hydrophilic properties can change. The reason for that lies in a changed curvature of the particle-water interface. Small-sized ENMs can, therefore, be hydrophobic whereas larger ENMs of the same chemical composition and same coating can be hydrophilic [44].

For the development of nanomedicines, it is important to notice that positively charged particles often form aggregates upon intravenous injection. This can cause, e.g., potentially lethal embolisms in the lung capillaries [45]. The surface charge has also indirectly an influence on the toxicity of ENMs as it influences the efficiency of the uptake of the particles by the cells, the uptake pathways and the cellular distribution [36, 39, 40].

12.3.2 Pharmacokinetics

The physicochemical characteristics of ENMs does not only has an influence on the toxicity of the particles but also on their pharmacokinetics. Pharmacokinetics, the knowledge and investigation of what the body does to a drug, follows the so-called ADME scheme studying the **A**dsorption, **D**istribution, **M**etabolism and **E**xcretion of a drug. Knowledge about pharmacokinetics is, therefore, very important for the development and fine-tuning of the efficacy, but also, toxicity of nanomedicines and plays a crucial role in health risk assessment for nanomedicines.

12.3.2.1 Absorption

In pharmacokinetics absorption is the process by which the drug crosses biological membranes and reaches the bloodstream. The absorption efficiency and involved absorption mechanisms of nanomedicines are greatly depending on how the nanomedicine is administered. Administration of nanomedicines can occur orally, via inhalation, dermally, intravenously, subcutaneously and intramuscularly. Obviously, the absorption process for a drug is bypassed if the drug is directly injected into the bloodstream. As most of the nanotoxicological studies focus on the unintentional exposure of workers and consumers to ENMs, absorption after subcutaneous and intramuscular administration is much less investigated and there is no conclusive information on the absorption process available at this point.

The effectiveness of the absorption process differs greatly with uptake through the skin (dermal) as the least effective and injection, either subcutaneously, intramuscular or intravenously as the most effective administration routes. Although research is still ongoing, most of the studies on dermal absorption of ENMs confirm that the human skin can normally not be penetrated by particles even when they are in the nanoscale range [23]. Neither titanium dioxide nanoparticles nor quantum dots were able to reach the bloodstream [21]. Even if titanium dioxide nanoparticles were found in the stratum corneum the particles never reached the lower layers of the dermis. The stratum corneum consists of multiple layers of dead keratinized cells, which are difficult to penetrate for ENMs. However, this might be the case if the skin is damaged by wounds, sunburn or skin diseases like eczema. In addition, movement and stretching of the skin, particle charge, follicular openings, gender, and age might affect the barrier function of the skin [21]. But as there are reports showing the absorption or at least penetration of the stratum corneum there is currently no consensus about the ability of ENMs to be absorbed through the skin [46].

Absorption of ENMs via inhalation is probably the best investigated exposure route and will also be discussed in more detail in the section "Pulmonary toxicity". The most important parameter for absorption of ENMs through inhalation is their size and aerodynamic diameter. Depending on these parameters particles will be deposited more or less deep in the respiratory tract. The main mechanism for particle deposition is diffusion due to displacement when the particles collide with the molecules of the air. Depending on the size of the ENMs, they will be deposited in the nasopharyngeal, tracheobronchial or alveolar region [13]. Alveoli are the deepest part of the lung and nanoscale particles have been found to reach this region. The approximate size limits for particle deposition are that particles with an aerodynamic diameter >50 μm do not enter the respiratory tract as they are filtered quite efficiently by the nose, particles >10 μm are deposited in the upper respiratory tract, particles between 2 and 10 μm can reach trachea, bronchi and bronchioles, and particles smaller than 1 μm can reach the alveoli [13]. ENMs that reach the alveoli are mainly cleared by alveolar macrophages [13]. However, if the particles persist in the alveoli they are able to access the pulmonary interstitium either through diffusion or, more likely, transcytosis through the alveolar epithelium. From there the particles can cross the endothelium of the capillary, enter the bloodstream and translocate to systemic sites. Alternatively, the particles could enter sensory nerve endings that are embedded in the airway epithelial, a mechanism that seems to be specific for nanoscaled material [23, 47].

Particles that have been inhaled may be cleared from the lungs via the mucociliary escalator and through this way reach the gastrointestinal tract (GIT) [13, 48]. In addition, the GIT is also an important absorption route for nanomedicines and nanocarriers are currently under development for a more effective oral uptake of drugs and vaccines [46]. Nanomedicines that are administered orally are ingested into the GIT and are absorbed by a process called persorption, the paracellular translocation through transitory leaks in the epithelial cell layer [49–51]. It is thought that loosened tight junctions in the mucosa allow undissolvable particles in nano- but also microscale to be transported in the epithelial cell layer. From there the transport occurs into the sub-epithelial region via the thoracic duct either through lymph tracts or through veins and reach the bloodstream. This process seems to be quite fast as within a few minutes particles are found in the peripheral blood [49–51]. However, newer studies on ENMs showed that the absorption through the GIT increases with decreasing particle sizes and that micro-sized particles are trapped within the Peyer's patches, which are organized lymphoid nodules that are found in the lower regions of the small intestine. Particles trapped there does not seem to reach the systemic circulation in high numbers [48]. Besides the size, the charge of the ENMs is important as positively charged particles are more efficiently absorbed than negative or neutral charged ENMs [46].

12.3.2.2 Distribution and Cellular Uptake

After absorption or direct administration by intravenous injection the nanomedicine is distributed in the body through the bloodstream. Within pharmacokinetics, distribution is the reversible transfer of drugs away from the site of absorption to other sites within the body, including the target site, into interstitial and intracellular fluids. Again, the physicochemical properties of the ENMs are important for the biodistribution as they influence the way ENMs interact with cells, body fluids and proteins. The binding of ENMs to proteins can influence the mobility of the particles. If these proteins promote cellular uptake of the ENMs in specific organs or immune cells the biodistribution of the particles might be limited.

A general rule is that smaller ENMs have a much greater biodistribution compared to larger ENMs. For example, intravenously administered 10 nm gold nanoparticles were found in liver, spleen, kidney, testis, thymus, heart, lung and brain whereas 50 and 250 nm gold nanoparticles were only found in liver and spleen. One explanation is that the 10 nm particles were too small to be efficiently recognized and internalized by professional phagocytes that normally will clear the blood for foreign particles. Therefore, the 10 nm particles were able to reach more organs compared to the larger particles [52]. Iron oxide nanoparticles with a size of 22 nm were shown to be quickly translocated to the bloodstream and distributed to liver, spleen, kidney and testis after intratracheally instillation of rats [53]. Although research is still ongoing it seems that in general particles with a small diameter (10–20 nm) or a positive charge are more easily translocated through the alveolar barrier of the lung [21]. However, the situation might be different when chronic exposure occurs or in the case of a pathological situation. For example, inflammation seems to increase the translocation of ENMs from the alveoli into the blood and has, thereby, an influence on the biodistribution of the particles [54, 55].

The cell membrane is the last barrier ENMs have to cross if they are used as carriers to transport the drug into the target cells. Due to the particulate or vesicular form of nanomedicines most of the cellular uptake of nanomedicines will occur via active transport mechanisms into the cell. These active transport mechanisms include internalization pathways like phagocytosis, (macro)pinocytosis and receptor-mediated endocytosis via clathrin coated pits or caveolae [21]. Which of these cellular uptake mechanisms apply is greatly dependent on the size of the particles and on their surface coating [21, 56–60]. If particles reach a size of larger then approximately 500 nm they are mainly taken up via phagocytosis by so-called professional phagocytes like neutrophils, monocytes, macrophages, dendritic and mast cells; smaller particles are primarily processed by endocytic pathways. An alternative for the uptake of larger aggregates (0.5–5 μm) might be micropinocytosis [61]. However, at this point the precise role and importance of this pathway for the uptake of ENMs is not very well investigated. The surface coating can, e.g., allow binding to specific receptors on the cell surface. This is exploited for targeted drug delivery by interaction with receptors that are exclusively expressed at the surface of the target cells. In addition, targeting a specific receptor will also define by which cellular uptake routes the particles enter the cell as some receptors are exclusively found in clathrin-coated pits or caveolae [21]. After entry into the cells ENMs are present in intracellular membrane-coated vesicles. Depending on the uptake pathway, vesicles can be, e.g., endosomes, lysosomes or caveosomes. Furthermore, ENMs were also found in mitochondria, the nucleus or just free in the cytosol [21].

12.3.2.3 Metabolism

Metabolism of drugs covers the biochemical modification or biotransformation of pharmaceutical active substances or xenobiotics, substances that are foreign to the organism. The goal of these biochemical modifications is to convert lipophilic substances into more readily excreted hydrophilic products. The metabolic pathways are the same as for detoxification of poisons and include usually specialized enzymatic systems like the cytochrome P450 oxidase protein family. These enzymes are involved in the first of three metabolic phases where they introduce reactive or polar groups. In phase II, transferase enzymes such as glutathione S-transferases are catalyzing the conjugation of these modified substances to polar compounds that are in some cases further processed in phase III before they are pumped out of the cells by efflux transporters. These reactions are a defense mechanism of the cells to detoxify foreign substances, however, in some cases metabolic intermediates of normally non-toxic compounds can themselves be toxic.

Whereas the metabolism or biotransformation of the pharmaceutical active substance of the nanomedicine is likely to be well investigated and known, the metabolism of their carriers, the ENMs, is generally not very well investigated and understood. So far, ENMs were predominantly found not to be metabolized but that is, of course, very much dependent on the chemical composition of the ENMs. Qdots, e.g., seem to have a very long half-life in the body of several weeks or month. In contrast to nanoparticles, nanoscaled liposomes are likely to be much easier degraded and metabolized if they are able to fuse with cellular membranes. However, generally, there is still very little information available about what happens to ENMs after they have been taken up by cells. And there are concerns that breakdown of the nanostructures can again lead to unique unpredictable molecular responses.

Nevertheless, metabolization of ENMs has to be considered a very important step for clearance of the body from the particles. If ENMs are not metabolized or degraded they might not be excreted and, therefore, accumulate in the cells of the body. This might especially be a problem for repeated long-term administration of nanomedicines. In addition, if the particles are non-biodegradable even a short-term exposure and low toxicity of the administered ENMs might lead to a cumulative toxic effect over time. This is of special concern if an interaction with DNA occurs which could result in carcinogenesis. Furthermore, it has to be kept in mind that theoretically all biological effects of ENMs can be enhanced if the particles persist within the body for several months, years or even through the entire residual lifetime.

Though, in some cases can the solubility of ENMs result in toxic effects. For example, some or all of the observed toxicity of ENMs consisting of ZnO, CuO or Ag has been attributed to the released metal ions [62, 63]. The dissolution of ENMs can either occur in body fluids but also intracellularly. Here, especially the acidic environment of endosomes and lysosomes are thought to contribute significantly to the degradation of ENMs and to the release of toxic metal ions [64]. However, the pH alone is not in all cases enough for the dissolution of particles. In case of silver nanoparticles the interaction with cellular proteins seem to play an important role for the degradation of the particles as well [65]. Nevertheless, it is important to notice that not all ENMs end up in endosomes and/or lysosomes and that there is a number of materials that either cannot be degraded in endosomes or lysosomes or are able to escape these compartments. Factors that are important for the dissolution rate of ENMs are the size of the particles, roughness, coating, and aggregation state [21, 66–69]. After dissolution of the ENMs the particle compounds might be available for biotransformation and subsequent excretion.

12.3.2.4 Excretion

The two major routes for excretion are through feces and urine and only to a lower degree via the lung and skin. When discussing the excretion of ENMs one has to differentiate between biodegradable and non-biodegradable particles. Biodegradable ENMs are digested and the metabolites excreted by the body through urine or feces, and does no longer pose a health threat. However, excretion of non-biodegradable ENMs might take very long or might even be not at all existing. In general, circulation of the particles in the blood is a prerequisite for their excretion. Also for non-biodegradable ENMs the major routes for excretion are via urine or feces.

For the excretion via urine the blood is filtered in the kidney through the renal glomerula and via this way particles with a size lower than 8 nm can be filtered out of the blood whereas particles that are larger in size will accumulate in the mononuclear phagocyte system (MPS) [21]. The MPS consists of phagocytic cells that are located in reticular connective tissue, which is found around the liver, kidney, spleen, and lymph nodes as well as in bone marrow. In the liver, this system is particularly well developed and the macrophages of the liver, the Kupffer cells, are responsible for clearance of the largest part of the particles. In the case of non-biodegradable ENMs, the particles accumulate and persist in the macrophages. In addition, hepatocytes are able to take up particles via endocytosis but if they can metabolize and secrete the particles into the bile is not known.

Although there is still a great demand for investigations on the fate of non-biodegradable ENMs it seems that particles that are administered intravenously are either rapidly cleared by the kidney or are taken up by the mononuclear phagocyte system and persist in the body [21]. Water-soluble single-walled carbon nanotubes have been shown to be excreted via the renal route in rats and mice, whereas, titanium dioxide nanoparticles accumulate in the liver and spleen for several weeks [21]. Independent of the physicochemical properties of ENMs, the highest accumulation of particles is in general found in the liver [52, 70].

If particles or agglomerated ENMs reach a size of larger then approximately 500 nm they are mainly cleared from the blood via phagocytosis by so-called professional phagocytes like neutrophils, monocytes, macrophages, dendritic and mast cells. Especially macrophages have been in focus for pharmacokinetics and toxicity studies as they quite efficiently can clear the bloodstream from foreign particles. This might be a great problem for the efficacy of a drug but can also lead to toxicological complications and inflammatory responses if the particles persist within the cells. Positively charged nanomaterials are cleared fast from the blood and their aggregates accumulate in the liver and lung. Neutral ENMs have a decreased rate of uptake by macrophages of the liver or spleen. Neutral surface charge increases, therefore, the half-life of ENMs in the blood and the availability for uptake by other organs. In addition, binding of opsonins leads to enhanced phagocytosis and clearance of the particles from the bloodstream.

12.3.3 Mechanisms of Toxicity of Nanomaterials

Due to the small size of ENMs, the particles enter the organs, tissues and cells of the human body much easier then their larger counterparts. However, one of the most important questions is if the particles induce a toxicological response in the body once they are absorbed and what happens if non-degradable or slowly degradable ENMs accumulate in the body. If talking about toxicity of agents or drugs, a number of terms and definitions are used depending on what is the focus and aim of the toxicological study. Toxicity can be described based on the route, number and duration of exposure, primary toxic effects (target organ), and mechanism of toxicity. Terms like local and systemic toxic effects; acute, subchronic, chronic toxicity; transient, persistent, cumulative, latent toxicity are briefly described in the next section.

12.3.3.1 Toxicity Terms

The toxic effects of a drug, no matter if these are desired therapeutic effects or undesired side effects, can be either local or systemic. Local effects are those harmful effects that occur at the site of the initial contact, e.g., contact dermatitis. Systemic effects occur after absorption of the agent and include toxic effects in organs or tissues that are distant from the side of the original exposure [71]. Local and systemic effects can occur after acute (single) and repeated exposure where the repeated exposure can either be short-term (5 % of lifespan), subchronic (5–20 % of lifespan) or chronic (majority or entire lifespan) [71]. The vast majority of studies on nanotoxicity are on short-term effects after acute exposure while long-term effects after chronic exposures are mainly unknown. Dependent on when and how long the toxic effects are arising, one distinguishes between transient, persistent and latent toxic effects. Transient effects are temporary and reversible whereas persistent effects are permanent and present during the complete residual lifetime. Latent toxic effects have a delayed onset and can appear days, weeks, month or even years after exposure. Latent toxic effects occur mainly after acute exposure, whereas, cumulative toxic effects are progressing effects after repeated exposure [71].

As nanomedicines are usually a mixture of different components, e.g., the pharmaceutically active drug and the carrier ENM, also other toxicological terms might be important that play a role especially for exposures to mixtures. The toxic effects of the different components of nanomedicines can be *additive* ($2+3=5$; the overall toxic effect is the sum of the toxicity of each component); *antagonistic* ($2+3<5$; at least one of the components antagonize the toxicity of the other); *potentiating* ($0+3>3$; one non-toxic component enhances the toxicity of another toxic component); or *synergistic* ($2+3\gg5$; two toxic components are increasing the overall toxicity much more than the sum of the toxicity of each component). Potentiating and synergistic toxic effects can easily be confused. However, in case of a potentiating toxic effect one of the components has to be non-toxic, in case of a synergistic effect both components have to be toxic.

Although it is not very well studied how the different components of a nanomedicine are affecting each other's toxicity, there are a few examples where the co-exposure with two different kinds of ENMs leads to a potentiating or synergistic toxic effect. For example, pure cobalt and carbide particles have no toxic effect whereas the combination of both components leads to hard metal lung disease caused by the release of reactive oxygen species [72]. Oxidative effects are also observed after co-exposure with carbon black and iron oxide nanoparticles that are not observed for either particle type alone [73].

12.3.3.2 Reactive Oxygen Species, Oxidative Stress and Inflammation

The toxicological effects of ENMs on cells include cytotoxicity and genotoxicity. One if not the most important underlying mechanism for these effects is the induction of oxidative stress in the cells [74, 75]. Oxidative stress is caused by an imbalance between the formation of reactive oxygen species and the antioxidant capacity of the cells [13, 76]. Reactive oxygen species are chemically reactive molecules, and as the name suggests, do contain oxygen. Examples for reactive oxygen species are oxygen itself, superoxide anion, peroxide, hydroxyl radicals and ions, and hydrogen peroxide. These molecules are always present in cells as they are natural byproducts of the oxygen metabolism but, e.g., cellular stress, infection or other environmental factors can lead to an excessive formation of reactive oxygen species. In addition to reactive oxygen species, reactive nitrogen species containing nitric oxide can also be involved in the induction of oxidative stress [77].

There are different mechanisms of how exposure to ENMs might lead to an increased formation of reactive oxygen species. One possibility is the generation of free radicals by ENMs in aqueous suspensions in vitro. Another possibility is an increased production of reactive oxygen species in mitochondria but also the depletion of antioxidants and the subsequent impairment of the antioxidant capacity have been discussed as possible mechanisms. Reactive oxygen species and oxygen-free radicals are mainly produced in the mitochondria and thereby the mitochondria themselves are a major target for oxidative stress and injury.

The existence of too high concentrations of reactive oxygen species within the cell induces lipid peroxidation, mitochondrial damage, damages to DNA, RNA and proteins and lead to the induction of redox sensitive pathways that are involved in pro-inflammatory responses, cell cycle/proliferation as well as apoptosis (programmed and targeted cell death) and necrosis (non-programmed cell death) [74, 75]. The increased formation of reactive oxygen species is thought to be involved in the inactivation of protein functions that are important for cellular DNA repair. Reactive oxygen species might directly attack DNA leading to modified DNA bases like, e.g., 8-oxo-7,8-dihydroguanine (8-oxoG) as the most abundant and best investigated DNA alteration [75, 78]. Impaired repair of DNA alterations such as modified nucleotides is associated with mutagenesis and carcinogenesis. Furthermore, oxidative stress is thought to be involved in a number of different diseases like, e.g., Parkinson's and Alzheimer's disease, and cardiovascular diseases like atherosclerosis and myocardial infarction [74]. Therefore, the increased formation of oxidative stress by most of the investigated ENMs can be linked to a number of diseases and is the reason for concerns about the health effects of ENMs. However, as mentioned before, epidemiological human studies on ENMs are still very rare and many studies are using unrealistic high particle concentrations or purely characterized ENMs. Therefore, definite and especially general conclusions cannot be drawn on the induction of reactive oxygen species by ENMs and the toxicity of ENMs at this point.

12.3.3.3 Genotoxicity and Carcinogenicity

The high surface reactivity of ENMs and the induction of oxidative stress upon exposure to ENMs has raised the concern that they might be genotoxic and carcinogenic. An agent is classified as genotoxic when it has a DNA damaging capacity. Genotoxic events are normally very efficiently repaired by the cellular DNA repair system unless the DNA damage is too extensive. If the latter is the case programmed cell death, apoptosis, is induced. Mutagenesis is the permanent change of the original genetic information and occurs only if the DNA damage leads to persistent mutations within the genome. These persistent mutations can eventually lead to uncontrolled cell growth in form of neoplasms, which in the worst case can be malignant. The formation of malignant neoplasms, also better known as cancer, is called carcinogenesis. Carcinogenesis is a multistep process and requires not only an initiation stage but also a promotion stage. In the initiation stage the cell is exposed to a genotoxic agent whereas in the promotion stage the initiated cell is exposed to a promoting agent. This multistep process is resulting in carcinogenic characteristics of the cell like evasion of apoptosis, uncontrolled cell growth and metastasis. The promoting agent does not need to be genotoxic itself but either persistent or repeated exposure is required for carconigenesis. Whereas the initiating stage has no threshold (one DNA damaging event is in theory enough at this stage to initiate the cell), the promoting activity of an agent may have a threshold. In principle, the initiating and promoting agent could be the same [71].

The genotoxicity of ENMs and the underlying mechanisms are, at the point of writing this book chapter, not very well understood. In 2012, 4346 articles had been published on nanotoxicology whereof 94 described in vitro and 22 in vivo genotoxicity studies [7]. Although this number has tripled since then, it shows that there still is limited information on the genotoxicity of ENMs available considering the large amount of different types of nanomaterials that has been developed. However, in vitro studies suggest that several ENMs may have genotoxic potential, e.g., carbon nanotubes, C60 fullerenes, titanium dioxide and silver nanoparticles [7, 9, 79–81]. However, the results are somewhat conflicting and often due to limited information on the physicochemical properties of the investigated ENMs or variations in the experimental settings hardly to compare. Despite of these limitations, several mechanisms and factors are currently discussed that could lead to genotoxicity and carcinogenicity of ENMs. These could either be direct primary mechanisms (direct interaction of the particles with the genome), indirect primary mechanisms (interaction with proteins involved in cell cycle, binding to mitotic spindle components, inhibition of antioxidant defense and DNA repair activity or release of toxic ions from soluble ENMs, formation of reactive oxygen species by mitochondria) or secondary mechanisms (formation of reactive oxygen species by inflammatory cells) [7, 82]. These genotoxic mechanisms can result in oxidative modifications of DNA bases, bulky DNA adducts, single and double strand breaks, structural changes of the DNA (deletions, duplications, inversion and translocation of chromosome segments) or changes in the number of chromosomes [7].

Importantly, information on genotoxic and carcinogenic effects of ENMs on humans is even more limited at this point. Epidemiological studies on workers that were exposed to titanium dioxide nanoparticles were inconclusive and could not show an association between exposure to these particles and an increased cancer risk [83]. However, indications for a genotoxic potential of especially non-biodegradable persistent ENMs should of course be taken seriously and require further and more detailed investigations.

12.3.3.4 Neurotoxicity

Neurotoxicological health effects are adverse effects on the brain and the central nervous system. Normally, the blood-brain barrier protects the brain from entry of foreign particles or other unwanted compounds. The passage even of small molecules is tightly regulated and efficient translocation of drugs through the blood-brain barrier is hard to achieve. However, depending on their physicochemical properties, especially their size and surface charge, nanoscale particles might be able to enter the brain or the central nervous system. ENMs that are 20–50 nm in size as well as hydrophobic particles might be able to enter the brain even if the blood-brain barrier is intact [23, 84–86]. However, the reports in this matter are conflicting and intravenous injection of ENMs like 40 nm gold nanoparticles did not lead to translocation of a detectable amount of particles into the brain [87]. In contrast, a recent study by Huang et al. showed an accumulation of intravenously administered lipid nanoparticles in the brain parenchyma of mice after 3 h. The particles persisted there for more than 24 weeks [88]. Furthermore, for polymeric nanoparticles the surfactants seem to be more important than the size of the particles [89]. Aging, injury or disease may limit the protective capacity of the blood-brain barrier and allow for an easier access [90]. Another possibility is the entry via the olfactory bulb where there is a connection between the nasal epithelium and olfactory neurons [47]. This has been shown for carbon nanotubes, gold nanoparticles, quantum dots and manganese oxide nanoparticles [13, 47, 91–94]. ENMs were found in the olfactory bulb but have also been found in the hippocampus [95]. The entry into the brain was shown to be associated with an inflammatory response [93, 95, 96]. Although animal studies have shown that ENMs can reach the brain through the olfactory bulb, it is not known which role this entry pathway might occur in humans, as humans have a significantly less developed olfactory bulb compared to rodents [21]. In addition, it is not known what the health effects of ENMs actually are after they reach the brain or central nervous system. Animal and in vitro studies suggest that the presence of ENMs in the brain can cause brain damage. Neutrophils and lymphocyte numbers as well as protein carbonyl levels were increased and oxidative stress, lipid peroxidation and inflammatory responses (glia activation) have been discussed to be induced as a result of the high surface area and reactivity of ENMs [47, 76, 88, 93, 97, 98]. In addition, it is more likely that neurotoxicity is of chronic nature than to be acute due to the difficulties of reaching the brain [90].

12.3.3.5 Pulmonary Toxicity

From studies on nanoscale particles in air pollution we know that their inhalation can have adverse health effects and is associated with increased risk to develop cardiovascular diseases, lung fibrosis and lung cancer. The adverse health effects that are associated with inhalation of particles are occurring due to the deposition of the particles in the lung. Although there has been a number of studies on nanoscale particles in air pollution on human health, there are still only limited data available on the health effects of ENMs. This is also due to the difficulties to separate exposure to ENMs from background exposure of ambient particles. In a study on workers that were exposed to polyacrylate nanoparticles it has been suggested that these particles induce pleural effusion, pulmonary fibrosis and granuloma [99]. The respiratory effects that have been described are mainly inflammation, oxidative stress and functional disturbances. The inflammatory response includes local invasion of leukocytes and release of cytokines [21]. As described before, the toxic effect of ENMs are dependent on the size and shape of the particles. Nanoscale particles have been shown to reach the alveoli, the deepest part of the lung. Alveoli have an extreme large surface area (estimates are between 30 and 100 m^2) but the distance between the surface of the alveoli and the bloodstream measures only 2 μm [21]. Therefore, this region is less protected against inhaled particles [100]. Macrophages are mainly responsible for clearance of particles in the lung via phagocytosis. However, the efficiency of this process is strongly dependent on the size of the particles [21, 101]. It seems that alveolar macrophages are unable to recognize particles as foreign and to phagocyte them when they are less than 70 nm [23]. In contrast, nanofibres with a length of more than 20 μm are too long for phagocytosis. In both cases, the particles are suspected to stay in the lung for month or even years resulting in nonspecific pulmonary inflammatory responses. These inflammatory processes might even spread systemically as, e.g., also an increased risk for cardiovascular diseases is associated with pulmonary exposure to ENMs [23].

12.3.3.6 Cardiovascular Toxicity

Cardiovascular toxicity of ENMs or other nanoscale particles, e.g., ultrafine particles in air pollution, has mainly been observed after exposure via inhalation. Several epidemiological studies have shown the association between particles in air pollution, especially the ultrafine fraction, and cardiovascular diseases [102]. The observed short-term health effects after exposure to nanoscale particles and particles in air pollution include arrhythmia, coagulation disturbances, thrombosis, blood pressure abnormalities and in the long perspective a generally increased risk for development of cardiovascular diseases. Although the reasons and mechanisms are still somewhat unclear, it is thought that deposited particles in the lung induce inflammatory responses and conditions that cause these health effects. The release of inflammatory and prothrombotic mediators from the site of exposure into the blood might cause the activation of immune cells leading to the development of these adverse conditions [103]. Especially if these inflammatory responses are chronic and become systemic they will, obviously, have stronger effects on the cardiovascular system. After uptake by alveolar macrophages ENMs might be translocated from the respiratory to the cardiovascular system where the particles can directly induce cardiovascular toxicity by induction of inflammatory responses through cellular stress and increased release of reactive oxygen species [21]. Although epidemiological data for ENMs are still rare, it is believed that ENMs underlie the same toxicological mechanisms and, thereby, have somewhat the same adverse effects on the cardiovascular system as ultrafine particles in air pollution. As several studies have shown that ENMs can induce pulmonary toxicity this is believed to be an indicator for potential cardiovascular damage due to the close association of pulmonary and cardiovascular toxicity [104].

12.3.3.7 Reproductive Toxicity

The reproductive toxicity, which includes adverse effects on the sexual function and the fertility of adult males and females as well as the developmental toxicity in the offspring, is probably the least investigated toxicological effects of ENMs. Some studies in mice have shown that titanium dioxide nanoparticles were able to cross the blood-testes barrier and reduced the sperm production in the offspring of the treated mice. In addition, it has been suggested that these particles might be able to affect the development of the central nervous system in the offspring as they affected the gene expression of genes involved in the development and function of the neural system [46]. However, since a possible accumulation of ENMs has been found, further studies are needed to exclude any reproductive and developmental effects due to an exposure to ENMs or nanomedicines. In addition, the potential of ENMs to cross the fetal-placental barrier has to be investigated as well [46].

12.4 Drug Safety Testing

Like conventional drugs, also nanomedicines have to be approved before used on patients. In the U.S.A. this is done by the U.S. Food and Drug Administration (FDA) and in countries of the European Union, the European Medicines Agency (EMA) is the responsible regulatory agency. The approval process involves several phases including preclinical studies as well as clinical trial phases. In these different phases the safety and efficacy of the drug is investigated by the applicant under the supervision of the responsible regulatory agency.

12.4.1 Preclinical Studies

Before entering the clinical trials drugs are tested in pre-clinical studies, normally by the drug-developing pharmaceutical company. The aim of these pre-clinical studies is to collect basic safety and efficacy data. Based on these data a plan for further testing of the drug on humans is developed and an application for clinical trials is submitted. Another important goal of these studies is to ensure that the drug is a promising candidate that justifies the enormous costs and efforts that are associated with a drug approval process. Therefore, these pre-clinical studies are quite extensive and include in vitro cell culture studies as well as in vivo animal studies to investigate the preliminary efficacy, toxicity and pharmacokinetics of the drug. In the case of nanomedicines, nanotoxicological aspects have to be considered in addition to standard toxicological investigations and evaluations.

12.4.2 *Clinical Trials*

The clinical trials involve three phases where the drug is tested on either healthy volunteers (phase I) or patients (phase II and III). The goal of phase I is the determination of the most frequent side effects of the drug and, frequently, the pharmacokinetics of the drug that gives information about how the drug is taken up, transported, metabolized and excreted. Typically, between 20 and 80 volunteers are involved in phase I and the safety of the drug is stressed. However, there are circumstances when patients have to be enrolled. This is the case when the drug is expected to cause severe side-effects in healthy individuals. After a successful phase I, approximately 100–300 patients are enrolled in phase II where the effectiveness of the drug is investigated. Normally, the effect of the drug on patients will be compared to patients receiving either a placebo or standard treatment. Phase III contain 1000 or more patients to further investigate the safety and effectiveness of the drug. Different dosages and the use of the drug in combination with other drugs are studied. Based on the results from the clinical trials the authorities decide whether the drug can be approved or not. However, as it is not possible to predict and determine all side-effects and especially long-term effects in the clinical trials, the drug will be further monitored to detect any adverse effects when on the market. Again, when investigating the safety of nanomedicines their specific nano-related characteristics and properties have to be taken into consideration in all phases of the drug approval process as special nanoscale related safety issues have to be addressed.

12.5 Risk Assessment of Engineered Nanomaterials

12.5.1 *Risk Assessment*

Toxicological investigations of ENMs are important not only for drug safety testing, but are also essential parts of the risk assessment of these agents. Nanomedicines and the used ENMs have to be manufactured and depending on the scale of this manufacturing process unintentional exposure of workers could occur. Risk assessment is also necessary to regulate the use of nanomedicines and ENMs properly. To cover all aspects of safety considerations for nanomedicines a short introduction into risk assessment will be given based on the WHO tool kit and IPCS harmonization project (WHO Human Health Risk Assessment Toolkit: Chemical Hazards; http://www.who.int/ipcs/methods/harmonization/areas/ra_toolkit/en/).

For the investigation of the adverse effects of a newly developed drug, a broad spectrum of methodologies are used that range from experimental in-vitro studies to animal studies and epidemiological investigations on whole populations. The identification of adverse effects form the basis for risk assessments of any given chemical, physical or biological agent. *Risk Assessment* is a process where it is evaluated to which degree, and with which probability, these agents affect human health and the environment. It is the primary objective to identify and characterize potential hazards, estimate exposure and assess the overall risks for humans or the environment. The assessment if an agent poses a risk is the first component of a risk analysis that also includes risk management and risk communication.

When talking about risk it is important to keep in mind that a hazardous substance does only pose a risk to humans but also the environment if there is a likelihood for an exposure. Risk is therefore defined as a function of hazard and exposure:

$$\text{Risk} = \text{f}(\text{Hazard; Exposure}) \quad (12.1)$$

In Eq. (12.1) is risk = zero if either hazard or exposure equals zero. Equation (12.1) is probably the most important risk assessment paradigm and can be very illustratively explained using the tiger in a cage example. Everybody will probably agree that it is a risk to visit a living tiger inside its cage, especially a hungry one. In this situation, we are exposed to a hazardous biological agent. However, when the tiger is separated from the visitor by a cage there is no risk to the visitor (there is a hazard but no exposure) just as a mounted tiger outside a cage poses no risk to the visitor (there is an exposure but no hazard). This example shows that for a thorough risk assessment the exposure assessment is just as important as the identification of the potential hazards of chemical, physical or biological agents.

A risk assessment will always begin with a problem formulation to establish the scope and objective of the risk assessment. The risk assessment itself consists then of four steps including *hazard identification, hazard characterization, exposure assessment* and *risk characterization.* In the following chapters a short overview over these four steps is given.

12.5.1.1 Hazard Identification

The first step in risk assessment is the hazard identification that is mainly based on the results from toxicological studies. These toxicological studies include human studies (mostly epidemiological studies), animal-based and in vitro toxicology studies as well as structure-activity studies. Although risk assessment and toxicology also include the investigation of adverse effects on the environment, the following will focus on the health hazards and effects on humans as nanomedicines are primarily intended to be used in humans being well aware of that the production of nanomedicines could pose environmental hazards and risks.

The purpose of the hazard identification is to identify (1) the specific hazard, (2) the type and nature this hazard may have to an individual or (sub)population and (3) investigate if exposure to the agent has the potential to be harmful.

The hazard identification begins with the identification of the chemical composition and, as nanomedicines consists of nanomaterials, particle characteristics of the nanomedicine. Identifying the chemical composition of a nanomedicine will give information if the parent material is already classified by the CLP regulation of the European Union (CLP stands for "Classification, Labelling and Packaging") and if it is already known to be hazardous. The health hazards that are CLP classified include acute toxicity, sensitization of respiratory tract and skin, skin corrosion and irritation, serious eye irritation and eye damage, reproductive toxicity, germ cell mutagenicity, carcinogenicity, specific target organ toxicity after single and repeated exposure, and aspiration toxicity.

In addition, as previously mentioned high aspect ratio nanomaterials like, e.g., nanotubes, nanofibers, nanowires and nanorods are generally considered hazardous when they are at the same time biopersistent, able to pass ciliated airways and able to initiate frustrated phagocytosis, which leads to the release of pro-inflammatory molecules.

If the nanomaterial is not categorized as high aspect ratio nanomaterials (HARN) there has to be investigated if the nanomaterial induces acute or chronic toxicity including genotoxicity, neurotoxicity, carcinogenicity, pulmonary, cardiovascular, or reproductive toxicity or if the nanomaterial accumulates in organs. If there cannot be excluded that the nanomaterial is potentially hazardous one proceeds with the hazard characterization.

12.5.1.2 Hazard Characterization

Whereas the hazard identification recognizes the type and nature of the hazard, the objective of the hazard characterization is to obtain a qualitative or quantitative description of the inherent properties of the agent that is potentially hazardous when one is exposed to it. A quantitative description will, wherever possible, include a dose-response assessment, identification a no-observed-adverse-effect level (NOAEL), no-observed-effect level (NOEL) or cancer potency factor and take uncertainty factors into account. In addition, based on dose-response assessments no effect levels (NEL) of an agent are derived.

The information on NOAEL and an eventual cancer potency factor are used to establish tolerable daily intake (TDI), acceptable daily intake (ADI) value as guidance values while including uncertainty factors like, e.g., interspecies and intraspecies variability, and data quality. TDI and ADI are both referring to a dose that is safe to consume for humans during an entire lifetime. ADI is used for, e.g., food additives, whereas, TDI is used for agents we are unintentionally exposed to like, e.g., air pollution, contaminants of water.

Depending on the uncertainty level of these data, e.g., if there has to be extrapolated from in vitro or animal studies to humans, if it is necessary to include susceptible population groups etc., uncertainty factors in the range of 10–10,000 are applied to cover also worst case scenarios and population groups. By applying an uncertainty factor the acceptable concentration for the exposure to an agent is reduced to a value where also the most susceptible population groups are not experiencing adverse health effects. Thereby it is avoided that parts of the human population might be unprotected.

Like for the hazard identification, data are obtained from human studies (mostly epidemiological studies), are animal-based or in vitro toxicology studies as well as structure-activity studies or combinations of these studies.

12.5.1.3 Exposure Assessment

The exposure assessment does not only include the investigation whether there is a contact with a potentially toxic agent. It determines also the concentration, route and duration of exposure. It has also the goal to establish safety margins and thresholds by evaluating the likelihood and level of exposure. An exhaustive exposure assessment requires that all possible exposure scenarios are taken into account and that includes the identification of particular susceptible population groups like children, pregnant woman, elderly and predisposed people. Another important part of the exposure assessment is the identification of the route and duration of exposure. In addition, this knowledge is important for the regulation and legislation of toxic agents but also nanomedicines. As described before, the exposure can occur orally, via inhalation, dermally, intravenously, subcutaneously and intramuscularly and the toxicity of a substance may be dependent on the route of exposure. For the estimation of exposures, either measurement or modelling approaches are used. In most cases when unintentional exposures occur the exact measurement and determination of an exposure is not available and a worst case scenario is modelled.

For nanomedicines the administered dose is exactly known. However, the internal dose is dependent on the absorption and excretion of a drug and can therefore vary from the administered dose. In some cases, the internal dose can be estimated using biomarkers. Biomarkers are measurable indicators for the presence of a substance in the body and can be measured in tissues or body fluids like blood, urine but also feces. The duration of an exposure can be acute, subacute, subchronic or chronic. The acute exposure has a duration of less than 24 h and is often a single exposure. Subacute exposure refers to repeated exposures with a duration of up to a month and subchronic exposure lasts for 1–3 month. If the exposure duration exceeds 3 month chronic exposure occurs.

12.5.1.4 Risk Characterization

The last step in the risk assessment process is the risk characterization. The aim of the risk characterization is, if possible, quantitative determination of the probability that known potential adverse health effects occur under defined exposure conditions. These exposure conditions might be actual or predicted exposures. Risk characterization includes the results that have been obtained from hazard identification and characterization and exposure assessment. Based on these results, risk quotients or margins of safety are calculated and exposure and no effect levels are compared to estimate the risks. However, there are no absolute measures of risks and the conclusion if and when a given agent comprises a risk might vary from scientist to scientist especially if they include implicit value judgments.

12.6 Regulatory Affairs

The regulation of the use of ENMs is still debated due to the relatively short time period ENMs have been in focus of toxicological investigations. Many toxicity and safety related uncertainties of ENMs have not been clarified and this gap in knowledge results of course in uncertainties about the safety of nanomedicines. Nevertheless, nanomedicines have been authorized by licensing agencies like the FDA and EMA for more than 30 years.

The primary regulatory bodies in the U.S. and European Union (EU) that are relevant for the regulation of nanomedicines are the FDA and EMA, respectively. The EMA is accompanied by several committees and groups, whereof, the Committee for Medicinal Products for Human Use (CHMP), the Innovation Task Force (ITF) and the New and Emerging Technologies (N&ET) Working group are relevant for the regulation of nanomedicine. In the case of the FDA the following centers and groups are relevant: Center for Drug Evaluation and Research (CDER), Center for Devices and Radiological Health (CDRH), Center for Biologics Evaluation and Research (NTIG) and Nanotechnology Task Force (NTF). In other countries, national agencies are responsible for the approval of nanomedicines.

12.6.1 Definition of Engineered Nanomaterials for Regulatory Purposes

For risk assessments and regulatory purposes a definition of the term nanomaterial is of utmost importance. However, the sheer number of different nanomaterials makes a definition much more complex than one perhaps first realizes. On one hand, the definition for nanomaterials has to be so comprehensive that it includes all nanomaterials but should, on the other hand, be also be simple and precise as possible. At the moment, there are no standardized definitions what a nanomaterial is and the definition varies between organizations and countries. One of these definitions was proposed by the International Organization for Standardization (ISO) in cooperation with the European Committee for Standardization (CEN). According to the ISO/TR 11360:2010 definition a nanomaterial is a material with any external dimension in the nanoscale or having internal or surface structure in the nanoscale with nanoscale (or nano range) defined as size range from approximately 1 nm to 100 nm (ISO/TR 11360:2010 Nanotechnologies—Methodology for the classification and categorization of nanomaterials https://www.iso.org/obp/ui/#iso:std:iso:tr:11360:ed-1:v1:en).

From a scientific point of view an implementation of a fixed size limit might not make sense and the approximate size range might be preferred. However, for regulatory purposes a fixed size limit is needed and, therefore, implemented by the EU Commission. This definition is based on the ISO definition, an opinion of the Scientific Committee on Emerging and Newly Identified Health Risks (SCENIHR) and a report of the Joint Research Centre (JCR). Nanomaterial means a "natural, incidental or manufactured material containing particles, in an unbound state or as an aggregate or as an agglomerate and where, for 50 % or more of the particles in the number size distribution, one or more external dimensions is in the size range 1–100 nm.

Fullerenes, graphene flakes and single wall carbon nanotubes with one or more external dimensions below 1 nm should be considered as nanomaterials.

In specific cases and where warranted by concerns for the environment, health, safety or competitiveness the number size distribution threshold of 50 % may be replaced by a threshold between 1 % and 50 %". In addition, the European Commission has acknowledged that a upper limit of 100 nm might not always be scientifically justified and that there are special circumstances prevailing in the pharmaceutical sector (EU Scientific Committee on Emerging and Newly Identified Health Risks. Scientific basis for the definition of the term 'Nanomaterial'. European Commission, Brussels, Belgium (2010)).

In the U.S.A. the FDA has another definition for nanomaterials and according to this a nanomaterial is defined to be any material with at least one dimension smaller than 1000 nm and a nanoparticle is an object with all three external dimensions in the size range from ~1 nm to 100 nm (http://www.fda.gov/RegulatoryInformation/Guidances/ucm257698.htm; accessed August 2014).

The lack of an adequate definition of nanomaterials becomes especially problematic when dealing with follow-up nanomedicines that are based on already approved medicines formerly not classified as nanomaterials or not registered to contain nanomaterials.

Another reason for the still ongoing debate on the definition of nanomaterials is the challenge of a comprehensive characterization of ENMs used as nanomedicines. The methods that are available for characterization are not necessarily applicable for ENMs in complex mixtures and sometimes only the primary material might be suitable for characterization. A too rigid definition and regulation might, therefore, lead to the reluctance of regulatory agencies to issue manufacturing licenses or marketing authorizations. Therefore, not only a definition for nanomaterials is needed but also a definition of standards for the characterization of nanomaterials [105].

12.6.2 Regulation of Nanomedicines

The exact definition of a nanomaterial is only one of many questions that have to be addressed for a proper regulative approach for nanomedicines. In addition, the situation might be even more complicated when it has to be decided if a nanomedicine that uses ENMs as carrier is a medicine (medical product) or a medical device? This classification is not unimportant as medical products and medical devises are regulated in different ways. Furthermore, before a new nanomedicine can be accepted for the use on patients it has to be decided, which regulatory regime is applicable. A medical device fulfills its function by physical means like mechanical or chemical action, whereas, a medical product exclusively fulfills its function by pharmacological, immunological or metabolic means [105]. Clarification on these matters is of course of uppermost importance not only for the safety of patients but also for the pharmaceutical industry that demands a greater harmonization in current nanomedicine regulatory framework. In the EU, the decision on the classification as a medical device or medicine is based on the EU Directive 2001/83/EC on human medicines, as amended, and the EU Directive 93/42/EEC on medical devices, as amended. According to these directives the decision is been made according to the principal mode of action of the nanomedicine. Especially for nanomedicines, which have a complex mode of action, this may prove difficult as their mode of action might involve and combine physicochemical and pharmacological properties. In addition, in some situations, when the nanomedicines are based on viable cells or tissues, they might also be classified as advanced therapy medicinal products and fall under the Regulation (EC) 1394/2007 on advanced therapy medicinal products [105].

Toxicological investigations are the basis for the safety and risk assessment of nanomedicines and as such crucial for their approval. For conventional medicines a battery of OECD approved methods are available for the investigation of eventual toxic effects of a drug. However, it is not clear if these tests are applicable for ENM based nanomedicines. For example, bacteria-based genotoxicity assays like the Ames test may not be appropriate as ENMs may not be able to penetrate the bacteria [105].

Taken together, it becomes clear that the regulation of ENMs and nanomedicines is still under development. Just recently as from 21 June to 13 September 2013 the European Commission had launched a public consultation on the modification of the REACH Annexes (REACH—Registration, Evaluation, Authorisation and Restriction of Chemical substances) on nanomaterials with the aim to improve the clarity on how nanomaterials are defined and their safety demonstrated. This consultation was open for the public and interested stakeholders. The European Commission states that “The REACH legislation must ensure a high level of health, safety and environmental protection. At the same time it should permit access to innovative products and promote innovation and competitiveness.” (http://ec.europa.eu/nanotechnology/policies_en.html).

12.7 Conclusion

Efficacy and toxicity of nanomedicines are inseparable interconnected as changes of the physicochemical properties can influence absorption, distribution, metabolism, and excretion of nanomedicines at the cellular but also organism level. For example, lipid particles or biodegradable ENMs might be less harmful than non-degradable inorganic ENMs and are most likely to be used in the clinic sooner. Therefore, an early implementation of toxicological investigations is fundamental when developing new nanomedicines.

Although there are no nano-specific directives and regulations at this time, it is important to point out that nanomedical products are not unregulated. Although it is not clear at this point if this procedure is adequate, the FDA and EMA are applying of course the existing legislation on medical products and devices, tissue engineering etc. that are relevant for nanomedicines. Despite the doubts, in 2013 as much as 247 nanomedicine products were listed in FDA registers as approved or to be in various stages of clinical trials most of them attended to be administered intravenously [106]. Hopefully, in time the safety and regulatory challenges that come with nanomedicines are solved to utilize the full potential of nanomedicines.

References

1. Project on Emerging Nanotechnologies. (2014). http://www.nanotechproject.org/
2. Chan VS (2006) Nanomedicine: an unresolved regulatory issue. Regul Toxicol Pharmacol 46:218–224
3. Donaldson K, Stone V (2003) Current hypotheses on the mechanisms of toxicity of ultrafine particles. Ann Ist Super Sanita 39:405–410
4. Schwarze PE, Ovrevik J, Lag M, Refsnes M, Nafstad P, Hetland RB et al (2006) Particulate matter properties and health effects: consistency of epidemiological and toxicological studies. Hum Exp Toxicol 25:559–579
5. Donaldson K, Stone V, Tran CL, Kreyling W, Borm PJ (2004) Nanotoxicology. Occup Environ Med 61:727–728
6. Maynard AD, Warheit DB, Philbert MA (2011) The new toxicology of sophisticated materials: nanotoxicology and beyond. Toxicol Sci 120(Suppl 1):S109–S129
7. Magdolenova Z, Collins A, Kumar A, Dhawan A, Stone V, Dusinska M (2014) Mechanisms of genotoxicity. A review of in vitro and in vivo studies with engineered nanoparticles. Nanotoxicology 8:233–278
8. Stone V, Nowack B, Baun A, van den Brink N, Kammer F, Dusinska M et al (2010) Nanomaterials for environmental studies: classification, reference material issues, and strategies for physico-chemical characterisation. Sci Total Environ 408:1745–1754
9. Karlsson HL (2010) The comet assay in nanotoxicology research. Anal Bioanal Chem 398:651–666
10. Worle-Knirsch JM, Pulskamp K, Krug HF (2006) Oops they did it again! Carbon nanotubes hoax scientists in viability assays. Nano Lett 6:1261–1268
11. Landsiedel R, Kapp MD, Schulz M, Wiench K, Oesch F (2009) Genotoxicity investigations on nanomaterials: methods, preparation and characterization of test material, potential artifacts and limitations—many questions, some answers. Mutat Res 681:241–258
12. Monteiro-Riviere NA, Inman AO, Zhang LW (2009) Limitations and relative utility of screening assays to assess engineered nanoparticle toxicity in a human cell line. Toxicol Appl Pharmacol 234:222–235
13. Oberdorster G, Oberdorster E, Oberdorster J (2005) Nanotoxicology: an emerging discipline evolving from studies of ultrafine particles. Environ Health Perspect 113:823–839
14. Landsiedel R, Ma-Hock L, Kroll A, Hahn D, Schnekenburger J, Wiench K et al (2010) Testing metal-oxide nanomaterials for human safety. Adv Mater 22:2601–2627

15. Kumar S, Verma MK, Srivastava AK (2013) Ultrafine particles in urban ambient air and their health perspectives. Rev Environ Health 28:117–128
16. Society of Toxicology SOT (2005). http://www.toxicology.org/ai/pub/si05/SI05_Define.asp. Ref Type: Online Source
17. Oberdorster G (2010) Safety assessment for nanotechnology and nanomedicine: concepts of nanotoxicology. J Intern Med 267:89–105
18. Zhu M, Nie G, Meng H, Xia T, Nel A, Zhao Y (2013) Physicochemical properties determine nanomaterial cellular uptake, transport, and fate. Acc Chem Res 46:622–631
19. Zhu M, Perrett S, Nie G (2013) Understanding the particokinetics of engineered nanomaterials for safe and effective therapeutic applications. Small 9:1619–1634
20. Foss Hansen S, Larsen BH, Olsen SI, Baun A (2007) Categorization framework to aid hazard identification of nanomaterials. Nanotoxicology 1:243–250
21. Baeza-Squiban A, Boland S, Hussain S, Marano F (2009) Health effects of nanoparticles. In: General, applied and systems toxicology. John Wiley & Sons, Ltd, Chichester
22. Borm PJ, Kreyling W (2004) Toxicological hazards of inhaled nanoparticles—potential implications for drug delivery. J Nanosci Nanotechnol 4:521–531
23. Gellein K, Syversen T (2009) Nanotoxicology—the toxicology of nanomaterials. In: General, applied and systems toxicology. John Wiley & Sons, Ltd, Chichester
24. Cho M, Cho WS, Choi M, Kim SJ, Han BS, Kim SH et al (2009) The impact of size on tissue distribution and elimination by single intravenous injection of silica nanoparticles. Toxicol Lett 189:177–183
25. Hussain S, Boland S, Baeza-Squiban A, Hamel R, Thomassen LC, Martens JA et al (2009) Oxidative stress and proinflammatory effects of carbon black and titanium dioxide nanoparticles: role of particle surface area and internalized amount. Toxicology 260:142–149
26. Stoeger T, Reinhard C, Takenaka S, Schroeppel A, Karg E, Ritter B et al (2006) Instillation of six different ultrafine carbon particles indicates a surface area threshold dose for acute lung inflammation in mice. Environ Health Perspect 114:328–333
27. Piret JP, Vankoningsloo S, Mejia J, Noel F, Boilan E, Lambinon F et al (2012) Differential toxicity of copper (II) oxide nanoparticles of similar hydrodynamic diameter on human differentiated intestinal Caco-2 cell monolayers is correlated in part to copper release and shape. Nanotoxicology 6:789–803
28. Tarantola M, Pietuch A, Schneider D, Rother J, Sunnick E, Rosman C et al (2011) Toxicity of gold-nanoparticles: synergistic effects of shape and surface functionalization on micromotility of epithelial cells. Nanotoxicology 5:254–268
29. World Health Organization (1997) Determination of airborne fibre number concentrations: a recommended method, by phase-contrast optical microscopy, membrane filter method. World Health Organization, Geneva
30. Feng W, Nie W, He C, Zhou X, Chen L, Qiu K et al (2014) Effect of pH-responsive alginate/chitosan multilayers coating on delivery efficiency, cellular uptake and biodistribution of mesoporous silica nanoparticles based nanocarriers. ACS Appl Mater Interfaces 6:8447–8460
31. Ogawara K, Furumoto K, Nagayama S, Minato K, Higaki K, Kai T et al (2004) Pre-coating with serum albumin reduces receptor-mediated hepatic disposition of polystyrene nanosphere: implications for rational design of nanoparticles. J Control Release 100:451–455
32. Cedervall T, Lynch I, Lindman S, Berggard T, Thulin E, Nilsson H et al (2007) Understanding the nanoparticle-protein corona using methods to quantify exchange rates and affinities of proteins for nanoparticles. Proc Natl Acad Sci U S A 104:2050–2055
33. Deng ZJ, Mortimer G, Schiller T, Musumeci A, Martin D, Minchin RF (2009) Differential plasma protein binding to metal oxide nanoparticles. Nanotechnology 20:455101
34. Vertegel AA, Siegel RW, Dordick JS (2004) Silica nanoparticle size influences the structure and enzymatic activity of adsorbed lysozyme. Langmuir 20:6800–6807
35. Casals E, Puntes VF (2012) Inorganic nanoparticle biomolecular corona: formation, evolution and biological impact. Nanomedicine (Lond) 7:1917–1930
36. Frohlich E (2012) The role of surface charge in cellular uptake and cytotoxicity of medical nanoparticles. Int J Nanomedicine 7:5577–5591

37. Magrez A, Kasas S, Salicio V, Pasquier N, Seo JW, Celio M et al (2006) Cellular toxicity of carbon-based nanomaterials. Nano Lett 6:1121–1125
38. Yang SH, Heo D, Park J, Na S, Suh JS, Haam S et al (2012) Role of surface charge in cytotoxicity of charged manganese ferrite nanoparticles towards macrophages. Nanotechnology 23:505702
39. Bhattacharjee S, Ershov D, Gucht J, Alink GM, Rietjens IM, Zuilhof H et al (2013) Surface charge-specific cytotoxicity and cellular uptake of tri-block copolymer nanoparticles. Nanotoxicology 7:71–84
40. Asati A, Santra S, Kaittanis C, Perez JM (2010) Surface-charge-dependent cell localization and cytotoxicity of cerium oxide nanoparticles. ACS Nano 4:5321–5331
41. Bhattacharjee S, de Haan LH, Evers NM, Jiang X, Marcelis AT, Zuilhof H et al (2010) Role of surface charge and oxidative stress in cytotoxicity of organic monolayer-coated silicon nanoparticles towards macrophage NR8383 cells. Part Fibre Toxicol 7:25
42. Xu M, Zhao Y, Feng M (2012) Polyaspartamide derivative nanoparticles with tunable surface charge achieve highly efficient cellular uptake and low cytotoxicity. Langmuir 28:11310–11318
43. Sayes CM, Liang F, Hudson JL, Mendez J, Guo W, Beach JM et al (2006) Functionalization density dependence of single-walled carbon nanotubes cytotoxicity in vitro. Toxicol Lett 161:135–142
44. Chiu CC, Moore PB, Shinoda W, Nielsen SO (2009) Size-dependent hydrophobic to hydrophilic transition for nanoparticles: a molecular dynamics study. J Chem Phys 131:244706
45. Ogris M, Brunner S, Schuller S, Kircheis R, Wagner E (1999) PEGylated DNA/transferrin-PEI complexes: reduced interaction with blood components, extended circulation in blood and potential for systemic gene delivery. Gene Ther 6:595–605
46. Zhao J, Castranova V (2011) Toxicology of nanomaterials used in nanomedicine. J Toxicol Environ Health B Crit Rev 14:593–632
47. Oberdorster G, Sharp Z, Atudorei V, Elder A, Gelein R, Kreyling W et al (2004) Translocation of inhaled ultrafine particles to the brain. Inhal Toxicol 16:437–445
48. Hagens WI, Oomen AG, De Jong WH, Cassee FR, Sips AJ (2007) What do we (need to) know about the kinetic properties of nanoparticles in the body? Regul Toxicol Pharmacol 49:217–229
49. Volkheimer G (1974) Passage of particles through the wall of the gastrointestinal tract. Environ Health Perspect 9:215–225
50. Volkheimer G, Schulz FH, Aurich I, Strauch S, Beuthin K, Wendlandt H (1968) Persorption of particles. Digestion 1:78–80
51. Volkheimer G, Schulz FH (1968) The phenomenon of persorption. Digestion 1:213–218
52. De Jong WH, Hagens WI, Krystek P, Burger MC, Sips AJ, Geertsma RE (2008) Particle size-dependent organ distribution of gold nanoparticles after intravenous administration. Biomaterials 29:1912–1919
53. Zhu MT, Feng WY, Wang Y, Wang B, Wang M, Ouyang H et al (2009) Particokinetics and extrapulmonary translocation of intratracheally instilled ferric oxide nanoparticles in rats and the potential health risk assessment. Toxicol Sci 107:342–351
54. Chen J, Tan M, Nemmar A, Song W, Dong M, Zhang G et al (2006) Quantification of extrapulmonary translocation of intratracheal-instilled particles in vivo in rats: effect of lipopolysaccharide. Toxicology 222:195–201
55. Meiring JJ, Borm PJ, Bagate K, Semmler M, Seitz J, Takenaka S et al (2005) The influence of hydrogen peroxide and histamine on lung permeability and translocation of iridium nanoparticles in the isolated perfused rat lung. Part Fibre Toxicol 2:3
56. Doherty GJ, McMahon HT (2009) Mechanisms of endocytosis. Annu Rev Biochem 78:857–902
57. Gould GW, Lippincott-Schwartz J (2009) New roles for endosomes: from vesicular carriers to multi-purpose platforms. Nat Rev Mol Cell Biol 10:287–292
58. Harush-Frenkel O, Rozentur E, Benita S, Altschuler Y (2008) Surface charge of nanoparticles determines their endocytic and transcytotic pathway in polarized MDCK cells. Biomacromolecules 9:435–443
59. Mayor S, Pagano RE (2007) Pathways of clathrin-independent endocytosis. Nat Rev Mol Cell Biol 8:603–612

60. Zhang LW, Monteiro-Riviere NA (2009) Mechanisms of quantum dot nanoparticle cellular uptake. Toxicol Sci 110:138–155
61. Zhao F, Zhao Y, Liu Y, Chang X, Chen C, Zhao Y (2011) Cellular uptake, intracellular trafficking, and cytotoxicity of nanomaterials. Small 7:1322–1337
62. Mortimer M, Kasemets K, Kahru A (2010) Toxicity of ZnO and CuO nanoparticles to ciliated protozoa Tetrahymena thermophila. Toxicology 269:182–189
63. Beer C, Foldbjerg R, Hayashi Y, Sutherland DS, Autrup H (2012) Toxicity of silver nanoparticles—nanoparticle or silver ion? Toxicol Lett 208:286–292
64. Xia T, Kovochich M, Liong M, Madler L, Gilbert B, Shi H et al (2008) Comparison of the mechanism of toxicity of zinc oxide and cerium oxide nanoparticles based on dissolution and oxidative stress properties. ACS Nano 2:2121–2134
65. Jiang X, Miclaus T, Wang L, Foldbjerg R, Sutherland DS, Autrup H et al (2014) Fast intracellular dissolution and persistent cellular uptake of silver nanoparticles in CHO-K1 cells: implication for cytotoxicity. Nanotoxicology 9:181–189
66. Borm P, Klaessig FC, Landry TD, Moudgil B, Pauluhn J, Thomas K et al (2006) Research strategies for safety evaluation of nanomaterials: Part V. Role of dissolution in biological fate and effects of nanoscale particles. Toxicol Sci 90:23–32
67. Ha ES, Choo GH, Baek IH, Kim JS, Cho W, Jung YS et al (2014) Dissolution and bioavailability of lercanidipine-hydroxypropylmethyl cellulose nanoparticles with surfactant. Int J Biol Macromol 72C:218–222
68. Martin MN, Allen AJ, MacCuspie RI, Hackley VA (2014) Dissolution, agglomerate morphology, and stability limits of protein-coated silver nanoparticles. Langmuir 30(38):11442–11452
69. Lanzl CA, Baltrusaitis J, Cwiertny DM (2012) Dissolution of hematite nanoparticle aggregates: influence of primary particle size, dissolution mechanism, and solution pH. Langmuir 28:15797–15808
70. Fraga S, Brandao A, Soares ME, Morais T, Duarte JA, Pereira L et al (2014) Short- and long-term distribution and toxicity of gold nanoparticles in the rat after a single-dose intravenous administration. Nanomedicine 10:1757–1766
71. Ballantyne B, Marrs TC, Syversen T (2009) Basic elements of toxicology. In: General, applied and systems toxicology. John Wiley & Sons, Ltd, Chichester
72. Lison D, Carbonnelle P, Mollo L, Lauwerys R, Fubini B (1995) Physicochemical mechanism of the interaction between cobalt metal and carbide particles to generate toxic activated oxygen species. Chem Res Toxicol 8:600–606
73. Guo B, Zebda R, Drake SJ, Sayes CM (2009) Synergistic effect of co-exposure to carbon black and Fe_2O_3 nanoparticles on oxidative stress in cultured lung epithelial cells. Part Fibre Toxicol 6:4
74. Sayes CM, Banerjee N, Romoser AA (2009) The role of oxidative stress in nanotoxicology. In: General, applied and systems toxicology. John Wiley & Sons, Ltd, Chichester
75. Singh N, Manshian B, Jenkins GJ, Griffiths SM, Williams PM, Maffeis TG et al (2009) NanoGenotoxicology: the DNA damaging potential of engineered nanomaterials. Biomaterials 30:3891–3914
76. Nel A, Xia T, Madler L, Li N (2006) Toxic potential of materials at the nanolevel. Science 311:622–627
77. Winterbourn CC (2008) Reconciling the chemistry and biology of reactive oxygen species. Nat Chem Biol 4:278–286
78. Cattley RC, Glover SE (1993) Elevated 8-hydroxydeoxyguanosine in hepatic DNA of rats following exposure to peroxisome proliferators: relationship to carcinogenesis and nuclear localization. Carcinogenesis 14:2495–2499
79. Ahamed M, Karns M, Goodson M, Rowe J, Hussain SM, Schlager JJ et al (2008) DNA damage response to different surface chemistry of silver nanoparticles in mammalian cells. Toxicol Appl Pharmacol 233:404–410
80. Foldbjerg R, Dang DA, Autrup H (2011) Cytotoxicity and genotoxicity of silver nanoparticles in the human lung cancer cell line, A549. Arch Toxicol 85:743–750

81. Jiang X, Foldbjerg R, Miclaus T, Wang L, Singh R, Hayashi Y et al (2013) Multi-platform genotoxicity analysis of silver nanoparticles in the model cell line CHO-K1. Toxicol Lett 222:55–63
82. Kumar A, Dhawan A (2013) Genotoxic and carcinogenic potential of engineered nanoparticles: an update. Arch Toxicol 87:1883–1900
83. Ng CT, Li JJ, Bay BH, Yung LY (2010) Current studies into the genotoxic effects of nanomaterials. J Nucleic Acids 2010:947859
84. Cheng Y, Dai Q, Morshed RA, Fan X, Wegscheid ML, Wainwright DA et al (2014) Blood-brain barrier permeable gold nanoparticles: an efficient delivery platform for enhanced malignant glioma therapy and imaging. Small 10:5137–5150
85. Hwang SR, Kim K (2014) Nano-enabled delivery systems across the blood-brain barrier. Arch Pharm Res 37:24–30
86. Cupaioli FA, Zucca FA, Boraschi D, Zecca L (2014) Engineered nanoparticles. How brain friendly is this new guest? Prog Neurobiol 119–120C:20–38
87. Sadauskas E, Wallin H, Stoltenberg M, Vogel U, Doering P, Larsen A et al (2007) Kupffer cells are central in the removal of nanoparticles from the organism. Part Fibre Toxicol 4:10
88. Huang JY, Lu YM, Wang H, Liu J, Liao MH, Hong LJ et al (2013) The effect of lipid nanoparticle PEGylation on neuroinflammatory response in mouse brain. Biomaterials 34:7960–7970
89. Voigt N, Henrich-Noack P, Kockentiedt S, Hintz W, Tomas J, Sabel BA (2014) Surfactants, not size or zeta-potential influence blood-brain barrier passage of polymeric nanoparticles. Eur J Pharm Biopharm 87:19–29
90. Gibson RM (2007) Understanding the potential neurotoxicology of nanoparticles. In: Nanotoxicology: characterization, dosing, and health effects. Informa Healthcare, pp 99–316
91. Mistry A, Stolnik S, Illum L (2009) Nanoparticles for direct nose-to-brain delivery of drugs. Int J Pharm 379:146–157
92. Mistry A, Glud SZ, Kjems J, Randel J, Howard KA, Stolnik S et al (2009) Effect of physicochemical properties on intranasal nanoparticle transit into murine olfactory epithelium. J Drug Target 17:543–552
93. Hopkins LE, Patchin ES, Chiu PL, Brandenberger C, Smiley-Jewell S, Pinkerton KE (2014) Nose-to-brain transport of aerosolised quantum dots following acute exposure. Nanotoxicology 8:885–893
94. Aschner M, Erikson KM, Dorman DC (2005) Manganese dosimetry: species differences and implications for neurotoxicity. Crit Rev Toxicol 35:1–32
95. Elder A, Gelein R, Silva V, Feikert T, Opanashuk L, Carter J et al (2006) Translocation of inhaled ultrafine manganese oxide particles to the central nervous system. Environ Health Perspect 114:1172–1178
96. Ze Y, Sheng L, Zhao X, Hong J, Ze X, Yu X et al (2014) TiO_2 nanoparticles induced hippocampal neuroinflammation in mice. PLoS One 9:e92230
97. Li XB, Zheng H, Zhang ZR, Li M, Huang ZY, Schluesener HJ et al (2009) Glia activation induced by peripheral administration of aluminum oxide nanoparticles in rat brains. Nanomedicine 5:473–479
98. Wang J, Liu Y, Jiao F, Lao F, Li W, Gu Y et al (2008) Time-dependent translocation and potential impairment on central nervous system by intranasally instilled TiO(2) nanoparticles. Toxicology 254:82–90
99. Song Y, Li X, Du X (2009) Exposure to nanoparticles is related to pleural effusion, pulmonary fibrosis and granuloma. Eur Respir J 34:559–567
100. Yang W, Peters JI, Williams RO III (2008) Inhaled nanoparticles—a current review. Int J Pharm 356:239–247
101. Geiser M, Kreyling WG (2010) Deposition and biokinetics of inhaled nanoparticles. Part Fibre Toxicol 7:2
102. Donaldson K, Tran L, Jimenez LA, Duffin R, Newby DE, Mills N et al (2005) Combustion-derived nanoparticles: a review of their toxicology following inhalation exposure. Part Fibre Toxicol 2:10

103. Simeonova PP, Erdely A (2009) Engineered nanoparticle respiratory exposure and potential risks for cardiovascular toxicity: predictive tests and biomarkers. Inhal Toxicol 21(Suppl 1): 68–73
104. Simeonova P, Erdely A, Li Z (2007) Carbon nanotube exposure and risk for cardiovascular effects. In: Nanotoxicology: characterization, dosing, and health effects. Informa Healthcare, pp 237–246
105. Kelly B (2010) Nanomedicines: regulatory challenges and risks ahead. Scrip Regulatory Affairs
106. Etheridge ML, Campbell SA, Erdman AG, Haynes CL, Wolf SM, McCullough J (2013) The big picture on nanomedicine: the state of investigational and approved nanomedicine products. Nanomedicine 9:1–14

Chapter 13
The Application of Nanotechnology for Implant Drug Release

Morten Østergaard Andersen

Abstract The use of medical implants is a cornerstone of modern medicine. All implants face, however, a number of challenges including infection and inflammation which cause many of them to fail. In addition, tissue engineering implants must also direct stem cell differentiation and tissue regeneration in order to work properly. These problems may be overcome using drugs that are delivered directly from the implant. For this to work the drugs have to be protected until they have performed their function, their release must be timed with when they are needed, they may have to affect specific regions of an implant only and some drugs must be delivered to specific sub-cellular locations in certain cells types. This chapter explores how various forms of nanotechnology may be employed to reach these goals and reviews many of the studies that have used nanotechnology for different implant mediated drug release applications.

Keywords Controlled drug release • Drug encapsulation • Drug adsorption • Local drug delivery • Stem cell differentiation • Implant associated infection • Inflammation control • Tissue engineering

13.1 Introduction

Nanotechnology plays an increasing role in improving drug release from medical implants, thereby, improving their ability to combat infections, modulate inflammation and promote tissue regeneration. Organ failure caused by trauma, congenital defects, infection or cancer has always been a major problem for human health. As a result, implants that support or replace our body parts have been in use for millennia as evidenced by ancient dental implants [1], wooden prosthetic limbs [2] and precious metal plates for use in cranioplasty after surgical trepanation [3]. Major

M.Ø. Andersen (✉)
The Maersk Mc-Kinney Moller Institute and Department of Chemical Engineering, Biotechnology and Environmental Technology, Faculty of Engineering, University of Southern Denmark, Niels Bohrs Allé 1, 5230 Odense M, Denmark
e-mail: moan@mmmi.sdu.dk

K.A. Howard et al. (eds.), *Nanomedicine*, Advances in Delivery Science and Technology, DOI 10.1007/978-1-4939-3634-2_13

new improvements in treating organ failure took place in the twentieth century including the first successful donor organ transplantations of kidneys in 1954 [4] and of hearts in 1967 [5] as well as the development of complex artificial electromechanical organs such as the Jarvik-7, the first working total artificial heart, implemented in 1982 [6]. Today the implantation of both donor organs and electromechanical implants form a critical part of modern medicine. However, the inherent limitations to both of these types of organ and tissue replacements have led to the development of a new research field: Tissue Engineering (TE). TE combines implant materials with cells and drugs to establish biological replacement organs. Unfortunately, donor organs, electromechanical implants and tissue engineered organs all face several major limitations which limit their clinical potential. Many of these problems arise from detrimental interactions between the implant and its biological surroundings such as implant associated infections, the immune system, and the peri-implant tissue. Various strategies have been devised to address these issues such as using different implant materials or changing surface chemistry or topography to prevent infection [7] or control stem cell differentiation [8]. The release of drugs from implants to modulate their surroundings has also been investigated extensively. The first of these devices such as drug eluting vascular metal stents [9] and antibiotics releasing orthopedic implants [10] have long been on the market and have helped reduce the occurrence of in-stent restenosis and implant associated infections, respectively. Yet many implant related challenges still exist. This chapter explores how nanotechnology is, and will be, used to improve implant functionality by modulating release of drugs. It is not possible to cover all studies on device-based drug delivery in this chapter; there are tens of thousands of publications already on this topic. Instead selected studies are reported that demonstrate different delivery/release principles, different nanotechnologies and different applications. Given the sheer number of studies that already exist, important studies may have been overlooked; however, the described studies do provide sufficient data to infer important generalized conclusions about device-based drug release. The following section introduces the three types of implants and some of the challenges they encounter, ending with an introduction into the concept of biocompatibility.

13.1.1 Donor Organ Transplantation

Organ transplants have successfully prolonged the life of millions of patients, with ~112.800 solid organ transplantations in 2012 carried out worldwide in order of frequency, kidney, liver, heart, lung, pancreas and intestinal transplants [11]. The benefits of living organ transplantation include patients receive a living implant that is fully capable of carrying out the multitude of functions that organs perform and that they can interact fully with the rest of the body, for example, through hormones and nerves. However, donor organs also have two severe limitations, the availability and their immunogenicity, as well as several other issues such as the risk of transferring undetected cancer or pathogens.

There are three main reasons for the lack of donor organs. First of all, a donor is needed, and with the exception of kidneys and small tissue grafts this means a recently deceased person who has consented to donate. Donor organs also need to be immunologically matched to the recipient otherwise the organ will be rejected [12]. Finally, donor organs must also have an appropriate size to fit into the patient which is a problem when either the donor or the recipient is a child [13]. The whole process is associated with several ethical issues such as the definition of being brain dead, the presumption of consent and the allocation of the organs [14]. As long term banking of living organs is not yet possible [15], patients have to wait until a suitable donor becomes available. In the USA alone [16], over 120,000 patients were on a waiting list for an organ at the start of 2014 and some 115,580 patients died while on a waiting list between 1995 and 2013. These numbers exclude even greater numbers of patients who could have benefited from an organ transplant but did not make it onto an official waiting list.

The second major limitation is the immunosuppressive medicine that organ recipients need to take for the rest of the life. Patients need to take these medicines to prevent organ rejection as perfect immunological matching of the donor is never achieved except in the rare cases where the organ is from an identical twin. Immunosuppression, unfortunately, leads to increased susceptibility to infections, diabetes, hypertension, cancer and several other diseases [17]. While implant survival and immunosuppressive therapy has improved, there is still a need for other treatment options where immunosuppression is not necessary and where the treatment can be carried out without waiting for a donor, until recently, that meant an artificial organ.

13.1.2 Artificial Electromechanical Implants

The main advantage of artificial electromechanical organs, such as total artificial hearts and ventricular assist devices, lies with their availability [18]; they are, thus, very useful in the cases where a suitable donor organ cannot be found in time. They also behave relatively reproducibly, compared to biological organs, even if the responses of different patients to identical implants differ. Devices also exist that can assist or replace tissue functions that are not amenable to organ transplantation; such devices include cochlear implants and deep brain stimulators. However, artificial implants also suffer from several major drawbacks and the clinical outcome of modern artificial implants are only beginning to match what is achieved with donor organs despite all their drawbacks [19]. Many human functions are too complex to completely emulate with current technology, examples include the signaling in the central nervous system, liver metabolism, kidney excretion and re-adsorption. Artificial organs also cannot communicate with the rest of the body for example via hormonal exchange, and, thus, cannot adapt to changing conditions else in the body, a total artificial heart is not regulated by thyroidal hormones nor does it secrete natriuretic hormones like a living tissue would. In contrast to many biological tissues, artificial organs cannot repair themselves when damaged which leads to mechanical fatigue, functional failure and the shedding of wear particles. The materials that artificial organs are composed can also be non-biocompatible and induce inflammatory foreign body responses, thrombus formation and/or lead to the destruction of the surrounding tissue due to mechanical strength incompatibilities. Mass produced, one size fits all, implants may inadequately fit patients and do not grow with the rest of their body; this is especially a problem for pediatric patients. Finally, electrically active implants depend on a power source as they cannot utilize the body's own energy carriers. Some of these limitations may be overcome by delivering drugs from the implant to modulate e.g. the foreign body response.

13.1.3 Tissue Engineering

While both donor organ transplantation and artificial organs continue to improve, their inherent drawbacks have led to a parallel effort being pursued where the aim is to engineer patient compatible living replacement organs, this research field is known as tissue engineering. The overall concept for tissue engineering was laid out by Langer and Vacanti in a 1993 paper [20]. The general approach consists of extracting cells from a patient, placing these cells on a scaffold material and then applying stimuli that induce the cells to form the correct tissue on this scaffold. The resulting tissue can then be re-implanted into the patient to replace a failing organ. This approach is illustrated in Fig. 13.1.

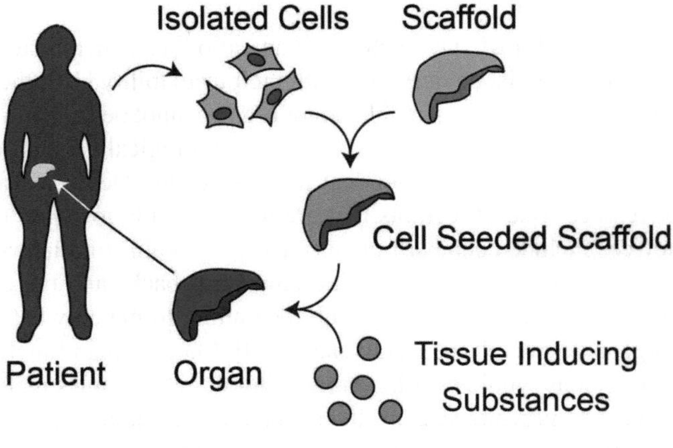

Fig. 13.1 Tissue engineering. Cells are first extracted from the patient and expanded in vitro, these are then seeded onto a scaffold were they are subjected to tissue inducing factors that may be delivered from the scaffold or an external source. When the desired tissue has formed it can be implanted into the patient

The first successful clinical transplantations have now been carried out using tissue engineered arteries in 2001 [21], heart valves in 2006 [22], bladders in 2006 [23] and trachea in 2008 [24]. Since then the long-term clinical success of tissue engineered organ transplantations has been proven [25]. The advantages of tissue engineering are plentiful. Ideally the organ becomes a living complete replacement organ fully capable of carrying out all normal functions and of communicating with the rest of the body. Since the tissue is derived from the patient's own cells no immunological rejection should take place. The organ shape and size is determined by the scaffold so they it can be made to fit the patient perfectly. No immediately deceased highly matched donor is required, and in the cases where a synthetic scaffold is used, no donor is required at all. Finally, since the graft is completely biological in origin it is ideally suited for pediatric patients [26]. The advantage in this case is that the transplant can grow with the patient obviating the need for repeated surgical interventions and reimplantations [27].

There are still many limitations to tissue engineering including cost, complexity, comparisons of different approaches and development of standardized best practices that enable studies from different groups to be compared. More animal studies are needed to determine the fate of the implants and potential problems they may cause. Clinical trials are also lacking. The lack of animal and clinical data causes concern [28] but that fact remains that some patients have been treated with tissue engineered organs with long-term success. This certainly indicates that tissue engineering will clearly be a treatment option in the future [29] even if the optimal approach has yet to be determined. However, the main tissues that have been developed so far have been relatively simple in their macro- and microstructure. To progress and develop more complex tissue engineered organs such as livers, kidneys, lungs and hearts (the organs that make up the majority of transplanted organs) there is a need to developed more complex scaffolds capable of inducing the formation of the intricate structures seen in these organs. This may potentially be achieved using scaffolds capable of advanced spatial and temporal control over drug release.

13.1.4 The Convergence of Artificial and Biological Implants

While tissue engineering offers limitless perfectly matched replacement organs that may be used to restore normal bodily function the possibility of merging artificial active implants with tissue engineered implants has been investigated. Mannoor et al., for example, additively manufactured a bionic ear composed of a chondrocyte filled alginate hydrogel and an inductive coil antenna made from conductive silver nanoparticles [30]. Such cyborg organs composed of biological and electronic components may provide capabilities far beyond what human tissues are normally capable of, and they could be the ultimate future implants. But even these implants will face problems that necessitate drug release to control infection, inflammation and differentiation.

13.1.5 *Biocompatibility*

Biocompatibility is one of the key issues in implant medicine [31]. Traditionally biocompatibility was seen as the absence of a host response as this led to the greatest success for implants. With the rise of tissue engineering where the ultimate goal is for the implant to merge completely with the body, this is no longer the case. Therefore, biocompatibility has been redefined as: "biocompatibility refers to the ability of a material to perform with an appropriate host response" [32]. Biocompatibility is clearly a complex and application dependent matter, but major determinants of biocompatibility and ultimately of implant success are implant associated inflammation and infection, as well as the interaction with the surrounding tissue. These are biological responses and, thus, amenable to pharmaceutical modulation.

Releasing drugs from implants has many advantages over systemic intervention such as greater local effect, lower side-effects at distant sites and smaller amounts of expensive drugs. The efficiency of drugs released from implants relies greatly on how they are interfaced with the implant in the first place. Drugs may, for example, be adsorbed or encapsulated into the implant, leading to fast and slow release, respectively. Poor drug stability and/or weak drug-implant interactions may prevent inclusion of drugs and in these cases, the drugs may be embedded in polymeric layers that are coated onto implants or in drug delivery particles that can be adsorbed onto or encapsulated into the implant surface. These then protect the drugs and determine how fast they are released. Nanotechnology is used in various ways to improve the functionality of these drug release system. Nanoparticles may, for example, be used to ensure maximum dispersion of the drug in the implant leading to a constant release; they may through their large surface area promote greater interaction with the implant matrix and they may facilitate delivery of bio-molecular drugs such as DNA and RNA to the intracellular sites of action. The following section describes how drug release strategies have been applied to various implants to improve their functionality in specific applications.

13.2 Implant Drug Release and Applications

13.2.1 *Nanotechnology, Drug Release and Implant Associated Infections*

It has long been known that implants provide extremely good breeding grounds for fungal and bacterial pathogens and that they may increase the virulence of bacteria more than 10^5-fold [33]. Implant associated infections are still a major problem [34]. Infections have traditionally been prevented by maintaining standard anti-septic operating conditions, thoroughly sterilizing implants and by giving patients antibiotics systemically after the surgery. However, even when sterile surgical conditions and implants are employed, a low number of pathogens may still find their way into the body. Furthermore, post-implantation, bacteria may arrive via the bloodstream and colonize otherwise sterile implants [35]. Interestingly, the implant material seems to have little effect on the infection rate, one study found that different metals gave roughly the same infection rate in vivo [36] and another that biological donor materials behaved like mechanical implants with regards to infection rates [37]. This is thought to be a consequence of the rapid formation of a serum protein film on the surface of all implant materials that subsequently masks the underlying material and provide good adhesion for microbial organisms [38]. When implant associated infection take place, the bacteria start to form biofilms on the implant surface [39]. These provide protection against the systemic treatment with antibiotics which cannot penetrate into the biofilm; they also deactivate the host's immune response by preventing the attachment of IgG and C3a [40]. As a result the implant usually has to be removed. To address this problem a wide range of new approaches have been developed [41, 42] these include changing the surface nano- and microstructure and its chemistry [7] as well as developing entirely new polymers [43]. However, immobilization [44] and/or release [45] of antimicrobial compounds such as organic molecules and silver are also being investigated often using nanotechnology to improve functionality. The progression of biofilm formation and treatment is illustrated on Fig. 13.2.

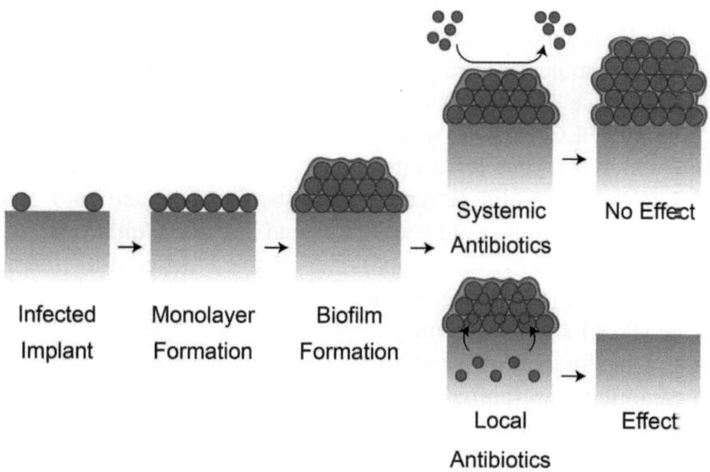

Fig. 13.2 Implant associated infections. Initially bacteria, implanted together with the implant either as free planktonic bacteria or already be attached to the implant, colonize the surface inside the body as a monolayer and then start to form a 3D multilayer biofilm. Systemic antibiotics cannot penetrate this biofilm but antibiotics released from the implant can due to their local high concentration

13.2.1.1 Controlled Release of Antibiotics from Implants

Chemical antibiotics have been in clinical use since the discovery of sulfonamides in the beginning of the twentieth century [46]. As antibiotics perform poorly when given systemically against implant associated infection due to biofilm formation, the local release of antibiotics from implants has been explored extensively and several products have reached the market [10]. The idea is that by delivering the drug beneath the biofilm it can locally accumulate to a much higher concentration and better penetrate the film. Liu et al. developed ethylcellulose particles encapsulating the antibiotic ceftazidime, which were encapsulated into a hydroxyapatite/polyurethane scaffold for bone tissue engineering [47]. The resulting scaffold released ceftazidime and could inhibit grow of *S. aureus*, additionally, the use of adsorbed ceftazidime encapsulating particles provided a longer release and bacterial inhibition than the simple inclusion of a "naked" ceftazidime in the scaffolds. Ma et al. demonstrated that nanosized cationic liposomes containing the antibiotic vancomycin could be incorporated into porous nano-hydroxyapatite/chitosan/konjac glucomannan tissue engineering scaffolds [48] which could then release liposomal vancomycin which destroyed *S. aureus*, the release rate could be tailored by the amount of konjac glucomannan that was incorporated into the scaffold. Similar to the study by Liu et al., this study also compared the release and antimicrobial activity of "naked" antibiotic with the formulated antibiotic and also found that encapsulating a formulated antibiotic led to prolonged release and anti-bacterial effect. Feng et al. showed that the antibiotic doxycycline could be incorporated into poly (lactic-co-glycolic) acid (PLGA) nanospheres than could then be placed on a porous nanofiber polylactic acid (PLA) scaffold [49]. This paper also found that doxycycline nanospheres provided a longer release and anti-bacterial activity than "naked" adsorbed doxycycline, with high molecular weight and lactic acid content in the PLGA indicative of longer release. Hong et al. developed an electrospun scaffold composed of polyurethane and PLGA fibers, where the PLGA fibers contained the antibiotic tetracycline which was then released destroying the *E. coli* [50]. Kim et al. demonstrated that the antibiotic cefoxitin could be incorporated into electrospun PLGA fibers where from it could be released and destroy *S. aureus*, introducing a polyethylene glycol (PEG)-PLA copolymer enabled better drug retention and prolonged the release [51]. Teo et al. blended the antibiotic gentamicin with polycaprolactone (PCL) and tricalcium phosphate (TCP) and deposited the blending into a wound healing mesh using fused deposition modelling, the mesh then released gentamicin and inhibited bacterial growth in vitro and in vivo where it promoted faster wound healing [52]. As an alternative to incorporating the antibiotic into the implant material, Li et al. demonstrated that the antibiotic cefazolin could be coated onto orthopedic implants using a layer-by-layer (LbL) technique; the resulting surfaces released the antibiotic and killed *S. aureus* [53]. These concepts are not limited to anti-bacterial compounds, Verreck et al. showed that the anti-fungal compound itraconazole could be encapsulated and released from electrospun polyurethane fibers [54].

These studies clearly indicate that by formulating antibiotics using nanotechnology, their activity, local retention and duration of effect can be improved. A major problem with using antibiotics to combat implant associated infection is, however, resistance to these drugs in bacterial communities including those found in implant associated infections [55–57]. This development has prompted a search for alternative means to control infection including implants that release non-organic compounds such as nitric oxide [58] and silver [59], amongst these approaches, silver has received by far the greatest interest.

13.2.1.2 Controlled Release of Silver from Implants

Silver has long been used as a bactericidal implant coating [60], and several commercial anti-infection products based on silver are on the market including Acticoat and Aquacel AG by Smith & Nephews and ConvaTec, respectively. Their effect is thought to be caused solely by the silver ion Ag^+ which silver surfaces release when oxidized [61]. Due to their large surface area which facilitates Ag^+ release and the ease of incorporating them into materials, silver nanoparticles are now used extensively to coat implants [62], where they seem to give better results than traditional silver formulations [63]. Those tested clinically include wound dressings [64], burn dressings [65], urinary catheters [66], ventricular drain catheters [67] and central venous catheters [68, 69] although the clinical effect varies. However, great caution must be exercised when using any drug for combating infections as they may have adverse effects on neighboring cells. Silver nanoparticles are known to be cytotoxic to a wide variety of cells [70] including stem cells commonly used for tissue engineering [71] and have been shown to be cytotoxic to osteoblasts and osteoclasts at concentrations lower than those needed for their antibacterial effect [72]. The use of nanotechnology such as silver nanoparticles to overcome one problem such as infection may, therefore, cause another problem such as local tissue damage. It is realistic; however, that further nanotechnological progress will enable us to use such potentially toxic particles in a controlled manner so that they do not cause harm, the key is to keep the local concentration of silver within the therapeutic window by controlling its release rate and local retention that can be achieved with enabling nanotechnologies.

This can be done by encapsulating the silver into a polymer or by attaching it onto the surface of the implant. Taglietti et al. used (3-aminopropyl)triethoxysilane to bind silver nanoparticles onto the surface of glass substrate where they inhibited microbial biofilm formation [73]. Gordon et al. developed a nanostructured coordination network composed of silver coordinating polymers and silver nanoparticles that could be coated onto titanium cages [74]. These cages effectively killed *S. epidermis* while the only damage to the mouse host was transient and minor damage to leukocytes. Wang et al. encapsulated silver nanoparticles into a PLGA electrospun coating that could be spun onto, and around, implants and showed that this coating was capable of killing *S. aureus* and *E. coli* while releasing concentrations far below those that caused acute toxicity [75]. Mohiti-Asli et al. showed that an electrospun nanofibrous PLA scaffold could be coated with proprietary formulation of $AgNO_3$ and polymer binder where after the scaffold became bactericidal [76]. Sheikh et al. showed that silk nanofibers containing hydroxyapatite and silver nanoparticles could be electrospun into a mat that inhibited *S. aureus* and *E. coli* but also induced cytotoxicity in fibroblasts [77]. Samberg et al. developed an electrospun PLCL skin tissue engineering scaffold that incorporated silver nanoparticles and showed that it could both support a confluent layer of fibroblasts as well as kill *S. aureus* and *E. coli* [78]. The use of silver is often promoted as a solution to drug resistance and does indeed kill bacteria that are resistant to antibiotics as shown by Cheng et al. who formed silver nanoparticles inside titania rods that had been created on the surface of titanium implants by anodization, these silver nanoparticles could subsequently kill multi resistant *S. aureus* (MRSA) [79]. However, it is important to note that microorganisms may also evolve genes that confer resistance to silver and the increasing use of silver in infection control may, therefore, trigger the evolution and spread of these genes rendering the silver approach no better than organic antibiotics [80].

13.2.2 Nanotechnology, Drug Release and Implant Associated Inflammation

13.2.2.1 The Foreign Body Response

When any implant is placed within the body a sequence of inflammatory events occurs that either resolves itself or leads to a foreign body response that causes implant failure [81]. During surgery, skin, connective tissue, blood vessels and possibly other tissue types are damaged and blood accumulates around the implant. The inflammation sequence starts when blood proteins adhere to the surface of the implant forming a provisional matrix composed of coagulated fibrin which encapsulates a range of blood borne bioactive proteins including mitogens, cytokines, complement factors, growth factors and chemokines. This matrix then acts as both a controlled release device shedding these bioactive proteins and as a cellular attachment scaffold. Neutrophils are among the first cells that attach to this matrix and they trigger the innate immune response. Mast cells are recruited and degranulate releasing histamine which, among other things, attracts macrophages. This acute inflammatory phase then either resolves itself typically within a week following normal wound healing or it progresses towards chronic inflammation. Chronic inflammation is characterized by the continual presence of lymphocytes and macrophages on the implant surface. If this persists, a granulation tissue is formed around the adhering immune cells by infiltrating fibroblasts. This granulation tissue may eventually progress and become a fibrous capsule that encapsulates and isolates the implant from the body causing the implant to fail. At the same time macrophages on the implant surface may fuse and become foreign body giant cells that form a large isolated pocket between themselves and the implant surface into which they secrete protons, reactive oxygen species and degradative enzymes which destroy the implant. These end results are known as the foreign body response. The direction that the inflammatory response takes (normal healing versus foreign body response) depends on the amount and type of proteins adsorbed onto the surface of the implant which depends on the surface chemistry and topography. A simplified overview of the foreign body response is shown in Fig. 13.3.

13.2.2.2 Wear Debris

On a longer term some implants shed nano- and micrometer sized debris particles [82]. This is the case when non-biodegradable metals, ceramics and polymers break down due to mechanical fatigue in areas where they are subjected to movement or loading e.g. in joint replacements. Wear particles then disperse into the surrounding tissue and can even travel to distant sites in the body. This is a major problem as

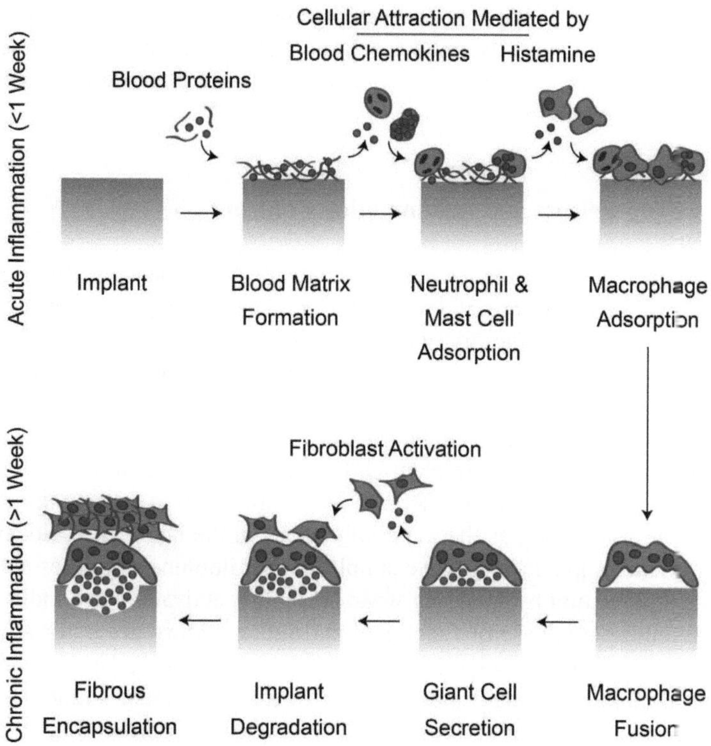

Fig. 13.3 The foreign body response. When an implant is inserted into the body a blood derived matrix forms attracting neutrophils, masts cells which then recruit macrophages. This acute inflammatory phase then either resolves itself, typically within 1 week, or it becomes chronic when macrophages start to fuse and degrade the implant while the implant is, in parallel, isolated by the formation of a fibrous capsule

macrophages engage these non-degradable particles. In the case of microparticles that are too big to be internalized, a frustrated phagocytosis response is triggered leading to giant cell formation and the production of reactive oxygen species. Some non-degradable wear debris nanoparticles are readily internalized by macrophages but since they cannot be broken down in the endo- and lysosomes, they end up destabilizing these [83]. Nanoparticles released by the wear of non-degradable polymer implants, such as those made of polyethylene, may also trigger inflammation through the TLR 1/2 receptor. In these cases the end result is the activation of the NALP3 inflammasome and the release of the pro-inflammatory cytokines IL-1, IL-6, IL-10, IL-12, TNF-α and IFN-γ. Additional pathways such as metal allergies may also contribute to the inflammatory response. The result is continual myelomonocyte infiltration and generation of additional active macrophages [84]. Eventually this cellular response causes peri-implant osteolysis where the bone tissue surrounding the implant is broken down due to three reasons: (1) Formation of bone-degrading osteoclasts triggered by inflammatory cytokines, (2) wear particle-induced apoptosis of bone-building osteoblasts and (3) degradation of the bone tissue which is caused by acidification of the tissue by bursting endosomes and the release of degradative enzymes such as collagenases and matrix metalloproteinases by the activated immune cells and osteoclasts. The end result is the degradation of the surrounding tissue and the loosening and eventual implant failure necessitating major surgical intervention.

13.2.2.3 Drug Release and Inflammation Control

Implant associated inflammation can be controlled by releasing drugs from the implants. However, there are two major problems with using drug release from to modulate inflammation. The first problem is that a certain degree of inflammatory response is absolutely critical to the success of most implants. The immune system controls implant associated infection, the immune system is involved in healing surgical tissue damage [85] and in the case of biodegradable implants, it is involved in degrading and clearing the implant and remodeling the resulting tissue to work together with the surrounding tissue. The second problem is that the inflammatory response may last as long as the material persists in the body. Non-biodegradable metal or polymer implants may, for example, trigger immune responses many years after their implantation by shedding wear debris that activates surrounding macrophages. Retaining a release of functional anti-inflammatory drugs for such a prolonged period is difficult. Despite these concerns, drug release has been studied to counteract implant associated inflammation in many different studies. These are increasingly making use of various nanotechnologies to improve functionality, for example, by achieving greater control over drug release.

To reduce early inflammation anti-inflammatory drugs may be released. Jayant et al. demonstrated that alginate microparticles containing and releasing the anti-inflammatory drugs dexamethasone or diclofenac could reduce the inflammatory response to a glucose sensor [86], in this case an LbL nanofilm was deposited on the surface of the particles to control drug release. Vacanti et al. incorporated dexamethasone into electrospun PCL or PLA nanofibers and demonstrated that drug loaded PLA nanofibers could inhibit inflammation and fibrous capsule formation in vivo [87]. Interestingly, dexamethasone release differed greatly between the two polymers, it was burst released from PCL but not PLA fibers and only drug loaded PLA fibers were capable of reducing the foreign body response.

Another approach is to target the late stages of inflammation such as the fibrous capsule formation, this leaves the acute response unaffected and able to clear infections and damaged tissue. Rujitanaroj et al. have developed a biodegradable electrospun polymer nanofiber scaffold that encapsulated different siRNA nanoparticle formulations targeted against Collagen 1 A1, a major component in the fibrous capsule [88]. These scaffolds were capable of silencing Collagen 1 A1 in cells and when placed in vivo fibrous capsule formation was inhibited. Additionally, release rates and duration of silencing could be altered by choosing different nanoparticle formulations. Takahashi et al. used a similar strategy, encapsulating in PEG based hydrogels, polyethylenimine formulated siRNA nanoparticles targeted towards mTOR, a gene involved in fibroblast proliferation and collagen production [89]. While they observed effective results in vitro no effect was observed in vivo.

Implant associated pain is another area where drug release may be used to benefit a patient. The implantation of an implant is a serious intervention which is often associated with great pain for the patient. Weldon et al. has, therefore, developed an electrospun PLGA fiber suture which incorporated bupivacaine, a local anesthetic which caused anesthesia around the implant in an animal model [90].

13.2.2.4 Inflammation and Neuroprosthetics

Tissue inflammation is also a major problem for neural implants [91] which typically consist of metal electrodes such as those used for deep brain stimulation. In the CNS, inflammation is mediated by reactive astrocytes and activated microglial cells and results in glial scar formation which removes the nerves from the implanted electrodes which cease to function as the impedance rises. Device based drug release has been pursued to inhibit the inflammatory process in some cases using nanotechnology [92]. Mercanzini et al. developed dexamethasone encapsulating polypropylene sulfide nanoparticles that could be coated onto electrode neuroprosthetics using a dissolvable polyethylene oxide (PEO) based coating [93]. When implanted in the motor cortex of rats the nanoparticles were released from the electrodes but stayed in the vicinity of the implant which resulted in lower inflammation and impedance. Similarly, Kim and Martin developed dexamethasone encapsulating PLGA nanoparticles which were coated onto neuroprosthetic implants using alginate hydrogels, these coatings also reduced inflammation and impedance when the implants were implanted in the auditory cortex of guinea pigs [94]. Interestingly, this study also showed that the release rate was much lower when the particles were embedded in an alginate hydrogel than when they were in a PBS solution.

13.2.2.5 Immunostimulatory Nanocoats

Interestingly, the opposite approach, local immunostimulation, has also been investigated with the aim of boosting inflammation to reduce implant associated infections [95]. Li et al. developed an immunostimulatory nanolayer that was deposited onto surgical steel K-wires by electrostatic LbL deposition of IL-12, BSA and PLL. When implanted the deposition of IL-12 led to a marked reduction in infection rates [96]. The concept has later been expanded by including chemokines such as MCP-1 in the coating to attract immune cells to the implant, this also led to a reduction in implant associated infection [97]. While this strategy may prove effective against bacteria that are resistant to antibiotics and silver, its use must be carefully balanced against increases in inflammation and the foreign body response.

13.2.3 Nanotechnology, Drug Release and Stem Cell Differentiation

Promoting tissue development is important in tissue engineering where a new tissue has to be created de novo within the implant. But it is also relevant for non-biodegradable artificial implants such as orthopedic and dental implants where a strong fixation in the bone is desired and bone development on the implant surface is, therefore, required, known as osseointegration. During embryonic development as well as in adult tissue repair, tissue development typically takes place when stem cells proliferate to generate more specialized cell types by asymmetric cell division [98]. These specialized cell types then undergo a succession of differentiation step which increasingly specialize the cell. Embryonic stem cells, for example, differentiate into endodermal, mesodermal and ectodermal stem cells during embryonic gastrulation, these then specialize further to produce the tissues our bodies contain. Likewise, mesenchymal stem cells can undergo differentiation into structural tissues such bone, fat, cartilage and muscle in a step wise manner by first producing precursor cells typically suffixed "blasts" which can then form the final cell types typically suffixed "cytes" as in osteoblasts and osteocytes, respectively. Often additional steps are present, commonly indicated by prefix "pre" as in a pre-osteoblast. During each of these differentiation steps the cell types perform different functions, pre-osteoblasts, for example, lay down the protein components of bone matrix, osteoblasts mineralize the matrix and osteocytes maintain the developed bone. Each differentiation step is typically started by exogenous factors such as the extracellular matrix or endocrine or paracrine signaling through hormones (often steroids and vitamins) and protein growth factors. These factors then act on transcription factors and microRNA which combine to down regulate the set of genes that define the current stage of differentiation as well as upregulate the expression of the set genes that are required for the next stage of development.

13.2.3.1 Drug Release and Stem Cell Differentiation

Different strategies exist to control the development of tissue on the surface of implants including changing the surface chemistry and topography [99]. However, since the body itself to a large extent controls stem cell differentiation through extra- and intracellular biomolecules it is only natural that much focus has been placed on delivering these biomolecules as well as synthetic molecules from implants to promote tissue development. Molecules delivered from implants to influence stem cell differentiation include proteins [100], vitamins [101], plasmid DNA [102] as well as various RNA molecules such as microRNA [103] and siRNA [104]. These molecules have been incorporated into scaffolds using a wide variety of methods. Wadagaki et al., for example, demonstrated that the synthetic molecule simvastatin could be incorporated into electrospun nanofibers that were deposited into a scaffold and that the subsequent release of simvastatin enhanced the formation of bone in vivo [105]. Shah et al. used a chitosan/poly aspartic acid LbL nanofilm system to load the osteoinductive bone morphogenetic protein 2 (BMP2) and hydroxyapaptite onto durable implants made of PEEK or titanium [106]. The inclusion of a hydrolytically cleavable molecule in the films enabled tuning the BMP2 release and the coatings promoted bone regeneration in vitro and in vivo. Nie and Wang formed chitosan nanoparticles containing plasmid DNA encoding BMP2, these were then either encapsulated or adsorbed into PLGA scaffolds, and a sample with adsorbed "naked" plasmid was also prepared [107]. When placed in phosphate buffered saline the "naked" DNA was released fastest and the encapsulated plasmid nanoparticles slowest, the encapsulation of the plasmid led to the best combination of transgene expression and cell viability.

13.2.3.2 Drug Release and Other Tissue Engineering Applications

Adequate cell seeding on a tissue engineering scaffold is of great importance for the formation of functional tissues once the scaffolds has broken down. While most tissue engineering strategies rely on cell seeding prior to implantation in order to achieve sufficient cell densities, other approaches utilize implant mediated release of chemoattractive molecules to stimulate the mobilization and invasion of desired cell types. Zhao et al. developed a hydrogel that released the chemoattractive protein hepatocyte growth factor (HGF) [108]. This hydrogel was then capable of attracting mesenchymal stem cells in an in vitro transwell invasion assay whereas non-loaded hydrogels were not. Similarly, Li et al., demonstrated that a HGF loaded hydrogel could attract neural stem cells in a transwell assay [109].

Diffusion of oxygen and nutrients from the bloodstream into tissues and is another area of critical importance to tissue function and where insufficient transport takes place necrosis occurs [110]. Diffusion is distance limited and can typically only supply cells that are within 100 μm of a capillary. Neovascularization is promoted automatically when cells within a scaffold suffers from a lack of nutrient or oxygen [111], but may take place too slowly to prevent necrosis in large implants. Releasing drugs that promote vascularization from scaffolds has, therefore, received much attention in tissue engineering. Nelson et al. formed polymeric nanoparticles containing siRNA against prolyl hydroxylase domain 2 (PHD2), an anti-angiogenic factor, and incorporated these into a biodegradable polymeric foam scaffold by mixing the nanoparticles with the scaffold foam prior to foaming [112]. These scaffolds were then capable of silencing PHD2 and promoting the ingrowth of more and denser vessels.

13.2.4 Other Applications of Implant Drug Release

Besides controlling implant associated infection, inflammation and tissue engineering, implant drug release is also being employed for other applications such as combating residual cancer cells around implants. When cancer tumors are removed a surgeon often has to balance conservative resection that retains tissue function but risk incomplete removal of all cancer cells against aggressive resection that more likely removes all cancer cells but which may destroy tissue function. Implants are often placed where tumors are removed to fill the void, therefore, the possibility of release drugs to kill residual cancer cells have been explored extensively [113]. Wie and Wang developed an electrospun PLGA fiber scaffold that encapsulated the chemotherapeutic paclitaxel and showed that its release rate could be tailored by controlling the fiber diameter with nanofibers releasing faster than microfibers [114], released paclitaxel was capable of killing glioma cells. Yohe et al. showed that the anti-cancer drugs CPT-11 and SN-38, campothecin analogues, could be incorporated into an electrospun PCL/ PGC-C18 meshes designed to bridge colorectal anastomosis, these meshes could the kill colorectal cancer cells [115]. Intriguingly this study uses included air as a means to control drug release.

13.3 Nanotechnology Strategies in the Development of Drug Releasing Implants

Drugs can be interfaced with implants using a variety of means such as encapsulation and adsorption which affects amongst other things, drug accessibility and release rate. These parameters can be additionally tuned by formulating the drugs in nanoparticles. Figure 13.4 shows some simple modes for drug/implant interfacing.

13.3.1 Drug Adsorption

The simplest method for incorporating drugs is to adsorb them onto the surface of the formed implant by drying or lyophilizing a solution of the drug onto the surface. In this case the release of the drug from the surface depends on the binding strength

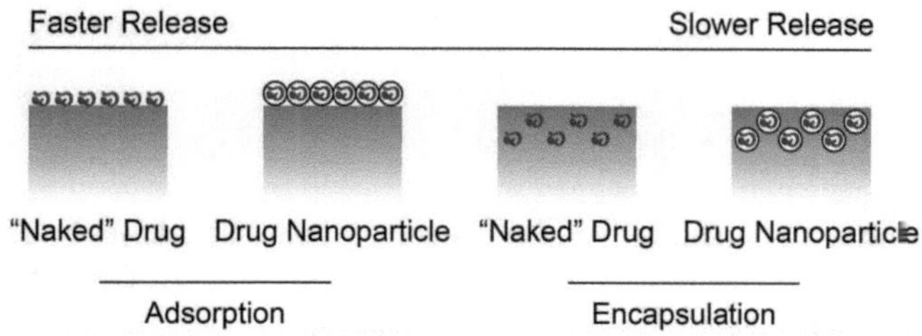

Fig. 13.4 Drugs can be adsorbed onto the surface of a polymeric implant after that surface has been formed (*far left*). If the drug is incorporated into a drug delivery vehicle before adsorption and that vehicle promote adhesion the release rate may be lowered (*center left*). If the drug is added to the polymer melt or solution before forming the implant surface the drug may be encapsulated which delays its release (*center right*). By incorporating the drug into a drug delivery system before its encapsulation the release may be further slowed if the system limits drug diffusion within the polymeric matrix (*far right*)

between the surface and the drug relative to the flow and content of the medium surrounding the implant. If this medium contains molecules that bind to the surface these may displace the drug. If slower release is desired, the drug may be incorporated into a drug delivery system that exhibits greater binding to the implant surface. A drug can also be covalently tethered to the surface, either permanently if the drug is only to affect adhering cells or temporally if the drug is to be released slowly to the implant surroundings. To increase surface loading LbL depositions may be used, in this case several alternately charged polymers are deposited as layers onto the implant surface, this provides a matrix wherein drugs may be loaded. Alternatively the implant surface can be modified so it provides a larger surface area by which the drug may bind stronger and to which more drug may be bound. Additionally surface cavities or tubes may be used to protect adsorbed drugs from the flow of the surrounding medium slowing their release. Çalışkan et al. recently showed that nanotubes could be formed on titanium surfaces using anodic oxidation and that these improved the loading and delayed the release of gentamicin; these surfaces, but not plain titanium, could kill *S. aureus* [116]. It is worth noting that adsorbed drugs are exposed to the surrounding aqueous medium and that many labile biological drugs may rapidly degrade in this release mode if unprotected. The typical adsorption modes are illustrated in Fig. 13.5.

13.3.2 Drug Encapsulation

If a slower release or greater drug protection is desired, the drug may be incorporated into the implant surface or the entire implant. This can be achieved by dissolving the drug in the polymer melt or solution prior to forming the implant, the drug then becomes encapsulated as the polymer matrix solidifies. Using this

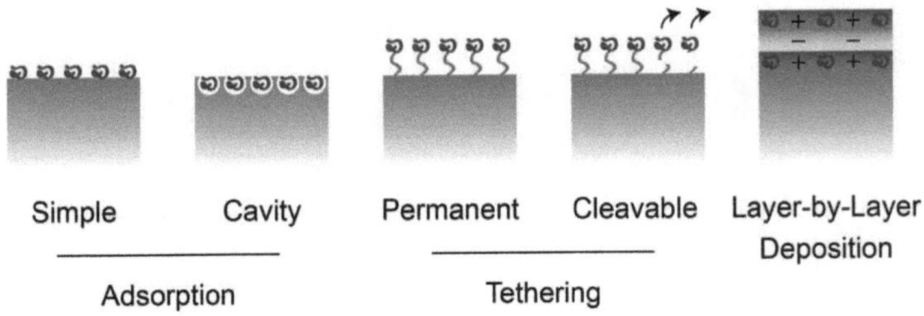

Fig. 13.5 Adsorption modes. There exist several different methods for adsorbing drugs onto an implant surface. Drugs may simply be adsorbed onto a flat surface but will in this case be susceptible to fast release. To slower the release rate, the drugs may be placed in cavities that reduce fluid motion and increase the surface area with which the drug may bind. Alternatively the drug may be covalently attached to the surface either permanently or by a linker that degrades slowly or fast in response to stimuli. Finally, drugs may be adsorbed as part of a layer-by-layer coating

method requires that the drug can tolerate the solvent or temperature used in the implant formation process which may not be the case for labile biological drugs, in this case a solvent/water emulsion may be used so the drugs can be kept in a aqueous phase prior to dehydration. Several studies show that compared to absorption, encapsulated drugs are released slower. The release rate depends on the drug diffusion rate through the polymer matrix and in the case of a biodegradable matrix, on the degraded volume as a function of time. Therefore, the desired release rate may be achieved by embedding the drug in a matrix with a tailored thickness or degradation rate. Alternatively the drug may be encapsulated into particles that are in turn encapsulated into the implant matrix, these particles may then retain the drug within the matrix and can also function as delivery vectors for drugs such as nucleic acids that need to reach the intracellular space of their target cells.

13.3.2.1 Drug Encapsulation by Co-Axial Electrospinning

Drugs can be incorporated onto or into electrospun polymeric implant the same ways as a normal polymeric implant by adsorption or encapsulation. Electrospinning, however, offers an additional mode of drug incorporation known as co-axial electrospinning, this method deposits a core-shell fiber containing an internal core phase and an outer shell phase which may be chemically different [117]. This allows the deposition of functional drugs into a drug compatible core phase even when harsh solvents are employed for the outer shell phase. In one example, Korehei and Kadla used co-axial electrospinning to deposit a T4 bacteriophage into hollow poly(ethylene oxide)/cellulose diacetate fibers and demonstrated that these fibers could subsequently release bacteriophages capable of killing *E. coli* in vitro [118]. The different delivery modes in electrospinning can be seen in Fig. 13.6.

Fig. 13.6 The drug incorporation modes commonly applied in electrospun implants. Adsorption provides faster release than encapsulation. Core-shell encapsulation may be used if the polymer processing solvent is incompatible with the drug to be encapsulated

13.3.2.2 Drug Release from Metal and Ceramic Implants

Metal and ceramic implants can be coated with drugs by adsorption the same way as with polymeric implants, indeed dip coating metal prosthesis in solutions of antibiotics prior to implantation is a simple solution that has been employed [119]. Unfortunately, such coated drugs are released fast. Drug encapsulation as a means to control the release is not as easy as with polymeric implants as most drugs will not survive forging or sintering temperatures or pressures. In these cases other approaches have to be tried. A solution is to coat the metal or ceramic surface with a polymeric or hydrogel layer encapsulating the drug for example using the LbL method [53]. Achieving a uniform layer using this method is, however, not simple on three dimensional implants. A different solution is to electrospin polymeric fibers around an implant, as was demonstrated by Wang et al. These three methods of drug functionalizing metal and ceramic implants are shown on Fig. 13.7.

13.3.3 Co-Release of Multiple Drugs

In many cases it would be beneficial to release more than one drug from an implant. In some cases it could be that different drugs accomplish different functions such as combating infection while promoting stem cell differentiation [120], whereas in other cases, drugs act synergistic to induce a desired phenomenon. Each step in stem cell differentiation, for example, is typically promoted by more than one growth factor and co-delivering several growth factors is, thus, necessary to recapitulate natural development completely [121].

We have shown one application where the anti-cancer drug doxorubicin was co-released with chitosan/siRNA nanoparticles from tissue engineering scaffolds to adjacent cancer cells wherein gene silencing and cell death was induced in lung cancer cells with the aim of their susceptibility to the chemotherapeutic [122]. Zheng et al. demonstrated that antimicrobial silver nanoparticles and the osteogenic growth factor BMP-2 could be co-released from a PLGA scaffold killing *S. aureus*

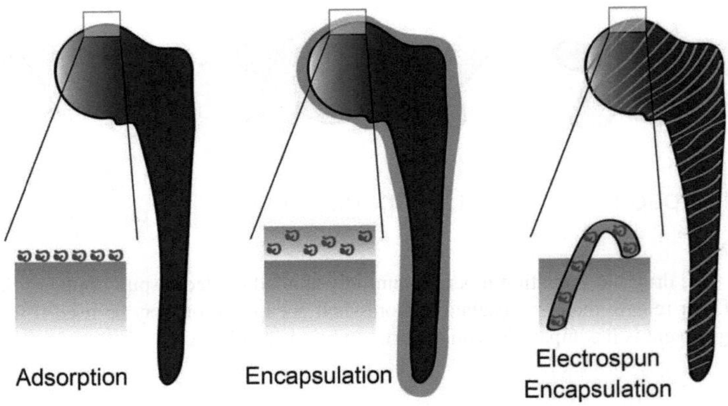

Fig. 13.7 Strategies for functionalizing metal and ceramic implants with drugs, illustrated with a hip prosthesis. The drug may either be adsorbed onto the implant surface (*left*) or be encapsulated in a polymeric layer that is coated onto the implant (*center*). A novel way of drug functionalizing metal and ceramic implants may be to wrap it in electrospun fibers that encapsulated drugs (*right*)

while promoting bone healing [123]. Similarly, Stevanovic et al. showed that silver nanoparticles and the anti-oxidant/osteogenic factor ascorbic acid could be co-incorporated into PLGA nanoparticles which could then kill *E. coli* and MRSA as well as promote bone formation [124], although in this case incorporation into medical devices was only suggested not tried. He et al. showed that polyethylenimine nanopolyplexes containing plasmids encoding vascular endothelial growth factor (VEGF) and fibroblast growth factor (FGF) could be incorporated into fibers scaffolds made of PEG-PLA [125]. When implanted into mice these plasmids synergized to promote ingrowth of denser mature vessels than either plasmid alone. Interestingly, the authors compared polyplex encapsulation with adsorption and showed that encapsulation led to greater expression of the encoded proteins as well as greater vessel density than adsorption.

13.3.4 Temporal Controlled Drug Release

Many implant related phenomenon are time dependent. Implant associated infections and acute inflammation take place immediately upon implantation, whereas, fibrous encapsulation occurs weeks after implantation and wear debris induced inflammation may not occur until years following implantation. The same is true in stem cell differentiation in tissue engineering where each differentiation step occurs in successions. Controlling drug release temporally is, thus, of great importance and can be achieved by choosing an appropriate delivery strategy that has suitable release kinetics for the biological function that needs to be modulated.

Drug adsorption, for example, leads to rapid release and may be suitable for combating infections or acute inflammation early on, whereas drug encapsulation leads to a slower release and may be more suited for modulating chronic inflammation or stem cell differentiation. By combining various delivery methods it is even possible to deliver different drugs from the same implant at different times. Clark et al. incorporated anti-inflammatory ketoprofen into a poly (β-amino ester) based hydrogel and BMP2 into PLGA microspheres, these two components were then sintered together to provide a composite scaffold [126]. When hydrated, the ketoprofen was released rapidly, whereas, the BMP2 was released slowly. Min et al. used an LbL technique to place BMP2 beneath a layer of the gentamicin on an implant [127]. Gentamicin was released fastest and was capable of killing *S. aureus*, subsequently BMP2 was released and was capable of promoting osteogenesis in pre-osteoblasts. Interestingly, the addition of clay layers in the layer by layer film was able to slow the release. In a different study by the same group, BMP2 was deposited underneath the angiogenic growth factor VEGF and demonstrated that VEGF was released first, followed by BMP2, and that their co-release led to denser bone formation than delivering BMP2 alone [128]. This was speculated to be due to the inclusion of fast releasing VEGF mediated the formation of an early vascular network through which cells penetrated the implant before dense bone formation was induced by slower releasing BMP2. Other strategies may also be used to control temporally controlled delivery of multiple drugs. Basmanav et al. incorporated BMP2 and BMP7 (BMP7 is another later acting osteoinductive growth factor) into crosslinked alginate microspheres with different alginate content and showed that the alginate determined the release rate [129]. When fast releasing BMP2 microspheres and slow releasing BMP7 microspheres were incorporated into a PLGA scaffold subsequently seeded with BMSCs bone formation was promoted. In a series of studies by the same group [130–132], Yilgor et al. incorporated BMP2 and BMP7 into nanocapsules composed of fast degrading PLGA or slow degrading PHBV, these particles were then either adsorbed onto or encapsulated into chitosan scaffolds where encapsulation led to a slower release. The authors then used these delivery modes to deliver BMP2 and BMP7 sequentially which led to greater bone formation than delivering them simultaneous or singularly.

13.3.5 Spatial Restricted Drug Release

Spatial restriction by nanofunctionalization is useful in the cases where a differential response is desired in different parts of an implant. This is relevant, for example, when complex tissue composed of multiple cell types has to be grown. Traditionally, such tissues have been grown by seeding different cell types onto different parts of an implant [133, 134] or by using different materials for each phase [135–138]. Sheehy et al., for example, created an osteochondral tissue by encapsulating chondrocytes and mesenchymal stem cells in separate regions of an alginate hydrogel implant whereas Filardo et al. repaired the knees in patients using a cell-free osteochondral implant consisting of collagen I layers with or with hydroxyapatite. Another strategy is to load each phase with different drugs typically loaded in drug vehicles, these drugs can then be released in a spatially restricted manner and induce the desired local effects [139, 140]. Dormer et al., for example, generated an osteochondral tissue using a bi-phasic scaffold composed of TGF-β and BMP2 loaded regions. Mohan et al. generated an osteochondral tissue by combining a hydroxyapatite gradient and a BMP2/TGF-β particle gradient in a biphasic scaffold [141]. Using a similar strategy [142], we have previously grown a bone/fat tissue using siRNA nanoparticles. In this case we formed a macroporous PCL scaffold with additional nanopores which we filled with nanoparticles containing siRNAs against either TRIB2, an inhibitor of adipogenesis, or BCL2L2, an inhibitor of osteogenesis. The nanoparticles adhered to the scaffold for days, presumably because of ionic interaction within the pores, and where able to silence genes locally when mesenchymal stem cells were seeded into the scaffold. When scaffolds containing two compartments with either of the siRNAs were implanted into mice, a spatial differentiated tissue development took place (Fig. 13.8).

Using the aforementioned techniques is only relevant to the engineering of "simpler" tissues where the different cells, materials or drugs can be arranged in geometries amenable to separate cell/drug seeding or material joining by hand. Most

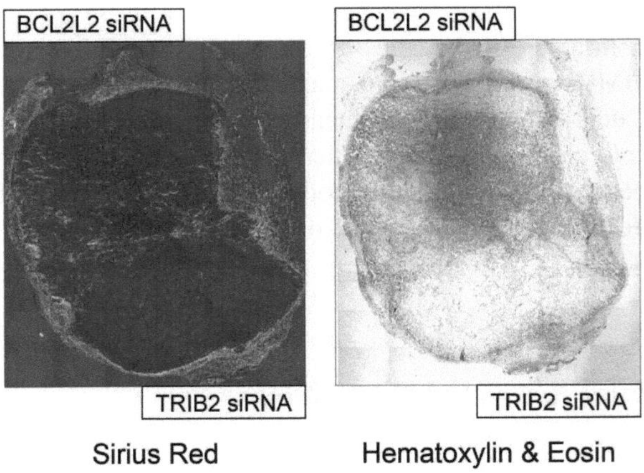

Fig. 13.8 A two-sided cylindrical scaffold containing adipogenic TRIB2 siRNA nanoparticles on one side and osteogenic BCL2L2 siRNA nanoparticles on the other side was seeded with mesenchymal stem cells and implanted sub cutaneously in a mouse. After 8 weeks pre-adipogenic tissue had developed in the TRIB2 side whereas pre-osteogenic tissue had developed in the BCL2L2 side. Figure modified from Andersen et al. [142]

internal organs and tissues, such as the kidney, liver, spleen, lungs and pancreas have more complex microstructures that necessitate different strategies. Towards engineering such tissues, various additive manufacturing strategies are being explored [143]. Many approaches rely neither on nanotechnology nor on drug release, instead, most research groups focus on printing different cell types into different volumes of an implant [144–147]. However, a different approach would be to 3D print different drugs into different compartments that correspond to different cell types in the desired tissue similar to the strategies that have worked well with simpler tissues [139–142]. Towards this, we have recently shown that nanoparticles containing different siRNAs can be 3D printed into different regions of a three component carbohydrate hydrogel using a patient derived CT-scan as the deposition guide [148], our approach is shown on Fig. 13.9. Once MSCs were seeded the siRNAs could then induce spatially restricted gene silencing. Co-printing cells and drug formulations can be combined, Xu et al. co-printed plasmid nanoparticles and endothelial cells into a collagen gel implants, during subsequent culture the endothelial cells became transfected by the plasmids [149].

13.4 Conclusion

The release of drugs from implants has clearly shown its clinical efficacy in fighting implant associated infections, modulating inflammation and in promoting tissue healing. It is, therefore, likely that most future implants will utilize drugs and that they will use nanotechnology to modulate their release and function. For artificial implants there will likely be a move towards employing multiple drugs to control infection, inflammation and other tissue responses. Nanotechnologies such as controlled release nanoparticles, responsive tethering and various layer encapsulation strategies will be used increasingly to ensure that effective but non-toxic concentrations of the different drugs are released at the specific times when each of these drugs are needed. For biological implants, there will likely be increased use of multiple drugs that promote different events in tissue regeneration. An implant will thus likely contain different stimulators of the progressive stages of stem cell differentiation as well as modulators of vascularization, reinnervation and cell survival, proliferation and migration. Many of these drugs will be labile biological macromolecules that are needed only at specific time points and locations. Nanotechnology will surely play a role in protecting the drugs until they are needed and in delivering the drugs to the right sub-cellular compartment in the targeted cells at the right time in the correct regions of the implant. This will be achieved by combining advanced targeted drug delivery systems with the advances in new scaffold fabrication technologies such as additive manufacturing and electrospinning.

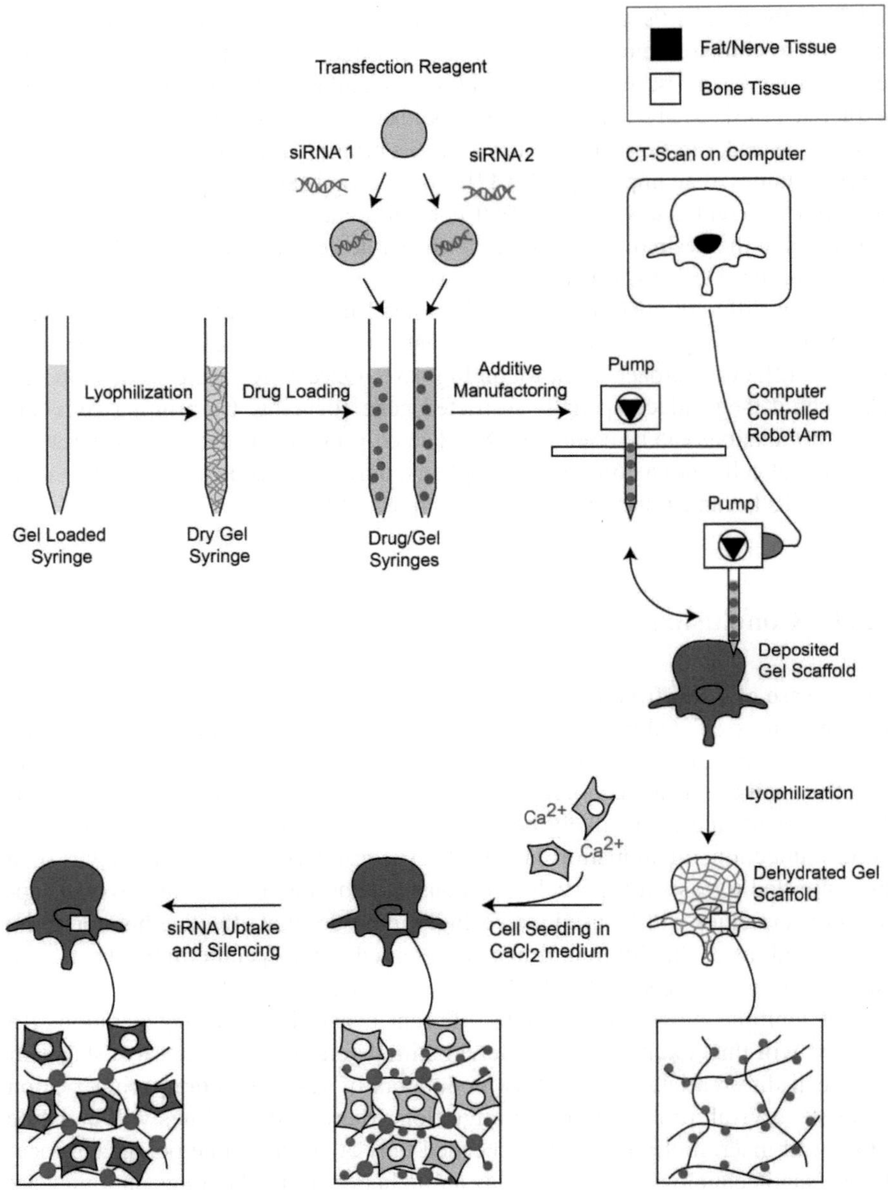

Fig. 13.9 Additive manufacturing of a complex implant guided by patient scanning data and using two hydrogels loaded with different siRNA particles

References

1. Iskander Z, Harris JE (1977) A skull with a silver bridge to replace a central incisor. Ann Serv Antiq Egypte 62:85–90
2. Thurston AJ (2007) Pare and prosthetics: the early history of artificial limbs. ANZ J Surg 77(12):1114–1119
3. Sanan A, Haines SJ (1997) Repairing holes in the head: a history of cranioplasty. Neurosurgery 40(3):588–603
4. Merrill JP, Murray JE, Harrison JH, Guild WR (1956) Successful homotransplantation of the human kidney between identical twins. JAMA 160(4):277–282
5. Barnard CN (1968) Human cardiac transplantation. An evaluation of the first two operations performed at the Groote Schuur Hospital, Cape Town. Am J Cardiol 22(4):584–596
6. DeVries WC, Anderson JL, Joyce LD, Anderson FL, Hammond EH, Jarvik RK, Kolff WJ (1984) Clinical use of the total artificial heart. N Engl J Med 310(5):273–278
7. Bazaka K, Jacob MV, Crawford RJ, Ivanova EP (2012) Efficient surface modification of biomaterial to prevent biofilm formation and the attachment of microorganisms. Appl Microbiol Biotechnol 95(2):299–311
8. Kolind K, Leong KW, Besenbacher F, Foss M (2012) Guidance of stem cell fate on 2D patterned surfaces. Biomaterials 33(28):6626–6633
9. Urban P, Gershlick AH, Guagliumi G, Guyon P, Lotan C, Schofer J, Seth A, Sousa JE, Wijns W, Berge C, Deme M, Stoll HP, e-Cypher Investigators (2006) Safety of coronary sirolimus-eluting stents in daily clinical practice: one-year follow-up of the e-Cypher registry. Circulation 113(11):1434–1441
10. Zilberman M, Elsner JJ (2008) Antibiotic-eluting medical devices for various applications. J Control Release 130(3):202–215
11. Those 2012 data are based on the Global Observatory on Donation and Transplantation (GODT) data, produced by the WHO-ONT collaboration
12. Buckley RH (2003) Transplantation immunology: organ and bone marrow. J Allergy Clin Immunol 111(2 Suppl):S733–S744
13. van Heurn E, de Vries EE (2009) Kidney transplantation and donation in children. Pediatr Surg Int 25(5):385–393
14. Freeman RB, Bernat JL (2012) Ethical issues in organ transplantation. Prog Cardiovasc Dis 55(3):282–289
15. Fahy GM, Wowk B, Wu J (2006) Cryopreservation of complex systems: the missing link in the regenerative medicine supply chain. Rejuvenation Res 9(2):279–291
16. The organ procurement and transplantation network. The OPTN Web site offers a wealth of information about transplantation (http://optn.transplant.hrsa.gov)
17. Burra P, De Bona M (2007) Quality of life following organ transplantation. Transpl Int 20(5):397–409
18. Timms D (2011) A review of clinical ventricular assist devices. Med Eng Phys 33(9): 1041–1047
19. Milla F, Pinney SP, Anyanwu AC (2012) Indications for heart transplantation in current era of left ventricular assist devices. Mt Sinai J Med 79(3):305–316
20. Langer R, Vacanti JP (1993) Tissue engineering. Science 260:920–926
21. Shin'oka T, Imai Y, Ikada Y (2001) Transplantation of a tissue-engineered pulmonary artery. N Engl J Med 344:532–533
22. Cebotari S, Lichtenberg A, Tudorache I, Hilfiker A, Mertsching H, Leyh R, Breymann T, Kallenbach K, Maniuc L, Batrinac A, Repin O, Maliga O, Ciubotaru A, Haverich A (2006) Clinical application of tissue engineered human heart valves using autologous progenitor cells. Circulation 114(1 Suppl):I132–I137
23. Atala A, Bauer SB, Soker S, Yoo JJ, Retik AB (2006) Tissue-engineered autologous bladders for patients needing cystoplasty. Lancet 367(9518):1241–1246

24. Macchiarini P, Jungebluth P, Go T, Asnaghi MA, Rees LE, Cogan TA, Dodson A, Martorell J, Bellini S, Parnigotto PP, Dickinson SC, Hollander AP, Mantero S, Conconi MT, Birchall MA (2008) Clinical transplantation of a tissue-engineered airway. Lancet 372(9655):2023–2030
25. Gonfiotti A, Jaus MO, Barale D, Baiguera S, Comin C, Lavorini F, Fontana G, Sibila O, Rombolà G, Jungebluth P, Macchiarini P (2013) The first tissue-engineered airway transplantation: 5-year follow-up results. Lancet 383(9913):238–244
26. Elliott MJ, De Coppi P, Speggiorin S, Roebuck D, Butler CR, Samuel E, Crowley C, McLaren C, Fierens A, Vondrys D, Cochrane L, Jephson C, Janes S, Beaumont NJ, Cogan T, Bader A, Seifalian AM, Hsuan JJ, Lowdell MW, Birchall MA (2012) Stem-cell-based, tissue engineered tracheal replacement in a child: a 2-year follow-up study. Lancet 380(9846): 994–1000
27. Ott HC (2012) Engineering tissues for children: building grafts that grow. Lancet 380(9846):957–958
28. Vogel G (2013) Trachea transplants test the limits. Science 340:266–268
29. Russell AJ (2014) The end of the beginning for tissue engineering. Lancet 383(9913): 193–195
30. Mannoor MS, Jiang Z, James T, Kong YL, Malatesta KA, Soboyejo WO, Verma N, Gracias DH, McAlpine MC (2013) 3D printed bionic ears. Nano Lett 13(6):2634–2639
31. Williams DF (2008) One the mechanism of biocompatibility. Biomaterials 29:2941–2953
32. Williams DF (1987) Definitions in biomaterials. Elsevier, Amsterdam
33. Elek SD, Conen PE (1957) The virulence of Staphylococcus pyogenes for man. A study of the problems of wound infection. Br J Exp Pathol 38(6):573–586
34. Montanaro L, Speziale P, Campoccia D, Ravaioli S, Cangini I, Pietrocola G, Giannini S, Arciola CR (2011) Scenery of Staphylococcus implant infections in orthopedics. Future Microbiol 6(11):1329–1349
35. Zimmerli W, Zak O, Vosbeck K (1985) Experimental hematogenous infection of subcutaneously implanted foreign bodies. Scand J Infect Dis 17(3):303–310
36. Hudetz D, Ursic Hudetz S, Harris LG, Luginbühl R, Friederich NF, Landmann R (2008) Weak effect of metal type and ica genes on staphylococcal infection of titanium and stainless steel implants. Clin Microbiol Infect 14(12):1135–1145
37. Hammermeister K, Sethi GK, Henderson WG, Grover FL, Oprian C, Rahimtoola SH (2000) Outcomes 15 years after valve replacement with a mechanical versus a bioprosthetic valve: final report of the Veterans Affairs randomized trial. J Am Coll Cardiol 36(4):1152–1158
38. Zimmerli W, Sendi P (2011) Pathogenesis of implant-associated infection: the role of the host. Semin Immunopathol 33:295–306
39. Götz F (2002) Staphylococcus and biofilms. Mol Microbiol 43(6):1367–1378
40. Kristian SA, Birkenstock TA, Sauder U, Mack D, Götz F, Landmann R (2008) Biofilm formation induces C3a release and protects Staphylococcus epidermidis from IgG and complement deposition and from neutrophil-dependent killing. J Infect Dis 197(7):1028–1035
41. Campoccia D, Montanaro L, Arciola CR (2013) A review of the biomaterials technologies for infection-resistant surfaces. Biomaterials 34(34):8533–8554
42. Ketonis C, Parvizi J, Jones LC (2012) Evolving strategies to prevent implant-associated infections. J Am Acad Orthop Surg 20:478–480
43. Hook AL, Chang CY, Yang J, Luckett J, Cockayne A, Atkinson S, Mei Y, Bayston R, Irvine DJ, Langer R, Anderson DG, Williams P, Davies MC, Alexander MR (2012) Combinatorial discovery of polymers resistant to bacterial attachment. Nat Biotechnol 30(9):868–875
44. Hickok NJ, Shapiro IM (2012) Immobilized antibiotics to prevent orthopaedic implant infections. Adv Drug Deliv Rev 64(12):1165–1176
45. Hetrick EM, Schoenfisch MH (2006) Reducing implant-related infections: active release strategies. Chem Soc Rev 35(9):780–789
46. Otten H (1986) Domagk and the development of the sulphonamides. J Antimicrob Chemother 17:689–696

47. Liu H, Zhang L, Shi P, Zou Q, Zuo Y, Li Y (2010) Hydroxyapatite/polyurethane scaffold incorporated with drug-loaded ethyl cellulose microspheres for bone regeneration. J Biomed Mater Res B Appl Biomater 95(1):36–46
48. Ma T, Shang BC, Tang H, Zhou TH, Xu GL, Li HL, Chen QH, Xu YQ (2011) Nano-hydroxyapatite/chitosan/konjac glucomannan scaffolds loaded with cationic liposomal vancomycin: preparation, in vitro release and activity against Staphylococcus aureus biofilms. J Biomater Sci Polym Ed 22(12):1669–1681
49. Feng K, Sun H, Bradley MA, Dupler EJ, Giannobile WV, Ma PX (2010) Novel antibacterial nanofibrous PLLA scaffolds. J Control Release 146(3):363–369
50. Hong Y, Fujimoto K, Hashizume R, Guan J, Stankus JJ, Tobita K, Wagner WR (2008) Generating elastic, biodegradable polyurethane/poly(lactide-co-glycolide) fibrous sheets with controlled antibiotic release via two-stream electrospinning. Biomacromolecules 9(4):1200–1207
51. Kim K, Luu YK, Chang C, Fang D, Hsiao BS, Chu B, Hadjiargyrou M (2004) Incorporation and controlled release of a hydrophilic antibiotic using poly(lactide-co-glycolide)-based electrospun nanofibrous scaffolds. J Control Release 98(1):47–56
52. Teo EY, Ong SY, Chong MS, Zhang Z, Lu J, Moochhala S, Ho B, Teoh SH (2011) Polycaprolactone-based fused deposition modeled mesh for delivery of antibacterial agents to infected wounds. Biomaterials 32(1):279–287
53. Li H, Ogle H, Jiang B, Hagar M, Li B (2010) Cefazolin embedded biodegradable polypeptide nanofilms promising for infection prevention: a preliminary study on cell responses. J Orthop Res 28(8):992–999
54. Verreck G, Chun I, Rosenblatt J, Peeters J, Dijck AV, Mensch J, Noppe M, Brewster ME (2003) Incorporation of drugs in an amorphous state into electrospun nanofibers composed of a water-insoluble, nonbiodegradable polymer. J Control Release 92(3):349–360
55. Karbach J, Callaway AS, Willershausen B, Wagner W, Al-Nawas B (2013) Multiple resistance to betalactam antibiotics, azithromycin or moxifloxacin in implant associated bacteria. Clin Lab 59(3–4):381–387
56. Rams TE, Degener JE, van Winkelhoff AJ (2014) Antibiotic resistance in human peri-implantitis microbiota. Clin Oral Implants Res 25(1):82–90
57. Rodriguez DJ, Afzal A, Evonich R, Haines DE (2012) The prevalence of methicillin resistant organisms among pacemaker and defibrillator implant recipients. Am J Cardiovasc Dis 2(2):116–122
58. Nablo BJ, Prichard HL, Butler RD, Klitzman B, Schoenfisch MH (2005) Inhibition of implant-associated infections via nitric oxide release. Biomaterials 26(34):6984–6990
59. Cao H, Liu X (2010) Silver nanoparticles-modified films versus biomedical device-associated infections. Wiley Interdiscip Rev Nanomed Nanobiotechnol 2(6):670–684
60. Schierholz JM, Lucas LJ, Rump A, Pulverer G (1998) Efficacy of silver-coated medical devices. J Hosp Infect 40(4):257–262
61. Xiu ZM, Zhang QB, Puppala HL, Colvin VL, Alvarez PJ (2012) Negligible particle-specific antibacterial activity of silver nanoparticles. Nano Lett 12(8):4271–4275
62. Chernousova S, Epple M (2013) Silver as antibacterial agent: ion, nanoparticle, and metal. Angew Chem Int Ed 52:1636–1653
63. Gravante G, Caruso R, Sorge R, Nicoli F, Gentile P, Cervelli V (2009) Nanocrystalline silver: a systematic review of randomized trials conducted on burned patients and an evidence-based assessment of potential advantages over older silver formulations. Ann Plast Surg 63(2):201–205
64. Lo SF, Hayter M, Chang CJ, Hu WY, Lee LL (2008) A systematic review of silver-releasing dressings in the management of infected chronic wounds. J Clin Nurs 17(15):1973–1985
65. Verbelen J, Hoeksema H, Heyneman A, Pirayesh A, Monstrey S (2014) Aquacel® Ag dressing versus Acticoat™ dressing in partial thickness burns: a prospective, randomized, controlled study in 100 patients. Part 1: Burn wound healing. Burns 40(3):416–427, pii: S0305-4179(13)00230-1

66. Rupp ME, Fitzgerald T, Marion N, Helget V, Puumala S, Anderson JR, Fey PD (2004) Effect of silver-coated urinary catheters: efficacy, cost-effectiveness, and antimicrobial resistance. Am J Infect Control 32(8):445–450
67. Lackner P, Beer R, Broessner G, Helbok R, Galiano K, Pleifer C, Pfausler B, Brenneis C, Huck C, Engelhardt K, Obwegeser AA, Schmutzhard E (2008) Efficacy of silver nanoparticles-impregnated external ventricular drain catheters in patients with acute occlusive hydrocephalus. Neurocrit Care 8(3):360–365
68. Kalfon P, de Vaumas C, Samba D, Boulet E, Lefrant JY, Eyraud D, Lherm T, Santoli F, Naija W, Riou B (2007) Comparison of silver-impregnated with standard multi-lumen central venous catheters in critically ill patients. Crit Care Med 35(4):1032–1039
69. Antonelli M, De Pascale G, Ranieri VM, Pelaia P, Tufano R, Piazza O, Zangrillo A, Ferrario A, De Gaetano A, Guaglianone E, Donelli G (2012) Comparison of triple-lumen central venous catheters impregnated with silver nanoparticles (AgTive®) vs conventional catheters in intensive care unit patients. J Hosp Infect 82(2):101–107
70. Beer C, Foldbjerg R, Hayashi Y, Sutherland DS, Autrup H (2012) Toxicity of silver nanoparticles—nanoparticle or silver ion? Toxicol Lett 208(3):286–292
71. Pauksch L, Hartmann S, Rohnke M, Szalay G, Alt V, Schnettler R, Lips KS (2014) Biocompatibility of silver nanoparticles and silver ions in primary human mesenchymal stem cells and osteoblasts. Acta Biomater 10(1):439–449
72. Albers CE, Hofstetter W, Siebenrock KA, Landmann R, Klenke FM (2013) In vitro cytotoxicity of silver nanoparticles on osteoblasts and osteoclasts at antibacterial concentrations. Nanotoxicology 7(1):30–36
73. Taglietti A, Arciola CR, D'Agostino A, Dacarro G, Montanaro L, Campoccia D, Cucca L, Vercellino M, Poggi A, Pallavicini P, Visai L (2014) Antibiofilm activity of a monolayer of silver nanoparticles anchored to an amino-silanized glass surface. Biomaterials 35(6):1779–1788
74. Gordon O, Slenters TV, Brunetto PS, Villaruz AE, Sturdevant DE, Otto M (2010) Silver coordination polymers for prevention of implant infection: thiol interaction, impact on respiratory chain enzymes, and hydroxyl radical induction. Antimicrob Agents Chemother 54:4208–4218
75. Wang H, Cheng M, Hu J, Wang C, Xu S, Han CC (2013) Preparation and optimization of silver nanoparticles embedded electrospun membrane for implant associated infections prevention. ACS Appl Mater Interfaces 5(21):11014–11021
76. Mohiti-Asli M, Pourdeyhimi B, Loboa EG (2014) Novel, silver-ion-releasing nanofibrous scaffolds exhibit excellent antibacterial efficacy without the use of silver nanoparticles. Acta Biomater 10(5):2096–2104, pii: S1742-7061
77. Sheikh FA, Woo Ju H, Mi Moon B, Jung Park H, Kim JH, Joo Lee O, Hum Park C (2014) Facile and highly efficient approach for the fabrication of multifunctional silk nanofibers containing hydroxyapatite and silver nanoparticles. J Biomed Mater Res A 102(10): 3459–3469
78. Samberg ME, Mente P, He T, King MW, Monteiro-Riviere NA (2013) In vitro biocompatibility and antibacterial efficacy of a degradable poly(L-lactide-co-epsilon-caprolactone) copolymer incorporated with silver nanoparticles. Ann Biomed Eng 42:1482–1493
79. Cheng H, Li Y, Huo K, Gao B, Xiong W (2013) Long-lasting in vivo and in vitro antibacterial ability of nanostructured titania coating incorporated with silver nanoparticles. J Biomed Mater Res A 102:3488–3499
80. Mijnendonckx K, Leys N, Mahillon J, Silver S, Van Houdt R (2013) Antimicrobial silver: uses, toxicity and potential for resistance. Biometals 26(4):609–621
81. Anderson JM, Rodriguez A, Chang DT (2008) Foreign body reaction to biomaterials. Semin Immunol 20(2):86–100
82. Cobelli N, Scharf B, Crisi GM, Hardin J, Santambrogio L (2011) Mediators of the inflammatory response to joint replacement devices. Nat Rev Rheumatol 7(10):600–608
83. Maitra R, Clement CC, Scharf B, Crisi GM, Chitta S, Paget D, Purdue PE, Cobelli N, Santambrogio L (2009) Endosomal damage and TLR2 mediated inflammasome activation by alkane particles in the generation of aseptic osteolysis. Mol Immunol 47(2–3):175–184

84. Ingham E, Fisher J (2005) The role of macrophages in osteolysis of total joint replacement. Biomaterials 26(11):1271–1286
85. Thomas MV, Puleo DA (2011) Infection, inflammation, and bone regeneration: a paradoxical relationship. J Dent Res 90(9):1052–1061
86. Jayant RD, McShane MJ, Srivastava R (2011) In vitro and in vivo evaluation of anti-inflammatory agents using nanoengineered alginate carriers: towards localized implant inflammation suppression. Int J Pharm 403(1–2):268–275
87. Vacanti NM, Cheng H, Hill PS, Guerreiro JD, Dang TT, Ma M, Watson S, Hwang NS, Langer R, Anderson DG (2012) Localized delivery of dexamethasone from electrospun fibers reduces the foreign body response. Biomacromolecules 13(10):3031–3038
88. Rujitanaroj PO, Jao B, Yang J, Wang F, Anderson JM, Wang J, Chew SY, Rujitanaroj PO, Jao B, Yang J, Wang F, Anderson JM, Wang J, Chew SY (2013) Controlling fibrous capsule formation through long-term down-regulation of collagen type I (COL1A1) expression by nanofiber-mediated siRNA gene silencing. Acta Biomater 9(1):4513–4524
89. Takahashi H, Wang Y, Grainger DW (2010) Device-based local delivery of siRNA against mammalian target of rapamycin (mTOR) in a murine subcutaneous implant model to inhibit fibrous encapsulation. J Control Release 147(3):400–407
90. Weldon CB, Tsui JH, Shankarappa SA, Nguyen VT, Ma M, Anderson DG, Kohane DS (2012) Electrospun drug-eluting sutures for local anesthesia. J Control Release 161(3):903–909
91. Polikov VS, Tresco PA, Reichert WM (2005) Response of brain tissue to chronically implanted neural electrodes. J Neurosci Methods 148(1):1–18
92. Yue Z, Moulton SE, Cook M, O'Leary S, Wallace GG (2013) Controlled delivery for neuro-bionic devices. Adv Drug Deliv Rev 65(4):559–569
93. Mercanzini A, Reddy ST, Velluto D, Colin P, Maillard A, Bensadoun JC, Hubbell JA, Renaud P (2010) Controlled release nanoparticle-embedded coatings reduce the tissue reaction to neuroprostheses. J Control Release 145(3):196–202
94. Kim DH, Martin DC (2006) Sustained release of dexamethasone from hydrophilic matrices using PLGA nanoparticles for neural drug delivery. Biomaterials 27(15):3031–3037
95. Li B, McKeague AL (2011) Emerging ideas: interleukin-12 nanocoatings prevent open fracture-associated infections. Clin Orthop Relat Res 469(11):3262–3265
96. Li B, Jiang B, Boyce BM, Lindsey BA (2009) Multilayer polypeptide nanoscale coatings incorporating IL-12 for the prevention of biomedical device-associated infections. Biomaterials 30(13):2552–2558
97. Li B, Jiang B, Dietz MJ, Smith ES, Clovis NB, Rao KM (2010) Evaluation of local MCP-1 and IL-12 nanocoatings for infection prevention in open fractures. J Orthop Res 28(1):48–54
98. Kolios G, Moodley Y (2013) Introduction to stem cells and regenerative medicine. Respiration 85(1):3–10
99. Tomisa AP, Launey ME, Lee JS, Mankani MH, Wegst UG, Saiz E (2011) Nanotechnology approaches to improve dental implants. Int J Oral Maxillofac Implants 26(Suppl):25–44, discussion 45–49
100. Luginbuehl V, Meinel L, Merkle HP, Gander B (2004) Localized delivery of growth factors for bone repair. Eur J Pharm Biopharm 58(2):197–208
101. Yoon SJ, Park KS, Kim MS, Rhee JM, Khang G, Lee HB (2007) Repair of diaphyseal bone defects with calcitriol-loaded PLGA scaffolds and marrow stromal cells. Tissue Eng 13(5):1125–1133
102. Storrie H, Mooney DJ (2006) Sustained delivery of plasmid DNA from polymeric scaffolds for tissue engineering. Adv Drug Deliv Rev 58(4):500–514
103. Andersen MØ, Dillschneider P, Kjems J (2013) The role of MicroRNAs in natural tissue development and application in regenerative medicine. In: RNA interference from biology to therapeutics. Advances in delivery science and technology, vol 15. Springer, New York, pp 57–78
104. Andersen MØ, Kjems J (2011) RNA interference enhanced implants. In: Active implants and scaffolds for tissue regeneration. Studies in mechanobiology, tissue engineering and biomaterials, vol 8. Springer, New York, pp 145–165

105. Wadagaki R, Mizuno D, Yamawaki-Ogata A, Satake M, Kaneko H, Hagiwara S, Yamamoto N, Narita Y, Hibi H, Ueda M (2011) Osteogenic induction of bone marrow-derived stromal cells on simvastatin-releasing, biodegradable, nano- to microscale fiber scaffolds. Ann Biomed Eng 39(7):1872–1881
106. Shah NJ, Hyder MN, Moskowitz JS, Quadir MA, Morton SW, Seeherman HJ, Padera RF, Spector M, Hammond PT (2013) Surface-mediated bone tissue morphogenesis from tunable nanolayered implant coatings. Sci Transl Med 5(191):191ra83
107. Nie H, Wang CH (2007) Fabrication and characterization of PLGA/HAp composite scaffolds for delivery of BMP-2 plasmid DNA. J Control Release 120(1–2):111–121
108. Zhao J, Zhang N, Prestwich GD, Wen X (2008) Recruitment of endogenous stem cells for tissue repair. Macromol Biosci 8(9):836–842
109. Li X, Liu X, Zhao W, Wen X, Zhang N (2012) Manipulating neural-stem-cell mobilization and migration in vitro. Acta Biomater 8(6):2087–2095
110. Radisic M, Malda J, Epping E, Geng W, Langer R, Vunjak-Novakovic G (2006) Oxygen gradients correlate with cell density and cell viability in engineered cardiac tissue. Biotechnol Bioeng 93(2):332–343
111. Wittenborn T, Nielsen T, Nygaard JV, Larsen EK, Thim T, Rydtoft LM, Vorup-Jensen T, Kjems J, Nielsen NC, Horsman MR, Falk E (2012) Ultrahigh-field DCE-MRI of angiogenesis in a novel angiogenesis mouse model. J Magn Reson Imaging 35(3):703–710
112. Nelson CE, Kim AJ, Adolph EJ, Gupta MK, Yu F, Hocking KM, Davidson JM, Guelcher SA, Duvall CL (2014) Tunable delivery of siRNA from a biodegradable scaffold to promote angiogenesis in vivo. Adv Mater 26(4):607–614
113. Wadee A, Pillay V, Choonara YE, du Toit LC, Penny C, Ndesendo VM, Kumar P, Murphy CS (2011) Recent advances in the design of drug-loaded polymeric implants for the treatment of solid tumors. Expert Opin Drug Deliv 8(10):1323–1340
114. Xie J, Wang CH (2006) Electrospun micro- and nanofibers for sustained delivery of paclitaxel to treat C6 glioma in vitro. Pharm Res 23(8):1817–1826
115. Yohe ST, Herrera VL, Colson YL, Grinstaff MW (2012) 3D superhydrophobic electrospun meshes as reinforcement materials for sustained local drug delivery against colorectal cancer cells. J Control Release 162(1):92–101
116. Calışkan N, Bayram C, Erdal E, Karahaliloğlu Z, Denkbaş EB (2014) Titania nanotubes with adjustable dimensions for drug reservoir sites and enhanced cell adhesion. Mater Sci Eng C Mater Biol Appl 35:100–105
117. Yarin AL (2011) Coaxial electrospinning and emulsion electrospinning of core-shell fibers. Polym Adv Technol 22(3):310–317
118. Korehei R, Kadla JF (2014) Encapsulation of T4 bacteriophage in electrospun poly(ethylene oxide)/cellulose diacetate fibers. Carbohydr Polym 100:150–157
119. Zhao L, Chu PK, Zhang Y, Wu Z (2009) Antibacterial coatings on titanium implants. J Biomed Mater Res B Appl Biomater 91(1):470–480
120. Wenke JC, Guelcher SA (2011) Dual delivery of an antibiotic and a growth factor addresses both the microbiological and biological challenges of contaminated bone fractures. Expert Opin Drug Deliv 8(12):1555–1569
121. Chen FM, Zhang M, Wu ZF (2010) Toward delivery of multiple growth factors in tissue engineering. Biomaterials 31(24):6279–6308
122. Chen M, Andersen MØ, Dillschneider P, Chang C-C, Gao S, Le DQS, Yang C, Hein S, Bünger C, Kjems J (2015) Co-delivery of siRNA and doxorubicin to cancer cells from additively manufactured implants. RSC Adv 5:101718–101725
123. Zheng Z, Yin W, Zara JN, Li W, Kwak J, Mamidi R, Lee M, Siu RK, Ngo R, Wang J, Carpenter D, Zhang X, Wu B, Ting K, Soo C (2010) The use of BMP-2 coupled—nanosilver-PLGA composite grafts to induce bone repair in grossly infected segmental defects. Biomaterials 31(35):9293–9300
124. Stevanović M, Uskoković V, Filipović M, Škapin SD, Uskoković D (2013) Composite PLGA/AgNpPGA/AscH nanospheres with combined osteoinductive, antioxidative, and antimicrobial activities. ACS Appl Mater Interfaces 5(18):9034–9042

125. He S, Xia T, Wang H, Wei L, Luo X, Li X (2012) Multiple release of polyplexes of plasmids VEGF and bFGF from electrospun fibrous scaffolds towards regeneration of mature blood vessels. Acta Biomater 8(7):2659–2669
126. Clark A, Milbrandt TA, Hilt JZ, Puleo DA (2014) Mechanical properties and dual drug delivery application of poly(lactic-co-glycolic acid) scaffolds fabricated with a poly(β-amino ester) porogen. Acta Biomater 10(5):2125–2132. doi:10.1016/j.actbio.2013.12.061, pii: S1742-7061(14)00011-7
127. Min J, Braatz RD, Hammond PT (2014) Tunable staged release of therapeutics from layer-by-layer coatings with clay interlayer barrier. Biomaterials 35(8):2507–2517
128. Shah NJ, Macdonald ML, Beben YM, Padera RF, Samuel RE, Hammond PT (2011) Tunable dual growth factor delivery from polyelectrolyte multilayer films. Biomaterials 32(26): 6183–6193
129. Basmanav FB, Kose GT, Hasirci V (2008) Sequential growth factor delivery from complexed microspheres for bone tissue engineering. Biomaterials 29(31):4195–4204
130. Yilgor P, Sousa RA, Reis RL, Hasirci N, Hasirci V (2010) Effect of scaffold architecture and BMP-2/BMP-7 delivery on in vitro bone regeneration. J Mater Sci Mater Med 21(11): 2999–3008
131. Yilgor P, Hasirci N, Hasirci V (2010) Sequential BMP-2/BMP-7 delivery from polyester nanocapsules. J Biomed Mater Res A 93(2):528–536
132. Yilgor P, Tuzlakoglu K, Reis RL, Hasirci N, Hasirci V (2009) Incorporation of a sequential BMP-2/BMP-7 delivery system into chitosan-based scaffolds for bone tissue engineering. Biomaterials 30(21):3551–3559
133. Sheehy EJ, Vinardell T, Buckley CT, Kelly DJ (2013) Engineering osteochondral constructs through spatial regulation of endochondral ossification. Acta Biomater 9(3):5484–5492
134. Keeney M, Pandit A (2009) The osteochondral junction and its repair via bi-phasic tissue engineering scaffolds. Tissue Eng Part B Rev 15(1):55–73
135. Cook JL, Kuroki K, Bozynski CC, Stoker AM, Pfeiffer FM, Cook CR (2014) Evaluation of synthetic osteochondral implants. J Knee Surg 27(4):295–302
136. Kon E, Filardo G, Robinson D, Eisman JA, Levy A, Zaslav K, Shani J, Altschuler N (2014) Osteochondral regeneration using a novel aragonite-hyaluronate bi-phasic scaffold in a goat model. Knee Surg Sports Traumatol Arthrosc 22(6):1452–1464
137. Filardo G, Kon E, Di Martino A, Busacca M, Altadonna G, Marcacci M (2013) Treatment of knee osteochondritis dissecans with a cell-free biomimetic osteochondral scaffold: clinical and imaging evaluation at 2-year follow-up. Am J Sports Med 41(8):1786–1793
138. Schleicher I, Lips KS, Sommer U, Schappat I, Martin AP, Szalay G, Hartmann S, Schnettler R (2013) Biphasic scaffolds for repair of deep osteochondral defects in a sheep model. J Surg Res 183(1):184–192
139. Wang X, Wenk E, Zhang X, Meinel L, Vunjak-Novakovic G, Kaplan DL (2009) Growth factor gradients via microsphere delivery in biopolymer scaffolds for osteochondral tissue engineering. J Control Release 134(2):81–90
140. Dormer NH, Singh M, Zhao L, Mohan N, Berkland CJ, Detamore MS (2012) Osteochondral interface regeneration of the rabbit knee with macroscopic gradients of bioactive signals. J Biomed Mater Res A 100(1):162–170
141. Mohan N, Dormer NH, Caldwell KL, Key VH, Berkland CJ, Detamore MS (2011) Continuous gradients of material composition and growth factors for effective regeneration of the osteochondral interface. Tissue Eng Part A 17(21–22):2845–2855
142. Andersen MØ, Nygaard JV, Burns JS, Raarup MK, Nyengaard JR, Bünger C, Besenbacher F, Howard KA, Kassem M, Kjems J (2010) siRNA nanoparticle functionalization of nanostructured scaffolds enables controlled multilineage differentiation of stem cells. Mol Ther 18(11):2018–2027
143. Melchels FPW, Domingos MAN, Klein TJ, Malda J, Bartolo PJ, Hutmacher DW (2012) Additive manufacturing of tissues and organs. Prog Polym Sci 37(8):1079–1104
144. Xu T, Zhao W, Zhu JM, Albanna MZ, Yoo JJ, Atala A (2013) Complex heterogeneous tissue constructs containing multiple cell types prepared by inkjet printing technology. Biomaterials 34(1):130–139

145. Jakab K, Norotte C, Marga F, Murphy K, Vunjak-Novakovic G, Forgacs G (2010) Tissue engineering by self-assembly and bio-printing of living cells. Biofabrication 2(2):022001
146. Kolesky DB, Truby RL, Gladman AS, Busbee TA, Homan KA, Lewis JA (2014) 3D bioprinting of vascularized, heterogeneous cell-laden tissue constructs. Adv Mater 26(19): 3124–3130
147. Lee JS, Hong JM, Jung JW, Shim JH, Oh JH, Cho DW (2014) 3D printing of composite tissue with complex shape applied to ear regeneration. Biofabrication 6(2):024103
148. Andersen MØ et al (2013) Spatially controlled delivery of siRNAs to stem cells in implants generated by multi-component additive manufacturing. Adv Funct Mater 23(45):5599–5607
149. Xu T, Rohozinski J, Zhao W, Moorefield EC, Atala A, Yoo JJ (2009) Inkjet-mediated gene transfection into living cells combined with targeted delivery. Tissue Eng Part A 15(1):95–101

Chapter 14
Guided Cellular Responses by Surface Cues for Nanomedicine Applications

Ryosuke Ogaki, Ole Zoffmann Andersen, and Morten Foss

Abstract Using surface cues to guide and ultimately control cellular responses is of paramount importance in numerous biomedical applications. Since cells react on feature sizes both on the micro and nanometer scale, these length scales are crucial parameters when designing new materials for applications in medicine e.g. for tissue engineering and drug delivery. Thus, variation of the features at the nanometer length scale is an integral part of nanomedicine research and development. In this chapter the interaction between biological systems and artificial materials will be addressed in general with focus on simplified model systems where only one or a few parameters are varied at two-dimensional (2D) surfaces. At first biomolecular adsorption/immobilization on surfaces will be addressed followed by a discussion of approaches to synthesize functionalized surfaces and the influence of such surfaces on cellular response. Some of the key parameters, which will be discussed in more detail, are topography, chemistry, and elastic modulus of the substrate. Even though there is a vast amount of published data it will become clear that there is still not a detailed understanding of the influence of these parameters on biosystems at the cellular level. Obviously, the degree of complexity increases when several of the surface cues are combined. The challenges in understanding the cellular responses in detail in real systems with the simultaneous variation of multiple parameters leads to the introduction of the field of high throughput screening of biomaterials.

Keywords 2D surfaces • Surface cues • Artificial niche • Cell behavior • Nano/micro fabrication

R. Ogaki • O.Z. Andersen • M. Foss (✉)
Interdisciplinary Nanoscience Center (iNANO), Aarhus University,
Gustav Wieds Vej 14, 8000 Aarhus C, Denmark
e-mail: foss@inano.au.dk

K.A. Howard et al. (eds.), *Nanomedicine*, Advances in Delivery Science and Technology, DOI 10.1007/978-1-4939-3634-2_14

14.1 Introduction

The search for routes to replace or repair malfunctioning tissue or organs in the human body has led to multiple breakthroughs within the field of medicine. Systems such as orthopedic implants, heart valves, intraocular lenses, stents and pacemakers are routinely used to sustain body function and to improve quality of life of patients worldwide. To enable continuous improvement of exiting systems and to enable further advances within the field of biomedical devices, a detailed understanding of the interaction at the nanometer scale between cells and their surroundings is needed. Understanding this interplay will allow researchers to tailor improved biomaterials to be used for e.g. guided stem cell differentiation and tissue engineering purposes. The research and development focused on novel materials with functionality depending on nanometer scale features is an integral part of the field of nanomedicine.

Fundamental cellular functions and behavior are governed by diverse and complex external stimuli from the surrounding microenvironment, known as the niche. The niche consists of extracellular matrix (ECM), soluble factors as well as other cells that are present and these provide a variety of physical and chemical factors. A multitude of these external stimuli essentially orchestrates the fate of cells by initiating a cascade of biochemical signals by interacting and consequently influencing the fundamental cellular functions including motility, proliferation, differentiation and apoptosis. A focal point of research has been aimed at deconstructing the complex relationship between external signals and cellular responses to increase our knowledge in understanding the cellular regulatory mechanisms. Since mammalian cell niches are challenging to control experimentally, the properties of niche regulatory components have been extensively explored in vitro using a *reductionistic approach* and a toolbox of artificial (bio) materials and fabrication strategies for creating the artificial cues. Here, the surface and interfacial properties of the materials become of paramount importance for studying the effect of physical, chemical and biomolecular 'cues' on cellular function and fate, since it is the surface that first comes in contact with the cells and regulatory processes are initiated at the interface between the material and the cells.

This chapter describes the surface and interfacial aspects of artificial materials for guiding cellular behavior. Current investigations are, to a large extend, based on trial and error but continuous research within this area holds promise to gradually expand the nanomedicine tool-box and ultimately enable rational design of surfaces for specific purposes. The focus is two-dimensional (2D) nanoscale structures thus excluding the field of porous materials/scaffolds (3D). The effect of various artificially constructed external material surface cues including chemical, biomolecular and physical properties as well as a combination of multiple cues that are influential to the cell behavior are highlighted. It is important to note that the cellular responses obtained from experiments using 2D and 3D systems do not always correlate with one another and recent growing evidences indicate that the 3D experimental platforms depicts more of a native cellular niche. Interested readers should refer to excellent reviews [1–7] which discuss the 2D and 3D culture systems and the impact on cell behavior. Nevertheless, 2D experimental platforms are highly advantageous as a simplified approach in determining the individual role of cues in a well-defined manner on cell behavior via a repertoire of material fabrication processes and commonly employed engineering strategies for generating 2D systems. Recent emergence of high throughput screening (HTS) platforms is also highlighted; as such platforms have now become a proven tool for rapidly and systematically probing the effect of multiple nanoscale niche components.

14.2 The Role of Surfaces in Biorecognition and Cell Responses

14.2.1 Biomolecules at Interfaces

Biorecognition by the cells interacting with an artificial material depends on the type, orientation and conformation of the biomolecules presented on the surface. The state of how biomolecules reside on the surface is influenced by the properties of the surface [8]. This successive interfacial event occurs as, for example, when a medical device is implanted into the body (Fig. 14.1). Among a variety of biomolecules, adsorption of proteins to surfaces in particular has been extensively studied since the cell adhesion to the surface and the subsequent cellular response depends on the availability of specific cell binding peptide sequences presented by the proteins. At constant temperature and pressure, protein adsorption to surfaces occurs by lowering the Gibbs energy, *G* of the system with the contributions from enthalpic (*H*) and entropic (*S*) effects; $\Delta_{ads}G = \Delta_{ads}H - T\Delta_{ads}S$. Numerous enthalpic and entropic effects including van der Waals and electrostatic forces, hydrogen bonding, hydrophobic interactions and surface packing restrictions via steric and excluded volume effects determine protein adsorption [9, 10]. The intricacy of the protein adsorption becomes further evident as some of these effects can be simultaneously altered by the external parameters of temperature, pH and ionic strength: [11, 12] Diffusion of proteins towards the surface increases with increasing temperature and entropic gain is expected to occur as the water molecules and ions are released upon protein adsorption. Changes in the pH or ionic strength alters the electrostatic state of the proteins [13]. At the isoelectric point and/or at high ionic strength, protein-protein repulsion is minimized leading to higher protein adsorption as a result of higher surface packing density, as well as causing the proteins to aggregate.

Protein adsorption on surfaces occurs in several consecutive steps and the concentration of proteins in solution differs from those on the surface with time, depending on the properties of the surface, types of protein involved and their environment [8, 14]. This means that the surface exposed to a single or a few protein types may be very different from those surfaces exposed to multiple types of proteins. In the mixed protein system (Fig. 14.2a), certain types of proteins initially

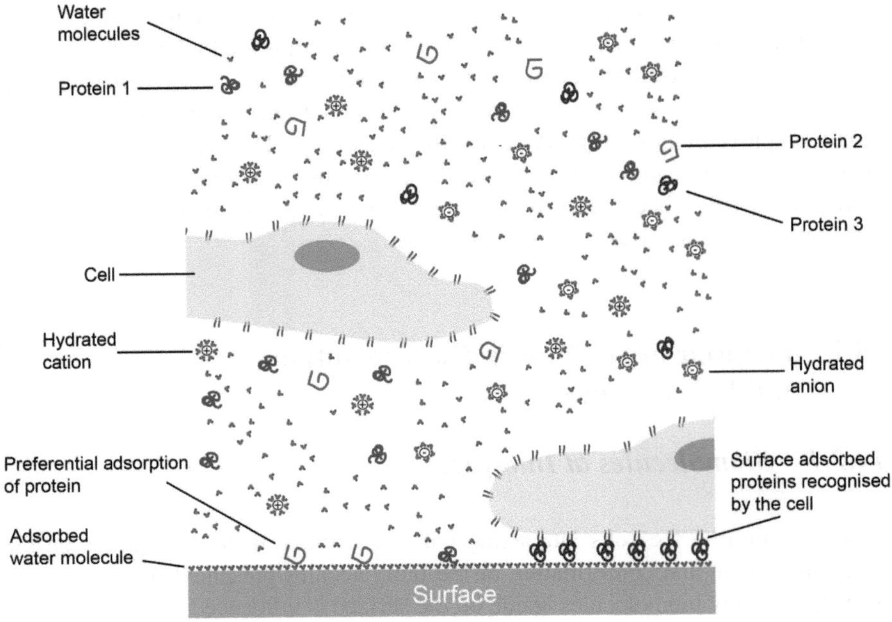

Fig. 14.1 Schematic diagram depicting the interface between the (implant) material surface and the biological milieu in vivo (adapted from Kasemo [8]). Successive surface adsorption events occur as water molecules first reach the surface, followed by proteins and finally the cells. The surface-cell interaction depends on the properties of the surface

adsorb onto the surface almost immediately after exposure. Depending on the state and type of the adsorbed proteins, displacement and reorganization by other types of protein may occur at the interface over longer adsorption time periods of minutes to days, a phenomenon known as the Vroman effect [14]. Once adsorbed, each protein on the surface could adopt orientational and conformational changes, depending on the biological and/or local environment (Fig. 14.2b and c). The final structure of the adsorbed protein layer also depends on the rate at which the proteins adsorbed onto the surface and the kinetics of protein adsorption is largely governed by the bulk concentration and protein size. The rate of adsorption is higher with higher bulk concentration and the neighboring proteins may sterically prevent conformational changes, whereas at a lower bulk concentration, the rate of adsorption is lower, leading to longer time for the given proteins to undergo conformational changes on the surface [14]. The size of the protein also contributes to the kinetics and arrangement of proteins on the surface, where a higher number of smaller proteins adsorb onto the surface compared to larger proteins because of higher rates of diffusion, but larger proteins typically possess higher binding affinity towards the surface due to having a larger potential contact area [12]. Preferential orientation of adsorbed proteins on the surface originates from its complex structure, with different regions of proteins presenting different properties depending on the local amino

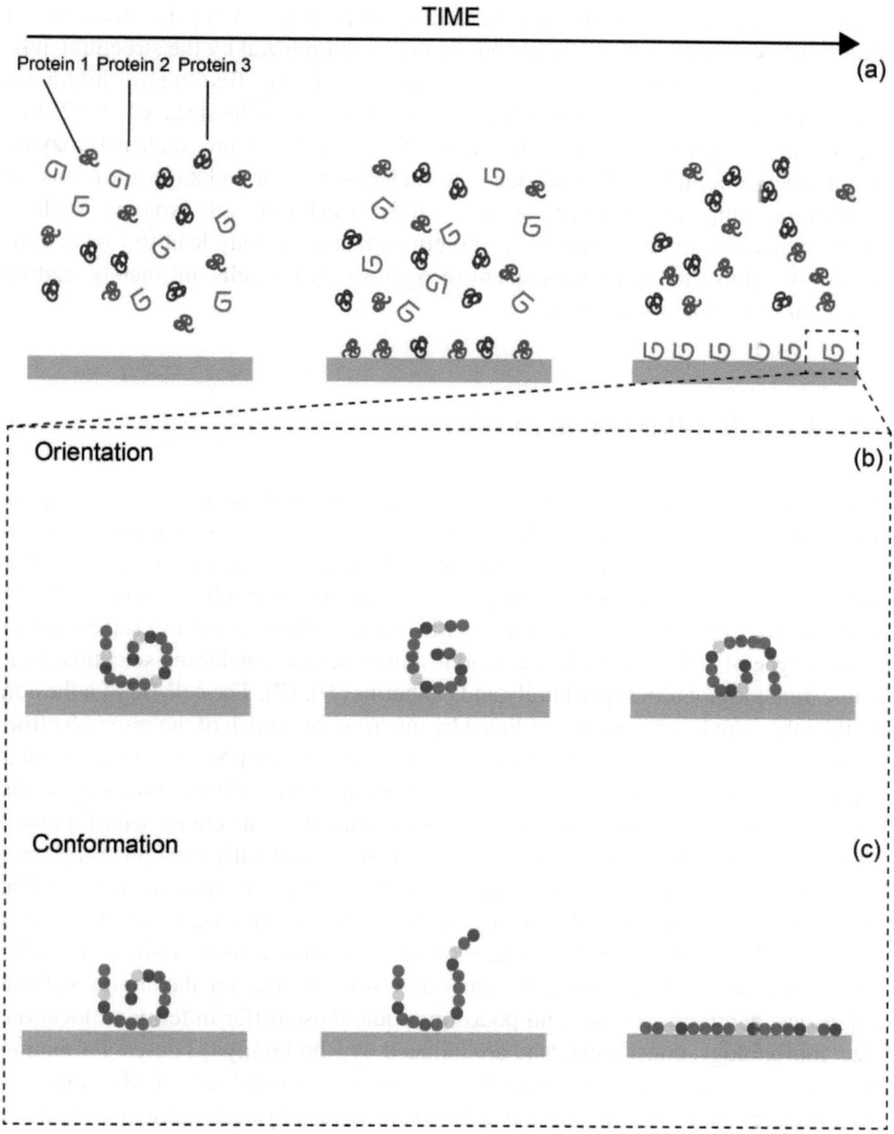

Fig. 14.2 Protein adsorption on a surface with time (adapted from Castner and Ratner [14]). In a mixed protein solution, displacement of the initially adsorbed protein by another type can occur over a prolonged period of time (**a**). Once a protein is adsorbed on a surface, orientation (**b**) and conformation (**c**) change may take place

acid compositions. If the adsorbed protein is structurally stable and the adsorption is driven by electrostatic interaction between the protein and the surface, the orientation of the protein could change further if a sufficient surface protein density is reached. High surface density leads to close proximity between the neighboring adsorbed proteins exposing the electrostatically repulsing domains leading to reorientation towards a less repulsive arrangement (Fig. 14.2c). Conformational change (denaturation) of proteins at adsorption is epitomized by the structural ability of the protein and the result of the difference in the free energy minimum between the proteins in solution to the proteins adsorbed on the surface [11, 12, 14]. Small and rigid proteins, classically referred to as 'hard proteins' such as lysozyme and β-lactoglobulin have less tendency to undergo structural change upon surface adsorption compared to larger and less tightly structured 'soft proteins' such as albumin and immunoglobulins [15]. Protein denaturation may lead to loss of biological function and hinder subsequent recognition by the cells, ultimately leading to e.g. an inflammatory response.

14.2.2 Cell Adhesion on Surfaces

In vivo most cells (except non-adherent cells such as blood and tumor cells) require adhesion to a surface of extracellular matrix (ECM) as well as surrounding cells in order to maintain and regulate the functionality and development of multicellular organisms. Cells interact with two types of surfaces in vivo; cell-cell and cell-ECM interactions. Cell-cell adhesion is a specific process where cells bind selectively to another type of cells mediated by cell adhesion molecules, including selectins, integrins, immunoglobulin super family and cadherins [16, 17]. The cell-ECM adhesion on the other hand, is primarily mediated by integrins and much of the research effort in the recent years has been focused on unraveling the composition and molecular architecture of integrin adhesions as well as its adhesion related signaling events [18–20]. Integrins are heterodimeric transmembrane proteins composed of α and β subunits (most common being $\alpha_5\beta_1$ and $\alpha_v\beta_3$) that specifically recognize different short amino acid chain (e.g. the often mentioned Arg-Gly-Asp sequence, often termed RGD) motifs present in many of the ECM proteins. Each set of subunits possess a distinct capacity for bi-directional transduction (outside-in and inside-out). Characteristic features in cell-matrix adhesions include focal complexes, focal adhesions, hemidesmosomes and podosomes and these differ in terms of location, size, morphology, constituents and are induced by Rho family GTPases. Formation of focal adhesion has been the most studied and characterized type of adhesions. An initial step in the formation of focal adhesion involves physical interaction between the ligand and integrin as its conformation changes into an active upright state, opening the intracellular domains of the integrin for the subsequent attachment to cytoplasmic proteins including talin, vinculin and focal adhesion kinase (FAK) that connects to the actin cytoskeleton [21]. This further leads to the assembly of focal adhesion by recruiting additional components leading to further clustering of integrin complexes promoted by the binding of vinculin to talin and reinforcement of the integrin-cytoskeleton bonds. Molecular complexity in the formation of focal adhesion is evident, with integrin-mediated adhesion alone consisting of ~160 distinct components including actin regulators, adaptor proteins and signaling molecules with ~700 links to each other [19].

14.3 Fabrication Strategies for the Generation of Artificial Surface Cues

14.3.1 Chemical and Biomolecular Patterning

In the context of conducting fundamental cell studies, chemical and biomolecular surface patterning involves precise positioning and presentation of chemical functional groups and biomolecules onto a substrate over micrometer to nanometer lengths scale. Generally the most straightforward approach is to fabricate a pattern consisting of areas containing a single chemical/biomolecular moiety which the cells can interact with, while the surrounding regions are non-fouling/bioresistant. Poly- and oligo-ethylene glycol (PEG [22, 23] and OEG [24, 25], respectively) containing materials are often used for creating such bioresistant regions, though other types of materials such as polysaccharides [26] and zwitterionic polymers [27, 28] have been found to provide robust bioresistance. Inspired by the microelectronics fabrication methods used for preparing e.g. microprocessors [29], MEMS [30], and NEMS [31], the early surface patterning approaches have been to employ 'top-down' instruments such as electron beam (e-beam) lithography [32] and photolithography [33]. In contrast, 'bottom-up' methods such as colloidal lithography (CL [34]), block copolymer lithography (BCL [35]) and block copolymer micelle lithography (BCML [36]) have been developed for the fabrication of micro and nanoscale surface patterns in an autologous manner typically via self-assembly processes. A number of specific advantages and drawbacks exist in both fabrication strategies; for example 'top-down' approaches possess flexibility over feature shape, pattern size and positioning. However, serial and multi-step fabrication as well as access to a clean room is often required which can be costly. On the other hand, 'bottom-up' approaches can be straightforwardly conducted 'on a bench' at a low cost and over large areas but suffer from a lack of versatility in the geometrical features that can be patterned as well as the frequent presence of pattern defects and order at a longer length-scale. In almost all cases, self-assembled monolayers (SAMs) such as thiols [37, 38], silanes [37, 39], phosphates [37, 40] and polyelectrolytes [41] are used in combination with lithographic techniques either; (1) directly, (2) indirectly or (3) via material guided self-assembly. The direct patterning of SAMs typically involves using SAMs as an 'ink' to spatially position SAMs onto a substrate and, subsequently, backfill the 'clean' non-printed areas with bio-resisting SAMs. Examples include micro- and nano-contact printing (μCP [42] and nCP [43]), dip and polymer pen nanolithography (DPN [44, 45] and PPN [46]) and ink jet printing [47]. The indirect patterning of SAMs involves temporary pre-patterning of surfaces with polymer resists to create a surface pattern of exposed/unexposed substrate surface. The surface is subsequently immersed in the solution with components for generating SAMs which will self-assemble onto the substrate-exposed areas and, finally, the polymers are subsequently removed and the surface passivated. The pre-patterns can be fabricated at sub-micro and nanometer scale using photolithography [41, 48], e-beam lithography [49] and nanoimprint lithography (NIL [50]) using polymer resists. In comparison to the above approaches, the material-guided self-assembly takes advantage of the material-specific self-assembling nature of SAMs (e.g. thiols on gold and silanes on oxide surfaces), where surfaces are first patterned with different materials and sequentially incubated in different types of SAMs. Patterning of different materials can be carried out by, for example, photolithography [51] and CL [52] using physical vapor deposition (PVD) and/or chemical etching to make e.g. a stable chemical contrast or height variation.

For biomolecular surface patterning, biomolecules can be immobilized either directly (i.e. physically adsorbed) onto a surface or via specific conjugation. Physical adsorption of biomolecules is commonly carried out via electrostatic interactions between biomolecules and substrates bearing opposite charge, or via hydrophobic interaction by rendering the substrate hydrophobic using SAMs, depending on the substrate wettability. Specific conjugation via covalent (e.g. N-hydroxyl succiimide or NHS coupling between primary amines of biomolecules to a carboxylic acid moiety) or affinity-ligand based interactions (e.g. biotin-streptavidin) permits presentation of biomolecules in its native state without denaturation or displacement, while orientation can be controlled by targeting a specific amino acid sequence on the protein for conjugation.

Although patterning and presentation of more than one type of chemical moieties can be conducted on a surface with relative ease, the patterning of multiple biomolecules may become a challenging affair, unless direct immobilization methods are employed. For generating multiple protein patterns via indirect or material guided self-assembly routes, a chemical moiety that can be couple specifically to one type of biomolecules must be patterned while minimizing/avoiding non-specific bioadsorption from another type of biomolecule which is to be patterned on the surface. This implies that the general covalent coupling method such as NHS cannot be straightforwardly used to pattern multiple proteins, thus different affinity coupling moieties with a non-fouling background is required [53–55]. For further details in chemical and biomolecular patterning, see further Refs. 55–60.

14.3.2 Topographical Patterning

Surface topography, or in other words the structural landscape of a surface, is a physical cue that has already found applications within the field of biomaterials. It is for instance well established that rough surfaces with features on the micro and nanometer scale are able to facilitate accelerated bone ingrowth for orthopedic and dental implants [61] as compared to smooth controls. Such surfaces are typically prepared by blasting the surface with an appropriate blasting media yielding a random structure composed of valleys and peaks. While random structures are of interest application wise, the trend with respect to more fundamental studies is moving towards employing a higher degree of structural coherence when investigating cellular responses. Such systems are frequently composed of ordered, semi-disordered or disordered structures with well-defined feature sizes in the micro and/or

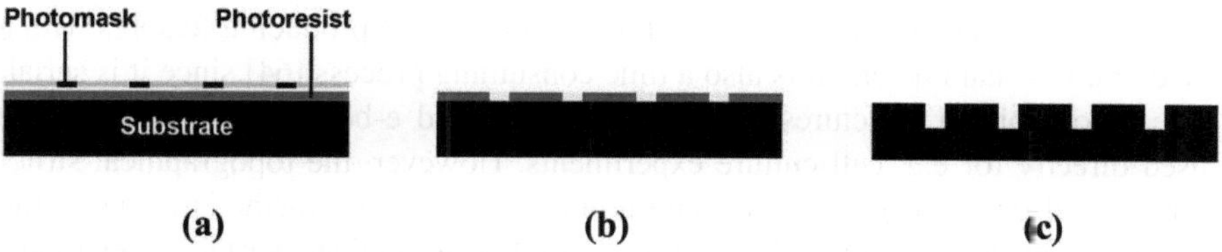

Fig. 14.3 Schematic overview of the photolithography process using a negative resist. (**a**) The substrate is coated with a thin layer of photoresist(s) (*red*) and a pattern from a pre-designed photomask is then transferred to the resist layer by irradiation with ultra violet (UV) light through the photomask. Molecules of the exposed regions of the resist react to form cross-linking bonds, which stabilize the material. (**b**) Following UV exposure, the unreacted resist is removed and the underlying substrate is exposed. The remaining resist protects the underlying substrate and it is now possible to selectively modify the substrate using different methods such as e.g. physical vapor deposition or reactive ion etching (RIE). (**c**) Final surface topography prepared using RIE (Color online)

nanometer size range [62, 63]. A wide range of methods can be utilized when synthesizing such cell culture substrates, most of which are based on the "top-down" approach as previously mentioned. One of the most frequently used techniques for producing well defined topographical features in the micrometer range is photolithography. The basic principle behind this method is illustrated in Fig. 14.3 and relies on using a photoactive polymer, known as a resist, to coat a substrate material, e.g. silicon.

The resist at the surface is subsequently selectively exposed to electromagnetic radiation through a photomask which screens part of the resist. Two main types of resist exist known as "negative" and "positive" which indicate if the resist is cross-linked or destabilize, respectively, by the exposure. Typically radiation in the UV or deep UV range is used as the short wavelengths allow for a higher ultimate resolution (diffraction limited) and, by that, higher feature resolution. Upon radiation the pattern transferred from the photo mask to the resist is developed, thus, exposing the underlying substrate at the intended areas. Subsequently, the underlying substrate can be modified and this process is typically based on either a reactive ion etching process, through which substrate material is removed, or on a physical vapor deposition process enabling addition of a material onto the substrate surface. While several modifications to standard photolithography has been developed, which allows for a very high lateral feature resolution of ~40 nm [64, 65], these are extremely costly and mainly employed in the semiconductor industry and most commercially available methods have a maximum resolution of approximately 250 nm. The advantage of photolithography is the high throughput nature of the process when the mask has been produced. Even though the production is relatively straight forward, access to a clean room is required. The mask does, however, represent a significant expense and if lateral changes to the structures are needed, a new mask must be obtained. If features in the lower nanometer size range are needed, the process of electron beam (e-beam) lithography can be employed. This method also utilizes a resist layer, but rather than using a mask to transfer a pattern to the substrate surface, it uses an electron beam to "write" the structures directly into the resist. This is known as a direct write method and while it is capable of producing features with a size of a few nanometers, it is also a time consuming process [64] since it is serial. The topographical structures produced by photo and e-beam lithography can be used directly for e.g. cell culture experiments. However, the topographical structures can also be utilized as masters for replica molding. This method belongs to the family of soft lithography and the basic principle is to replicate surface topographical features using a polymeric material. Several different polymeric materials can be used for this process but the material most frequently employed for cell culture studies is based on polydimethylsiloxane (PDMS). Typically the liquid polymer material is cast onto the surface structures and set to cure. Subsequently, the polymer material can be separated from the master, resulting in a negative replica of the surface structures. As the master structure can be reused, replica molding is a very cost effective route for obtaining a large number of samples. With respect to the structural integrity of the transferred pattern, the elastic modulus of the material used for the process greatly affects the lower limit for the minimum feature resolution. Utilizing a chemically modified PDMS composition, topographical replication of carbon nanotubes with a vertical dimension of ~2 nm has been achieved [66]. This special material is referred to as "hard" PDMS and has an elastic modulus of ~9 MPa [67].

14.3.3 Modulation of Mechanical Surface Properties

When designing substrate surfaces with varying mechanical properties, e.g. viscoelasticity, it is important to keep in mind that the simultaneous introduction of any chemical changes also has the potential to affect the cellular response. With respect to elucidating the fundamental mechanisms involved in cellular mechano-sensing, it is therefore important to employ model systems that present varying mechanical properties with a minimal degree of change to other properties such as e.g. surface chemistry, wetting properties and topography. In the literature there are two main approaches that have found applications for research purposes. These are; (1) varying the degree of cross-linking of polymeric materials and (2) geometric design of the surface topography of elastic materials. With respect to the approach of varying the chemistry of the substrates, the two systems that are most frequently employed are based on polyacrylamide hydrogels or polydimethylsiloxane (PDMS) elastomers.

Polyacrylamide hydrogels are prepared by copolymerization of the two components; acrylamide and N,N′-Methylenebisacrylamide (bis-acrylamide). During this reaction the acrylamide monomer will form polyacrylamide chains which are cross-linked to each other by the bis-acrylamide cross-linker component, thereby forming an interlocking polymer network [68]. Depending on the amount of bis-acrylamide used in the preparation the number of cross-linking events can be varied which in turn determines the elastic properties of the final hydrogel [69]. Increasing the concentration of the bis-acrylamide component will increase the elastic modulus of the material. The viscoelastic properties of the polyacrylamide system can also be designed by varying the ratio between the acrylamide monomer and the bis-acrylamide cross-linker. Using this, it is possible to vary the loss modulus (a materials ability to retain a specific strain as a function of time) while keeping the storage modulus (elastic properties) of the material constant between the different substrate formulations [70]. With respect to the PDMS system the classical method for modulating the mechanical properties is also to vary the amount of cross-linker used in the preparation, similarly to the acrylamide system. This will result in varying degrees of cross-linking and produces materials with varying elastic modulus. The use of alternating curing periods or curing temperature have also been employed for the production of cell culture substrates [71]. This approach is, however, linked to the risk of time dependent alterations to the mechanical properties as the curing reaction for many PDMS formulations proceeds at room temperature. With respect to PDMS based systems, Fu et al., have described an alternative for varying the effective mechanical properties without affecting the chemical composition of the material. In their study, a geometrical approach, where closely spaced pillars were prepared by replica molding, is described. By keeping the chemical composition constant but varying the height of these pillars, substrates were generated showing varying pillar deflections in response to cellular traction forces [72]. Similarly to the challenges associated with varying chemical compositions, varying pillar deflections between the different substrates will ultimately change the local topographical appearance of the substrate surrounding the individual cells. This, in turn, might also hold the potential to affect the cellular response.

14.4 Surface Cues for Controlling Cell Behavior

Enormous complexity of the cell-matrix interaction is evidently present in a typical in vivo nanomedicine application; it is far from a trivial task in deconstructing numerous direct and indirect signaling interactions that can activate or inhibit targeting molecules which ultimately determines the cellular fate. One might, as Geiger et al. [19] expressed, attempt two approaches for unraveling this multifarious adhesion signaling events: (1) investigate individual cellular adhesion mechanism in detail, integrate and respond to the signal given by the cells, or (2) initiate specific cellular response by altering the properties of the cell interacting surfaces. With substantial progress made in the areas of materials science and nanotechnology in the recent decades, reconstruction and mimicry of in vivo niches down to the nanometer length scale has become possible for measuring cell-surface interactions. Furthermore, discovery of numerous synthetic materials have allowed investigating beyond the in vivo mimicry for identifying and seeking new cellular behavior on novel materials that are absent in vivo. Single and often multiple cues can be presented on a surface with varying physical, chemical and biomolecular properties as described in the previous sections. Physical properties may include; surface roughness/topography, viscoelasticity as well as shape, size, and spacing of topographical features. Chemical and biomolecular properties include; composition (i.e. basic chemical moieties, bare material composition) and material-specific surface chemistry (i.e. metals, polymers and their associated surface functionality), amino acids, proteins, carbohydrates and lipids. Experimental conditions and the corresponding cell response to the interfacial parameters should be prudently designed and accounted for. For example, topographical effects on cellular behavior could be interpreted in terms of wettability (i.e. surface energy) or the roughness of the surface investigated. In addition, cell culturing conditions such as presence/absence of serum and the concentration, type of growth media, cell seeding density, durations of study and type of cells/passage number may have a dramatic effect on the cell behavior. The following sections will provide critical examples of studies conducted in the past and recent and their importance in determining relationships between artificial instructive cues and cellular behavior.

14.4.1 Chemical and Biomolecular Cues for Guided Cellular Behavior

14.4.1.1 Chemical Cues

The early studies into the effect of surface chemical properties on cellular behavior have been conducted directly on material surfaces such as polymers to determine the relationship between the polymer surface properties and cell behavior. For example, van Wachem et al. studied adhesion and proliferation of human endothelial cells on a range of polymers with varying wettability and surface charge and observed polymers with moderate wettability showed highest number of cells adhered [73, 74]. The concentration of functional groups within a polymer can be readily controlled by copolymerization [74, 75] and the surface chemical groups can be quantitatively determined by surface analytical techniques such as X-ray photoelectron spectroscopy (XPS) and Fourier-transform infrared spectroscopy (FTIR). While polymer surfaces provide an ideal platform for biomaterial research and plays an important role in elucidating material-specific interaction with cells, it is difficult to readily obtain precise control over other surface parameters (e.g. topography) that may also contribute to cellular behavior depending on the type of polymers used. To determine the sole effect of chemical groups on cellular behavior, self-assembled monolayers (SAMs) of thiols or silanes have been typically employed on flat surfaces. For studies using a single type of functionalized alkane thiols, general trends in the cell number has been observed regardless of different cell types, where, for example the growth of bovine aortic endothelial cells (BAEC) [76] increased in the order of $OH < COOCH_3 < CH_3 < COOH$, glioma cell line C6 [77] generally increased in the order of $SH < CH_3 \leq OH < NH_2 < COOH$, and MCF-7 cancer cells [78] increased in the order of $SH \leq OH \leq CH_3 < NH_2 \leq COOH$. These studies have been conducted with the presence of serum, where some studies have highlighted the correlation between chemical functional groups and the level of elutability of adsorbed albumin and adsorption of cell adhesive proteins such as fibronectin [76, 79, 80]. Similar trends in the cell behavior such as cell proliferation and spreading on various functional groups have also been observed using functional silanes for a range of cell types including fibroblasts [81–83] and stem cells [84, 85]. In particular, positively charged NH_2 surfaces led to highest viable human mesenchymal stem cells (hMSCs) adhesion and increase in the osteogenic differentiation pathway, while negatively charged COOH surfaces showed rounded morphology with chondrocytic differentiation [84]. As for neural stem cells (NSCs), SO_3H surfaces showed highest cell contact area and favored oligodendrocyte differentiation, while COOH, NH_2, SH and CH_3 guided NSCs to differentiate into neurons, astrocytes and oligodendrocytes [85].

Limited studies into the sole direct effect of micro-patterned chemical moieties on cell behavior have been reported to date. Basic studies into the baby hamster kidney (BHK) cell adhesion and proliferation were carried out on hydrophilic/hydrophobic silane patterns on a quartz surface using photolithography showed higher cell number and proliferation on the patterned hydrophilic regions [86]. Similar cell adhesion results were obtained on osteoblasts [87] using alkyl thiols with different terminal functional groups patterned via photolithography. Preferential cell growth were observed with hierarchical preference of $CH_3 < OH < COOH$. However, the study involved pre-adsorption of fibronectin onto these moieties and the results indicated that the focal contact formation by the cells varied in the same order and depend on the adsorbed protein density and cell-cell interactions.

The direct effect of nanoscale chemical patterns on cells has been studied more extensively. In particular, several cell adhesive short peptide motifs such as RGD [88] have been patterned in nanoscale to study cell adhesion, spreading, focal adhesion assembly, migration and polarization using block copolymer micelle lithography (BCML) and MC3T3 osteoblasts [89–92]. These studies have demonstrated that cyclic RGDfK peptide-functionalized gold nanoparticles that can only interact with a single integrin per particle spaced less than 73 nm showed cell adhesion and spreading, whereas spacings larger than 73 nm or with a random peptide sequence have shown limited adhesion and spreading. This critical finding of the ligand spacing threshold of ~70 nm also applied to ordered vs disordered pattern, where nanoscale disordering with global average ligand spacing of larger than 70 nm inhibited cell adhesion [92]. Controlled differentiation of human mesenchymal stem cells (hMSCs) have been demonstrated on nanoarrays of COOH, NH_2, CH_3 and OH with feature size of ~70 nm having defined spacings ranging from 140 to 1000 nm [93]. Compared to standard tissue culture polystyrene (TCPS), CH_3 nanopatterned arrays enhanced the expression of CD29, CD73, CD90, CD105 and CD166 and maintained an enhanced expression of phenotypic markers after 28 days, highlighting that simple chemical nanoarrays can promote and maintain MSC phenotypes in vitro without exogenous biological factors and heavily supplemented cell media.

14.4.1.2 Biomolecular Cues

The effect of biomolecular cues on cellular behavior has been researched intensely over the last several decades and continues to be an important area in the field of nanomedicine. Analogous to the studies conducted on simple chemical moieties/cues as described in the previous section, the early studies into the effect of biomolecules on cellular behavior were carried out on flat surfaces. The studies in the 1970s and 1980s in particular involved making basic observations of adhesion, motility and growth of different types of cells on ECM components physically adsorbed on surfaces. With a repertoire of surface patterning strategies becoming readily accessible in the 1990s, the research focus shifted towards investigating the micro and nano-scale effect of surface patterned biomolecules. One of the notable studies into the effect of geometric shapes patterned with adhesive ECM protein was demonstrated by Chen et al. [94, 95], where the behavior of endothelial cells were switched from growth to apoptosis by changing the size of fibronectin adhesive islands. Micro meter scale ECM geometrical limits on cell spreading and adhesion were studied further by Lehnert et al. [96] using mouse B16F1 melanoma cells, buffalo rat liver cells and NIH-3T3 fibroblasts. They discovered that all of the studied cell types regulated the number of focal adhesions formed to the range of surface coverage and the spacing of the fibronectin pattern, with maximum distance of 25 μm between the adhesive surfaces with fibronectin coating for cells to achieve half-maximal spreading. Dot area of as low as 0.1 μm^2 was capable of inducing focal adhesion formation but did not support spreading when the inter-dot spacing was increased more than 5 μm. The effect of pattern shape was also investigated on stem cells. McBeath et al. [97] employed similar approach from the previous study [94] to investigate the lineage commitment of human mesenchymal stem cells (hMSCs). The study have importantly found that the cell-adhesive pattern that allowed cells to spread on the surface, steered hMSCs into the osteogenic lineage, but the pattern that restricted the spreading led hMSCs into the adipogenic lineage and the cell shape, cytoskeletal tension and RhoA signaling were found to be integral to the lineage commitment of the hMSCs.

Tseng et al. [98] investigated how ECM proteins affect the spatial organization of intercellular junctions by constraining the location of epithelial cell-fibronectin adhesion, by patterning fibronectin in various shapes. It was found that intercellular junctions were displaced due to a large tensional force when they were close to fibronectin, whereas the junctions remained stable with low tensional force in the area without fibronectin, highlighting the importance of ECM in morphogenesis. Micro-patterning has also been used to study cell motility. Bailly et al. [99] investigated the regulation of lamellipodial protrusion during chemotaxis by using 10 μm line patterns of vitronectin surrounded by non-cell adhesive regions. The cells only adhered on the vitronectin lines but upon exposure to epidermal growth factor (EGF), the cells were able to protrude into the non-cell adhesive regions. However, a stable protrusion was only achieved by establishing a firm contact to the substrate. Kumar et al. [100] and Kushiro et al. [101] demonstrated controlled directional migration of fibroblast, endothelial and epithelial cells by generating asymmetric 'tear drop' shaped adhesive islands of fibronectin. Although the same 'tear drop' micro-patterns were used in both studies, differing directional bias were observed due to the disparate migration properties of the cell types and the difference in the environmental cues presented to the cells. The importance of cell-cell interactions in cell motility were further highlighted by micro-patterning a combination of cell-ECM (collagen IV) and cell-cell (E-cadherin-Fc) adhesion cues to determine the motile behavior of epithelial cells on alternating micro-patterned lines [102]. While collagen IV was found to be essential for cell migration and E-cadherin-Fc was not, when both biomolecules were present, the direction of cell migration as a result of reduced lamellipodial activity was promoted by E-cadherin-Fc in a concentration-dependent manner without affecting the migration rate, indicating the importance of crosstalk between cell-cell and cell-ECM adhesion complexes upon migration. Beyond motility, microcontact printing was used to investigate the effect of spatial distribution of ECM on cell shape anisotropy and cell polarity in the context of cell division. The study concluded that ECM controls the location of intracellular actin dynamics at the membrane which in turn affect the orientation of the cell divisional axis [103].

The impact of nanoscale effect of biomolecules on cell behavior has also been studied in recent years in attempt to gain further understanding of the underlying mechanisms of cell function and fate. Colloidal lithography has been used to identify the nanoscale geometric effect of cell-ECM and cell-cell interaction processes by quantifying fundamental cellular processes including cell adhesion, spreading, focal adhesion formation and maturation and adherence junction formation [104–106]. In particular, the focal adhesion formation and development was found to vary depending on the ECM protein type, where a single focal adhesion was able to bridge across the gap between patches of vitronectin with sizes of 200 nm but not for fibronectin patches, until the patch size reached 500 nm. This may be due to the differences in the force which cells can exert on these proteins, with fibronectin being more prone to unfolding and mechanically more pliable than vitronectin [106]. Such approach was extended to study the effect of nanoscale geometry of the adhesive patches on epidermal stem cell differentiation, where it was found that such nano-scale patterns dictate the stem cell fate decision via the control of cell shape and AP-1 transcription activity, an important transcription factor controlling the expression of epidermal terminal transcription factors [107]. Giam et al. [108] also studied the effect of fibronectin feature size on human mesenchymal stem cell (hMSC) differentiation with and without the presence of osteogenic inducing media. By monitoring the FAK phosphorylation level, it was found that 300 nm fibronectin patterned features without osteogenic inducing media was more effective at inducing osteogenesis than the unpatterned surface with osteogenic inducing media, indicating the importance of programmed nanoscale feature sizes of ECM proteins for controlling the fate of hMSCs. Recent studies have further highlighted the importance of nanoscale ligand geometry in immune system responses [109, 110]. For example, using the block copolymer micelle lithography (BCML) approach, Delcassian et al. [109] generated CD3 and CD16 antibody patterns to monitor the effect of ligand spacing on the activation levels of T cells and Natural Killer (NK) cells. The activations were assessed in terms of membrane-localized phosphotyrosine for T cells and the size of contact area for the NK cells, and the study showed decreased cell response with increasing ligand spacing in both cell types, with the threshold activation at the ligand spacing being 69 nm and 104 nm for T cells and NK cells, respectively. All together, these studies conclusively epitomize the importance of nanoscale biomolecular patterned cues for controlling cellular behavior in many biological settings.

14.4.2 Topographical Cues for Guided Cellular Behavior

It has been known for many years that topographical structures influence cellular response. Early on, already in the 1940s, Weiss demonstrated that cells are guided by directed structures and the term "contact guidance" was established [111]. Fabrication of topographical structures in the micrometer range for cell culture followed where the influence of topographical patterning on several cell types was examined, pioneered by the group of Curtis et al. [112, 113]. In 1964 it was shown that the cells align along the direction of a few micrometer wide grooves. Since then, a vast number of articles and reviews have addressed the influence of topography on cellular response. Following the expanding range of possibilities to synthesize and characterize structures with features at the nanometer scale (nanostructures), the focus has shifted to exploring this length-scale in the recent years [114, 115]. One focus area has been application of such nanostructures in nanomedicine.

One of the key issues has been to distinguish between the influence of topography on the adsorbing biolayer (e.g. denaturation and orientation of proteins) as already discussed in this chapter from the direct influence of topography on the attached cells. Multiple studies have shown that the amount [116] and the orientation/denaturation of the proteins adsorbed [117, 118] depend critically on nanoscale features. The change in conformation may change the presentation of binding sites of the protein either exposing or hiding such sites [119] thereby potentially also changing the cellular response. The field is further complicated by the fact that the actual roughness of a surface relevant to a given biomolecule is dependent on the size of the molecule making lateral size-scale dependent roughness measurements desirable [120]. Thus, it is still expected that features at the micrometer length-scale is less dependent on the detailed protein adsorption as already concluded 40 years ago [113]. With respect to nanostructured artificial surfaces, detailed knowledge about the mechanisms guiding the cellular response, including attachment, spreading and proliferation, is still to be obtained. An additional challenge concerning the characterization of the cell-substrate system is the choice of a specific protocol, including cell type(s). Different cell types have shown to give different response making comparison between the published results even more challenging. An early example was shown by Turner et al. [121], where preferential attachment of a primary cell type and a cell line was observed either on nanostructured silicon 'grass' or flat control, respectively. Another apparent conflict has been observed in studies on surfaces consisting of TiO_2 nanotubes of varying diameter. In 2007 Park et al. published results showing that osteogenic differentiation of rat bone marrow derived MSCs was promoted on small diameter nanotubes (15 nm), while differentiation was inhibited on larger diameter tubes (100 nm) [122]. In contrast, Oh et al. showed that culturing human MSCs on TiO_2 nanotubes with a diameter of 30 nm promoted proliferation, while 100 nm nanotubes selectively differentiated the MSCs along the osteogenic lineage [123]. Several reasons for this dispute have been discussed including the origin (species) of the cells [124, 125].

Lately, much focus has been on stem cell behavior on artificial materials for e.g. stem cell therapy and ultimately production of artificial tissue/organs [63, 126, 127]. For these applications MSCs is an often used adult stem cell source. Topographies on the micrometer scale direct mesenchymal stem cells [128], and it has also been demonstrated that nano-topography as the sole factor is an important cue to induce a specific cellular response, e.g. osteogenic differentiation or proliferation. A range of artificial surface materials has been employed including TiO_2 [123] and polydimethylsiloxane (PDMS) [129]. In general cell types do react to nanotopographies. For example, Teixeira et al. [130] examined human corneal epithelial cells on nano-grooved silicon oxide surfaces. It was observed that the cells aligned along grooves with 70 nm wide ridges and a 400 nm pitch, while the cells were rounded on the flat control.

However, not only the features of the nanoscale topographies play an important role. Also the detailed *arrangement* of nano-features is of importance. It has been shown that, using nanoscale disorder (perturbing a 50 nm offset laterally) of a perfect array of 100 nm deep and 120 nm diameter pits with average pitch spacing of 300 nm, it is possible to stimulate hMSCs towards osteogenic differentiation in vitro [62]. Later it was further demonstrated by the same group that the same nanotopographical features with reduced value of offset and thereby a more "perfect" square lattice resulted in a switch of the hMSC behavior from osteogenic differentiation to a state more suitable for expansion of the hMSCs, still retaining the characteristics of the cell type (hMSC markers and multipotency) [131].

Apart from the structures with long-range order as discussed previously in this section, stochastic nano-scale structures have an impact on cellular behavior as well. For example, silica nanoparticles of different size have been used to create surfaces with varying nano-roughness. Such surfaces have been shown to affect fibroblastic morphology, cell adhesion and spreading for periods of up to 7 weeks [132]. For human endothelial cells, a significant reduction in the adhesion and proliferation was observed as compared to flat control silica surfaces [117].

The complex behavior of cells on topographically structured surfaces has led to a "learning by nature" approach. This route has been used for creating biological materials with functional integration [133] e.g. for creating superhydrophobic and oleophobic surfaces. New directions in research e.g. synthesizing nanoscale topographical features mimicking virus surface topography has been proven successful as well. In a recent study it has been demonstrated that cellular uptake efficacy depends on the nanoscale roughness meaning that the performance of delivery vehicles can be improved by topography [134].

As described above, the influence of topographical cues on cellular behavior is a complex affair involving both biomolecular adsorption and direct mechanotransduction. Furthermore, various cell types respond differently even to the same surface cues and small differences in experimental protocols may lead to very different conclusions. Therefore, more systematic testing is needed and several approaches to implement "high-throughput screening" in biomaterials have been reported the last decade. This is discussed further in Sect. 14.4.5 below.

14.4.3 Mechanical Cues for Guided Cellular Behavior

The types of tissue comprising the human body present a wide range of mechanical properties from the soft brain tissue, with an elastic modulus of approximately 0.2 kPa, to the hard mineralized bone tissue with an elastic modulus in the GPa range [135, 136]. The mechanical properties of bone, in particular, vary greatly with respect to specific anatomical sites, degree of mineralization and also among individuals [136]. Numerous studies have indicated the existence of a correlation between cellular behavior and the mechanical properties of the cell culture substrate. However, the major challenge preventing researchers from obtaining direct evidence of such correlation, was the biochemical complexity of the systems used for the investigations. This complexity did not allow for generating substrates with varying mechanical properties while retaining an identical ligand composition. The study by Pelham and Wang, 1997 was the first that described a culture system allowing for elucidating the direct effect of the mechanical properties. This system was based on polyacrylamide hydrogels which were mechanically modulated by varying the amount of bis-acrylamide cross-linker added to the material during preparation. The acrylamide substrates were subsequently modified by immobilizing type I collagen on the surface by chemical cross-linking. Using this biochemical reductionistic approach, the effect of substrate stiffness on motility and cellular morphology of kidney cells and fibroblasts was investigated. The results for both cell types showed a reduction in the motility of cells cultured on the soft substrates and that these cells were less spread and showed a more ruffled cell morphology, compared to the cells on the more rigid substrates [69]. In a later study, the effect of a mechanical gradient was examined, also using an acrylamide substrate. This study was conducted using fibroblasts and it was found that cells preferentially migrated towards the more rigid area of the substrate. Moreover, the cell spreading area increased as the cell encountered the rigid area of the substrate. Similarly it was found that local deformation of the substrate, either by increasing or decreasing the local tension of the substrate, would either cause the cell to change its direction of migration towards the deformation site in response to increased tension, or away from the site, in the case of decreased substrate tension [137].

Mechanical properties have also been found to affect the differentiation of adult stem cells. In a study by Engler et al., 2006 it was shown that mesenchymal stem cells preferentially differentiated towards the phenotypic lineages that matched the elastic modulus of the cell culture substrate. This study utilized a system similar to the one developed by Pelham and Wang, 1997 in order to generate cell culture substrates with elastic modulus in the range of brain tissue (0.1–1 kPa (soft)), muscle (8–17 kPa (intermediate)) and collagenous bone (25–40 kPa (stiff)). In this study, the cells cultured on different substrates were found to exhibit morphological characteristics of their in vivo lineage counterparts. Thus, soft substrates induced cell morphologies with branched protrusions, intermediate induced spindle like cell morphologies, and stiff substrates induced highly spread cells with polygonal characteristics. These characteristics were already appearing after 4 hours of culture and became increasingly pronounced in the period spanning the culture period. Microarray profiling of cellular mRNA transcripts revealed that the cellular expression of phenotypic markers was also affected by the mechanical properties of the substrate and that this expression followed the morphological appearance of the cells [135].

Engler et al. further investigated the stability of the differentiated cellular phenotypes by treating cells that had been pre-cultured on the mechanical modulated substrates with chemical differentiation factors intended to drive the cells towards alternate phenotypes. These investigations revealed the guiding potency of the flexible substrates, as the cells retained their acquired phenotypic characteristics and the differentiated state was found to be highly stable [135].

Later, a study by Cameron et al., 2011 showed that the cellular response was not only determined by the elastic modulus, but was also influenced by the viscoelastic properties of the culture substrate. In this study polyacrylamide substrates was developed having similar storage moduli but with loss modulus properties spanning from 1 to 130 Pa. This setup allowed for investigating the effect of the materials ability to dissipate the energy through viscous flow, via cellular traction forces, on the behavior of mesenchymal stem cells. Thus, cells cultured on substrates with a high loss modulus continuously experience that the traction forces, exerted by their focal adhesions, would decrease as time progressed. The findings of this study showed that substrates with a high loss modulus (high degree of energy loss) increased cell area and proliferation while the size and maturity of focal adhesions were found to be reduced. These findings indicate that the initial attachment of the cell, to the high loss modulus substrates, allows the cell to exert traction forces to the substrate but that the loss modulus dissipates the energy, thus, forcing the cell to spread out to regain traction. As this process is continuous, focal adhesion development is inhibited which in turn yields immature complexes. The study also found that cells cultured on the high loss modulus substrates showed increased expression of smooth muscle cell markers compared to the cells on the lower loss modulus substrates. Moreover, the use of adipogenic and osteogenic differentiation media, to push the cells towards the specific lineages, was most effective for cells cultured on the high loss modulus substrates. Chemical inhibition of the cells, ability to exert traction forces to the different substrates induced responses from the cells cultured on the low loss modulus substrate that were similar to the response found for cells on the high loss modulus substrates, indicating the importance of cytoskeletal tension forces in relation to guiding cellular responses [70].

Contrary to the chemically modified substrates described above, Fu et al., 2011 has demonstrated the feasibility of generating cell culture substrates where effective elastic modulus was varied purely by geometrical factors. This was achieved by generating PDMS substrates comprising closely spaced pillars of varying heights. In order to ensure that the cells would only adhere to the top of the pillars, these were coated with fibronectin via a micro contact printing approach. In the study, the differentiation of hMSCs was investigated using bi-potent differentiation medium for simultaneous induction of adipogenic and osteogenic differentiation. It was found that the cells cultured on the short pillars, showing higher rigidity, preferentially differentiated towards the osteogenic lineage. On the other hand, preferential differentiation towards the adipogenic linage was found for the taller, more flexible pillars [72].

The different studies investigating the effect of mechanical properties on cellular behavior has shown the guiding potency of this physical cue, both with respect to regulating cellular migration, proliferation and differentiation. These studies have paved the road for future investigations, thus enabling the scientific community to further elucidate the mechanisms involved in, and the implications of, the phenomenon of cellular mechano-sensing.

14.4.4 Combined and Asymmetrical Cues for Guided Cellular Behavior

While approaches investigating the effect of single guidance parameters on cells in vitro are important for understanding the fundamental processes resulting in specific cellular responses, the synergy of cues presented to cells in vivo necessitates studies designed to elucidate the combined effects of multiple cues. While replicating the exact conditions of most in vivo situations implies that 3D cultures are utilized, such systems are beyond the scope of the current text. There are, however, several studies that have begun to shed light at the interplay between various cues, using two dimensional culture approaches.

The study by Rowlands et al. (2008) has investigated the combined effect of substrate modulus and tissue specific adhesion ECM protein ligands on the differentiation of mesenchymal stem cells (MSCs) [138]. The findings of this study showed that the combination of a flat substrate with elastic modulus in the range of collagenous bone (>34 kPa), in combination with collagen I, favored bone formation relative to the same substrate modified with fibronectin. On the other hand, the use of fibronectin as ligand for substrates with an elastic modulus in the range of muscle tissue (8–17 kPa) favored myogenic differentiation, relative to similar collagen I modified substrates [138]. The study by Lee et al. (2013) investigated the effect of different adhesion ligands and geometrical restraints presented on substrates having an elastic modulus of approximately 0.6 kPa. Cellular differentiation was assessed with respect to the adipogenic and neurogenic lineages. Initially, the study investigated the individual effects of coating substrates with fibronectin, laminin or collagen, either in small circular patches, limiting cell spreading, or as continuous distributions, allowing the cells to spread freely. Generally, cells confined to the circular patches showed a higher tendency to differentiate along the adipogenic lineage compared to cells that were allowed to spread freely. Moreover, substrates modified with fibronectin were observed to favor adipogenic differentiation while collagen modified substrates generally favored neurogenesis. Taking the study one step further, the group investigated the influence of varying the shape and size of the ligand pattern on which the cells were allowed to attach and spread. More specifically, round, elongated and four-branched star structures were examined. The study found that the star shaped adhesion areas, favored neurogenesis compared to round areas, having the same nominal surface area [139].

The migration of fibroblasts in response to different configurations of mechanically modulated hydrogel gradients, functionalized to yield an opposing collagen I gradient has also been investigated. The substrates used for this study were designed so that the areas showing a high elastic modulus had a low collagen concentration and vice versa. The study found that the fibroblasts preferentially migrated towards increasing ligand concentrations and thus were less prone to be affected by the modulus of the underlying substrate [140].

While the various findings of the studies investigating combined an asymmetrical cues, first of all, highlights the complex interplay of various cues in relation to guiding cells towards a specific lineage, more importantly, it illustrates the complexity of designing and optimizing the biomaterials of the future. Carefully tailoring the interplay between parameters such as mechanical, chemical and structural properties can greatly increase the performance of a biomaterial for specific applications. The complexity of the interplay of the different cues has led to the development of high throughput screening approaches as discussed in the next section.

14.4.5 High Throughput Screening

Understanding the biomaterial-cell interaction is of utmost importance for the development of novel biomedical surfaces with tailored properties for specific applications in e.g. nanomedicine. In the preceding sections different cues from surfaces contributing to the overall guiding of cellular response has been discussed; both chemical/biomolecular, topographical, and mechanical properties has been shown to influence cellular behavior. Already with this span of surface parameters, it is extremely challenging to deduce guidelines. However, the cellular output obtained from these cues are more complex, as also such parameters as cell shape [141], cell colony size [142], and even the combined synergistic effect of the factors above [138] are of importance. Realizing that even looking for a needle in a haystack is simple compared to understanding the interplay between all the factors, additional routes are necessary to speed up the development of biomedical products which can benefit the individual patients.

The biomedical material research approaches have typically either been testing "off-the-shelf" materials in in vivo models or using a reductionistic approach focusing on a single parameter in in vitro experiments. Both routes have proven valuable and necessary parts of the research field. What would add further value would be a more direct route to obtain design guidelines for materials properties to specifically tailor cellular (and tissue) responses.

In the last decade, a new field in biomaterials has developed, sometimes termed "Materiomics" [143]. An important aspect of the field of Materiomics is high-throughput screening (HTS). Here, the detailed understanding of the cellular mechanisms is less important than the actual cellular response. One of the first demonstrations of the HTS approach was published by Anderson et al. [144]. In this study the influence of chemical composition of polymer mixtures on human embryonic stem cell (hESC) differentiation was addressed: 576 (24 × 24) combinations of different acrylate, diacrylate, dimethacrylate and triacrylate monomers were deposited in an array in triplicates on a standard glass slide with a non-fouling background. After crosslinking the monomers, the hESCs were seeded on the arrays and it was shown that specific polymer compositions which promote cytokeratin-positive cells could be identified. Since the detailed chemical composition of each individual spot was known, it was possible to produce hit arrays. A follow up study using the same platform addressed human MSC–biomaterial composite interactions [145]. Later studies included a more thorough surface characterizations including water contact-angle, surface topography, surface chemistry and indentation elastic modulus [146].

Screening of topographical structures at the micro and nanometer scale has been addressed by several groups. In an early demonstration, a combinatorial array was created by standard photolithography comprising 169 (13 × 13) individual microstructures with lateral feature sizes from 1 μm to 6 μm. The approach proved to give topographical guidelines for enhancing osteogenic expression and mineralization using murine pre-osteoblasts [147]. Using the same array, it was also possible to optimize structures promoting mineralization by human dental pulp stem cells [128], human fibroblastic proliferation [63] and enhance non-viral transfection efficiency in primary human fibroblasts [148]. Another approach was developed by de Boer et al., where the topographical structures were designed by mathematical algorithms to obtain nonbiased, random surface features. The resulting array was reproduced on a 2 × 2 cm^2 area, called a "TopoChip", which consisted of poly(lactic acid) with 2176 different topographies [149, 150].

Platforms involving combinatorial mixtures of extracellular matrix (ECM) proteins and soluble growth factors have been developed as well using robotic spotting technology. Such platform has been applied to study the effects of 32 different combinations of five extracellular matrix molecules (collagen I, collagen III, collagen IV, laminin and fibronectin) on cellular expansion/differentiation [151, 152].

It is evident that a 2D surface is far from the typical 3D environment experienced by cells in vivo. Thus it would be desirable to be able to do advanced HTS of 3D microenvironments with varying compositions. An attractive route is to use hydrogels mixed with cells. Several groups have tested cellular response inside various miniaturized cell-laden gels in a combinatorial manner, e.g. for osteogenic differentiation of human mesenchymal stem cells [153]. Recently, a method to create a 3D microarray platform using robotic nanolitre liquid-dispensing technology to create more than 1000 unique microenvironments has been published. In this study the combined effects of matrix elasticity, proteolytic degradability of the surrounding hydrogel and three distinct classes of signaling proteins (extracellular matrix proteins, cell–cell interaction proteins and soluble factors) has been addressed [154].

These newly developed methods to screen and identify surfaces or micro-niche environments with specific cell guiding properties may enable development of novel technologies and products for nanomedicine applications. One area with a huge potential is the expansion of multi- and pluripotent cells on a clinically relevant scale for novel cell-based therapies and tissue engineering.

14.5 Summary and Outlook

In this chapter the different types of cues from surfaces of artificial materials on cellular behavior has been discussed. It has been demonstrated that controlling a number of material parameters is necessary to control the cell-material system e.g. the protein adsorption, surface chemistry, surface topography and mechanical properties and combinations thereof. All these material parameters are of importance to control when designing materials for biological applications at the nanoscale e.g. for nanomedicine applications. The reader might feel that it is an enormous challenge and maybe even an obstacle to address this issue. Furthermore, the whole engineering aspect of transferring the obtained 2D results to real in vivo 3D applications is not straight forward. In some ways this is true. However, one might also view the parameters and the possible synergistic effects as a hitherto almost untapped source of possibilities to control cellular response to, and maybe ultimately even tissue interaction with, artificial materials. Furthermore, even though 2D surfaces are simple models of a real 3D system, all surfaces e.g. at and inside 3D scaffolds are at a certain level 2D. Thus it is expected that the knowledge gained by the examination of parameters in 2D systems may provide guidelines for the construction of 3D systems as well. With the advance of well-defined cell substrates and particularly high-throughput screening of single and multiple parameters the chances are that novel combinations of cues will be identified to optimize cellular/tissue responses and thus be of importance in the field of nanomedicine.

References

1. Abbott A (2003) Cell culture: biology's new dimension. Nature 424:870–872
2. Keung AJ, Kumar S, Schaffer DV (2010) Presentation counts: microenvironmental regulation of stem cells by biophysical and material cues. Annu Rev Cell Dev Biol 26:533–556

3. Lutolf MP, Gilbert PM, Blau HM (2009) Designing materials to direct stem-cell fate. Nature 462:433–441
4. Saha K, Pollock JF, Schaffer DV, Healy KE (2007) Designing synthetic materials to control stem cell phenotype. Curr Opin Chem Biol 11:381–387
5. Tibbitt MW, Anseth KS (2009) Hydrogels as extracellular matrix mimics for 3D cell culture. Biotechnol Bioeng 103:655–663
6. Vazin T, Schaffer DV (2010) Engineering strategies to emulate the stem cell niche. Trends Biotechnol 28:117–124
7. Watt FM, Huck WTS (2013) Role of the extracellular matrix in regulating stem cell fate. Nat Rev Mol Cell Biol 14:467–473
8. Kasemo B (2002) Biological surface science. Surf Sci 500:656–677
9. Haynes CA, Norde W (1994) Globular proteins at solid/liquid interfaces. Colloids Surf B: Biointerfaces 2:517–566
10. Hlady V, Buijs J (1996) Protein adsorption on solid surfaces. Curr Opin Biotechnol 7:72–77
11. Nakanishi K, Sakiyama T, Imamura K (2001) On the adsorption of proteins on solid surfaces, a common but very complicated phenomenon. J Biosci Bioeng 91:233–244
12. Rabe M, Verdes D, Seeger S (2011) Understanding protein adsorption phenomena at solid surfaces. Adv Colloid Interf Sci 162:87–106
13. Gray JJ (2004) The interaction of proteins with solid surfaces. Curr Opin Struct Biol 14:110–115
14. Castner DG, Ratner BD (2002) Biomedical surface science: foundations to frontiers. Surf Sci 500:28–60
15. Malmsten M (1998) Formation of adsorbed protein layers. J Colloid Interface Sci 207:186–199
16. Cooper GM (2000) The cell: a molecular approach, 2nd edn. ASM Press, Sunderland, USA
17. Alberts B, Bray D, Lewis J, Raff M, Roberts K, Watson JD (1994) Molecular biology of the cell. Garland Publishing, New York, USA
18. Geiger B, Bershadsky A, Pankov R, Yamada KM (2001) Transmembrane crosstalk between the extracellular matrix and the cytoskeleton. Nat Rev Mol Cell Biol 2:793–805
19. Geiger B, Spatz JP, Bershadsky AD (2009) Environmental sensing through focal adhesions. Nat Rev Mol Cell Biol 10:21–33
20. Zamir E, Geiger B (2001) Molecular complexity and dynamics of cell-matrix adhesions. J Cell Sci 114:3583–3590
21. Parsons JT, Horwitz AR, Schwartz MA (2010) Cell adhesion: integrating cytoskeletal dynamics and cellular tension. Nat Rev Mol Cell Biol 11:633–643
22. Kingshott P, Griesser HJ (1999) Surfaces that resist bioadhesion. Curr Opinion Solid State Mater Sci 4:403–412
23. Otsuka H, Nagasaki Y, Kataoka K (2001) Self-assembly of poly(ethylene glycol)-based block copolymers for biomedical applications. Curr Opin Colloid Interface Sci 6:3–10
24. Li L, Chen S, Zheng J, Ratner BD, Jiang S (2005) Protein adsorption on oligo(ethylene glycol)-terminated alkanethiolate self-assembled monolayers: the molecular basis for non-fouling behavior. J Phys Chem B 109:2934–2941
25. Harder P, Grunze M, Dahint R, Whitesides GM, Laibinis PE (1998) Molecular conformation in oligo(ethylene glycol)-terminated self-assembled monolayers on gold and silver surfaces determines their ability to resist protein adsorption. J Phys Chem B 102:426–436
26. Morra M, Cassineli C (1999) Non-fouling properties of polysaccharide-coated surfaces. J Biomater Sci Polym Ed 10:1107–1124
27. Lowe AB, McCormick CL (2002) Synthesis and solution properties of zwitterionic polymers†. Chem Rev 102:4177–4190
28. Jiang S, Cao Z (2010) Ultralow-fouling, functionalizable, and hydrolyzable zwitterionic materials and their derivatives for biological applications. Adv Mater 22:920–932
29. Ito T, Okazaki S (2000) Pushing the limits of lithography. Nature 406:1027–1031
30. Romankiw LT (1997) A path: from electroplating through lithographic masks in electronics to LIGA in MEMS. Electrochim Acta 42:2985–3005

31. Craighead HG (2000) Nanoelectromechanical Systems. Science 290:1532–1535
32. Broers AN, Molzen WW, Cuomo JJ, Wittels ND (1976) Electron-beam fabrication of 80-Å metal structures. Appl Phys Lett 29:596–598
33. Wu B, Kumar A (2007) Extreme ultraviolet lithography: a review. J Vac Sci Technol B 25:1743–1761
34. Yang S-M, Jang SG, Choi D-G, Kim S, Yu HK (2006) Nanomachining by colloidal lithography. Small 2:458–475
35. Park M, Harrison C, Chaikin PM, Register RA, Adamson DH (1997) Block copolymer lithography: periodic arrays of ~1011 holes in 1 square centimeter. Science 276: 1401–1404
36. Roman G, Martin M, Joachim PS (2003) Block copolymer micelle nanolithography. Nanotechnology 14:1153
37. Ulman A (1996) Formation and structure of self-assembled monolayers. Chem Rev 96:1533–1554
38. Love JC, Estroff LA, Kriebel JK, Nuzzo RG, Whitesides GM (2005) Self-assembled monolayers of thiolates on metals as a form of nanotechnology. Chem Rev 105:1103–1170
39. Haensch C, Hoeppener S, Schubert US (2010) Chemical modification of self-assembled silane based monolayers by surface reactions. Chem Soc Rev 39:2323–2334
40. Hofer R, Textor M, Spencer ND (2001) Alkyl phosphate monolayers, self-assembled from aqueous solution onto metal oxide surfaces. Langmuir 17:4014–4020
41. Falconnet D, Koenig A, Assi F, Textor M (2004) A combined photolithographic and molecular-assembly approach to produce functional micropatterns for applications in the biosciences. Adv Funct Mater 14:749–756
42. Xia Y, Whitesides GM (1998) Soft lithography. Annu Rev Mater Sci 28:153–184
43. Li H-W, Muir BVO, Fichet G, Huck WTS (2003) Nanocontact printing: a route to sub-50-nm-scale chemical and biological patterning. Langmuir 19:1963–1965
44. Salaita K, Wang Y, Mirkin CA (2007) Applications of dip-pen nanolithography. Nat Nano 2:145–155
45. Salaita K, Wang Y, Fragala J, Vega RA, Liu C, Mirkin CA (2006) Massively parallel dip–pen nanolithography with 55 000-pen two-dimensional arrays. Angew Chem 118:7378–7381
46. Huo F, Zheng Z, Zheng G, Giam LR, Zhang H, Mirkin CA (2008) Polymer pen lithography. Science 321:1658–1660
47. Pardo L, Wilson WC, Boland T (2002) Characterization of patterned self-assembled monolayers and protein arrays generated by the ink-jet method†. Langmuir 19:1462–1466
48. Ryan D, Parviz BA, Linder V, Semetey V, Sia SK, Su J et al (2004) Patterning multiple aligned self-assembled monolayers using light. Langmuir 20:9080–9088
49. Zhang G-J, Tanii T, Zako T, Hosaka T, Miyake T, Kanari Y et al (2005) Nanoscale patterning of protein using electron beam lithography of organosilane self-assembled monolayers. Small 1:833–837
50. Hoff JD, Cheng L-J, Meyhöfer E, Guo LJ, Hunt AJ (2004) Nanoscale protein patterning by imprint lithography. Nano Lett 4:853–857
51. Michel R, Lussi JW, Csucs G, Reviakine I, Danuser G, Ketterer B et al (2002) Selective molecular assembly patterning: a new approach to micro- and nanochemical patterning of surfaces for biological applications. Langmuir 18:3281–3287
52. Ogaki R, Cole MA, Sutherland DS, Kingshott P (2011) Microcup arrays featuring multiple chemical regions patterned with nanoscale precision. Adv Mater 23:1876–1881
53. Christman KL, Schopf E, Broyer RM, Li RC, Chen Y, Maynard HD (2008) Positioning multiple proteins at the nanoscale with electron beam cross-linked functional polymers. J Am Chem Soc 131:521–527
54. Dubey M, Emoto K, Takahashi H, Castner DG, Grainger DW (2009) Affinity-based protein surface pattern formation by ligand self-selection from mixed protein solutions. Adv Funct Mater 19:3046–3055
55. Ganesan R, Kratz K, Lendlein A (2010) Multicomponent protein patterning of material surfaces. J Mater Chem 20:7322–7331

56. Cretich M, Damin F, Pirri G, Chiari M (2006) Protein and peptide arrays: recent trends and new directions. Biomol Eng 23:77–88
57. Rusmini F, Zhong Z, Feijen J (2007) Protein immobilization strategies for protein biochips. Biomacromolecules 8:1775–1789
58. Ogaki R, Alexander M, Kingshott P (2010) Chemical patterning in biointerface science. Mater Today 13:22–35
59. Blawas AS, Reichert WM (1998) Protein patterning. Biomaterials 19:595–609
60. Ekblad T, Liedberg B (2010) Protein adsorption and surface patterning. Curr Opin Colloid Interface Sci 15:499–509
61. Wennerberg A, Albrektsson T (2010) On implant surfaces: a review of current knowledge and opinions. Int J Oral Maxillofac Implants 25:63–74
62. Dalby MJ, Gadegaard N, Tare R, Andar A, Riehle MO, Herzyk P et al (2007) The control of human mesenchymal cell differentiation using nanoscale symmetry and disorder. Nat Mater 6:997–1003
63. Kolind K, Dolatshahi-Pirouz A, Lovmand J, Pedersen FS, Foss M, Besenbacher F (2010) A combinatorial screening of human fibroblast responses on micro-structured surfaces. Biomaterials 31:9182–9191
64. Pimpin A, Srituravanich W (2012) Review on micro- and nanolithography techniques and their applications. Eng J 16:37–55
65. Gates BD, Xu Q, Stewart M, Ryan D, Willson CG, Whitesides GM (2005) New approaches to nanofabrication: molding, printing, and other techniques. Chem Rev 105:1171–1196
66. Gates BD, Whitesides GM (2003) Replication of vertical features smaller than 2 nm by soft lithography. J Am Chem Soc 125:14986–14987
67. Odom TW, Love JC, Wolfe DB, Paul KE, Whitesides GM (2002) Improved pattern transfer in soft lithography using composite stamps. Langmuir 18:5314–5320
68. Dasgupta BR, Weitz DA (2005) Microrheology of cross-linked polyacrylamide networks. Phys Rev E 71:021504
69. Robert J, Pelham J, Wang Y-L (1997) Cell locomotion and focal adhesions are regulated by substrate flexibility. Proc Natl Acad Sci U S A 94:13661–13665
70. Cameron AR, Frith JE, Cooper-White JJ (2011) The influence of substrate creep on mesenchymal stem cell behaviour and phenotype. Biomaterials 32:5979–5993
71. Fuard D, Tzvetkova-Chevolleau T, Decossas S, Tracqui P, Schiavone P (2008) Optimization of poly-di-methyl-siloxane (PDMS) substrates for studying cellular adhesion and motility. Microelectron Eng 85:1289–1293
72. Fu J, Wang Y-K, Yang MT, Desai RA, Yu X, Liu Z et al (2011) Mechanical regulation of cell function with geometrically modulated elastomeric substrates. Nat Methods 7:733–739
73. van Wachem PB, Beugeling T, Feijen J, Bantjes A, Detmers JP, van Aken WG (1985) Interaction of cultured human endothelial cells with polymeric surfaces of different wettabilities. Biomaterials 6:403–408
74. van Wachem PB, Hogt AH, Beugeling T, Feijen J, Bantjes A, Detmers JP et al (1987) Adhesion of cultured human endothelial cells onto methacrylate polymers with varying surface wettability and charge. Biomaterials 8:323–328
75. Lee JH, Jung HW, Kang I-K, Lee HB (1994) Cell behaviour on polymer surfaces with different functional groups. Biomaterials 15:705–711
76. Tidwell CD, Ertel SI, Ratner BD, Tarasevich BJ, Atre S, Allara DL (1997) Endothelial cell growth and protein adsorption on terminally functionalized, self-assembled monolayers of alkanethiolates on gold. Langmuir 13:3404–3413
77. Xu S-J, Cui F-Z, Yu X-L, Kong X-D (2013) Glioma cell line proliferation controlled by different chemical functional groups in vitro. Front Mater Sci 7:69–75
78. Hongji Y, Song Z, Jin H, Yanbin Y, Xiumei W, Xiongbiao C et al (2013) Self-assembled monolayers with different chemical group substrates for the study of MCF-7 breast cancer cell line behavior. Biomed Mater 8:035008
79. Barrias CC, Martins MCL, Almeida-Porada G, Barbosa MA, Granja PL (2009) The correlation between the adsorption of adhesive proteins and cell behaviour on hydroxyl-methyl mixed self-assembled monolayers. Biomaterials 30:307–316

80. Arima Y, Iwata H (2007) Effect of wettability and surface functional groups on protein adsorption and cell adhesion using well-defined mixed self-assembled monolayers. Biomaterials 28:3074–3082
81. Faucheux N, Schweiss R, Lützow K, Werner C, Groth T (2004) Self-assembled monolayers with different terminating groups as model substrates for cell adhesion studies. Biomaterials 25:2721–2730
82. Webb K, Hlady V, Tresco PA (1998) Relative importance of surface wettability and charged functional groups on NIH 3T3 fibroblast attachment, spreading, and cytoskeletal organization. J Biomed Mater Res 41:422–430
83. Faucheux N, Tzoneva R, Nagel M-D, Groth T (2006) The dependence of fibrillar adhesions in human fibroblasts on substratum chemistry. Biomaterials 27:234–245
84. Curran JM, Chen R, Hunt JA (2005) Controlling the phenotype and function of mesenchymal stem cells in vitro by adhesion to silane-modified clean glass surfaces. Biomaterials 26:7057–7067
85. Ren Y-J, Zhang H, Huang H, Wang X-M, Zhou Z-Y, Cui F-Z et al (2009) In vitro behavior of neural stem cells in response to different chemical functional groups. Biomaterials 30:1036–1044
86. Britland S, Clark P, Connolly P, Moores G (1992) Micropatterned substratum adhesiveness: a model for morphogenetic cues controlling cell behavior. Exp Cell Res 198:124–129
87. Scotchford CA, Gilmore CP, Cooper E, Leggett GJ, Downes S (2002) Protein adsorption and human osteoblast-like cell attachment and growth on alkylthiol on gold self-assembled monolayers. J Biomed Mater Res 59:84–99
88. Pierschbacher MD, Ruoslahti E (1984) Cell attachment activity of fibronectin can be duplicated by small synthetic fragments of the molecule. Nature 309:30–33
89. Arnold M, Cavalcanti-Adam EA, Glass R, Blümmel J, Eck W, Kantlehner M et al (2004) Activation of integrin function by nanopatterned adhesive interfaces. ChemPhysChem 5:383–388
90. Arnold M, Hirschfeld-Warneken VC, Lohmüller T, Heil P, Blümmel J, Cavalcanti-Adam EA et al (2008) Induction of cell polarization and migration by a gradient of nanoscale variations in adhesive ligand spacing. Nano Lett 8:2063–2069
91. Cavalcanti-Adam EA, Volberg T, Micoulet A, Kessler H, Geiger B, Spatz JP (2007) Cell spreading and focal adhesion dynamics are regulated by spacing of integrin ligands. Biophys J 92:2964–2974
92. Huang J, Gräter SV, Corbellini F, Rinck S, Bock E, Kemkemer R et al (2009) Impact of order and disorder in RGD nanopatterns on cell adhesion. Nano Lett 9:1111–1116
93. Curran JM, Stokes R, Irvine E, Graham D, Amro NA, Sanedrin RG et al (2010) Introducing dip pen nanolithography as a tool for controlling stem cell behaviour: unlocking the potential of the next generation of smart materials in regenerative medicine. Lab Chip 10:1662–1670
94. Chen CS, Mrksich M, Huang S, Whitesides GM, Ingber DE (1997) Geometric control of cell life and death. Science 276:1425–1428
95. Chen CS, Mrksich M, Huang S, Whitesides GM, Ingber DE (1998) Micropatterned surfaces for control of cell shape, position, and function. Biotechnol Prog 14:356–363
96. Lehnert D, Wehrle-Haller B, David C, Weiland U, Ballestrem C, Imhof BA et al (2004) Cell behaviour on micropatterned substrata: limits of extracellular matrix geometry for spreading and adhesion. J Cell Sci 117:41–52
97. McBeath R, Pirone DM, Nelson CM, Bhadriraju K, Chen CS (2004) Cell shape, cytoskeletal tension, and rhoa regulate stem cell lineage commitment. Dev Cell 6:483–495
98. Tseng Q, Duchemin-Pelletier E, Deshiere A, Balland M, Guillou H, Filhol O et al (2012) Spatial organization of the extracellular matrix regulates cell–cell junction positioning. Proc Natl Acad Sci 109:1506–1511
99. Bailly M, Yan L, Whitesides GM, Condeelis JS, Segall JE (1998) Regulation of protrusion shape and adhesion to the substratum during chemotactic responses of mammalian carcinoma cells. Exp Cell Res 241:285–299

100. Kumar G, Ho CC, Co CC (2007) Guiding cell migration using one-way micropattern arrays. Adv Mater 19:1084–1090
101. Kushiro K, Chang S, Asthagiri AR (2010) Reprogramming directional cell motility by tuning micropattern features and cellular signals. Adv Mater 22:4516–4519
102. Borghi N, Lowndes M, Maruthamuthu V, Gardel ML, Nelson WJ (2010) Regulation of cell motile behavior by crosstalk between cadherin- and integrin-mediated adhesions. Proc Natl Acad Sci 107:13324–13329
103. Thery M, Racine V, Pepin A, Piel M, Chen Y, Sibarita J-B et al (2005) The extracellular matrix guides the orientation of the cell division axis. Nat Cell Biol 7:947–953
104. Kristensen SH, Pedersen GA, Nejsum LN, Sutherland DS (2012) Nanoscale E-cadherin ligand patterns show threshold size for cellular adhesion and adherence junction formation. Nano Lett 12:2129–2133
105. Malmström J, Christensen B, Jakobsen HP, Lovmand J, Foldbjerg R, Sørensen ES et al (2010) Large area protein patterning reveals nanoscale control of focal adhesion development. Nano Lett 10:686–694
106. Malmström J, Lovmand J, Kristensen S, Sundh M, Duch M, Sutherland DS (2011) Focal complex maturation and bridging on 200 nm vitronectin but not fibronectin patches reveal different mechanisms of focal adhesion formation. Nano Lett 11:2264–2271
107. Gautrot JE, Malmström J, Sundh M, Margadant C, Sonnenberg A, Sutherland DS (2014) The nanoscale geometrical maturation of focal adhesions controls stem cell differentiation and mechanotransduction. Nano Lett 14:3945–3952
108. Giam LR, Massich MD, Hao L, Shin Wong L, Mader CC, Mirkin CA (2012) Scanning probe-enabled nanocombinatorics define the relationship between fibronectin feature size and stem cell fate. Proc Natl Acad Sci 109:4377–4382
109. Delcassian D, Depoil D, Rudnicka D, Liu M, Davis DM, Dustin ML et al (2013) Nanoscale ligand spacing influences receptor triggering in T cells and NK cells. Nano Lett 13:5608–5614
110. Sekula S, Fuchs J, Weg-Remers S, Nagel P, Schuppler S, Fragala J et al (2008) Multiplexed lipid dip-pen nanolithography on subcellular scales for the templating of functional proteins and cell culture. Small 4:1785–1793
111. Weiss P (1945) Experiments on cell and axon orientation in vitro: the role of colloidal exudates in tissue organization. J Exp Zool 100:353–386
112. Curtis AS (2004) Small is beautiful but smaller is the aim: review of a life of research. Eur Cell Mater 8:27–36
113. Curtis ASG, Varde M (1964) Control of cell behavior: topological factors. J Natl Cancer Inst 33:15–26
114. Dalby MJ, Gadegaard N, Oreffo ROC (2014) Harnessing nanotopography and integrin-matrix interactions to influence stem cell fate. Nat Mater 13:558–569
115. Lord MS, Foss M, Besenbacher F (2010) Influence of nanoscale surface topography on protein adsorption and cellular response. Nano Today 5:66–78
116. Rechendorff K, Hovgaard MB, Foss M, Zhdanov VP, Besenbacher F (2006) Enhancement of protein adsorption induced by surface roughness. Langmuir 22:10885–10888
117. Lord MS, Cousins BG, Doherty PJ, Whitelock JM, Simmons A, Williams RL et al (2006) The effect of silica nanoparticulate coatings on serum protein adsorption and cellular response. Biomaterials 27:4856–4862
118. Roach P, Farrar D, Perry CC (2006) Surface tailoring for controlled protein adsorption: effect of topography at the nanometer scale and chemistry. J Am Chem Soc 128:3939–3945
119. Sutherland DS, Broberg M, Nygren H, Kasemo B (2001) Influence of nanoscale surface topography and chemistry on the functional behaviour of an adsorbed model macromolecule. Macromol Biosci 1:270–273
120. Rechendorff K, Hovgaard MB, Chevallier J, Foss M, Besenbacher F (2005) Tantalum films with well-controlled roughness grown by oblique incidence deposition. Appl Phys Lett 87:073105

121. Turner S, Kam L, Isaacson M, Craighead HG, Shain W, Turner J (1997) Cell attachment on silicon nanostructures. J Vac Sci Tech B 15:2848–2854
122. Park J, Bauer S, von der Mark K, Schmuki P (2007) Nanosize and vitality: TiO2 nanotube diameter directs cell fate. Nano Lett 7:1686–1691
123. Oh S, Brammer KS, Li YSJ, Teng D, Engler AJ, Chien S et al (2009) Stem cell fate dictated solely by altered nanotube dimension. Proc Natl Acad Sci 106:2130–2135
124. Oh S, Brammer KS, Li YSJ, Teng D, Engler AJ, Chien S et al (2009) Reply to von der Mark et al.: looking further into the effects of nanotube dimension on stem cell fate. Proc Natl Acad Sci 106:E61
125. von der Mark K, Bauer S, Park J, Schmuki P (2009) Another look at "Stem cell fate dictated solely by altered nanotube dimension". Proc Natl Acad Sci 106:E60
126. Kulangara K, Leong KW (2009) Substrate topography shapes cell function. Soft Matter 5:4072–4076
127. Yang Y, Leong KW (2010) Nanoscale surfacing for regenerative medicine. Wiley Interdiscip Rev Nanomed Nanobiotechnol 2:478–495
128. Kolind K, Kraft D, Bøggild T, Duch M, Lovmand J, Pedersen FS et al (2014) Control of proliferation and osteogenic differentiation of human dental-pulp-derived stem cells by distinct surface structures. Acta Biomater 10:641–650
129. Teo BKK, Wong ST, Lim CK, Kung TYS, Yap CH, Ramagopal Y et al (2013) Nanotopography modulates mechanotransduction of stem cells and induces differentiation through focal adhesion kinase. ACS Nano 7:4785–4798
130. Teixeira AI, Abrams GA, Bertics PJ, Murphy CJ, Nealey PF (2003) Epithelial contact guidance on well-defined micro- and nanostructured substrates. J Cell Sci 116:1881–1892
131. McMurray RJ, Gadegaard N, Tsimbouri PM, Burgess KV, McNamara LE, Tare R et al (2011) Nanoscale surfaces for the long-term maintenance of mesenchymal stem cell phenotype and multipotency. Nat Mater 10:637–644
132. Cousins BG, Doherty PJ, Williams RL, Fink J, Garvey MJ (2004) The effect of silica nanoparticulate coatings on cellular response. J Mater Sci Mater Med 15:355–359
133. Liu K, Jiang L (2011) Bio-inspired design of multiscale structures for function integration. Nano Today 6:155–175
134. Niu Y, Yu M, Hartono SB, Yang J, Xu H, Zhang H et al (2013) Nanoparticles mimicking viral surface topography for enhanced cellular delivery. Adv Mater 25:6233–6237
135. Engler AJ, Sen S, Sweeney HL, Discher DE (2006) Matrix elasticity directs stem cell lineage specification. Cell 126:677–689
136. Zysset PK, Guo XE, Hoffler CE, Moore KE, Goldstein SA (1999) Elastic modulus and hardness of cortical and trabecular bone lamellae measured by nanoindentation in the human femur. J Biomech 32:1005–1012
137. Lo CM, Wang HB, Dembo M, Wang YL (2000) Cell movement is guided by the rigidity of the substrate. Biophys J 79:144–152
138. Rowlands AS, George PA, Cooper-White JJ (2008) Directing osteogenic and myogenic differentiation of MSCs: interplay of stiffness and adhesive ligand presentation. Am J Physiol Cell Physiol 295:1037–1044
139. Lee J, Abdeen AA, Zhang D, Kilian KA (2013) Directing stem cell fate on hydrogel substrates by controlling cell geometry, matrix mechanics and adhesion ligand composition. Biomaterials 34:8140–8148
140. Hale NA, Yang Y, Rajagopalan P (2010) Cell migration at the interface of a dual chemical-mechanical gradient. ACS Appl Mater Interfaces 2:2317–2324
141. Kilian KA, Bugarija B, Lahn BT, Mrksich M (2010) Geometric cues for directing the differentiation of mesenchymal stem cells. Proc Natl Acad Sci 107:4872–4877
142. Hwang Y-S, Chung BG, Ortmann D, Hattori N, Moeller H-C, Khademhosseini A (2009) Microwell-mediated control of embryoid body size regulates embryonic stem cell fate via differential expression of WNT5a and WNT11. Proc Natl Acad Sci 106:16978–16983
143. Cranford SW, de Boer J, van Blitterswijk C, Buehler MJ (2013) Materiomics: an -omics approach to biomaterials research. Adv Mater 25:802–824

144. Anderson DG, Levenberg S, Langer R (2004) Nanoliter-scale synthesis of arrayed biomaterials and application to human embryonic stem cells. Nat Biotech 22:863–866
145. Anderson DG, Putnam D, Lavik EB, Mahmood TA, Langer R (2005) Biomaterial microarrays: rapid, microscale screening of polymer–cell interaction. Biomaterials 26:4892–4897
146. Mei Y, Saha K, Bogatyrev SR, Yang J, Hook AL, Kalcioglu ZI et al (2010) Combinatorial development of biomaterials for clonal growth of human pluripotent stem cells. Nat Mater 9:768–778
147. Lovmand J, Justesen J, Foss M, Lauridsen RH, Lovmand M, Modin C et al (2009) The use of combinatorial topographical libraries for the screening of enhanced osteogenic expression and mineralization. Biomaterials 30:2015–2022
148. Adler AF, Speidel AT, Christoforou N, Kolind K, Foss M, Leong KW (2011) High-throughput screening of microscale pitted substrate topographies for enhanced nonviral transfection efficiency in primary human fibroblasts. Biomaterials 32:3611–3619
149. Baker M (2011) Trying out topographies. Nat Meth 8:900
150. Unadkat HV, Hulsman M, Cornelissen K, Papenburg BJ, Truckenmüller RK, Carpenter AE et al (2011) An algorithm-based topographical biomaterials library to instruct cell fate. Proc Natl Acad Sci 108:16565–16570
151. Flaim CJ, Chien S, Bhatia SN (2005) An extracellular matrix microarray for probing cellular differentiation. Nat Meth 2:119–125
152. Flaim CJ, Teng D, Chien S, Bhatia SN (2008) Combinatorial signaling microenvironments for studying stem cell fate. Stem Cells Dev 17:29–40
153. Dolatshahi-Pirouz A, Nikkhah M, Gaharwar AK, Hashmi B, Guermani E, Aliabadi H et al (2014) A combinatorial cell-laden gel microarray for inducing osteogenic differentiation of human mesenchymal stem cells. Sci Rep 4:3896
154. Ranga A, Gobaa S, Okawa Y, Mosiewicz K, Negro A, Lutolf MP (2014) 3D niche microarrays for systems-level analyses of cell fate. Nat Commun 5:4324

Index

K.A. Howard et al. (eds.), *Nanomedicine*, Advances in Delivery Science and Technology, DOI 10.1007/978-1-4939-3634-2

MIX
Papier aus verantwortungsvollen Quellen
Paper from responsible sources
FSC® C105338

If you have any concerns about our products,
you can contact us on
ProductSafety@springernature.com

In case Publisher is established outside the EU,
the EU authorized representative is:
Springer Nature Customer Service Center GmbH
Europaplatz 3, 69115 Heidelberg, Germany

Printed by Libri Plureos GmbH
in Hamburg, Germany